THE
FUTURE

BY AL GORE

The Future

The Assault on Reason

An Inconvenient Truth

An Inconvenient Truth: Adapted for a New Generation

Earth in the Balance: Ecology and the Human Spirit

*Joined at the Heart: The Transformation of the
American Family* (with Tipper Gore)

The Spirit of Family (with Tipper Gore)

Our Choice: A Plan to Solve the Climate Crisis

Our Purpose

*From Red Tape to Results: Creating a Government
That Works Better and Costs Less*

Common Sense Government

Businesslike Government: Lessons Learned from America's Best Companies
(illustrated by Scott Adams)

AL GORE

THE FUTURE

WH
ALLEN

2 4 6 8 10 9 7 5 3 1

First published in the United States in 2013 by Random House,
an imprint of The Random House Publishing Group,
a division of Random House, Inc., New York

First published in the UK in 2013 by WH Allen

This edition published in 2014 by WH Allen, an imprint of Ebury Publishing

A Random House Group Company

www.randomhouse.co.uk

Addresse und at:

A

Co n.

Book design by Susan Turner

Printed and bound by CPI Group (UK) Ltd, Croydon, CR0 4YY

ISBN: 9780753540503

To buy books by your favourite authors and register for offers, visit www.randomhouse.co.uk

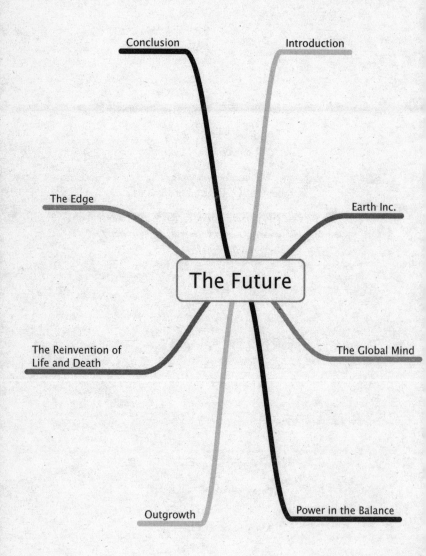

Conclusion

Introduction

The Edge

Earth Inc.

The Future

The Reinvention of
Life and Death

The Global Mind

Outgrowth

Power in the Balance

CONTENTS

INTRODUCTION   xiii

1
EARTH INC.   4

2
THE GLOBAL MIND   44

3
POWER IN THE BALANCE   92

4
OUTGROWTH   142

5
THE REINVENTION OF LIFE AND DEATH   204

6
THE EDGE   280

CONCLUSION   361

ACKNOWLEDGMENTS   375
BIBLIOGRAPHY   379
NOTES   389
INDEX   535

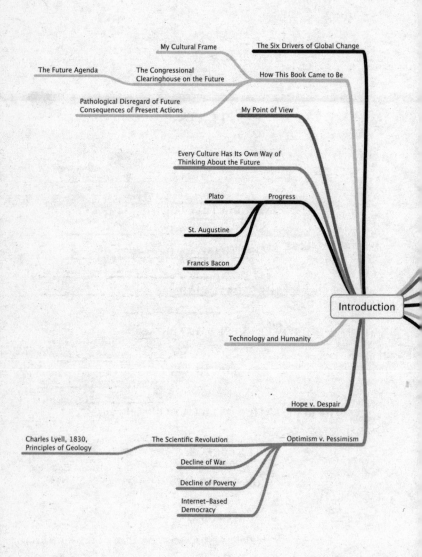

My Cultural Frame

The Six Drivers of Global Change

The Future Agenda

The Congressional
Clearinghouse on the Future

How This Book Came to Be

Pathological Disregard of Future
Consequences of Present Actions

My Point of View

Every Culture Has Its Own Way of
Thinking About the Future

Plato

Progress

St. Augustine

Francis Bacon

Introduction

Technology and Humanity

Hope v. Despair

Charles Lyell, 1830,
Principles of Geology

The Scientific Revolution

Optimism v. Pessimism

Decline of War

Decline of Poverty

Internet-Based
Democracy

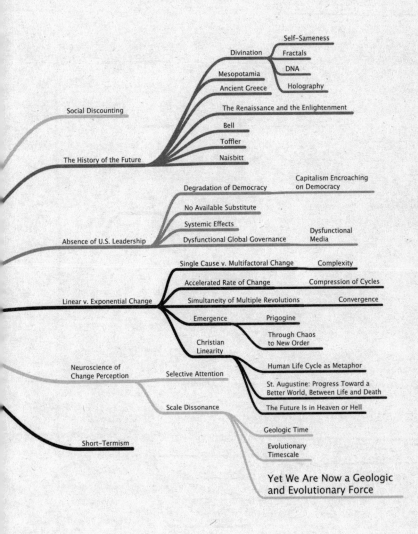

Social Discounting

The History of the Future
- Divination
  - Self-Sameness
  - Fractals
  - DNA
  - Holography
- Mesopotamia
- Ancient Greece
- The Renaissance and the Enlightenment
- Bell
- Toffler
- Naisbitt

Absence of U.S. Leadership
- Degradation of Democracy — Capitalism Encroaching on Democracy
- No Available Substitute
- Systemic Effects
- Dysfunctional Global Governance — Dysfunctional Media

Linear v. Exponential Change
- Single Cause v. Multifactoral Change — Complexity
- Accelerated Rate of Change — Compression of Cycles
- Simultaneity of Multiple Revolutions — Convergence
- Emergence — Prigogine
- Christian Linearity
  - Through Chaos to New Order
  - Human Life Cycle as Metaphor
  - St. Augustine: Progress Toward a Better World, Between Life and Death
  - The Future Is in Heaven or Hell

Neuroscience of Change Perception
- Selective Attention

Scale Dissonance
- Geologic Time
- Evolutionary Timescale

Short-Termism

Yet We Are Now a Geologic and Evolutionary Force

INTRODUCTION

LIKE MANY FULFILLING JOURNEYS, THIS BOOK BEGAN NOT WITH AN-swers but with a question. Eight years ago, when I was on the road, someone asked me: "What are the drivers of global change?" I listed several of the usual suspects and left it at that. Yet the next morning, on the long plane flight home, the question kept pulling me back, demanding that I answer it more precisely and accurately—not by relying on preconceived dogma but by letting the emerging evidence about an emerging world take me where it would. The question, it turned out, had a future of its own. I started an outline on my computer and spent several hours listing headings and subheadings, then changing their rank order and relative magnitude, moving them from one category to another and filling in more and more details after each rereading.

As I spent the ensuing years raising awareness about climate change and pursuing a business career, I continued to revisit, revise, and sharpen the outline until finally, two years ago, I concluded that it would not leave me alone until I dug in and tried to thoroughly answer the question that had turned into something of an obsession.

What emerged was this book, a book about the six most important drivers of global change, how they are converging and interacting with one another, where they are taking us, and how we as human beings—and as a global civilization—can best affect the way these changes unfold. In order to reclaim control of our destiny and shape the future, we must think freshly and clearly about the crucial choices that confront us as a result of:

- The emergence of a deeply interconnected global economy that increasingly operates as a fully integrated holistic entity with a completely new and different relationship to capital flows, labor, consumer markets, and national governments than in the past;

- The emergence of a planet- wide electronic communications grid connecting the thoughts and feelings of billions of people and linking them to rapidly expanding volumes of data, to a fast grow- ing web of sensors being embedded ubiquitously throughout the world, and to increasingly intelligent devices, robots, and thinking machines, the smartest of which already exceed the capabilities of humans in performing a growing list of discrete mental tasks and may soon surpass us in manifestations of intelligence we have always assumed would remain the unique province of our species;

- The emergence of a completely new balance of political, eco- nomic, and military power in the world that is radically different from the equilibrium that characterized the second half of the twentieth century, during which the United States of America provided global leadership and stability— shifting influence and initiative from West to East, from wealthy countries to rapidly emerging centers of power throughout the world, from nation- states to private actors, and from political systems to markets;

- The emergence of rapid unsustainable growth— in population; cities; resource consumption; depletion of topsoil, freshwater sup- plies, and living species; pollution flows; and economic output that is measured and guided by an absurd and distorted set of universally accepted metrics that blinds us to the destructive con- sequences of the self- deceiving choices we are routinely making;

- The emergence of a revolutionary new set of powerful biological, biochemical, genetic, and materials science technologies that are enabling us to reconstitute the molecular design of all solid mat- ter, reweave the fabric of life itself, alter the physical form, traits, characteristics, and properties of plants, animals, and people, seize active control over evolution, cross the ancient lines dividing spe- cies, and invent entirely new ones never imagined in nature; and

- The emergence of a radically new relationship between the aggregate power of human civilization and the Earth's ecological systems, including especially the most vulnerable— the atmosphere and climate balance upon which the continued flourishing of humankind depends— and the beginning of a massive global transformation of our energy, industrial, agricultural, and construction technologies in order to reestablish a healthy and balanced relationship between human civilization and the future.

This book is data-driven and is based on deep research and reporting— not speculation, alarmism, naïve optimism, or blue-sky conjecture. It represents the culmination of a multiyear effort to investigate, decipher, and present the best available evidence and what the world's leading experts tell us about the future we are now in the process of creating.

There is a clear consensus that the future now emerging will be extremely different from anything we have ever known in the past. It is a difference not of degree but of kind. There is no prior period of change that remotely resembles what humanity is about to experience. We have gone through revolutionary periods of change before, but none as powerful or as pregnant with the fraternal twins—peril and opportunity—as the ones that are beginning to unfold. Nor have we ever experienced so *many* revolutionary changes unfolding simultaneously and converging with one another.

This is not a book primarily about the climate crisis, though the climate crisis is one of the six emergent changes that are quickly reshaping our world, and its interaction with the other five drivers of change has revealed to me new ways to understand it. Nor is it primarily about the degradation of democracy in the United States and the dysfunctionality of governance in the world community—though I continue to believe that these leadership crises must be resolved in order for humankind to reclaim control of our destiny. Indeed all six of these emergent revolutionary changes are threatening to overtake us at a moment in history when there is a dangerous vacuum of global leadership.

Neither is this a manifesto intended to lay the groundwork for some future political campaign. I have run for political office often enough in the past. The joke I often use to deflect questions about whether I have finally surrendered any intention to do so again is actually as close to the truth as any words I can summon in describing my attitude toward politics: I am a recovering politician and the chances of a relapse have been diminishing for long enough to increase my confidence that I will not succumb to that temptation again. In the Conclusion, however, you will find a recommended agenda for action that is based on the analysis in this book.

A NEW LAW OF NATURE

As a young freshman member of the U.S. House of Representatives elected in 1976, I joined a new bipartisan group of congressmen and senators known as the Congressional Clearinghouse on the Future, founded by the late Charlie Rose of North Carolina.* In my second term, Rose asked me to succeed him as chair of the group. We organized workshops on the implications of new technologies and scientific discoveries and met with leaders in business and science. Among our other initiatives, we persuaded all 200 subcommittees in the Congress to publish a list of the most important issues they expected to emerge over the following twenty years and published it as "The Future Agenda." Most of all, we studied emerging trends and met regularly with the leading thinkers about the future: Daniel Bell, Margaret Mead, Buckminster Fuller, Carl Sagan, Alvin Toffler, John Naisbitt, Arno Penzias, and hundreds of others.

The visiting scholar who made perhaps the biggest impression on me was a short and balding scientist born in Russia a few months before the 1917 Revolution but educated in Belgium: Ilya Prigogine, who had just won the Nobel Prize in Chemistry for his discovery of a major corollary to the Second Law of Thermodynamics.

Entropy, according to the Second Law, causes all isolated physical systems to break down over time and is responsible for irreversibility in nature. For a simple example of entropy, consider a smoke ring: it begins as a coherent donut with clearly defined boundaries. But as the molecules separate from one another and dissipate energy into the air, the ring falls apart and disappears. All so-called closed systems are subject to the same basic process of dissolution; in some, entropy operates quickly, while in others the process takes more time.

---

* The Congressional Clearinghouse on the Future had a very able executive director, Anne Cheatham.

Prigogine's discovery was that an open system—that is, a system that imports flows of energy from outside the system into it, through it, and out again—not only breaks down, but as the flow of energy continues, the system then *reorganizes itself* at a higher level of complexity. In a sense, the phenomenon described by Prigogine is the opposite of entropy. Self-organization, as a law of nature and as a process of change, is truly astonishing. What it means is that complex new forms can *emerge* spontaneously through *self-organization*.

Consider the increased flows of information throughout the world following the introduction of the Internet and the World Wide Web. Elements of the old information pattern began to break down. Many newspapers went bankrupt, readership sharply declined in most others, bookstores consolidated and closed. Many business models became obsolete. But the new emergent pattern led to the self-organization of thousands of new business models, and volumes of online communication dwarfing those that characterized the world of the printing press.

The Earth itself, when viewed as a whole, is also an open system. It imports energy from the sun that flows into and through the elaborate patterns of energy transfer that make up the Earth system, including the oceans, the atmosphere, the various geochemical processes—and life itself. The energy then flows from the Earth back into the universe surrounding it as heat energy in the form of infrared radiation.

The essence of the emergent crisis of global warming is that we are importing enormous amounts of energy from the crust of the Earth and exporting entropy (that is, progressive disorder) into the previously stable, though dynamic, ecological systems upon which the continued flourishing of civilization depends. These new flows of energy, originally imported to the Earth from the sun ages ago, have been stabilized underground for millions of years as inert deposits of carbon.

By mobilizing them and injecting the waste products from their combustion into the atmosphere, we are breaking down the stable climate pattern that has persisted since not long after the end of the last Ice Age ten millennia ago. This was not long before the first cities and the beginning of the Agricultural Revolution, which began to spread in the valleys of the Nile, Tigris, Euphrates, Indus, and Yellow rivers 8,000 years ago after Stone Age women and men patiently picked and selectively bred the plant varieties on which our modern diet still depends. In the process, we are forcing the emergence of a new climate pattern very different from the one to which our entire civilization is tightly configured and within which we have thrived.

While Prigogine's discovery of this new law of nature may seem arcane, its implications for the way we should think about the future are profound. The modern meaning of the word "emergence," and the entire field of knowledge known as complexity theory, are both derived from Prigogine's work. The motivation for his exploration of emergence was his passion for understanding how the future becomes irreversibly different from the past. He wrote that, "given my interest in the concept of time, it was only natural that my attention was focused on . . . the study of irreversible phenomena, which made so manifest the 'arrow of time.'"

THE HISTORY OF THE FUTURE

The way we think about the future has a past. Throughout the history of human civilization, every culture has had its own idea of the future. In the words of an Australian futurist, Ivana Milojević, "Although the conception of time and the future exist universally, they are understood in different ways in different societies." Some have assumed that time is circular and that past, present, and future are all part of the same recurring cycle. Others have believed that the only future that matters is in the afterlife.

The crushing disappointments that are so often part of the human condition have sometimes led to crises of confidence in the future, replacing hope with despair. But most have learned from their life experiences and the stories told by their elders that what we do in the present, when informed by knowledge of the past, can shape the future in objectively better ways.

Anthropologists tell us of evidence dating back almost 50,000 years of humans trying to divine the future with the help of oracles or mediums. Some attempted to see into the future by reading clues to the unfolding patterns of life in the entrails of animals sacrificed to the gods, by studying the movements of fish, by interpreting marks on the Earth, or in any of a hundred other ways. Some still read the patterns of palms or Tarot cards for the same purpose. The implicit assumption in such searches is that all reality is of one fabric encompassing past, present, and future, according to a design whose meaning can be divined from particular portions of the whole and applied to other parts of the fabric in order to interpret the unfolding future.

Doctors and scientists now divine clues about the future of individuals from the pattern of DNA that is found in every cell. Mathematicians discern the nature of fractal equations—and the geometric forms derived from them—by observing the "self-sameness" of the patterns they manifest at every level of resolution. Holographic images are contained in their entirety in each molecule of the gaseous cylinders onto which the emergent larger image is projected.

According to historians, astrologers of ancient Babylon used a double clock—one for measuring the timescale of human affairs, and another for tracking the celestial movements they believed had an influence on earthly events. In divining our own future, we too must now pay due attention to a double clock. There is the one that measures our hours and days, and the other that measures the centuries and millennia over which our disruptions of the Earth's natural systems will continue to occur.

Even as teams of scientists race against the clock to compete with other teams in making new genetic discoveries that may cure diseases and lay the foundation for multibillion-dollar products, we must consult another clock that measures the timescales over which evolution operates—because the emergent capabilities bursting forth from the revolutionary advances in the life sciences are about to make us the principal agent of evolution.

Because of the new power that seven billion of us collectively wield with our new technologies, voracious consumption, and outsized economic dynamism, some of the ecological changes that we are setting in motion are going to unfold, the scientists tell us, in geologic time, measured by a planetary clock that tracks timespans that strain the limits of human imagination. Roughly a quarter of the 90 million tons of global warming pollution we put into the atmosphere each day will still linger there—still trapping heat—more than 10,000 years from now.

Consequently, in reconciling the difference between what "is" and what "*ought* to be," we are faced with an existential conundrum. Though we have great difficulty conceiving of geologic time, we have nevertheless become a geologic force; though we cannot imagine evolutionary timescales, we are nevertheless becoming the chief force behind evolution.

The idea that human history is characterized by progress from one era to the next is not, as some have long thought, an invention of the Enlightenment. The explosion of philosophy in ancient Greece marked the beginning of recorded contemplations about the future of humankind. In the fourth century BCE, Plato wrote about progress as "a continuous process, which improves the human condition from its original state of nature to higher and higher levels of culture, economic organization and political structure towards an ideal state. Progress flows from the growing complexity of society and the need to enlarge knowledge, through the development of sciences and arts."

In the fourth century CE, St. Augustine, who frequently quoted Plato, wrote, "The education of the human race, represented by the people of God, has advanced, like that of an individual, through certain epochs, or, as it were, ages, so that it might gradually rise from earthly to heavenly things, and from the visible to the invisible."

Nor is progress exclusively a Western invention. Many interpret the Tao of ancient China as a guide for those who wish to progress as they make their way forward in the world—though its conception of progress is very different from what emerged in the West. The eleventh-century Islamic philosopher Muhammad al-Ghazali wrote that Islam teaches that "Sincere accomplished work towards progress and development is, therefore, an act of religious worship and is rewarded as such. The end result will be a serious, scrupulous and perfect work, true scientific progress and hence actual achievement of balanced and comprehensive development."

At the beginning of the Renaissance, the rediscovery of the Aristotelian branch of ancient Greek philosophy—which had been preserved in Alexandria in Arabic and reintroduced to Europe in Al-Andalus—contributed to a fascination with the physical as well as the philosophical legacies of both Athens and Rome. The legacies of that recovered past nourished dreams that would find fruition in the Enlightenment, when a strong consensus emerged that secular progress is the dominant pattern in human history.

The discoveries of Copernicus, Galileo, Descartes, Newton, and the others who launched the Scientific Revolution helped to ignite a belief that, whatever God's role or plan, the growth of knowledge made progress in human societies inevitable. Francis Bacon, who more than any other emphasized the word "progress" in describing humanity's journey into the future, was also among the first to write about human progress with a special emphasis on subduing, dominating, and controlling nature—as if we were as separate from nature as Descartes believed the mind was separate from the body.

Centuries later, this philosophical mistake is still in need of correction. By tacitly assuming our own separateness from the ecological system of the planet, we are frequently surprised by phenomena that emerge from our inextricable connections to it. And as the power of our civilization grows exponentially, these surprises are becoming increasingly unpleasant.

The cultural legacy that still influences the scientific method is reductionist—that is, by dividing and endlessly subdividing the objects of our research and analysis, we separate interconnected phenomena and processes to develop specialized expertise. But the focusing of attention on ever narrower slices *of* the whole often comes at the expense of attention *to* the whole, which can cause us to miss the significance of emergent phenomena that spring unpredictably from the interconnections and interactions among multiple processes and networks. That is one reason why linear projections of the future are so often wrong.

A NEW VISION OF THE PAST AND THE FUTURE

The invention of powerful new tools and the development of potent new insights—and the discovery of rich new continents—led to exciting new ways of seeing the world and expansive optimism about the future. In the seventeenth century, the father of microbiology, Antonie van Leeuwenhoek, fashioned new lenses for the microscope (which itself had been invented in Holland less than a century earlier), and by looking through them discovered cells and bacteria. Simultaneously, his close friend in Delft, Johannes Vermeer, revolutionized portraiture with the use (most art historians agree) of the camera obscura, made possible by the new understanding of optics.

As the Scientific Revolution accelerated and the Industrial Revolution began, the idea of progress shaped prevailing conceptions of the future. In the years before his death, Thomas Jefferson wrote about the progress he had witnessed in his life and noted, "And where this progress will stop no one can say. Barbarism has, in the meantime, been receding before the steady step of amelioration, and will in time, I trust, disappear from the earth."

Four years after Jefferson's death, the publication by Charles Lyell of his masterwork, *Principles of Geology*, in 1830, profoundly disrupted the long prevailing view of humanity's relationship to time. In the Judeo-Christian world especially, most had assumed that the Earth was only a few thousand years old, and that humans were created not long after the planet itself, but Lyell amply proved that the Earth was not thousands, but at the very least millions of years old (4.5 billion, we now know). In reshaping the past, he also reshaped the idea of the future. And he provided the temporal context for the discovery by Charles Darwin of the principles of evolution. Indeed, as a young man Darwin took Lyell's books with him during his voyage on the *Beagle*.

The previously unimaginable longevity of the past revealed by Lyell inspired symmetrical dreams of distant futures in which the progress of man might reach limitless heights. In the generation that followed Lyell, Jules Verne conjured a future with rockets landing on the moon, a submarine traversing the oceans' depths, and men traveling to the center of the Earth.

The exuberant optimism of the nineteenth century was dampened for many by the excesses of the Second Industrial Revolution, but was revivified during the first decade of the twentieth century with the birth of a political movement based on the belief that progress required governmental policy interventions and social changes in order to ameliorate the problems accompanying industrialization and consolidate its obvious benefits. As the scientific and technological revolution brought some of the visions conjured by Verne and his successors into reality, optimism about the future gained further momentum.

But the balance of the twentieth century brought two world wars and the murder of millions by totalitarian dictators of the left and right to serve their own twisted conceptions of progress—and our view of the future began to change. The malignant nightmare of the Thousand Year Reich, the Holocaust, and the cruelties of Stalin and Mao came to be emblematic of the potential for emergent evil emanating from the use of any means, however horrific, in an effort to impose grand designs for the future of humanity that conformed to the visions of twisted men with too much power.

In the aftermath of World War II, the lingering dismay at the way totalitarian governments had used the wondrous new communications technologies of radio and film to persuade millions to suppress their better instincts and conform their lives to an evil design—coupled with the deep emotional and spiritual impact of the atomic sword of Damocles that the emergence of the nuclear arms race left hanging over civilization— reawakened concerns that new inventions might be double-edged. The uneasiness in the popular mind that powerful technologies—whatever their benefits—might also magnify the innate human vulnerability to hubris deepened for many the loss of their confidence that progress was a reliable guiding star.

The prophecies of Jules Verne were replaced by those of Aldous Huxley, George Orwell, and H. G. Wells, and popular movies about destructive monsters from the ancient past—awakened by nuclear testing or dangerous creatures modified by genetic engineering gone awry—and malevolent robots from the distant future or distant planets, all seemingly bent on ravaging humanity's future.

AND NOW MANY wonder: who are we? Aristotle wrote that the end of a thing defines its essential nature. If we are forced to contemplate the possibility that we might become the architects of our own demise as a civilization, then there are necessarily implications for how we answer the question: what is our essential nature as a species? As a scientist once reframed the question: is the combination of an opposable thumb and a neocortex viable as a sustainable form of life on Earth?

Our natural and healthy preference for optimism about the future is difficult to reconcile with the gnawing concerns expressed by many that all is not well, and that left to its own devices the future may be unfolding in ways that threaten some of the human values we most cherish. The future, in other words, now casts a shadow upon the present. It may be comforting, but of little practical use, to say, "I am an optimist!" Optimism is a form of prayer. Prayer does, in my personal view, have genuine spiritual power. But I also believe, in the words of the old African saying, "When you pray, move your feet." Prayer without action, like optimism without engagement, is passive aggression toward the future.

Even those who understand the different dangers we are facing and are committed to taking action often feel stymied by a sense of powerlessness. On the issue of climate, for example, they change their own behaviors and habits, reduce their impact on the environment, speak out and vote, but still feel they are having precious little impact, because the powerful momentum of the global machine we have built to give us progress seems almost independent of human control. Where are the levers to pull, the buttons to push? Is there a steering mechanism? Do our hands have enough strength to operate the controls?

More than a decade before writing *Faust*, Goethe wrote his well-known poem "The Sorcerer's Apprentice" about a young trainee who, left to his own devices, dared to use one of his master's magic spells in order to bring to life the broom he was supposed to be using to clean the workshop. But once animated, the broom could not be stopped. Growing desperate to halt the broom's increasing frenzy of activity, the apprentice split the broom with an axe—which caused it to self-replicate, with each half growing into another new animated broom. Only when the master returned was the process brought back under control.

DEMOCRATIC CAPITALISM AND ITS DISCONTENTS

The idea of making truly meaningful collective decisions in democracy that are aimed at steering the global machinery we have set in motion is naïve, even silly, according to those who have long since placed their faith in the future not in human hands, but in the invisible hand of the marketplace. As more of the power to make decisions about the future flows from political systems to markets, and as ever more powerful technologies magnify the strength of the invisible hand, the muscles of self-governance have atrophied.

That is actually a welcome outcome for some who have found ways to accumulate great fortunes from the unrestrained operations of this global machinery. Indeed, many of them have used their wealth to reinforce the idea that self-governance is futile at best and, when it works at all, leads to dangerous meddling that interferes with both markets and technological determinism. The ideological condominium formed in the alliance between capitalism and representative democracy that has been so fruitful in expanding the potential for freedom, peace, and prosperity has been split asunder by the encroachment of concentrated wealth from the market sphere into the democracy sphere.

Though markets have no peer in collecting, processing, and utilizing massive flows of information to allocate resources and balance supply with demand, the information in markets is of a particularly granular variety. It is devoid of opinion, character, personality, feeling, love, or faith. It's just numbers. Democracy, on the other hand, when it operates in a healthy pattern, produces from the interactions of people with different perspectives, predispositions, and life experiences emergent wisdom and creativity that is on a completely different plane. It carries dreams and hopes for the future. By tolerating the routine use of wealth to distort, degrade, and corrupt the process of democracy, we are depriving ourselves of the opportunity to use the "last best hope" to find a sustainable path for humanity through the most disruptive and chaotic changes civilization has ever confronted.

In the United States, many have cheered the withering of self-governance and have celebrated the notion that we should no longer even try to control our own destiny through democratic decision making. Some have recommended, only half in jest, that government should be diminished to the point where it can be "drowned in the bathtub." They have enlisted politicians in the effort to paralyze the ability of government to serve any interests other than those of the global machine, recruited a fifth column in the Fourth Estate, and hired legions of lobbyists to block any collective decisions about the future that serve the public interest. They even seem to sincerely believe, as many have often written, that there is no such thing as "the public interest."

The new self-organized pattern of the Congress serves the special interests that are providing most of the campaign money with which candidates—incumbents and challengers alike—purchase television commercials. It no longer responds to any but the most emotional concerns of the American people. Its members are still "representatives," but the vast majority of them now represent the people and corporations who donate money, not the people who actually vote in their congressional districts.

The world's need for intelligent, clear, values-based leadership from the United States is greater now than ever before—and the absence of any suitable alternative is clearer now than ever before. Unfortunately, the decline of U.S. democracy has degraded its capacity for clear collective thinking, led to a series of remarkably poor policy decisions on crucially significant issues, and left the global community rudderless as it faces the necessity of responding intelligently and quickly to the implications of the six emergent changes described in this book. The restoration of U.S. democracy, or the emergence of leadership elsewhere in the world, is essential to understanding and responding to these changes in order to shape the future.

One of the six drivers of change described in this book—the emergence of a digital network connecting the thoughts and feelings of most people in every country of the world—offers the greatest source of hope that the healthy functioning of democratic deliberation and collective decision making can be restored in time to reclaim humanity's capacity to reason together and chart a safe course into the future.

Capitalism—if reformed and made sustainable—can serve the world better than any other economic system in making the difficult but necessary changes to the relationship between the human enterprise and the ecological and biological systems of the Earth. Together, sustainable capitalism and healthy democratic decision making can empower us to save the future. So we have to think clearly about how both of these essential tools can be repaired and reformed.

The structure of these decision-making systems and the ways in which we measure progress—or the lack thereof—toward the goals we decide are important have a profound influence on the future we actually create. By making economic choices in favor of "growth," it matters a lot which definition of growth we use. If the impact of pollution is systematically removed from the measurement of what we call "progress," then we start to ignore it and should not be surprised when much of our progress is accompanied by lots of pollution.

If the systems we use for recognizing and measuring profit are based on a narrow definition—for example, quarterly projections of earnings per share, or quarterly unemployment statistics that don't include people who have given up looking for work, those who have been forced to take large pay cuts in order to continue working, or those who are flipping hamburgers instead of using higher-value skills hard won with education or prior experience—then what we are seeing is an imperfect and partial representation of a much larger reality. When we become accustomed to making important choices about the future on the basis of distorted and misleading information, the results of those decisions are more likely to fall short of our expectations.

Psychologists and neuroscientists have studied a phenomenon called selective attention—a tendency on the part of people who are so determined to focus intensely on particular images that they become oblivious to other images that are present in the field of vision.

We select the things to which we pay attention not only by curiosity, preference, and habit, but also through our selection of the observational tools, technologies, and systems we rely on in making choices. And these tools implicitly mark some things as significant and obscure others to the point that we completely ignore them. In other words, the tools we use can have their own selective attention distortions.

For example, the system of economic value measurement known as gross domestic product, or GDP, includes some values and arbitrarily excludes others. So when we use GDP as a lens through which to observe economic activity, we pay attention to that which is measured and tend to become oblivious to those things that are not measured at all. British mathematician and philosopher Alfred North Whitehead called the obsession with measurements "the fallacy of misplaced concreteness."

Here is a metaphor to illustrate the point: the electromagnetic spectrum is often portrayed as a long thin horizontal rectangle divided into differently colored segments that represent the different wavelengths of electromagnetic energy—usually ranging from very low frequency wavelengths like those used for radio on the left, extending through microwaves, infrared, ultraviolet, X-rays, and the like, to extreme high frequency gamma radiation at the right end of the rectangle.

Somewhere near the middle of this rectangle is a very thin section representing visible light—which is, of course, the only part of the entire spectrum that can be seen with the human eye. But since the human eye is normally the only "instrument" with which most of us attempt to "see" the world around us, we are naturally oblivious to all of the information contained in the 99.9 percent of the spectrum that is invisible to us.

By supplementing our natural vision with instruments capable of "seeing" the rest of the spectrum, however, we are able to enhance our understanding of the world around us by collecting and interpreting much more information. During the eight years I worked in the White House, I started every day, six days a week, with a lengthy briefing from the intelligence community on all the issues affecting national security and vital U.S. interests, and it routinely contained information collected from almost all parts of the electromagnetic spectrum. It was, as a result, a much more complete and accurate picture of a very complex reality.

One of the current realities in the business world that has been most surprising to me is the near consensus that markets are "short on long and long on short"—that is, there is an unhealthy focus on very short-term goals, to the exclusion of long-term goals. If the incentives routinely provided for business leaders—and political leaders—are focused on extremely short-term horizons, then no one should be surprised if the decisions they make in pursuit of the rewards to be gained are also focused on the short term—at the expense of any consideration of the future. Compensation and incentive structures reinforce these biases and penalize most CEOs and businesses that dare to focus on more sustainable longer-term strategies. "Short-termism" has long since become a frequently used buzzword in business circles. In both business and politics, short-term decision making is dominant.

"Quarterly capitalism" is a phrase some use to describe the prevailing practice of managing businesses from one three-month period to the next, and focusing budgets and strategies on the constant effort to ensure that each quarter's earnings per share report never fails to meet projections or the market's expectations. When investors and CEOs focus on a definition of "growth" that excludes the health and well-being of the communities where businesses are located, the health of the employees who do most of the work, and the impact of the businesses' operations on the environment, they are tacitly choosing to ignore material facts with the potential to make real growth unsustainable.

Similarly, the dominance of money in modern politics—particularly in the United States—has now led to what might be described as "quarterly democracy." Every ninety days, incumbent officeholders running for reelection and challengers in political contests are required to publicly report their fundraising totals for the previous ninety days. At the end of each of these quarters, there is a flurry of fundraising events, email solicitations, and fundraising telephone calls to maximize the amount that can be reported—much as a puffer fish increases its perceived size in the presence of another puffer fish encroaching on its territory.

Our evolutionary heritage has made us vulnerable to numerous stimuli that trigger short-term thinking. Though we also have the capacity for long-term thinking, of course, it requires effort, and neuroscientists tell us that distractions, stress, and fear easily disrupt the processes by which we focus on the longer term. When elected officials are under constant systemic stress to focus intently on short-term horizons, the future gets short shrift.

This is particularly dangerous during a period of rapid change. Some of the trends now under way are so well documented by observations in the past that projections of those same trends into the future can be made with a very high degree of confidence. The rate of advancement in computer chips, to pick a well-known example, is understood more than well enough to justify predictions that computer chips will continue to advance rapidly in the future.

The speedy drop in the cost of sequencing DNA has occurred for reasons that are understood more than well enough to justify predictions that this trend too will continue to shape our future. The accumulation of greenhouse gases in the past and the rise in global temperatures they have caused is also understood more than well enough to justify predictions of what will happen to global temperatures if we continue to increase emissions at the same rate in the future—and what the consequences of much higher global temperatures would be.

Other changes, however, burst upon the world seemingly fully formed: a brand-new pattern that represents a sudden shift from an older pattern that persisted for as far back in the past as humans can recall. In our own lives, we are accustomed to gradual, linear change. But sometimes the potential for change builds up without being visibly manifested until the inchoate pressure for change reaches a critical mass powerful enough to break through whatever systemic barriers have held the change back. Then suddenly one pattern gives way to another that is entirely new. This "emergence" of systemic change is often difficult to predict, but does occur frequently both in nature and in complex systems designed by human beings.

MANY WHO WERE once fascinated and excited about the possibilities of the future are now focused solely on the implications of the future's potential for the business, political, and security strategies of the present. As the Scientific Revolution accelerated in the last decades of the twentieth century, corporate planners and military strategists began to devote considerably more attention to the study of alternative futures, motivated by a concern that the potency of new scientific and technological discoveries could threaten the strategic interests—or even survival—of business models and the balance of power among nations.

What is our present conception of the future? How does our image of the future affect the choices we are making in the present? Do we still believe that we have the power to shape our collective future on Earth and choose from among the alternative futures one that preserves our deepest values and makes life better than it is in the present? Or do we have our own crisis of confidence in humanity's future?

If the spectrum of past, present, and future were displayed as a long thin rectangle similar to that used to portray the electromagnetic spectrum, the birth of Planet Earth 4.5 billion years ago would be at the far left end. Moving to the right, we would see the emergence of life 3.8 billion years ago, the appearance of multicellular life 2.8 billion years ago, the appearance of the first plant life on land 475 million years ago, the first vertebrates more than 400 million years ago, and the first primates 65 million years ago. Then, moving all the way to the right end of the rectangle, the death of the sun would appear 7.5 billion years from now.

The narrow slice of time to the left of the midpoint in this spectrum—the one that represents the history of the human species—is an even narrower slice of the spectrum of time than is visible light of the electromagnetic spectrum. The thoughts we devote to these vast stretches of time in the past and future are often fleeting at best.

There are ample reasons for optimism about the future. For the present, war seems to be declining. Global poverty is declining. Some fearsome diseases have been conquered and others are being held at bay. Lifespans are lengthening. Standards of living and average incomes—at least on a global basis—are improving. Knowledge and literacy are spreading. The tools and technologies we are developing—including Internet-based communication—are growing in power and efficacy. Our general understanding of our world, indeed, our universe (or multiverse!) has been growing exponentially. There have been periods in the past when limits to our growth and success as a species appeared to threaten our future, only to be transcended by new advances—the Green Revolution of the second half of the twentieth century, for example.

So the positive and negative sets of trends are occurring simultaneously. The fact that some are welcome and others are not has an effect on our perception of them. The unwelcome trends are sometimes ignored, at least in part because they are unpleasant to think about. Any uncertainty about them that can be conjured to justify inaction is often seized upon with enthusiasm, while new hard evidence establishing their reality is often resisted with even stronger denial of the reality the evidence supports.

Just as naïve optimism can amount to self-deception, so too can a predisposition to pessimism blind us to bases for legitimate hope that we can find a path that leads around and through the dangers that lie ahead. Indeed, I *am* an optimist—though my optimism is predicated on the hope that we will find ways to see and think clearly about the obvious trends that are even now gaining momentum, that we will reason together and attend to the dangerous distortions in our present ways of describing and measuring the powerful changes that are now under way, that we will actively choose to preserve human values and protect them, not least against the mechanistic and destructive consequences of our baser instincts that are now magnified by technologies more powerful than any that those in previous generations, even Jules Verne, could have imagined. I have tried my best to describe what I believe the evidence shows is more likely than not to present us with important choices that we must consciously make together. I do so not out of fear, but because I believe in the future.

THE
FUTURE

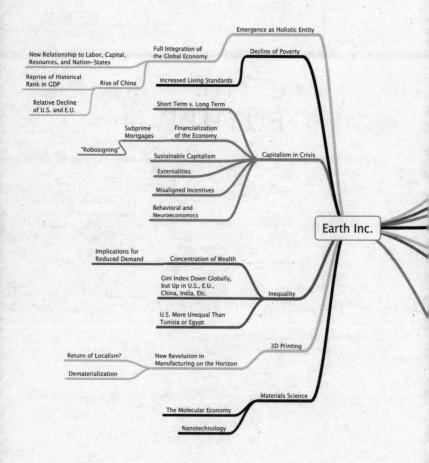

Emergence as Holistic Entity

Full Integration of
the Global Economy

New Relationship to Labor, Capital,
Resources, and Nation-States

Decline of Poverty

Reprise of Historical
Rank in GDP   Rise of China

Increased Living Standards

Relative Decline
of U.S. and E.U.

Short Term v. Long Term

Subprime   Financialization
Mortgages   of the Economy

"Robosigning"

Sustainable Capitalism   Capitalism in Crisis

Externalities

Misaligned Incentives

Behavioral and
Neuroeconomics

Earth Inc.

Implications for
Reduced Demand   Concentration of Wealth

Gini Index Down Globally,
but Up in U.S., E.U.,
China, India, Etc.   Inequality

U.S. More Unequal Than
Tunisia or Egypt

3D Printing

Return of Localism?   New Revolution in
                             Manufacturing on the Horizon

Dematerialization

Materials Science

The Molecular Economy

Nanotechnology

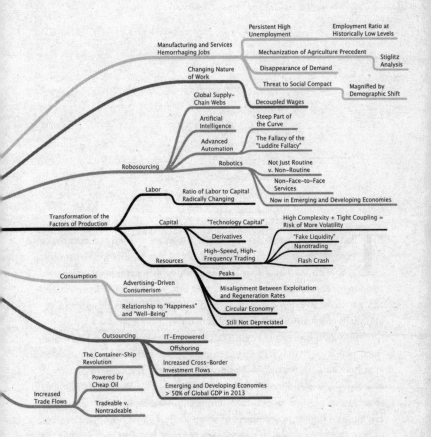

Persistent High Unemployment

Employment Ratio at Historically Low Levels

Manufacturing and Services Hemorrhaging Jobs

Mechanization of Agriculture Precedent

Stiglitz Analysis

Changing Nature of Work

Disappearance of Demand

Threat to Social Compact

Magnified by Demographic Shift

Global Supply-Chain Webs

Decoupled Wages

Artificial Intelligence

Steep Part of the Curve

Advanced Automation

The Fallacy of the "Luddite Fallacy"

Robosourcing

Robotics

Not Just Routine v. Non-Routine

Labor

Non-Face-to-Face Services

Ratio of Labor to Capital Radically Changing

Now in Emerging and Developing Economies

Transformation of the Factors of Production

Capital

"Technology Capital"

High Complexity + Tight Coupling = Risk of More Volatility

Derivatives

"Fake Liquidity"

Nanotrading

High-Speed, High-Frequency Trading

Resources

Flash Crash

Consumption

Peaks

Advertising-Driven Consumerism

Misalignment Between Exploitation and Regeneration Rates

Relationship to "Happiness" and "Well-Being"

Circular Economy

Still Not Depreciated

Outsourcing

IT-Empowered

Offshoring

The Container-Ship Revolution

Increased Cross-Border Investment Flows

Powered by Cheap Oil

Emerging and Developing Economies > 50% of Global GDP in 2013

Increased Trade Flows

Tradeable v. Nontradeable

EARTH INC.

T HE GLOBAL ECONOMY IS BEING TRANSFORMED BY CHANGES FAR greater in speed and scale than any in human history. We are living with, and in, Earth Inc.:* national policies, regional strategies, and long accepted economic theories are now irrelevant to the new realities of our new hyper-connected, tightly integrated, highly interactive, and technologically revolutionized economy.

Many of the most successful large enterprises in the world now produce goods in "virtual global factories," with intricate spiderwebs of supply chains connecting to hundreds of other enterprises in dozens of countries. More and more markets for goods—and increasingly services that do not require face-to-face interaction—are now global in nature. Higher and higher percentages of wage earners must now compete not only with wage earners in every other country, but also with intelligent machines interconnected with other machines and computer networks.

The digitization of work and the dramatic and relatively sudden metastasis of what used to be called automation are driving two massive changes simultaneously:

---

* This term was first coined by Buckminster Fuller in 1973, but he used it to convey a completely different meaning.

1. The outsourcing of jobs from industrial economies to developing and emerging economies with large populations and lower wages; and

2. The robosourcing of jobs from human beings to mechanized processes, computer programs, robots of all sizes and shapes, and still rudimentary versions of artificial intelligence that are improving in their efficacy, utility, and power with each passing year.

The transformation of the global economy is best understood as an emergent phenomenon—that is, one in which the whole is not only greater than the sum of its parts, but very different from the sum of its parts in important and powerful ways. It represents something new—not just a more interconnected collection of the same national and regional economies that used to interact with one another, but a completely new entity with different internal dynamics, patterns, momentum, and raw power than what we have been familiar with in the past. There are limits to cross-border flows of people, of course, and trade flows are stronger among countries that are close to one another, but the entire global economy has been knit together much more tightly than ever before.

Just as the thirteen American colonies in North America *emerged* as a unified whole in the last quarter of the eighteenth century—and just as the ancient walled city-states of Italy eventually became a unified nation in the second half of the nineteenth century—the world as a whole has now *emerged* as a single economic entity that is moving quickly toward full integration. At least that is the reality in the world of commerce and industry, in the world of science, and in the rapid spreading of most new technologies to centers of commerce throughout the world.

In the world of politics and governmental policy, nation-states remain the dominant players. Psychologically, emotionally, and in the ways we frame our identity, most of us still think and act as if we are still living in the world we knew when we were young. In fact, however, where the economic realities of life are concerned, that world is receding from view.

This powerful driver of global change—sometimes loosely and inadequately referred to as "globalization"—marks not only the end of one era in history and the beginning of another, it marks the emergence of a completely new reality with which we as human beings must come to grips.

OUTSOURCING AND ROBOSOURCING have typically been seen as two separate and distinct phenomena—studied and discussed by different groups of economists, technologists, and policy experts. Yet they are deeply intertwined and represent two aspects of the same mega-phenomenon.

The tectonic shift toward robosourcing *and* IT-empowered outsourcing dramatically changes the ratio of capital inputs to labor inputs and weakens the ability of working people to demand higher wages in industrial countries.

The political battles over labor rights in the first half of the twentieth century were fought to determine the relative distribution of income from labor and capital in enterprises where workers were organized. But technology-driven changes are now playing a much larger role in determining the future of work and what people earn in return for it. Arguments that used to occur in a zero-sum context no longer seem as relevant or persuasive when employers have the readily available options to: (a) simply close the factory or business and replicate it in a low-wage country, or (b) replace the labor with robots and automated systems.

From the standpoint of factory workers in the United States or Europe whose jobs are eliminated, the impact of automation and outsourcing is essentially the same. From the standpoint of the factory owner, productivity figures typically go up as a result of both offshoring *and* robosourcing—whether the new technology is deployed in the existing facility or in some foreign country.

Policymakers often count the result as a success because increased productivity is regarded as equivalent to the Holy Grail of progress. Yet they are often blind to the full impact of this process on employment in the country where the companies credited with productivity growth are nominally located, even though the trend is now accelerating to the point where the fundamental role of labor in the economy of the future is being called into question.

One manifestation of how the accelerating interconnection of the global economy drives both outsourcing *and* robosourcing simultaneously is that robosourcing is also occurring more and more rapidly in emerging and developing economies, and is beginning to eliminate a growing percentage of the jobs that were so recently outsourced from the advanced industrial economies.

There is a big difference between the investment of money in an offshore factory to replicate the same jobs that used to be located in the

West, and the provision of what economists are beginning to label "technological capital"—investments that not only increase the productivity of business and industry, but over time eliminate large numbers of jobs both in the countries that originally lose the factories as well as in the countries to which they are relocated.

The workers in lower-wage countries initially benefit from the new employment opportunities—until the improved living standards they help to produce lead them to demand higher wages themselves. Then they too become vulnerable to being replaced when the factory owners are able to purchase ever improved—and ever cheaper—robots and automated processes with the new profits they have freshly earned as a result of outsourcing from the West. One Chinese consumer electronics manufacturer, Foxconn, announced in 2012 that it would soon deploy one million new robots within two years.

A positive feedback loop has emerged between Earth Inc.'s increasing integration on the one hand, and the progressive introduction of interconnected intelligent machines on the other. In other words, both of these trends—increased robosourcing and the interconnectedness of the global economy driven by trade and investment—reinforce one another.

The impact of robosourcing on employment is sometimes misunderstood as a process in which entire categories of employment are completely eliminated when a technological breakthrough suddenly results in the replacement of people with intelligent interconnected machines. Far more common, however, is that the intelligent networked machines replace a significant percentage of the jobs while greatly enhancing the productivity of the smaller number of the employees remaining by empowering them to leverage the efficiency of the machines that are now part of the production process alongside them.

The jobs that remain sometimes command higher wages in return for the new skills required to work with the new technology. And this pattern reinforces our tendency to misunderstand the aggregate impact of this new acceleration of robosourcing and see it as part of the long familiar pattern by which old jobs are eliminated and replaced by new and better jobs.

But what is different today is that we are beginning to climb the steep part of this technology curve, and the aggregate impact of this same process occurring in multiple businesses and industries simultaneously produces a large decline in employment. Moreover, many employees lack skills (in decimal arithmetic, for example, which is necessary to operate many robots) that they need to fill the new jobs.

New companies have emerged to connect online workers with jobs that can be cheaply and efficiently outsourced over the Internet. Gary Swart, the CEO of one of the more successful online job brokerages, oDesk, said he is seeing increased demand across the board, including for "lawyers, accountants, financial executives, even managers." And robosourcing is beginning to have an impact on journalism. Narrative Science, a robot reporting company founded by two directors of Northwestern University's Intelligent Information Laboratory, is now producing articles for newspapers and magazines with algorithms that analyze statistical data from sporting events, financial reports, and government studies. One of the cofounders, Kristian Hammond, who is also a professor at the Medill School of Journalism, told me that the business is expanding rapidly into many new fields of journalism. The CEO, Stuart Frankel, said the few human writers who work for the company have become "meta-journalists" who design the templates, frames, and angles into which the algorithm inserts data. In this way, he said, they "can write millions of stories as opposed to a single story at a time."

THE CUMULATIVE EFFECT of the accelerating introduction of machine intelligence and the relocation of work to low-wage countries is also creating much greater inequality of incomes and net worth—not only in developed countries, but in the emerging economies as well. Those who lose their jobs have less income, while those who benefit from the increasing relative value of technological capital have increased income.

THE GLOBAL WEALTH GAP

As this shift in the relative value of technology to labor continues to accelerate, so too will the levels of inequality. This phenomenon is not in the realm of theory. It is happening right now on a large scale. As technological capital becomes more and more important compared to the value of labor, more and more of the income derived from productive activities is becoming more and more concentrated in the hands of fewer and fewer elites, while a much larger number of people suffer the harm of lost income.

There is a growing concentration of wealth at the top of the income ladder in almost every industrial country and emerging nations like China and India. Latin America is the rare exception. Globally, technological offshoring has at least temporarily improved the equality of income, because of the massive transfer of industrial—and now service—jobs to lower-wage countries as a group. On a nation by nation basis, though, inequality of income distribution—and of net worth—is increasing even faster in China and India than in the U.S or Europe. And income inequality reached a twenty-year high in 2012 in thirty-two developing countries surveyed by the global NGO Save the Children.

Over the past quarter century, the Gini coefficient—which measures inequality of income nation by nation on a scale from 0 to 100 (from everyone having the same income at 0 to one person having all the nation's income at 100)—has risen in the United States from 35 to 45, in China from 30 to the low 40s, in Russia from the mid 20s to the low 40s, and in the United Kingdom from 30 to 36. These nationwide numbers can obscure even more dramatic impacts within the wage ladder. For example, according to the OECD, the Organisation for Economic Cooperation and Development, the top 10 percent of wage earners in India now make more than twelve times what the bottom 10 percent make compared to six times just two decades ago.

The growing inequality of income and net worth in the United States has also been driven by changes in tax laws that favor those in higher-income brackets, including the virtual elimination of inheritance taxes and especially the taxation of investment income at the lowest tax rate of all—15 percent. When the tax rate imposed on income from capital investments is significantly lower than the tax rates imposed on income earned in return for labor or from those who sell the natural resources used in the process, then the ratio of income flowing to those providing the capital naturally increases.

In the United States 50 percent of all capital gains income goes to the top one thousandth of one percent. The current political ideology that supports this distribution of income refers to these wealthy investors as "job creators," but with robosourcing and outsourcing, the cumulative impact of the capital they provide is, whatever its beneficial effects, negative in terms of jobs.

It is interesting to note that the United States now has more inequality than either Egypt or Tunisia. The Occupy Wall Street movement caught fire because of a broad awakening to the dramatic increase in the concentration of wealth held by the top one percent, who now have more wealth than the people in the bottom 90 percent. The wealthiest 400 Americans—all of them billionaires—have more wealth as a group than the 150 million Americans in the bottom 50 percent. The five children and one daughter-in-law of Sam and Bud Walton (the founders of Walmart) have more wealth than the bottom 30 percent of Americans.

In terms of annual income, the top one percent now receive almost 25 percent of all U.S. income annually, up from 12 percent just a quarter century ago. While the after-tax income of the average American climbed only 21 percent over the last twenty-five years, the income of the top 0.1 percent increased over the same period by 400 percent.

Now that many jobs in services as well as manufacturing and agriculture are all subject to progressive dislocation by the innovation and productivity curves that measure the accelerating impact of the underlying technology revolution, the need for income replacement is becoming acute.

By 2011, the cumulative investment by industrial countries in the rest of the world had increased eightfold over the previous thirty years, in the process growing from 5 to 40 percent of the GDP in developed countries. While overall world GDP is projected to increase by almost 25 percent in the next five years, cross-border capital flows are expected to continue increasing three times faster than GDP.

The cumulative investment by the rest of the world in advanced economies is also growing—though not by as much. Stocks of foreign direct investment in industrialized countries like the United States increased from 5 to 30 percent of GDP from 1980 to 2011. Partly as a result, these global trends have not only eliminated jobs in the U.S. but also created many new ones. Foreign-owned automobile companies, for example, now employ almost a half million people in the United States, paying them wages that are 20 percent higher than the national average.

Overall, foreign-majority-owned companies now provide jobs for more than five million U.S. citizens. And many other jobs have been created in companies that serve as suppliers and subcontractors to foreign companies. For example, even though China now dominates the manufacture of solar panels, the United States has a positive balance of trade with China in the solar sector—because of U.S. exports to China of processed polysilicon and advanced manufacturing equipment.

Nevertheless, the impacts of this global economic revolution are already producing a tectonic reordering of the relative roles of the United

States, Europe, China, and other emerging economies. China's economy, one third the size of the United States' economy only ten years ago, will surpass the U.S. as the largest economy in the world within this decade. Indeed, China has already moved beyond America in manufacturing output, new fixed investment, exports, steel consumption, energy consumption, CO_2 emissions, car sales, new patents granted to residents, and mobile phones. It now has twice the number of Internet users. China's rise has become the most powerful symbol of the new pattern in the global economy quickly supplanting the one long associated with U.S. dominance.

The consequences of this transformation in the global economy are beginning to be manifested in unusually high rates of persistent unemployment and underemployment—and a slowdown in the demand for goods and services in consumer-oriented economies. The loss of middle-income jobs in industrial countries can no longer be blamed primarily on the business cycle—the alternating periods of recession and recovery that bring jobs in and out like the tide. Cyclical factors still account for considerable job gains and losses, but virtually all industrial countries seem perplexed and powerless in their efforts to create jobs with adequate wages, and are struggling with how to replace consumer demand for goods and services to reignite and/or solidify another recovery phase in the business cycle.

In the United States, the last ten years represents the only decade since the Great Depression when there have been zero net jobs added to the economy. During the same ten years, productivity growth has been higher than in any decade since the 1960s. Along with productivity, corporate profits have resumed healthy rates of increase while unemployment has barely declined. U.S. business spending on equipment and software increased by almost 30 percent while spending on private sector jobs increased by only 2 percent. Significantly, orders for new industrial robots in North America increased 41 percent.

Overall, the technology-enhanced integration of the global economy is lifting the relative economic strength of developing and emerging countries. This year (2013) the GDP of this group of countries (as measured by their purchasing power) will surpass the combined GDP of advanced economies for the first time in the modern era. The potential incapacity of these countries to maintain political and social stability and to deal with governance and corruption challenges may yet interrupt this trend. But the technological drivers of their ascent are powerful and are likely to prevail in consolidating and increasing a dramatic and truly fundamental change in the balance of global economic power. Already, in the aftermath of the Great Recession, it is the emerging economies that have become the principal engines of global growth. As a group they are growing much faster than the developed countries. Some analysts doubt the sustainability of these growth rates. But whatever their rate of growth, it is only a matter of time before these economies experience the same hemorrhaging of jobs to intelligent machines that is well under way in the West.

MOST PEOPLE AND political leaders in advanced industrial countries still attribute the disappearance of middle-income jobs simply to offshoring, without focusing on the underlying cause: the emergent reality of Earth Inc., and the deep interconnection between outsourcing and robosourcing. This misdiagnosis has led in turn to divisive debates over proposals to cut wages, impose trade restrictions, drastically change the social compact between old and young and rich and poor, and cut taxes on wealthy investors to encourage them to build more factories in the West.

These distracting and almost pointless arguments over labor policies are echoed in similarly misguided debates over the impact of national policies on financial flows in the age of Earth Inc. The nature and volume of capital movements in the ever more tightly interconnected global economy are being transformed by supercomputers and sophisticated software algorithms that now handle the vast majority of financial transactions with a destructive emphasis on extremely short-term horizons. One consequence of this change is a new level of volatility and contagion in the global economy *as a whole*. Major market disruptions are occurring with greater frequency and are reverberating more widely throughout the world.

THE NEED FOR SPEED

The sudden disruption in credit markets that began in 2008, and the global recession it triggered, resulted in the loss of 27 million jobs worldwide. When the period of weak recovery began one year later, global output started to increase again but the number of jobs restored— particularly in industrial countries—lagged far behind. Many economists attributed the jobless nature of the recovery to a new eagerness by employers to introduce new technology instead of hiring back more people.

Exotic, computer-driven "manufactured financial products" like the ones that led to the Great Recession now represent capital flows with a notional value twenty-three times larger than the entire global GDP. These so-called derivatives are now traded every day in volumes forty times larger than all of the daily trades in all of the world's stock markets put together. Indeed, even when the larger market in bonds is added to the market in stocks, the estimated value of derivatives is now thirteen times larger than the combined value of every stock and every bond on Earth.

The popular image of trading floors is still one where people yell at one another while making hand signals, but human beings have a much smaller role in the flows of capital in global markets now that they are dominated by high-speed, high-frequency trades made by supercomputers. In the United States, high-speed, high-frequency trading represented more than 60 percent of all trades in 2009. By 2012, in Europe as well as the U.S., it represented more than 60 percent of all trades. Indeed, stock exchanges now compete with one another with propositions like one from the London Exchange, which recently advertised its ability to complete a transaction in 124 microseconds (millionths of a second). More advanced algorithms will soon make trades in nanoseconds (billionths of a second), which according to some experts will further increase the risks of market disruptions.

An academic expert in automated trading at the University of Bristol, John Cartlidge, said recently that the result of the increasing speed of trades "is that we now live in a world dominated by a global financial market of which we have virtually no sound theoretical understanding." In the first week of October 2012, a single "mystery algorithm" accounted for 10 percent of the bandwidth allowed for trading on the U.S. stock market—and 4 percent of all traffic in stock quotes. Experts suspected the motivation was to slow down data speeds in order to enhance the advantage for the high-speed computer trader.

Advantages in the speed of information flow have played an important role in markets for at least 200 years, since the Rothschild bank used carrier pigeons to get early word of Napoleon's defeat at Waterloo, enabling them to make a fortune by shorting French bonds. Fifty years later, an American investor chartered faster sailboats to gain earlier knowledge of key battles in the U.S. Civil War and make a similar fortune by shorting bonds from the Confederacy. But the emphasis on speed has now reached absurd levels. Trading firms routinely place their supercomputers adjacent to their trading floors—because even at the speed of light, the amount of time it takes for the information to cross the street from another building would confer a competitive disadvantage.

A few years ago, a business friend in Silicon Valley told me about an opportunity to invest in an unusual project to build a straight-as-an-arrow fiber optic cable from Chicago's trading center in the inner Loop to the New York Stock Exchange's trading center in Mahwah, New Jersey. The inherent value of the project—since completed—came from its promise to shave three milliseconds off the time it took to transmit information over the 825 miles (from 16.3 to 13.3 milliseconds). Traders at the other end of the cable gain such a significant advantage with a three-millisecond head start over their competitors that access to this new cable is being sold at premium prices. An even newer microwave system with even faster data speeds (though less reliability in bad weather) is now being built along the same route.

The melting of the North Polar Ice Cap has led to the start of a new project to connect markets in Tokyo and New York with faster financial information flows via a fiber cable along the bottom of the Arctic Ocean. Three other projects have commenced to link Japan and Europe under the Arctic, and a new transatlantic cable being built for another $300 million is expected to increase the speed of data flows between New York and London by 5.2 milliseconds.

The spending of $300 million to save a few milliseconds in the flow of information is but one tiny example of how much of the wealth that used to be allocated to inherently productive activities is now diverted instead toward what many economists call the financialization of the economy. The share of the American economy now devoted to the financial sector has doubled from around 4 percent in 1980 to more than 8 percent at present.

Part of this startling increase reflected the large investments that financed the information technology explosion up to April of 2000, and part represented the rapid growth in mortgages that accompanied the buildup of the housing bubble up to 2008. Yet even after the bursting of the dot-com bubble, and later, the housing bubble, the financial services sector continued to gain a larger share of GDP. The driving force behind this historic shift has been the application of powerful supercomputers and algorithms to the manufacture of exotic financial derivatives—and the capitulation by government in the face of lobbying by the financial services industry for the relaxation of regulatory standards that used to impede the marketing of such instruments.

An estimated 82 percent of derivatives are exotic instruments based on interest rates, almost 11 percent are based on foreign exchange contracts, and roughly 6 percent are based on credit derivatives. Less than one percent are based on the value of actual commodities. But the overall flows are so incredibly large that, to pick one example, the value of oil derivatives traded on a typical day is an astonishing fourteen times the value of all the actual barrels of oil traded on that same day.

In theory, these high-volume, high-frequency computer-driven flows are justified by the assertion that they improve the liquidity and efficiency of markets. Many economists and bankers hold the view that the large flows of capital represented by derivatives actually add to the stability of markets and do not increase systemic risk, in part because banks hold collateral equal to a large percentage of what they are trading.

Others, however, point out that this view is based on the now obsolete assumption that more liquidity is always better—an assumption that is in turn based on two theories about markets that are part of the long discredited "standard model": that markets tend toward equilibrium (they don't), and that "perfect information" is implicitly reflected in the collective behavior found in the market (it isn't). Nobel Prize–winning economist Joseph Stiglitz says that high-speed trading produces only "fake liquidity."

THE CHALLENGE OF COMPLEXITY

Unlike trades in the stock and bond markets, derivatives trades are almost completely and totally unregulated. That adds to the risk of increased volatility in markets, especially when the daily volume of electronic transfers of capital now exceeds the combined total of all of the reserves in the central banks of all advanced countries. In practice, the progressive displacement of human decision making from the process and the explosion in the trading of artificial financial instruments in volumes that dwarf the transactions of real value in the global economy has contributed to the increased frequency of major dislocations in the role of capital as a reliable and efficient factor of production. Some of the artificial instruments now being traded in high volumes are difficult to distinguish from gambling.

There are two factors that explain the underlying reason why the management of global capital flows by supercomputers in microsecond intervals creates new systemic risks in markets: extreme complexity and tight coupling. And they work in combination. First, the complexity of the system sometimes produces large and troublesome anomalies caused by a form of "algorithmic harmonics" (essentially computer programs reacting to one another's simultaneous operations rather than underlying market realities). This complexity means that the problems thus introduced into the system's operation can be extremely difficult for any actual human being to understand without taking a considerable amount of time to get to the bottom of what has gone wrong. Second, the tight coupling of multiple supercomputers ensures that no one will have the luxury of time to figure out what's gone wrong, much less the time to address it.

One example: on May 6, 2010, the value of the New York Stock Exchange fell a thousand points and rebounded almost as much—all in the time span of sixteen minutes—for no apparent reason. There was no market-sensitive news of the kind normally associated with a sharp drop of that magnitude in such a short timespan. As *The New York Times* reported the following day, "Accenture fell more than 90 percent to a penny, P&G plunged to $39.37 from more than $60 within minutes." The *Times* quoted one trader as saying "it was almost like 'the Twilight Zone.'"

It required five months of intensive work by specialists to understand what had happened to cause this so-called Flash Crash, which they eventually found was the result of the complex interaction between automatic trading algorithms used by a large number of supercomputers in a way that created, in effect, an algorithmic echo chamber that caused prices to suddenly crash.

When one of these experts, Joseph Stiglitz, recommended remedies to prevent the recurrence of the Flash Crash, he suggested a new rule to require that offers to buy or sell must remain open for *one second*. The captains of finance in charge of the financial companies with the most at stake in the current pattern of business, however, reacted with horror to this proposal, claiming that the one-second requirement would bring the global economy to its knees. The proposal was then rejected.

The global market crisis of 2008 was primarily caused by a particular kind of derivatives: securitized subprime mortgages hedged with an exotic form of insurance that turned out to be illusory. Supercomputers sliced and diced the subprime mortgages into derivatives that were so complex that no human being could possibly understand them. And once again, the robosourcing of these exotic financial instruments aided and abetted the marketing of those same products to buyers throughout the global economy.

When the actual quality and real value of the mortgages in question were belatedly examined, they were suddenly repriced on a mass basis—triggering the credit crisis and bursting the housing bubble in the United States. The fact that they had been linked to a complex web of other computer-driven financial transactions (collateralized debt obligations, or CDOs) led to the credit crisis, a massive disruption in the availability of capital as a basic factor of production in the global economy—essentially, a global run on banks. This led, in turn, to the Great Recession, the effects of which we are still struggling to escape.

Incidentally, after the manufacturing of these derivatives gained momentum and scale, virtually the only remaining role for human beings in the process resulted from the legal requirement for a signature on each underlying mortgage by someone with the responsibility for reviewing the integrity of each mortgage that had been sliced, diced, securitized, and rubber-stamped with a AAA rating by corrupted and captive ratings agencies, then sold around the world.

As the subsequent lawsuits revealed, this requirement for signatures by actual human beings could not keep up with the speed of the supercomputers—so low-wage employees were hired to forge the signatures of loan officers a hundred times a minute, without the slightest attention to the substance and meaning of the documents they were signing—a practice that's been popularly labeled "robosigning." Though no robots were involved, the very term illustrates the intertwining of robosourcing and outsourcing.

Until the crisis of 2008, the volume of trading in derivatives had been increasing since 2000 at an average of 65 percent per year. Since banks in the U.S. have been earning roughly $35 billion per year off these derivatives trades, there is no reason to believe that the growth in volumes will not once again resume; and no reason to expect that the banks will not continue to use their lobbying power and campaign contributions to prevent them from being regulated.

GLOBAL INTEGRATION

The causes of this unprecedented acceleration in the integration of the global economy have included several factors simultaneously: the collapse of communism and the introduction of more market-oriented policies in the former communist bloc countries; the opening and modernization of China under Deng Xiaoping (a process that has also continued to accelerate with the rapid rise of China's economic strength); and revolutionary changes in transportation, communications, and information technology.

Perhaps most significantly, trade barriers were lowered in the liberalization process that began with the General Agreement on Tariffs and Trade (GATT) at the end of World War II (a process that has accelerated in the years since). International trade flows have increased tenfold over the last thirty years—from $3 trillion annually to $30 trillion annually—and are continuing to grow at a rate half again faster than global production.

There have, of course, been previous periods when new surges of global trade resulted in significant changes in the pattern of the global economy. The famous though brief voyages of the legendary Chinese eunuch Admiral Zheng He during the first three decades of the fifteenth century to East Africa prefigured the Voyages of Discovery by Christopher Columbus to the New World, by Vasco da Gama around the Cape of Good Hope, and by Cortés, Pizarro, and all the others that linked Europe to the New World and to Asia.

Prior to the development of intercontinental ocean trade routes, the establishment of the Mongol Empire in the thirteenth century and the Pax Mongolica that followed opened land routes for then unprecedented trade flows between China, India, Central Asia, Russia, and Eastern Europe. Following the Black Death in the mid-fourteenth century, and the weakening of Mongol rule, the closing of overland routes between Europe and Asia once again created a bottleneck that flowed through the Middle East, trade flows that were largely controlled by Venice and Egypt.

It was, in part, the intense economic pressure in Western Europe that contributed to the daring effort to find an ocean route to India and China. The influx of gold and silver from the New World to Europe—and not long after, the sharp gains in agricultural productivity that accompanied the introduction of maize (corn) and other New World food crops into Europe and Africa—revolutionized the old pattern of the global economy, such as it was.

Economic historians also remind us that China and India together accounted for half or more of world GDP from at least the year 1 through the beginning of the Second Industrial Revolution midway through the nineteenth century. China's economy was the single largest in the world in 1500 and again in the early nineteenth century prior to the First Opium War, which began in 1839.

Seen in that perspective, the dominance of the United States and Europe in the global economy over the last 150 years was the interruption of a much longer period of Asian dominance in the share of world GDP. That century-and-a-half period represented a breakout by those nations that first embraced the Industrial Revolution—the United Kingdom, then the United States and northwestern Europe—while four fifths of the world's population was left behind. In the modern era, it appears that China and other emerging and developing economies are the ones breaking out. Prior to the nineteenth century, the distribution of wealth in the world roughly correlated with population, but the surge in productivity enhancement that accompanied the Industrial Revolution and the Scientific and Technology Revolution led to much faster accumulations of wealth in the West. Then, when the East gained more access to the new technologies, the older pattern began to reassert itself.

Some economic experts attribute the rise of China and its imminent displacement of the United States as the world's largest economy to advantages inherent in their system of state-guided capitalism, which they claim is superior to the much freer form of capitalism in the United States. If that were truly the explanation, the United States could take comfort from the fact that similar warnings about the advantages of an allegedly superior form of economic organization turned out to be false alarms in the late 1950s (when the Soviet Union was seen as an economic as well as military threat) and the 1970s and 1980s (when Japan Inc. was feared as a new economic hegemon).

However, if the emergence of Earth Inc. is more responsible for this phenomenon, as I believe it is, this time really is different. All over the developing world, nations like India that have long been mired in poverty are now beginning to unlock their vast potential as young entrepreneurs connect to their counterparts in countries throughout Earth Inc. and discover and develop innovations, large and small.

IN THE PAST, centers of expertise in a particular technology or industry usually emerged in specific locations where a cluster of people with similar skills and experience developed a local network of connections with one another, learned from one another, and improved one another's innovations with incremental advances, sometimes called "tweaks." British-Canadian journalist Malcolm Gladwell, writing in *The New Yorker*, gives a powerful example of this phenomenon:

> In 1779, Samuel Crompton, a retiring genius from Lancashire, invented the spinning mule, which made possible the mechanization of cotton manufacture. Yet England's real advantage was that it had Henry Stones, of Horwich, who added metal rollers to the mule; and James Hargreaves, of Tottington, who figured out how to smooth the acceleration and deceleration of the spinning wheel; and William Kelly, of Glasgow, who worked out how to add water power to the draw stroke; and John Kennedy, of Manchester, who adapted the wheel to turn out fine counts; and, finally, Richard Roberts, also of Manchester, a master of precision machine tooling—and the tweaker's tweaker. He created the "automatic" spinning mule: an exacting, high-speed, reliable rethinking of Crompton's original creation. Such men, the economists argue, provided the "micro inventions necessary to make macro inventions highly productive and remunerative."

When the Industrial Revolution gained momentum in the United Kingdom during the eighteenth century, there was a proximate connection between the inventors, tinkerers, blacksmiths, and engineers who contributed to the improvement of a large cluster of technologies that later spread throughout the world. The revolution they started was at first confined to one country and then, slowly at first, spread throughout the North Atlantic region.

It's true that technology clusters still matter. Silicon Valley, in Northern California, is one of the premier examples. Face-to-face, personal interactions among cutting-edge experts focused on the same set of technologies is still one of the most powerful ways to advance innovation. Yet global connectivity is speeding up the application of new technologies to ever more fields of endeavor, simultaneously pointing the way toward ever more frequent macro- and micro-inventions that accelerate the replacement of human jobs by connected intelligent machines. And seemingly small improvements in automation and efficiency often have outsized consequences for the overall efficiency and productivity in a particular sector.

SMALL CHANGES, BIG IMPACTS

To illustrate this point, consider two examples: one from the late stages of the mechanization of agriculture, in the 1950s, and the second a seemingly mundane but highly significant example from the late stages of the global Transportation Revolution, also from the 1950s, that marked a significant empowerment of much higher levels of connectivity in the global economy.

When I was a boy spending my summers on our family farm, I sometimes helped retrieve eggs from the chicken coop—one by one—when the hens had left the coop for their morning chicken feed. I remember being slightly amazed less than twenty years later when my father automated this process by building two new large chicken houses, each one containing 5,000 chickens, according to a design that was then spreading quickly on many American farms with chickens. In each house, the chickens roamed on wire mesh and then retreated to the only dark and inviting place where they could lay their eggs—which happened to be located directly above a conveyer belt. All of the eggs thus automatically collected were then funneled to a relatively simple sorting machine near the front of the building where the eggs rolled precisely into cartons. When each carton was filled, the next moved automatically into place for the collection of its designated allocation of eggs.

Out of deference to the chickens' need for a rudimentary social life—so that they would remain sufficiently contented to lay eggs every day—heavily drugged roosters were placed approximately every fifteen square feet inside each chicken house. When they recovered from their stupor, they each established rule over their respective segments of the roost, and the hens in their immediate vicinity were happy. It also turned out that placing all the chickens in a confined area also conferred the operator of the chicken houses with a new ability (mildly disturbing to me at the time) to make the sun rise more than once per day—with artificial lighting—and thereby stimulate a greater production of eggs. (Note to PETA: I no longer have any connection to chicken houses.)

But what was most startling to me was that one employee was all that was needed to collect the daily output of eggs from 10,000 chickens. It was amazing that a single person could gather so many eggs, but why was a person involved at all? Sometimes an egg would be cracked and would have to be removed from the carton; sometimes a mechanical problem would interrupt the process and would require human intervention; it required a person to coordinate the transfer of the cartons to the truck that would regularly pick them up, to keep track of the total number of cartons per day, and so forth.

But it's easy to see how the introduction of rudimentary layers of intelligence into the machinery and the connection over the Internet of the chicken house and its various components to quality control programs, computers scheduling the delivery trucks, and mechanics on call to respond to the rare interruptions in the process could easily displace that sole remaining job.

Is it possible to imagine any set of government policies that could protect the jobs lost in this process? Consider the earliest efforts to stem the loss of agricultural jobs: at the beginning of the second half of the nineteenth century in the United States, the loss of jobs on farms was already well under way, but few could imagine the transformation that was in store during the decades that followed. In a speech prior to becoming president, Abraham Lincoln noted on September 30, 1859: "farmers, being the most numerous class, it follows that their interest is the largest interest. It also follows that that interest is most worthy of all to be cherished and cultivated—that if there be inevitable conflict between that interest and any other, that other should yield."

By the time of his inauguration, the percentage of all jobs represented by farm jobs had steadily declined from 90 percent at the beginning of the republic in 1789 to a little under 60 percent. The following year, in the spring of 1862, President Lincoln established the U.S. Department of Agriculture, and six weeks later signed the Morrill Land Grant College Act, providing public land for states to establish colleges of agriculture and the mechanical arts. Every state did so.

The crowding of cities with farm hands looking for work in factories led to a wholesale transformation of the nature of work for the vast majority in the U.S. The reforms of the Progressive Era, and later the New Deal, were introduced to address the human consequences of this transformation and, in part, to replace the lost flows of income with income transfer systems such as unemployment compensation and Social Security and disability payments.

When I became vice president in 1993, there were, on average, four different offices representing the Department of Agriculture located in every one of the 3,000 counties in the United States—yet the percentage of total jobs represented by farm jobs had declined to 2 percent. In other words, a determined and expensive national policy to promote agriculture for a century and a half did little or nothing to prevent the massive loss of employment opportunities on farms, although these policies arguably contributed to the massive increase in agricultural productivity. But the larger point is that many systemic technology-driven changes are simply too powerful for any set of policies to hold back.

Today, in fact, what is now referred to as factory farming has led to the mass introduction of partially automated systems for raising chickens, cattle, pigs, and other livestock—and for producing eggs. Over the last forty years, global production of eggs has increased by 350 percent. (China is by far the largest producer of eggs, with 70 million tons annually—four times the production of the United States.) Global trade in poultry meat has increased over the same period by more than 3,200 percent.

Here is a second example of a seemingly mundane advance that led to truly revolutionary progress in the efficiency of an entire industrial sector: the containership revolution began on October 4, 1957—on the very day that the first space satellite, Sputnik, was launched by the Soviet Union. Malcom McLean, a businessman who owned a trucking company in North Carolina, had wondered for almost twenty years why the cargo coming into U.S. ports from foreign countries was carried in boxes and enclosures of every size, shape, and description, which then had to be lifted and sorted individually onto the dock and moved from there to whatever conveyance was available to deliver each box to its ultimate destination—rather than packaged into enclosed symmetrical containers of the exact same size that could be lifted from each ship onto trains and trailer trucks and then transported to their destination.

In the spring of 1956, McLean experimented with his revolutionary idea by equipping one special deck on a ship bound from Newark, New Jersey, to Houston, Texas, with the bodies of fifty-eight trailer truck units that had been detached from the cab and chassis and loaded into slots on the ship. The experiment was so successful that eighteen months later he made history by outfitting an entire ship to carry 226 containers that were sent from port Newark and offloaded a week later in Houston onto the chassis of 226 trucks waiting to carry them to their destinations. The "containership revolution" that began in the fall of 1957 has had such an impact on global trade that in 2013 more than 150 million trailer-truck-sized containers will carry goods from one country to another.

The progressive introduction of intelligence and networking is accelerating this same process in almost all areas of manufacturing. High-quality large-screen television sets, for example, have come down more than 5 percent in price each year and are now in surplus supply (much as food grains were a few decades ago). The first color television set was sold in 1953 at a price that, in today's dollars, would be $8,000. The cheapest color television sets for sale today—with the same or larger screen size, much greater picture clarity, and the ability to play hundreds of channels instead of only three—are available for as little as $50—or approximately one half of one percent of the cost in return for a product of much higher quality and much higher capacity.

We take such dramatic price reductions (and simultaneous quality improvements) for granted these days, but on a cumulative basis the impact for the world of work can no longer be ignored. Indeed, many consumer products that were once described as high-tech are now referred to by economists as commodities. The massive increase in world trade, combined with outsourcing, robosourcing, the new flows of information and investment connecting virtually all locations in the world to one another have all reinforced each other in a massive global feedback loop.

ROBOSOURCING

This pattern of progressive improvement in the effectiveness and utility of machine intelligence is under way in thousands of industries and it is the cumulative impact that is driving the global change in the nature and purpose of work in the world. Look, for example, at the coal industry in the United States. In the last quarter century, production has increased by 133 percent, even as jobs have decreased by 33 percent.

To take another example, jobs in the U.S. copper mining industry have declined precipitously in the last half century even as output increased significantly over much of that period. As is often the case when new technology replaces jobs, the pattern was not an even and steady decline, but a decline that lurched downward from one plateau to the next as new innovations became available and were implemented. In one six-year period—from 1980 to 1986—the number of hours of labor required to produce a ton of copper fell by 50 percent. In that same decade, one of the largest companies, Kennecott, increased labor productivity in one of its largest mines by 400 percent.

Looking more closely at this industry as an illustrative example of the broad trend, the new technologies that replaced jobs included much larger trucks and shovels, much broader use of computers for micromanaging the schedule of the trucks and the operation of the mills, much more efficient crushers connected to better conveyer belts, and the introduction of new chemical and electrochemical processes to automatically separate the pure copper from the ore.

The copper mining industry in the United States also illustrates changes from robosourcing and outsourcing that impact the third classical factor of production—resources. As technology increased labor productivity and the number of tons of copper produced year by year, the industry eventually reached a tipping point when the available supplies of economically recoverable copper ore began to diminish. New sources of copper were developed in other countries, principally Chile. Sharp increases in the efficiency of production, coupled with increasing consumption rates driven by population growth and increased affluence, are driving many industries toward constraints in the supplies of natural resources essential to their production processes.

In a process that is further reducing jobs and demand in the industrial world, robosourcing and IT-empowered outsourcing are now also beginning to have a major impact on jobs in the largest category of employment: services. Consider the impact of intelligent programs for legal and document research in law firms. Some studies indicate that with the addition of these programs, a single first-year associate can now perform with greater accuracy the volume of work that used to be done by 500 first-year associates.

Indeed, many predict that the impact of robosourcing will be even more pronounced in services than in manufacturing. Much has been written about Google's success in developing self-driving automobiles, which have now traveled 300,000 miles in all driving conditions without an accident. If this technology is soon perfected—as many predict—consider the impact on the 373,000 people employed in the United States alone as taxi drivers and chauffeurs. Already, some Australian mining companies have replaced high-wage truck drivers with driverless trucks.

Where services are concerned, we are also seeing a third trend, which might be called "self-sourcing": individual consumers of services, empowered with laptops, smartphones, tablets, and other productivity-enhancing devices, are interacting with intelligent programs to effectively partner with machine intelligence to effectively replace many of the people who used to be employed in service jobs. Many airline travelers routinely make their own reservations, pick their own seats, and print their boarding passes. Many supermarkets and other stores enable shoppers to handle the checkout and payment process on their own. Banks began to provide cash with ATM machines and now offer extensive online banking services. Customers of many businesses now routinely deal with computers on the telephone. National postal services in many countries, including the U.S., are being progressively disintermediated (that is, their "middleman" role is being made obsolete) by email and social media.

This self-sourcing trend is still in its early stages and will accelerate dramatically as artificial intelligence improves year by year. One obvious problem is that there is no compensation for all the new work done by individuals, even as the compensation formerly paid to those in firms who lost their jobs is also lost to the economy as a whole. The enhanced convenience associated with self-sourcing improves efficiency and saves time, to be sure, but on an aggregate basis, the overall reduction in income for middle-income wage earners is beginning to have a noticeable impact on aggregate demand—particularly in consumer-oriented societies.

ON A GLOBAL basis, offshoring and robosourcing are together pushing the economy toward a simultaneous weakening of demand and surplus of production. The use of Keynesian stimulus policies—that is, government borrowing to finance temporary increases in aggregate demand—may become less effective over time as the secular, systemic shift to an economy with far fewer jobs relative to production represents a larger cause of declining incomes, and thus declining consumption and demand. In addition, as I'll detail later, unprecedented demographic shifts include a larger proportion of older, retired people in industrial countries whose incomes are already replaced by programs such as Social

Security—thereby limiting the ability of governments to replace income indefinitely to working age people.

Unless the lost income of the unemployed and underemployed factory workers in industrial countries can somehow be replaced, global demand for the products of the new highly automated factories will continue to decline. The industrial economies, after all, continue to provide the greatest share of global demand and consumption. Higher wages paid to workers in developing and emerging economies are far more likely—in part for cultural reasons—to go into savings instead of consumption. While both labor and capital have been globalized, the bulk of consumption in the world economy remains in wealthy industrial countries. This results in a mismatch between the distribution of income and the central role of consumption in driving global economic growth.

RETHINKING RESOURCES

These accelerating changes will therefore require us to reimagine the now central role of consumption in our economy and simultaneously replace the flows of income to workers that presently empower consumption. The current connection between ever rising levels of consumption and the health of the global economy is increasingly unstable in any case.

The accelerating technology revolution is not only transforming the role of labor and capital as factors of production in the global economy, it is also transforming the role of resources. The new technologies of molecular manipulation have led to revolutionary advances in the materials sciences and brand-new hybrid materials that possess a combination of physical attributes far exceeding those of any materials developed through the much older technologies of metallurgy and ceramics. As Pierre Teilhard de Chardin predicted more than sixty years ago, "In becoming planetized, humanity is acquiring new physical powers which will enable it to super-organize matter."

The new field of advanced materials science involves the study, manipulation, and fabrication of solid matter with highly sophisticated tools, almost on an atom-by-atom basis. It involves many interdisciplinary fields, including engineering, physics, chemistry, and biology. The new insights being developed into the ways that molecules control and direct basic functions in biology, chemistry, and the interaction of atomic and subatomic processes that form solid matter is speeding up the emergence of what some experts are calling the molecular economy.

Significantly, the new molecules and materials created need not be evaluated through the traditional, laborious process of trial and error. Advanced supercomputers are now capable of simulating the way these novel creations interact with other molecules and materials, allowing the selection of only the ones that are most promising for experiments in the real world. Indeed, the new field known as computational science has now been recognized as a third basic form of knowledge creation—alongside inductive reasoning and deductive reasoning—and combines elements of the first two by simulating an artificial reality that functions as a much more concrete form of hypothesis and allows detailed experimentation to examine the new materials' properties and analyze how they interact with other molecules and materials.

The properties of matter at the nanometer scale (between one and 100 nanometers) often differ significantly from the properties of the same atoms and molecules when they are clustered in bulk. These differences have allowed technologists to use nanomaterials on the surfaces of common products in order to eliminate rust, enhance resistance to scratches and dents, and in clothes to enhance resistance to stains, wrinkles, and fire. The single most common application thus far is the use of nanoscale silver to destroy microbes—a use that is particularly important for doctors and hospitals guarding against infections.

The longer-term significance that attaches to the emergence of an entirely new group of basic materials with superior properties is reflected in the names historians give to the ages of technological achievement in human societies: the Stone Age, the Bronze Age, and the Iron Age. As was true of the historical stages of economic development that began with the long hunter-gatherer period, the first of these periods—the Stone Age—was by far the longest.

Archaeologists disagree on when and where the reliance on stone tools gave way to the first metallurgical technologies. The first smelting of copper is believed to have taken place in eastern Serbia approximately 7,000 years ago, though objects made of cast copper emerged in numerous locations in the same era.

The more sophisticated creation of bronze—which is much less brittle and much more useful for many purposes than copper—involves a process in which tin is added to molten copper, a technique that combines high temperatures and some pressurization. Bronze was first created 5,000 years ago in both Greece and China, and more than 1,000 years later in Britain.

Though the first iron artifacts date back 4,500 years ago in northern Turkey, the Iron Age began between 3,000 and 3,200 years ago with the development of better furnaces that achieved higher temperatures capable of heating iron ore into a malleable state from which it could be made into tools and weapons. Iron, of course, is much harder and stronger than bronze. Steel, an alloy made from iron, and often other elements in smaller quantities, depending upon the properties desired, was not made until the middle of the nineteenth century.

The new age of materials created at the molecular level is leading to a historic transformation of the manufacturing process. Just as the Industrial Revolution was launched a quarter of a millennium ago by the marriage of coal-powered energy with machines in order to replace many forms of human labor, nanotechnology promises to launch what many are calling a Third Industrial Revolution based on molecular machines that can reassemble structures made from basic elements to create an entirely new category of products, including:

- Carbon nanotubes invested with the ability to store energy and manifest previously unimaginable properties;
- Ultrastrong carbon fibers that are already replacing steel in some niche applications; and
- Ceramic matrix nanocomposites that are expected to have wide applications in industry.

The emerging Nanotechnology Revolution, which is converging with the multiple revolutions in the life sciences, also has implications in a wide variety of other human endeavors. There are already more than 1,000 nanotechnology products available, most of them classified as incremental improvements in already known processes, mostly in the health and fitness category. The use of nanostructures for the enhancement of computer processing, the storage of memory, the identification of toxics in the environment, the filtration and desalination of water, and other uses are still in development.

The reactivity of nanomaterials and their thermal, electrical, and optical properties are among the changes that could have significant commercial impact. For example, the development of graphene—a form of graphite only one atom thick—has created excitement about its unusual interaction with electrons, which opens a variety of useful applications.

Considerable research is under way on potential hazards of nanoparticles. Most experts now minimize the possibility of "self-replicating nanobots," which gave rise to serious concerns and much debate in the first years of the twenty-first century, but other risks—such as the accumulation of nanoparticles in human beings and the possibility of consequent cell damage—are taken more seriously. According to David Rejeski, director of the Science and Technology Innovation Program at the Woodrow Wilson International Center for Scholars, "We know very little about the health and environmental impacts [of nanomaterials] and virtually nothing about their synergistic impacts."

In a sense, nanoscience has been around at least since the work of Louis Pasteur, and certainly since the discovery of the double helix in 1953. The work of Richard Smalley on buckminsterfullerene molecules ("buckyballs") in 1985 triggered a renewed surge of interest in the application of nanotechnology to the development of new materials. Six years later, the first carbon nanotubes offered the promise of electrical conductivity exceeding that of copper and the possibility of creating fibers with 100 times the strength and one sixth the weight of steel.

The dividing line between nanotechnology and new materials sciences is partly an arbitrary one. What both have in common is the recent development of new more powerful microscopes, new tools for guiding the manipulation of matter at nanoscales, the development of new more powerful supercomputer programs for modeling and studying new materials at the atomic level, and a continuing stream of new basic research breakthroughs on the specialized properties of nanoscale molecular creations, including quantum properties.

THE RISE OF 3D PRINTING

Humankind's new ability to manipulate atoms and molecules is also leading toward the disruptive revolution in manufacturing known as 3D printing. Also known as additive manufacturing, this new process builds objects from a three-dimensional digital file by laying down an ultrathin layer of whatever material or materials the object is to be made of, and then adds each additional ultrathin layer—one by one—until the object is formed in three-dimensional space. More than one different kind of material can be used. Although this new technology is still early in its development period, the advantages it brings to manufacturing are difficult to overstate. Already, some of the results are startling.

Since 1908, when Henry Ford first used identical interchangeable parts that were fitted together on a moving assembly line to produce the Model T, manufacturing has been dominated by mass production. The efficiencies, speed, and cost savings in the process revolutionized industry and commerce. But many experts now predict that the rapid development of 3D printing will change manufacturing as profoundly as mass production did more than 100 years ago.

The process has actually been used for several decades in a technique known as rapid prototyping—a specialized niche in which manufacturers could produce an initial model of what they would later produce en masse in more traditional processes. For example, the designs for new aircraft are often prototyped as 3D models for wind tunnel testing. This niche is itself being disrupted by the new 3D printers; one Colorado firm, LGM, that prototypes buildings for architects, has already made dramatic changes. The company's founder, Charles Overy, told *The New York Times*, "We used to take two months to build $100,000 models." Instead, he now builds $2,000 models and completes them overnight.

The emerging potential for using 3D printing is illuminating some of the inefficiencies in mass production: the stockpiling of components and parts, the large amount of working capital required for such stockpiling, the profligate waste of materials, and of course the expense of employing large numbers of people. Enthusiasts also contend that 3D printing often requires only 10 percent of the raw material that is used in the mass production process, not to mention a small fraction of the energy costs. It continues and accelerates a longer-term trend toward "dematerialization" of manufactured goods—a trend that has already kept the total tonnage of global goods constant over the past half century, even as their value has increased more than threefold.

In addition, the requirement for standardizing the size and shape of products made in mass production leads to a "one size fits all" approach that is unsatisfactory for many kinds of specialized products. Mass production also requires the centralization of manufacturing facilities and the consequent transportation costs for delivery of parts to the factory and finished products to distant markets. By contrast, 3D printing offers the promise of transmitting the digital information that embodies the design and blueprint for each product to widely dispersed 3D printers located in all relevant markets.

Neil Hopkinson, senior lecturer in the Additive Manufacturing Research Group at Loughborough University, said, "It could make offshore manufacturing half way round the world far less cost effective than doing it at home, if users can get the part they need printed off just round the corner at a 3D print shop on the high street. Rather than stockpile spare parts and components in locations all over the world, the designs could be costlessly stored in virtual computer warehouses waiting to be printed locally when required."

At its current stage of development, 3D printing focuses on relatively small products, but as the technique is steadily improved, specialized 3D printers for larger parts and products will soon be available. One company based in Los Angeles, Contour Crafting, has already built a huge 3D printer that travels on a tractor-trailer to a construction site and prints an entire house in only twenty hours (doors and windows not included)! In addition, while the 3D printers now available have production runs of one item up to, in some cases, 1,000 items, experts predict that within the next few years these machines will be capable of turning out hundreds of thousands of identical parts and products.

There are many questions yet to be answered about the treatment of intellectual property in a 3D printing era. The three-dimensional design will make up the lion's share of the value in a 3D printing economy, but copyright and patent law were developed without the anticipation of this technology and will have to be modified to account for the new emerging reality. In general, "useful" physical objects often do not have protection against replication under copyright laws.

Although there are skeptics who question how fast this new technology will mature, engineers and technologists in the United States, China, and Europe are working hard to exploit its potential. Its early use in printing prosthetics and other devices with medical applications is gaining momentum rapidly. Inexpensive 3D printers have already found their way into the hobbyist market at prices as low as $1,000. Carl Bass, the CEO of Autodesk, which has invested in 3D printing, said in 2012, "Some people see it as a niche market. They claim that it can't possibly scale. But this is a trend, not a fad. Something seismic is going on." Some advocates of more widespread gun ownership are promoting the

3D printing of guns as a way to circumvent regulations on gun sales. Opponents have expressed concern that any such guns used in crimes could be easily melted down to avoid any effort by law enforcement authorities to use the guns as evidence.

THE WAVE OF automation that is contributing to the outsourcing and robo-sourcing of jobs from developed countries to emerging and developing markets will soon begin to displace many of the jobs so recently created in those same low-wage countries. 3D printing could accelerate this process, and eventually could also move manufacturing back into developed countries. Many U.S. companies have already reported that various forms of automation have enabled them to bring back at least some of the jobs they had originally outsourced to low-wage countries.

CAPITALISM IN CRISIS

The emergence of Earth Inc. and its disruption of all three factors of production—labor, capital, and natural resources—has contributed to what many have referred to as a crisis in capitalism. A 2012 Bloomberg Global Poll of business leaders around the world found that 70 percent believe capitalism is "in trouble." Almost one third said it needs a "radical reworking of the rules and regulations"—though U.S. participants were less willing than their global counterparts to endorse either conclusion.

The inherent advantages of capitalism over any other system for organizing economic activity are well understood. It is far more efficient in allocating resources and matching supply to demand; it is far more effective at creating wealth; and it is far more congruent with higher levels of freedom. Most fundamentally, capitalism unlocks a larger fraction of the human potential with ubiquitous organic incentives that reward effort and innovation. The world's experimentation with other systems—including the disastrous experiences with communism and fascism in the twentieth century—led to a nearly unanimous consensus at the beginning of the twenty-first century that democratic capitalism was the ideology of choice throughout the world.

And yet publics around the world have been shaken by a series of significant market dislocations over the last two decades, culminating in the Great Recession of 2008 and its lingering aftermath. In addition, the growing inequality in most large economies in the world and the growing concentration of wealth at the top of the income ladder have caused a crisis of confidence in the system of market capitalism as it is presently functioning. The persistent high levels of unemployment and underemployment in industrial countries, added to unusually high levels of public and private indebtedness, have also diminished confidence that the economic policy toolkit now being used can produce a recovery that is strong enough to restore adequate vitality.

As Nobel Prize–winning economist Joseph Stiglitz put it in 2012:

> It is no accident that the periods in which the broadest cross sections of Americans have reported higher net incomes—when inequality has been reduced, partly as a result of progressive taxation—have been the periods in which the U.S. economy has grown the fastest. It is likewise no accident that the current recession, like the great Depression, was preceded by large increases in inequality. When too much money is concentrated at the top of society, spending by the average American is necessarily reduced—or at least it will be in the absence of an artificial prop. Moving money from the bottom to the top lowers consumption because higher-income individuals consume, as a fraction of their income, less than lower-income individuals do.

While developing and emerging economies are seeing increases in productivity, jobs, incomes, and output, inequality within these countries is also increasing. And of course, many of them still have significant numbers of people experiencing extreme poverty and deprivation. More than one billion people in the world still live on less than $2 a day, and almost 900 million of them still live in "extreme poverty"—defined as having an income less than $1.25 per day.

Most important of all, among the failures in the way the global market system is operating today is its almost complete refusal to include any recognition of major externalities, starting with its failure to take into account the cost and consequences of the 90 million tons of global warming pollution spewed every twenty-four hours into the planet's atmosphere. The problem of externalities in market theory is well known but has never been so acute as now. Positive externalities are also routinely ignored, leading to chronic underinvestment in education, health care, and other public goods.

In many countries, including the United States, the growing concentration of wealth in the hands of the top one percent has also led to distortions in the political system that now limit the ability of governments to consider policy changes that might benefit the many at the (at least short-term) expense of the few. Governments have been effectively paralyzed and incapable of taking needed action. This too has undermined public confidence in the way market capitalism is currently operating.

With the tightly coupled and increasingly massive flows of capital through the global economy, all governments now feel that they are hostage to the perceptions within the global market for capital. There are numerous examples—Greece, Ireland, Italy, Portugal, and Spain, to name a few—of countries' confronting policy choices that appear to be mandated by the perceptions of the global marketplace, not by the democratically expressed will of the citizens in those countries. Many have come to the conclusion that the only policies that will prove to be effective in restoring human influence over the shape of our economic future will be ones that address the new global economic reality on a global basis.

SUSTAINABLE CAPITALISM

Along with my partner and cofounder of Generation Investment Management, David Blood, I have advocated a set of structural remedies that would promote what we call Sustainable Capitalism. One of the best-known problems is the dominance of short-term perspectives and the obsession with short-term profits, often at the expense of the buildup of long-term value. Forty years ago, the average holding period for stocks in the United States was almost seven years. That made sense because roughly three quarters of the real value in the average business builds up over a business cycle and a half, roughly seven years. Today, however, the average holding period for stocks is less than seven months.

There are many reasons for the increasing reliance on short-term thinking by investors. These pressures are accentuated by the larger trends in the transformed and now interconnected global economy. As one analyst noted in 2012, "our banks, hedge funds and venture capitalists are geared toward investing in financial instruments and software companies. In such endeavors, even modest investments can yield extraordinarily quick and large returns. Financing brick-and-mortar factories, by contrast, is expensive and painstaking and offers far less potential for speedy returns."

This short-term perspective on the part of investors puts pressure on CEOs to adopt similarly short-term perspectives. For example, a premier business research firm in the United States (BNA) conducted a survey of CEOs and CFOs a few years ago in which it asked, among other things, a hypothetical question: You have the opportunity to make an investment in your company that will make the firm more profitable and more sustainable, but if you do so, you will slightly miss your next quarterly earnings report; under these circumstances, will you make the investment? Eighty percent said no.

A second well-known problem in the way capitalism currently operates is the widespread misalignment of incentives. The compensation of most investment managers—the people that make most of the daily decisions on the investment of capital—is calculated on a quarterly, or at most annual, basis. Similarly, many executives running companies are compensated in ways that reward short-term results. Instead, compensation should be aligned temporally with the period over which the maximum value of firms can be increased, and should be aligned with the fundamental drivers of long-term value.

In addition, companies should be encouraged to abandon the default practice of providing quarterly earnings guidance. These short-term metrics capture so much attention that they end up heavily penalizing firms that try to build sustainable value, and fail to take into account the usefulness of investments that pay for themselves handsomely over longer periods of time.

THE CHANGING NATURE OF WORK

One thing is certain: the transformation of the global economy and the emergence of Earth Inc. will require an entirely new approach to policy in order to reclaim humanity's role in shaping our own future. What we are now going through bears little relation to the problems inherent in the business cycle or the kinds of temporary market disruptions to which global business has become accustomed. The changes brought about by the emergence of Earth Inc. are truly global, truly historic, and are still accelerating.

Although the current changes are unprecedented in speed and scale, the pattern of productive activity for the majority of human beings has of course undergone several massive changes throughout the span of human history. Most notably, the Agricultural and Industrial revolutions both led to dramatic changes in the way the majority of people in the world spent their days.

The first known man-made tools, including spear points and axes, were associated with a hunting and gathering pattern that lasted, according to anthropologists, almost 200 millennia. The displacement of that dominant pattern by a new one based on agriculture (beginning not long after the last Ice Age receded) took less than eight millennia, while the Industrial Revolution required less than 150 years to reduce the percentage of agricultural jobs in the United States from 90 to 2 percent of the workforce. Even when societies still based on subsistence agriculture are included in the global calculation, less than half of all jobs worldwide are now on farms.

The plow and the steam engine—along with the complex universe of tools and technologies that accompanied the Agricultural and Industrial revolutions respectively—undermined the value of skills and expertise that had long been relied upon to connect the meaning of people's lives to the provision of subsistence and material gains for themselves, their families, and communities. Nevertheless, in both cases, the disappearance of old patterns was accompanied by the emergence of new ones that, on balance, made life easier and retained the link between productive activity and the meeting of real needs.

To be sure, the transformation of work opportunities required large changes in social patterns, including mass internal migrations from rural areas to cities, and the geographic separation of homes and workplaces, to mention only two of the most prominent disruptions. But the net result was still consistent with the hopeful narrative of progress and was accompanied by economic growth that increased net incomes dramatically and sharply reduced the amount of work necessary to meet basic human needs: food, clothing, shelter, and the like. In both cases, formerly common pursuits became obsolete while new ones emerged that called for new skills and a reconception of what it meant to be productive.

Both of these massive transformations occurred over long periods of time covering multiple generations. In both revolutions, new technologies opened up new opportunities for reorganizing the human enterprise into a new dominant pattern that was in each case disruptive and, for many, disorienting—but produced massive increases in productivity, large increases in the number of jobs, higher average incomes, less poverty, and historic improvements in the quality of life for most people.

Consider again the larger pattern traced in the history of these three epochs: the first lasted 200,000 years, the next lasted 8,000 years, and the Industrial Revolution took only 150 years. Each of these historic changes in the nature of the human experience was more significant than its predecessor and occurred over a radically shorter time span. All were connected to technological innovations.

Taken together, they trace the long gestation, infancy, and slow development of a technology revolution that eventually grew to play a central role in the advance of human civilization—then gradually but steadily gained speed and momentum in each of the last four centuries, jolted into a higher gear, and began to accelerate at an ever faster rate until it seemed to take on a life of its own. It is now carrying us with it at a speed beyond our imagining toward ever newer technologically shaped realities that often appear, in the words of Arthur C. Clarke, "indistinguishable from magic."

Because the change under way is one not only of degree but of kind, we are largely unprepared for what's happening. The structure of our brains is not very different from that of our ancestors 200,000 years ago. Because of the radical changes induced by technology in the way we live our lives, however, we are forced to consider making adaptations in the design of our civilization more rapidly than seems possible or even plausible.

We have difficulty even perceiving and thinking clearly about the pace of change with which we are now confronted. Most of us struggle with the practical meaning of exponential change—that is, change that is not only increasing but is increasing at a steadily faster pace. Consider the basic shape of all exponential curves. The pattern of change measured by such curves is slow at first, and then ascends at a gradually but ever increasing rate as the angle of ascent steepens. The steep phase of the curve drives changes at a far more rapid rate than the flat part of the curve—and it is this phase that has consequences not only of degree but of kind. As explained by Moore's Law, the fourth-generation iPad now has more computing power than the most powerful supercomputer in the world thirty years ago, the Cray-2.

The implications of this new period of hyper-change are not just mathematical or theoretical. They are transforming the fundamental link between how we play a productive role in life and how we meet our needs. What people do—their work, their careers, their opportunities to exchange productive activity for income to meet essential human needs and provide a sense of well-being, security, honor, dignity, and a sense of belonging as a member of the community: this basic exchange at the center of our lives is now changing on a global scale and at a speed with no precedent in human history.

In modern societies we have long since used money and other tangible symbols of credit and debit as the principal means of measuring and keeping track of this ongoing series of exchanges. But even in older forms of society where money was not the medium of exchange, productive work also was connected to the ability to meet one's needs, with a tacit recognition by the community of those who contributed to the needs of the group, and whose needs were then met partly by others in the group. It is that basic connection at the heart of human societies that is beginning to be radically transformed.

Many economists comfort themselves with the idea that this is actually an old and continuing story that they know and understand well— a story that has generated unnecessary alarm since Ned Ludd, a weaver, smashed the new knitting frames invented in the late eighteenth century, which he realized were making the jobs of weavers obsolete. The "Luddite fallacy"—a phrase coined to describe the mistaken belief that new technologies will result in a net reduction of good jobs for people— was validated on a large scale when the mechanization of farming eliminated all but a tiny fraction of farm-related jobs, and yet the new jobs that emerged in factories not only outnumbered those lost on farms but produced higher incomes, even as farms became far more productive and food prices sharply declined. Until recently, the large-scale automation of industry seemed to be repeating the same pattern again: routine, repetitive, and often arduous jobs were eliminated, while better jobs with higher wages more than replaced those that were eliminated.

Yet what we believe we learned during the early stages of this technology revolution may no longer be relevant to the new hyper-accelerated pace of change. The introduction of networked machine intelligence— and now artificial intelligence—may soon put a much higher percentage of employment opportunities at risk in ever larger sectors of the global economy. In order to adapt to this new emergent reality we may soon have to reimagine the way we as human beings exchange our productive potential for the income necessary to meet our needs.

Many scholars who have specialized in the study of technology's interaction with the pattern of society, including Marshall McLuhan, have described important new technologies as "extensions" of basic human capacities. The automobile, in the terms of this metaphor, is an extension of our capacity for locomotion. The telegraph, radio, and television are, in the same way, described as extensions of our ability to speak with one another over a greater distance. Both the shovel and the steam shovel are extensions of our hands and our ability to grasp physical objects. New technologies such as these made some jobs obsolete, but on balance created more new ones—often because the new technologically enhanced capacities had to be operated or used by people who could think clearly enough to be trained to use them effectively and safely.

In this context, the emergence of new and powerful forms of artificial intelligence represents not just the extension of yet another human capacity, but an extension of the dominant and uniquely human capacity to think. Though science has established that we are not the only sentient living creatures, it is nevertheless abundantly obvious that we as a species have become dominant on Earth because of our capacity to make mental models of the world around us and manipulate those models through thought to gain the power to transform our surroundings and exert dominion over the planet. The technological extension of the ability to think is therefore different in a fundamental way from any other technological extension of human capacity.

As artificial intelligence matures and is connected with all the other technological extensions of human capacity—grasping and manipulating physical objects, recombining them into new forms, carrying them over distance, communicating with one another utilizing flows of information of far greater volume and far greater speed than any humans are capable of achieving, making their own abstract models of reality, and learning in ways that are sometimes superior to the human capacity to learn—the impact of the AI revolution will be far greater than that of any previous technological revolution.

One of the impacts will be to further accelerate the decoupling of gains in productivity from gains in the standard of living for the middle class. In the past, improvements in economic efficiency have generally led to improvements in wages for the majority, but when the substitution of technology capital for labor creates the elimination of very large numbers of jobs, a much larger proportion of the gains go to those who provide the capital. The fundamental relationship between technology and employment is being transformed.

This trend is now nearing a threshold beyond which so many jobs are lost that the level of consumer demand falls below the level necessary to sustain healthy economic growth. In a new study of the Great Depression, Joseph Stiglitz has argued that the massive loss of jobs in agriculture that accompanied the mechanization of farming led to a similar contraction of demand that was actually a much larger factor in causing the Depression than has been previously recognized—and that we may be poised for another wrenching transition with the present ongoing loss of manufacturing jobs.

New jobs can and must be created, and one of the obvious targets for new employment is the provision of public goods in order to replace the income lost by those whose employment is being robosourced and outsourced. But elites who have benefited from the emergence of Earth Inc. have thus far effectively used their accumulated wealth and political influence to block any shift of jobs to the public sector. The good news is that even though the Internet has facilitated both outsourcing and robosourcing, it is also providing a new means to build new forms of political influence not controlled by elites. This is a major focus of the next chapter.

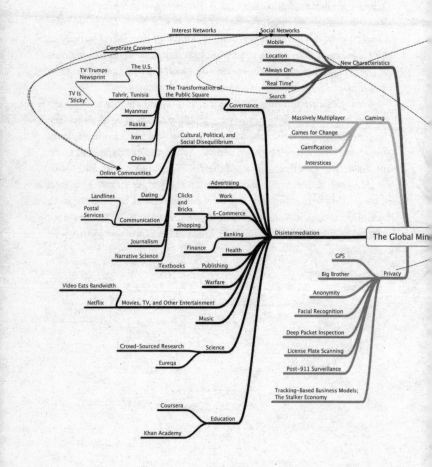

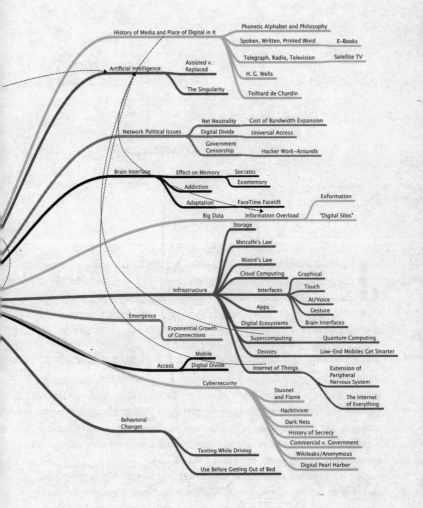

History of Media and Place of Digital in It
- Phonetic Alphabet and Philosophy
- Spoken, Written, Printed Word — E-Books
- Telegraph, Radio, Television — Satellite TV
- H. G. Wells
- Teilhard de Chardin

Artificial Intelligence
- Assisted v. Replaced
- The Singularity

Network Political Issues
- Net Neutrality — Cost of Bandwidth Expansion
- Digital Divide — Universal Access
- Government Censorship — Hacker Work-Arounds

Brain Interface
- Effect on Memory — Socrates
- Exomemory
- Addiction
- Adaptation — FaceTime Facelift
- Big Data — Information Overload — Exformation
- "Digital Silos"
- Storage

Infrastructure
- Metcalfe's Law
- Moore's Law
- Cloud Computing
- Interfaces — Graphical
- Touch
- AI/Voice
- Apps — Gesture
- Digital Ecosystems — Brain Interfaces
- Supercomputing — Quantum Computing
- Devices — Low-End Mobiles Get Smarter

Emergence
- Exponential Growth of Connections
- Mobile
- Access — Digital Divide
- Internet of Things — Extension of Peripheral Nervous System

Cybersecurity
- Stuxnet and Flame — The Internet of Everything
- Hacktivism
- Dark Nets
- History of Secrecy
- Commercial v. Government
- Wikileaks/Anonymous
- Digital Pearl Harbor

Behavioral Changes
- Texting While Driving
- Use Before Getting Out of Bed

THE GLOBAL MIND

J UST AS THE SIMULTANEOUS OUTSOURCING AND ROBOSOURCING OF PRO-
ductive activity has led to the emergence of Earth Inc., the simultaneous deployment of the Internet and ubiquitous computing
power have created a planet-wide extension of the human nervous system that transmits information, thoughts, and feelings to and from billions of people at the speed of light.

We are connecting to vast global data networks—and to one
another—through email, text messaging, social networks, multiplayer
games, and other digital forms of communication at an unprecedented
pace. This revolutionary and still accelerating shift in global communication is driving a tsunami of change forcing disruptive—and creative—
modifications in activities ranging from art to science and from collective
political decision making to building businesses.

Some familiar businesses are struggling to survive: newspapers,
travel agencies, bookstores, music, video rental, and photography stores
are among the most frequently recognized early examples of businesses
confronted with a technologically driven mandate to either radically
change or disappear. Some large institutions are also struggling: national
postal services are hemorrhaging customers as digital communication

displaces letter writing, leaving the venerable post office to serve primarily as a distribution service for advertisements and junk mail.

At the same time, we are witnessing the explosive growth of new business models, social organizations, and patterns of behavior that would have been unimaginable before the Internet and computing: from Facebook and Twitter to Amazon and iTunes, from eBay and Google to Baidu, Yandex.ru, and Globo.com, to a dozen other businesses that have started since you began reading this sentence—all are phenomena driven by the connection of two billion people (thus far) to the Internet. In addition to people, the number of digital devices connected to other devices and machines—with no human being involved—already exceeds the population of the Earth. Studies project that by 2020, more than 50 billion devices will be connected to the Internet and exchanging information on a continuous basis. When less sophisticated devices like Radiofrequency Identification (RFID) tags capable of transmitting information wirelessly or transferring data to devices that read them are included, the number of "connected things" is already much larger. (Some school systems, incidentally, have begun to require students to wear identification tags equipped with RFID tags in an effort to combat truancy, generating protests from many students.)

TECHNOLOGY AND THE "WORLD BRAIN"

Writers have used the human nervous system to describe electronic communication since the invention of the telegraph. In 1851, only six years after Samuel Morse received the message "What hath God wrought?" Nathaniel Hawthorne wrote: "By means of electricity, the world of matter has become a great nerve vibrating thousands of miles in a breathless point of time. The round globe is a vast brain, instinct with intelligence." Less than a century later, H. G. Wells modified Hawthorne's metaphor when he offered a proposal to develop a "world brain"—which he described as a commonwealth of all the world's information, accessible to all the world's people as "a sort of mental clearinghouse for the mind: a depot where knowledge and ideas are received, sorted, summarized, digested, clarified and compared." In the way Wells used the phrase "world brain," what began as a metaphor is now a reality. You can look it up right now on Wikipedia or search the World Wide Web on Google for some of the estimated one trillion web pages.

Since the nervous system connects to the human brain and the brain gives rise to the mind, it was understandable that one of the twentieth century's greatest theologians, Teilhard de Chardin, would modify Hawthorne's metaphor yet again. In the 1950s, he envisioned the "planetization" of consciousness within a technologically enabled network of human thoughts that he termed the "Global Mind." And while the current reality may not yet match Teilhard's expansive meaning when he used that provocative image, some technologists believe that what is emerging may nevertheless mark the beginning of an entirely new era. To paraphrase Descartes, "It thinks; therefore it is."*

The supercomputers and software in use have all been designed by human beings, but as Marshall McLuhan once said, "We shape our tools, and thereafter, our tools shape us." Since the global Internet and the billions of intelligent devices and machines connected to it—the Global Mind—represent what is arguably far and away the most powerful tool that human beings have ever used, it should not be surprising that it is beginning to reshape the way we think in ways both trivial and profound—but sweeping and ubiquitous.

In the same way that multinational corporations have become far more efficient and productive by outsourcing work to other countries and robosourcing work to intelligent, interconnected machines, we as individuals are becoming far more efficient and productive by instantly connecting our thoughts to computers, servers, and databases all over the world. Just as radical changes in the global economy have been driven by a positive feedback loop between outsourcing and robosourcing, the spread of computing power and the increasing number of people connected to the Internet are mutually reinforcing trends. Just as Earth Inc. is changing the role of human beings in the production process, the Global Mind is changing our relationship to the world of information.

---

* There is considerable debate and controversy over when—and even whether—artificial intelligence will reach a stage of development at which its ability to truly "think" is comparable to that of the human brain. The analysis presented in this chapter is based on the assumption that such a development is still speculative and will probably not arrive for several decades at the earliest. The disagreement over whether it will arrive at all requires a level of understanding about the nature of consciousness that scientists have not yet reached. Supercomputers have already demonstrated some capabilities that are far superior to those of human beings and are effectively making some important decisions for us already—handling high-frequency algorithmic trading on financial exchanges, for example—and discerning previously hidden complex relationships within very large amounts of data.

The change being driven by the wholesale adoption of the Internet as the principal means of information exchange is simultaneously disruptive and creative. The futurist Kevin Kelly says that our new technological world—infused with intelligence—more and more resembles "a very complex organism that often follows its own urges." In this case, the large complex system includes not only the Internet and the computers, but also us.

Consider the impact on conversations. Many of us now routinely reach for smartphones to find the answers to questions that arise at the dinner table by searching the Internet with our fingertips. Indeed, many now spend so much time on their smartphones and other mobile Internet-connected devices that oral conversation sometimes almost ceases. As a distinguished philosopher of the Internet, Sherry Turkle, recently wrote, we are spending more and more time "alone together."

The deeply engaging and immersive nature of online technologies has led many to ask whether their use might be addictive for some people. The *Diagnostic and Statistical Manual of Mental Disorders* (DSM), when it is updated in May 2013, will include "Internet Use Disorder" in its appendix for the first time, as a category targeted for further study. There are an estimated 500 million people in the world now playing online games at least one hour per day. In the United States, the average person under the age of twenty-one now spends almost as much time playing online games as they spend in classrooms from the sixth through twelfth grades. And it's not just young people: the average online social games player is a woman in her mid-forties. An estimated 55 percent of those playing social games in the U.S.—and 60 percent in the U.K.—are women. (Worldwide, women also generate 60 percent of the comments and post 70 percent of the pictures on Facebook.)

OF MEMORY, "MARKS," AND THE GUTENBERG EFFECT

Although these changes in behavior may seem trivial, the larger trend they illustrate is anything but. One of the most interesting debates among experts who study the relationship between people and the Internet is over how we may be adapting the internal organization of our brains—and the nature of consciousness—to the amount of time we are spending online.

Human memory has always been affected by each new advance

in communications technology. Psychological studies have shown that when people are asked to remember a list of facts, those told in advance that the facts will later be retrievable on the Internet are not able to remember the list as well as a control group not informed that the facts could be found online. Similar studies have shown that regular users of GPS devices began to lose some of their innate sense of direction.

The implication is that many of us use the Internet—and the devices, programs, and databases connected to it—as an extension of our brains. This is not a metaphor; the studies indicate that it is a literal reallocation of mental energy. In a way, it makes sense to conserve our brain capacity by storing only the meager data that will allow us to retrieve facts from an external storage device. Or at least Albert Einstein thought so, once remarking: "Never memorize what you can look up in books."

For half a century neuroscientists have known that specific neuronal pathways grow and proliferate when used, while the disuse of neuron "trees" leads to their shrinkage and gradual loss of efficacy. Even before those discoveries, McLuhan described the process metaphorically, writing that when we adapt to a new tool that extends a function previously performed by the mind alone, we gradually lose touch with our former capacity because a "built-in numbing apparatus" subtly anesthetizes us to accommodate the attachment of a mental prosthetic connecting our brains seamlessly to the enhanced capacity inherent in the new tool.

In Plato's dialogues, when the Egyptian god Theuth tells one of the kings of Egypt, Thamus, that the new communications technology of the age—writing—would allow people to remember much more than previously, the king disagreed, saying, "It will implant forgetfulness in their souls: they will cease to exercise memory because they rely on that which is written, calling things to remembrance no longer from within themselves, but by means of external marks."*

So this dynamic is hardly new. What is profoundly different about the combination of Internet access and mobile personal computing devices is that the instantaneous connection between an individual's brain

---

* The memory bank of the Internet is deteriorating through a process that Vint Cerf, a close friend who is often described as a "father of the Internet" (along with Robert Kahn, with whom he co-developed the TCP/IP protocol that allows computers and devices on the Internet to link with one another), calls "bit rot"—information disappears either because newer software can't read older, complex file formats or because the URL that the information is linked to is not renewed. Cerf calls for a "digital vellum"—a reliable and survivable medium to preserve the Internet's memory.

and the digital universe is so easy that a habitual reliance on external memory (or "exomemory") can become an extremely common behavior. The more common this behavior becomes, the greater one comes to rely on exomemory—and the less one relies on memories stored in the brain itself. What becomes more important instead are the "external marks" referred to by Thamus 2,400 years ago. Indeed, one of the new measures of practical intelligence in the twenty-first century is the ease with which someone can quickly locate relevant information on the Internet.

Human consciousness has always been shaped by external creations. What makes human beings unique among, and dominant over, life-forms on Earth is our capacity for complex and abstract thought. Since the emergence of the neocortex in roughly its modern form around 200,000 years ago, however, the trajectory of human dominion over the Earth has been defined less by further developments in human physical evolution and more by the evolution of our relationship to the tools we have used to augment our leverage over reality.

Scientists disagree over whether the use of complex speech by humans emerged rather suddenly with a genetic mutation or whether it developed more gradually. But whatever its origin, complex speech radically changed the ability of humans to use information in gaining mastery over their circumstances by enabling us for the first time to communicate more intricate thoughts from one person to others. It also arguably represented the first example of the storing of information outside the human brain. And for most of human history, the spoken word was the principal "information technology" used in human societies.

The long hunter-gatherer period is associated with oral communication. The first use of written language is associated with the early stages of the Agricultural Revolution. The progressive development and use of more sophisticated tools for written language—from stone tablets to papyrus to velum to paper, from pictograms to hieroglyphics to phonetic alphabets—is associated with the emergence of complex civilizations in Mesopotamia, Egypt, China and India, the Mediterranean, and Central America.

The perfection by the ancient Greeks of the alphabet first devised by the Phoenicians led to a new way of thinking that explains the sudden explosion in Athens during the fourth and fifth centuries BCE of philosophical discourse, dramatic theater, and the emergence of sophisticated concepts like democracy. Compared to hieroglyphics, pictographs, and cuneiform, the abstract shapes that made up the Greek alphabet—like those that make up all modern Western alphabets—have no more inherent meaning in themselves than the ones and zeros of digital code. But when they are arranged and rearranged in different combinations, they can be assigned gestalt meanings. The internal organization of the brain necessary to adapt to this new communications tool has been associated with the distinctive difference historians find in the civilization of ancient Greece compared to all of its predecessors.

The use of this new form of written communication led to an increased ability to store the collective wisdom of prior generations in a form that was external to the brain but nonetheless accessible. Later advances—particularly the introduction of the printing press in the fourteenth century (in Asia) and the fifteenth century (in Europe)—were also associated with a further expansion of the amount of knowledge stored externally and a further increase in the ease with which a much larger percentage of the population could gain access to it. With the introduction of print, the exponential curve that measures the complexity of human civilization suddenly bent upward at a sharply steeper angle. Our societies changed; our culture changed; our commerce changed; our politics changed.

Prior to the emergence of what McLuhan described as the Gutenberg Galaxy, most Europeans were illiterate. Their relative powerlessness was driven by their ignorance. Most libraries consisted of a few dozen hand-copied books, sometimes chained to the desks, written in a language that for the most part only the monks could understand. Access to the knowledge contained in these libraries was effectively restricted to the ruling elites in the feudal system, which wielded power in league with the medieval church, often by force of arms. The ability conferred by the printing press to capture, replicate, and distribute en masse the collected wisdom of preceding ages touched off the plethora of advances in information sharing that led to the modern world.

Less than two generations after Gutenberg's press came the Voyages of Discovery. When Columbus returned from the Bahamas, eleven print editions of the account of his journey captivated Europe. Within a quarter century sailing ships had circumnavigated the globe, bringing artifacts and knowledge from North, South, and Central America, Asia, and previously unknown parts of Africa.

In that same quarter century, the mass distribution of the Christian

Bible in German and then other popular languages led to the Protestant Reformation (which was also fueled by Martin Luther's moral outrage over the print-empowered bubble in the market for indulgences, including the exciting new derivatives product: indulgences for sins yet to be committed). Luther's Ninety-Five Theses, nailed to the door of the church in Wittenberg in 1517, were written in Latin, but thousands of copies distributed to the public were printed in German. Within a decade, more than six million copies of various Reformation pamphlets had been printed, more than a quarter of them written by Luther himself.

The proliferation of texts in languages spoken by the average person triggered a series of mass adaptations to the new flow of information, beginning a wave of literacy that began in Northern Europe and moved southward. In France, as the wave began to crest, the printing press was denounced as "the work of the Devil." But as popular appetites grew for the seemingly limitless information that could be conveyed in the printed word, the ancient wisdom of the Greeks and Romans became accessible. The resulting explosion of thought and communication stimulated the emergence of a new way of thinking about the legacy of the past and the possibilities of the future.

The mass distribution of knowledge about the world of the present began to shake the foundations of the feudal order. The modern world that is now being transformed by kind rather than degree rose out of the ruins of the civilization that we might say was creatively destroyed by the printing press. The Scientific Revolution began less than a hundred years after Gutenberg's Bible, with the publication of Nicolaus Copernicus's *Revolution of the Spheres* (a copy of which he received fresh from the printer on his deathbed). Less than a century later Galileo confirmed heliocentrism. A few years after that came Descartes's "Clockwork Universe." And the race was on.

Challenges to the primacy of the medieval church and the feudal lords became challenges to the absolute rule of monarchs. Merchants and farmers began to ask why they could not exercise some form of self-determination based on the knowledge now available to them. A virtual "public square" emerged, within which ideas were exchanged by individuals. The Agora of ancient Athens and the Forum of the Roman Republic were physical places where the exchange of ideas took place, but the larger virtual forum created by the printing press mimicked important features of its predecessors in the ancient world.

Improvements to the printing press led to lower costs and the pro-liferation of printers looking for material to publish. Entry barriers were very low, both for obtaining the printed works of others and for con-tributing one's own thoughts. Soon the demand for knowledge led to modern works—from Cervantes and Shakespeare to journals and then newspapers. Ideas that found resonance with large numbers of people attracted a larger audience still—in the manner of a Google search today.

In the Age of Enlightenment that ensued, knowledge and reason became a source of political power that rivaled wealth and force of arms. The possibility of self-governance within a framework of representative democracy was itself an outgrowth of this new public square created within the information ecosystem of the printing press. Individuals with the freedom to read and communicate with others could make decisions collectively and shape their own destiny.

At the beginning of January in 1776, Thomas Paine—who had mi-grated from England to Philadelphia with no money, no family connec-tions, and no source of influence other than an ability to express himself clearly in the printed word—published *Common Sense*, the pamphlet that helped to ignite the American War of Independence that July. The theory of modern free market capitalism, codified by Adam Smith in the same year, operated according to the same underlying principles. Individuals with free access to information about markets could freely choose to buy or sell—and the aggregate of all their decisions would constitute an "in-visible hand" to allocate resources, balance supply with demand, and set prices at an optimal level to maximize economic efficiency. It is fitting that the first volume of Gibbon's *Decline and Fall of the Roman Empire* was also published in the same year. Its runaway popularity was a counter-point to the prevailing exhilaration about the future. The old order was truly gone; those of the present generation were busy making the world new again, with new ways of thinking and new institutions shaped by the print revolution.

It should not surprise us, then, that the Digital Revolution, which is sweeping the world much faster and more powerfully than the Print Revolution did in its time, is ushering in with it another wave of new societal, cultural, political, and commercial patterns that are beginning to make our world new yet again. As dramatic as the changes wrought by the Print Revolution were (and as were those wrought earlier by the introduction of complex speech, writing, and phonetic alphabets), none of these previous waves of change remotely compares with what we are now beginning to experience as a result of today's emergent combination of nearly ubiquitous computing and access to the Internet. Computers have been roughly doubling in processing power (per dollar spent) every eighteen to twenty-four months for the last half-century. This remarkable pattern—which follows Moore's Law—has continued in spite of periodic predictions that it would soon run its course. Though some experts believe that Moore's Law may now finally be expiring over the next decade, others believe that new advances such as quantum computing will lead to continued rapid increases in computing power.

Our societies, culture, politics, commerce, educational systems, ways of relating to one another—and our ways of thinking—are all being profoundly reorganized with the emergence of the Global Mind and the growth of digital information at exponential rates. The annual production and storage of digital data by companies and individuals is 60,000 times more than the total amount of information contained in the Library of Congress. By 2011, the amount of information created and replicated had grown by a factor of nine in just five years. (The amount of digital storage capacity did not surpass analog storage until 2002, but within only five years the percentage of information stored digitally grew to 94 percent of all stored information.) Two years earlier, the volume of data transmitted from mobile devices had already exceeded the total volume of all voice data transmitted. Not coincidentally, from 2003 to 2010, the average telephone call grew shorter by almost half, from three minutes to one minute and forty-seven seconds.

The number of people worldwide connected to the Internet doubled between 2005 and 2010 and in 2012 reached 2.4 billion users globally. By 2015, there will be as many mobile devices as there are people in the world. The number of mobile-only Internet users is expected to increase 56-fold over the next five years. Aggregate information flow using smartphones is projected to increase 47-fold over the same period. Smartphones already have captured more than half of the mobile phone market in the United States and many other developed countries.

But this is not just a phenomenon in wealthy countries. Although computers and tablets are still more concentrated in advanced nations, the reduction in the cost of computing power and the proliferation of smaller, more mobile computing devices is spreading access to the Global Mind throughout the world. More than 5 billion of the 7 billion people in the world now have access to mobile phones. In 2012, there were 1.1 billion active smartphone users worldwide—still under one fifth of the global market. While smartphones capable of connecting to the Internet are still priced beyond the reach of the majority of people in developing countries, the same relentless cost reductions that have characterized the digital age since its inception are now driving the migration of smart features and Internet connectivity into affordable versions of low-end smartphones that will soon be nearly ubiquitous.

Already, the perceived value of being able to connect to the Internet has led to the labeling of Internet access as a new "human right" in a United Nations report. Nicholas Negroponte has led one of two competing global initiatives to provide an inexpensive ($100 to $140) computer or tablet to every child in the world who does not have one. This effort to close the "information gap" also follows a pattern that began in wealthy countries. For example, the United States dealt with concerns in the 1990s about a gap between "information haves" and "information have-nots" by passing a new law that subsidized the connection of every school and library to the Internet.

The behavioral changes driven by the digital revolution in developed countries also have at least some predictive value for the changes now in store for the world as a whole. According to a survey by Ericsson, 40 percent of smartphone owners connect to the Internet immediately upon awakening—even before they get out of bed. And that kick-starts a behavioral pattern that extends throughout their waking hours. While they are driving to work in the morning, for example, they encounter one of the new hazards to public health and safety: the use of mobile communications devices by people who email, text, play games, and talk on the phone while simultaneously trying to operate their cars and trucks.

In one extreme example of this phenomenon, a commercial airliner flew ninety minutes past its scheduled destination because both the pilot and copilot were absorbed with their personal laptops in the cockpit, oblivious as more than twelve air traffic controllers in three different cities tried to get their attention—and as the Strategic Air Command readied fighter jets to intercept the plane—before the distracted pilots finally disengaged from their computers.

The popularity of the iPhone and the amount of time people communicate over its videoconferencing feature, FaceTime, has caused a few to actually modify the appearance of their faces in order to adapt to the new technology. Plastic surgeon Robert K. Sigal reported that "patients come in with their iPhones and show me how they look on FaceTime. The angle at which the phone is held, with the caller looking downward into the camera, really captures any heaviness, fullness and sagging of the face and neck. People say, 'I never knew I looked like that! I need to do something!' I've started calling it the 'FaceTime Facelift' effect. And we've developed procedures to specifically address it."

THE RISE OF "BIG DATA"

Just as we have extended our consciousness into the Global Mind, we are now extending our *peripheral* nervous system into the Internet of Things, which operates almost entirely below the level of consciousness and controls functions important to maintaining the efficiency of Earth Inc. It is this part of the global Internet that is proliferating most rapidly, generating far more data than people themselves produce, and evolving toward what some call the "Internet of Everything."

The emerging field labeled "Big Data," one of the exciting new frontiers of information science, is based on the development of new algorithms for supercomputers to sift through voluminous new quantities of data that have not previously been seen as manageable. More than 90 percent of the information collected by Landsat satellites has been sent directly to electronic storage without ever firing a single neuron in a human brain, and without being processed by computers for patterns and meaning. This and other troves of unutilized data may now finally be analyzed.

Similarly, most of the data now being collected during the operation of industrial processes by embedded systems, sensors, and tiny devices such as actuators has been disposed of soon after it is collected. With the plummeting cost of data storage and the growing sophistication of Big Data, some of this information is now being kept and analyzed and is already producing a flood of insights that promote efficiency in industry and business. To take another example, some commercial vehicles mount a small video camera on the windshield that collects data continuously but only saves twenty seconds at a time; in the event of an accident, the information collected during the seconds prior to and during the accident is saved for analysis. The same is true of black boxes on airplanes and most security cameras in buildings. The data collected is constantly erased to make room for newer information. Soon, most all of this information will be kept, stored, and processed by Big Data algorithms for useful insights.

Plans for gathering—and analyzing—even larger amounts of information are now under way throughout the world. IBM is working with the Netherlands Institute of Radio Astronomy to develop a new generation of computer technology to store and process the data soon to be captured by the Square Kilometre Array, a new radio telescope that will collect each day twice the amount of information presently generated on the entire World Wide Web.

Virtually all human endeavors that routinely produce large amounts of data will soon be profoundly affected by the use of Big Data techniques. To put it another way, just as psychologists and philosophers search for deeper meanings in the operations of the human subconscious, cutting-edge supercomputers are now divining meaningful patterns in the enormous volumes of data collected on a continuous basis not only on the Internet of Things but also by analyzing patterns in the flood of information exchanged among people—including in the billions of messages posted each day on social networks like Twitter and Facebook.

The U.S. Geological Survey has established a Twitter Earthquake Detector to gather information on the impact and location of shaking events more quickly, particularly in populated areas with few seismic instruments. And in 2009, U.N. Secretary General Ban Ki-moon launched the Global Pulse program to analyze digital communications in order to detect and understand economic and social shocks more quickly. The pattern with which people add money to their mobile phone accounts is an early warning of job loss. Online food prices can be surveyed to help predict price spikes and food shortages. Searches for terms like "flu" and "cholera" can give warnings of disease outbreaks.

The intelligence community is using the techniques of Big Data analysis to search for patterns in vast flows of communication to predict social unrest in countries and regions of particular interest. Some new businesses are now using similar techniques to analyze millions of messages or tweets in order to predict how well Hollywood—and Bollywood—movies will perform at the box office.

DEMOCRACY IN THE BALANCE

As always, the imperatives driving commerce and national security adapt quickly to the emergence of new technologies, but what about democracy in this new age? The rapid and relentless rise of Internet-based communication is surely a hopeful sign for the renewed health of self-governance, largely because the structural characteristics of the Internet are so similar to the world of the printing press: individuals have extremely low entry barriers and ever easier access. As was true in the age of the printing press, the quality of ideas conveyed over the Internet can be at least partially assessed by the number of people with whom they resonate. And as more people find resonance with particular expressions, more still have their attention directed to the expressions whose popularity is rising.

The demand for content on the Internet is also linked to a significant rise in reading—a faint echo of the "big bang" of literacy that accompanied the creation of the Gutenberg Galaxy. In fact, after reading declined following the introduction of television, it has now tripled in just the last thirty years because the overwhelmingly dominant content on the Internet is printed words.

With democracy having fallen on hard times due to the current dominance over the public interest in so many countries by wealth and corporate power—and in others by the entrenched power of authoritarian dictatorships—many supporters of democratic self-governance are placing their hopes on the revival of robust democratic discourse in the age of the Internet.

Already, revolutionary political movements—from the Tahrir Square protesters in Cairo to Los Indignados in Spain to Occupy Wall Street to the surprisingly massive crowds of election protesters in Moscow—are predominantly shaped by the Internet. Facebook and Twitter have played a particularly important role in several of these movements, along with email, texting, and instant messaging. Google Earth has also been significant in spotlighting the excesses of elites, in Bahrain for example—and in the Libyan revolution, Google Earth was actually used by rebels in Misrata to guide their mortars. (Google Earth also, by the way, triggered a small border dispute and brief armed standoff between Nicaragua and Costa Rica, when it mistakenly attributed a tiny portion of Costa Rica to the national territory of Nicaragua.)

Thus far, however, reformist and revolutionary movements that have begun on the Internet have mostly followed the same pattern: enervation and excitement followed by disappointment and stasis. It is still an open question whether these Internet-inspired reform movements will gain a second wind and, after a period of simmering, reemerge and ultimately reach their goals.

One of the first revolutionary movements in which the Internet played a key igniting role was the 2007 Saffron Revolution in Myanmar. Activists took extreme personal risks to spread their messages urging democratic reforms by using the World Wide Web with false names from Internet cafés and by smuggling thumb drives across the border to collaborators in the diaspora living in Thailand. Unfortunately, the authoritarian government in Myanmar was able to smother and shut down the Saffron Revolution, but only at the cost of completely blacking out the Internet inside the country's borders.

Nevertheless, the revolutionary fires lit before the Internet was shut down continued to smolder in Myanmar and continued to burn brightly in other parts of the world where the forces of conscience had been awakened to the abuses and injustices of the Myanmar dictatorship. (Diasporas, particularly educated and wealthy diasporas in Western countries, have been newly empowered by the Internet to play significant roles in fostering and sustaining reform movements in their countries of origin.) A few years later, the government of Myanmar was pressured to loosen its controls on political dialogue and release the leader of the reform movement, Aung San Suu Kyi, from her long house arrest, and in March 2012 she was triumphantly elected to the Parliament amidst many signs that the popular movement that had begun on the Internet was reemerging as a force for change that seemed destined to take control of the government.

In many other authoritarian countries, however, the ferocious resistance to reform has been more effective in snuffing out Internet-based dissent movements. In 2009, Iran's Green Revolution began as a popular protest against the fraudulent presidential election. Although Western sympathizers had the impression that Twitter played a key role in igniting and sustaining the protest movement, in actuality social media played a much smaller role inside than outside Iran because the Iranian government was successful in largely controlling Internet use by the protesters. While it is true that YouTube videos documented government excesses (most famously, the tragic death of Neda Agha-Soltan), the more potent social media sites that would have enabled dissenters to build a larger protest movement were almost completely shut down. Indeed, during the election campaign itself, when the principal opposition candidate, Mir-Hossein Moussavi, began to gain momentum by organizing on Facebook, the government simply blacked it out.

Worse still, the Iranian security forces gave the world a demonstration of what a malignant authoritarian government can do to its citizens by using the knowledge it gains from their Internet connections and social graphs to identify and track down dissenters, read their private communications, and effectively stifle any effective resistance to the dictatorship's authority. The entire episode was a chilling alarm that underscored the extent to which the lack of privacy on the Internet can potentially increase the power of government over the governed more easily than it can empower reform and revolution.

China, in particular, has introduced by far the most sophisticated measures to censor content on the Internet and exercise control over its potential for fostering reformist or revolutionary fervor. The "Great Firewall of China" is the largest effort at Internet control in the world today. (Iran and the retro-Stalinist dictatorship of Belarus are the other two countries that have attempted such efforts.) China's connection to the global Internet is monopolized by state-run operators that carefully follow a system of protocols that effectively turn the Internet within China into a national intranet. In 2010, even an interview with the then premier of China, Wen Jiabao, in which he advocated reforms, was censored and made unavailable to the people of China.

In 2006 the Chinese plan to control content on the Internet collided with the open values of the world's largest search engine, Google. As one who participated in the company's deliberations at the time, I saw firsthand how limited the options were. After searching for ways to reconcile its commitment to full openness of information with China's determined effort to block any and all content it found objectionable, Google made the principled decision to withdraw from China and instead route its site through Hong Kong, which still maintains a higher level of freedom, albeit within constraints imposed from Beijing. Facebook, by the way, has never been allowed into China. The cofounder of Google, Sergey Brin, said in 2012 that China had been far more effective in controlling the Internet than he had expected. "I thought there was no way to put the genie back in the bottle," Brin noted, "but now it seems in certain areas the genie *has* been put back in the bottle."

The much admired Chinese artist Ai Weiwei expressed a different view: "[China] can't live with the consequences of that. . . . It's hopeless to try to control the Internet." China now has the largest number of Internet users of any country in the world—more than 500 million people, 40 percent of its total population. As a result, most observers believe it is only a matter of time before more open debate—even on topics controversial in the eyes of the Communist Party—will become uncontrollable inside China. Already, a number of Chinese leaders have found it necessary to take to the Internet themselves in order to respond to public controversies. In neighboring Russia, former president Dmitri Medvedev also felt the pressure to engage personally on the Internet.

As the role played by the Internet and connected computing devices becomes more prominent and pervasive generally, authoritarian governments may find it increasingly difficult to exert the same degree of control. When the Arab Spring began in Tunisia, it was partly due to the fact that four out of every ten Tunisians were connected to the Internet, with almost 20 percent of them on Facebook (80 percent of the Facebook users were under the age of thirty).

So even though Tunisia was one of the countries cited by Reporters Without Borders as censoring political dissent on the Internet, the largely nonviolent revolution gained momentum with startling speed, and the pervasive access to the Internet within Tunisia made it difficult for the government to control the digital blossoming of public defiance. The man who set himself on fire in protest, Mohamed Bouazizi, was not the first to do so, but he was the first to be *video-recorded* doing so. It was the downloaded video that ignited the Arab Spring.

In Saudi Arabia, Twitter has facilitated public criticism of the government, and even of the royal family. As the number of tweets grew faster there in 2012 than in any other country, a thirty-one-year-old lawyer, Faisal Abdullah, told *The New York Times*, "Twitter for us is like a parliament, but not the kind of parliament that exists in this region. It's a true parliament, where people from all political sides meet and talk freely."

But experts in the region argue that it is important to look carefully at the interplay between the Internet and other significant factors in the Arab Spring—including some that were at least as important as the Internet in bringing about this sociopolitical explosion. The combination of population growth, the growing percentage of young people, economic stagnation, and rising food prices created the conditions for unrest. When governments in the region first promised economic and political reforms, then appeared to backtrack, the frustrations reached a boiling point.

The change that many analysts believe was most important in sowing the seeds of the Arab Spring was the introduction in 1996 of the feisty and relatively independent satellite television channel Al Jazeera. Al Jazeera was soon followed by approximately 700 other satellite television channels that were easily accessed with small, cheap satellite dishes—even in countries where they are technically illegal. Several governments attempted to control the proliferation of small dishes, but the result was an incredible outburst of political discussion, including on topics that had not been debated openly before. By the time the Arab Spring erupted in Cairo's Tahrir Square, both access to satellite television *and* the Internet had spread throughout Egypt and the region. Sociologists and political scientists have had a difficult time parsing the relative influence of these two new electronic media in causing and feeding the Arab Spring, but most believe that Al Jazeera and its many siblings were the more important factor. In 2004, when then Egyptian president Hosni Mubarak paid a visit to Al Jazeera's headquarters in Qatar, he said, "All that trouble from this little matchbox?" Perhaps both were necessary but neither was sufficient.

Like Tunisia, Egypt found it difficult to shut down access to the Internet in the way Myanmar and Iran had. By 2011 it was so pervasive that when the government blocked all of the Internet access points entering the country, the public's reaction was so strong that the fires of revolt grew even hotter. The determination of the protesters ultimately succeeded in forcing Mubarak to step down, but their cohesion faded during the political struggle that followed.

Some analysts, including Malcolm Gladwell, have argued that online connections are inherently weak and often temporary because they do not support the stronger relationships formed when mass movements rely upon in-person gatherings. In Egypt, for example, the crowds of Tahrir Square actually represented a tiny fraction of Egypt's huge population— and those in the rest of the country who sympathized with their complaints against the Mubarak government did not remain aligned with the protesters when the time came to form a new political consensus around what kind of government would follow Mubarak. The Egyptian military soon asserted its control of the government, and in the elections that followed, Islamist forces prevailed in establishing a new regime based on principles far different from those advocated by most of the Internet-inspired reformers who predominated in Tahrir Square.

Indeed, not only in Egypt but also in Libya, Syria, Bahrain, Yemen, and elsewhere—including Iran—the same pattern has unfolded: an emergent reform movement powered by a new collective political consciousness born on the Internet has stimulated change, but failed to consolidate its victory. The forces of counterrevolution have tightened control of the media and have reestablished their dominance.

The unique history of communications technology in the Middle East and North Africa offers one of the reasons for the failure by reformers to consolidate their gains. The emergent political consciousness that accompanied the Print Revolution in Europe, and later North America, bypassed the Middle East and North Africa when the Ottoman Empire banned the printing press for Arabic-speaking peoples. This contributed to the isolation of the Ottoman-ruled lands from the rapid advances (such as the Scientific Revolution) that the printing press triggered in Europe. Two centuries later, when Arab Muslims first asked the historic question "What went wrong?" part of the answer was that they had deprived themselves of the fruits of the Print Revolution.

As a result, the institutions that emerged in the West to embody representative democracy never formed in the Middle East. Centuries later, therefore, the new political consciousness born on the Internet could not easily be *embodied* in formal structures that could govern according to the principles articulated by the reformers. Yet the forces of authoritarianism could easily embody their desire to control society and the economy in the institutions that were already present—including the military, the national police, and the bureaucracies of autocratic rule.

Other analysts have connected the disappointment in the wake of Tahrir Square to what they regard as yet another example of "techno-optimism," in which an exciting new technology is endowed with unrealistic hopes, while overlooking the simple fact that all technologies can be used for good or ill, depending on how they are used and who uses them to greatest effect. The Internet can be used not only by reformers, but also by opponents of reform. Still, the exciting promise of Internet-based reform—both in the delivery of public goods and, more crucially, in the revitalization of democracy—continues to inspire advocates of freedom, precisely because it enables and fosters the emergence of a new collective political consciousness within which individuals can absorb political ideas, contribute their own, and participate in a rapidly evolving political dialogue.

This optimism is further fueled by the fact that some governments providing services to individuals are making dramatic improvements in their ability to communicate important information on the Internet and engage in genuinely productive two-way communication with citizens. Some nations—most notably, Estonia—have even experimented with Internet voting in elections and referenda. In neighboring Latvia, two laws have already been passed as a result of proposals placed by citizens on a government website open to suggestions from the public. Any idea attaining the support of 10,000 people or more goes directly into a legislative process. In addition, many cities are using computerized statistics and sophisticated visual displays to more accurately target the use of resources and achieve higher levels of quality in the services they deliver. Some activists promoting Internet-based forms of democracy, including NYU professor Clay Shirky, have proposed imaginative ways to use open source programming to link citizens together in productive dialogues and arguments about issues and legislation.

In Western countries, however, the potential for Internet-based reform movements has been blunted. Even in the United States, in spite of the prevailing hopes that the Internet will eventually reinvigorate democracy, it has thus far failed to do so. In order to understand why, it is important to analyze the emerging impact of the Internet on political consciousness in the broader context of the historic relationship between communications media and governance—with particular attention to the displacement of print media by the powerful mass medium of television.

In the politics of many countries—including the United States—we find ourselves temporarily stuck in a surprisingly slow transition from the age of television to the age of the Internet. Television is still by far and away the dominant communications medium in the modern world. More people even watch Internet videos on television screens than on computer screens. Eventually, bandwidth limitations on high-quality video will become less of a hindrance and television will, in the words of novelist William Gibson, "be appropriated into the realm of the digital." But until it does, broadcast, cable, and satellite television will continue to dominate the public square. As a result, both candidates and leaders of reform movements will continue to face the requirement of paying a king's ransom for the privilege of communicating effectively with the mass public.

Well before the Internet and computer revolution was launched, the introduction of electronic media had already begun transforming the world that had been shaped by the printing press. In a single generation, television displaced print as the dominant form of mass communication. Even now, while the Internet is still in its early days, Americans spend more time watching television than in any other activity besides sleeping and working. The average American now watches television more than five hours per day. Largely as a result, the average candidate for Congress spends 80 percent of his or her campaign money on thirty-second television advertising.

To understand the implications for democracy that flow from the continuing dominance of television, consider the significant differences between the information ecosystem of the printing press and the information ecosystem of television. First of all, access to the virtual public square that emerged in the wake of the print revolution was extremely cheap; Thomas Paine could walk out of his front door in Philadelphia and easily find several low-cost print shops.

Access to the public square shaped by television, though, is extremely expensive. The small group of corporations that serve as gatekeepers controlling access to the mass television audience is now more consolidated than ever before and continues to charge exorbitant sums for that access. If a modern-day Thomas Paine walked to the nearest television station and attempted to broadcast a televised version of *Common Sense*, he would be laughed off the premises if he could not pay a small fortune. By contrast, paid pundits whose views reflect the political philosophy of the corporations that own most networks are given many hours each week to promulgate their ideology.

So long as commercial television dominates political discussion, candidates will find it necessary to solicit large and ever growing sums of money from wealthy individuals, corporations, and special interests to gain access to the only public square that matters when the majority of voters spend the majority of their free time staring at television screens. This requirement, in turn, has led to the obscene dominance of decision making in American democracy by these same wealthy contributors—

especially corporate lobbies. Because recent Supreme Court decisions—especially the *Citizens United* case—have overturned long-standing prohibitions against the use of corporate funds to support candidates, this destructive trend is likely to get much worse before it gets better. It is, in a very real sense, a slow-motion corporate coup d'état that threatens to destroy the integrity and functioning of American democracy.

Although the political systems and legal regimes of countries vary widely, the relative roles of television and the Internet are surprisingly similar. It is notable that in both China and Russia, television is much more tightly controlled than the Internet. In the Potemkin democracy that has been constructed in Vladimir Putin's Russia, the government is choosing to tolerate a much freer, more robust freedom of speech on the Internet than on television. Mikhail Kasyanov, one of the prime ministers who served under Putin (and whose candidacy for president against Putin's handpicked successor, Dmitri Medvedev, was derailed when Putin ordered him removed from the ballot), told me that when he was prime minister under Putin he was given clear instructions that debate on the Internet mattered little so long as the government exercised tight control over what appeared on Russian television.

Four years later, in the spring of 2012, the Internet-inspired protest movement challenging the obviously fraudulent process used in the first round of the elections (in which Putin was ultimately victorious, as expected), one Russian analyst said, "The old people come and the old people come and the old people come and all vote for one candidate—for Putin. Why are they voting for Putin? Watch TV. There is one face: Putin." And indeed, one of the many reasons for television's dominance in the political media landscape of almost every country is that older people both simultaneously vote in higher percentages *and* watch television more hours per day than any other age group. In the U.S., people aged sixty-five and older watch, on average, almost seven hours per day.

In many nations, institutions important to the rise and survival of democracy, like journalism, have also been profoundly affected by the historic transformation of communications technology. Newspapers have fallen on hard times. They used to be able to bundle together revenue from subscriptions, commercial advertising, and classified advertising to pay not only for the printing and distribution of their papers but also the salaries of professional reporters, editors, and investigative journalists. With the introduction of television—and particularly with the launch of evening television news programs—the afternoon newspapers in most major cities that people used to read upon returning home from work were the first to go bankrupt. The loss of increasing amounts of commercial advertising to television and radio also began to hurt the morning newspapers. Then, when classified advertising migrated en masse to the Internet and the widespread availability of online news sources led many readers to stop their subscriptions to newspapers, the morning newspapers began to go bankrupt as well.

Eventually, Internet-based journalism will begin to thrive. In the U.S., digital news stories already reach more people than either newspapers or radio. As yet, however, a high percentage of quality journalism available on the Internet is still derived from the repurposing of articles originally prepared for print publications. And there are as yet few business models for journalism originating on the Internet that bundle together enough revenue to support the salaries of reporters engaged in the kind of investigative journalism essential to provide accountability in a democracy.

Like the journalism essential to its flourishing, democracy itself is now stuck in this odd and dangerous transition era that falls between the waning age of the printing press and the still nascent maturation of effective democratic discourse on the Internet. Reformers and advocates of the public interest are connecting with one another in ever larger numbers over the Internet and are searching with ever greater intensity for ways to break through the quasi-hypnotic spell cast over the mass television audience—day after day, night after night—by constant, seductive, expensive, and richly produced television programming.

Virtually all of this programming is punctuated many times each hour by slick and appealing corporate messages designed to sell their products and by corporate issue advertising designed to shape the political agenda. During election years, especially in the United States, television viewers are also deluged with political advertisements from candidates who—again because of the economics of the television medium—are under constant and unrelenting pressure from wealthy and powerful donors to adopt the donors' political agendas—agendas that are, unsurprisingly, congruent with those contained in the corporate issue advertising.

Public goods—such as education, health care, environmental protection, public safety, and self-governance—have not yet benefited from the new efficiencies of the digital age to the same extent as have private goods. The power of the profit motive has been more effective at driving the exploitation of new opportunities in the digital universe. By contrast, the ability of publics to insist upon the adoption of new, more efficient, digital models for the delivery of public goods has been severely hampered by the sclerosis of democratic systems during this transition period when digital democracy has yet to take hold.

EDUCATION AND HEALTH CARE IN A NEW WORLD

The crisis in public education is a case in point. Our civilization has barely begun the necessary process of adapting schools to the tectonic shift in our relationship to the world of knowledge. Education is still too frequently based on memorizing significant facts. Yet in a world where all facts are constantly at our fingertips, we can afford to spend more time teaching the skills necessary to not only learn facts but also learn the connections among them, evaluate the quality of information, discern larger patterns, and focus on the deeper meaning inherent in those patterns. Students accustomed to the rich and immersive experience of television, video games, and social media frequently find the experience of sitting in desks staring at chalk on a blackboard to be the least compelling and engaging part of their day.

There is clearly a great potential for the development of a new curriculum, with tablet-based e-books and search-based, immersive, experiential, and collaborative online courses. E. O. Wilson's new, enhanced digital textbook *Life on Earth* is a terrific example of what the future may hold. In higher education, a new generation of high-quality ventures has emerged—including Coursera, Udacity, Minerva, and edX—that is already beginning to revolutionize and globalize world-class university-level instruction. Most of the courses are open to all, for free!

The hemorrhaging of government revenues at the local, state, and national level—caused in part by the lower wages and persistent high unemployment associated with the outsourcing and robosourcing in Earth Inc., and declining property values in the wake of the global economic crisis triggered in part by computer-generated subprime mortgages—is leading to sharp declines in budgets for public education at the very time when reforms are most needed. In addition, the aging of populations in developed countries and the declining percentage of parents of school-aged children have diminished the political clout wielded by advocates for increasing these budgets.

Even though public funding for education has been declining, many creative teachers and principals have found ways to adapt educational materials and routines to the digital age. The Khan Academy is a particularly exciting and innovative breakthrough that is helping many students. Nevertheless, in education as in journalism, no enduring model has yet emerged with enough appeal to replace the aging and decaying model that is now failing to meet necessary standards. And some online, for-profit ventures—like the University of Phoenix and Argosy University Online—appear to have taken advantage of the hunger for college-level instruction on the Internet without meeting their responsibility to the students who are paying them. One online college, Trinity Southern University, gave an online degree in business administration to a cat named Colby Nolan, which happened to be owned by an attorney general. The school was later prosecuted and shut down.

Health care, like education, is struggling to adapt to the new opportunities inherent in the digital universe. Crisis intervention, payment for procedures, and ridiculously expensive record keeping required by insurance companies and other service providers still dominate the delivery of health care. We have not yet exploited the new ability inherent with smartphones and purpose-built digital health monitors to track health trends in each individual and enable timely, cost-effective interventions to prevent the emergence of chronic disease states that account for most medical problems.

More sophisticated information-based strategies utilizing genomic and proteomic data for each individual could also clearly improve health outcomes dramatically at much lower cost. Epidemiological strategies—such as the monitoring of aggregate Internet searches for flu symptoms—are beginning to improve the allocation and deployment of public health resources. While interesting experiments have begun in these and other areas, however, there has as yet been no effectively focused public pressure or sustained political initiative to implement a comprehensive new Internet-empowered health care strategy. Some insurance companies have begun to use data mining techniques to scour social media and databases aggregated by marketing companies in order to better assess the risk of selling life insurance to particular individuals. At least two U.S. insurance companies have found the approach so fruitful that they even waive medical exams for customers whose data profiles classify them as low-risk.

THE SECURITY CONUNDRUM

With all of the exciting potential for the Internet to improve our lives, why have the results been so mixed thus far? Perhaps because of human nature, it is common for us to overemphasize the positive impacts of any important new technology when it is introduced and first used. It is also common, unfortunately, for us to give short shrift to the risks of new technologies and underestimate unintended side effects.

History teaches, of course, that *any* tool—the mighty Internet included—can and will be used for both good and ill. While the Internet may be changing the way we organize our thinking, and while it is changing the way we organize our relationships with one another, it certainly does not change basic human nature. And thus the age-old struggle between order and chaos—and dare I say good and evil—will play out in new ways.

More than four centuries ago, when the explosion of information created by the printing press was just beginning, the legend of Doctor Faust first appeared. Some historians claim that Faust was based on the financier and business partner of Gutenberg, Johann Fust, who was charged in France with witchcraft because of the seemingly magical process by which thousands of copies of the same text could be replicated perfectly.

In the Faust legend, which has appeared in varying forms over the centuries, the protagonist makes a deal with the devil in which he exchanges his soul for "unlimited knowledge and worldly pleasures." Ever since then, as the scientific and technological revolution accelerated, many new breakthroughs, like nuclear power and stem cell technology, among others, have frequently been described as "Faustian bargains." It is a literary shorthand for the price of power—a price that is often not fully comprehended at the beginning of the bargain.

In our time, when we adapt our thinking processes to use the Internet (and the devices and databases connected to it) as an extension of our own minds, we enter into a kind of "cyber-Faustian bargain"—in which we gain the "unlimited knowledge and worldly pleasures" of the Internet. Unless we improve privacy and security safeguards, however, we may be risking values more precious than worldly wealth.

For individuals, the benefits of this bargain—vastly increased power to access and process information anywhere and anytime, a greatly increased capacity to communicate and collaborate with others—are incredibly compelling. But the price we pay in return for these incalculable benefits is a significant loss of control over the security and privacy of the thoughts and information that we send into this extended nervous system. Two new phrases that have crept into our lexicon— "the death of distance" and "the disappearance of privacy"—are intimately connected, each to the other. Most who use the Internet are tracked by many websites that then sell the information. Private emails can be read by the government without a warrant, without permission, and without notification. And hacking has become easy and widespread.

The same cyber-Faustian bargain has been made by corporations and governments. Like individuals, they are just beginning to recognize the magnitude of the cybersecurity price that apparently has to be paid on an ongoing basis. And to be clear, virtually no one argues with the gains in efficiency, power, productivity, and convenience that accompany this revolutionary change in the architecture of the information economy. What is not yet clear is how the world can resolve—or at least manage—the massive new threats to security and privacy that accompany this shift.

Internet and software companies are themselves also making the same bargain, with a historic and massive shift from software, databases, and services located within computers themselves to "the cloud"— which means, essentially, using the Internet and the remote servers and databases connected to it as extensions of the memory, software, and processing power that used to be primarily contained within each computer. The growing reliance on the cloud creates new potential choke points that may have implications for both data security and reliability of service. In late 2012, several popular Internet companies in the U.S. that rely on Amazon.com's cloud services were all knocked out of commission when problems shut down Amazon's data centers in Virginia.

The world's historic shift onto the Internet confronts us with a set of dilemmas that are inherent in the creation of a planet-wide nervous system connecting all of us to the global brain. Some of these dilemmas have arisen because digital information is now recognized—and valued—as the key strategic resource in the twenty-first century.

Unlike land, iron ore, oil, or money, information is a resource that you can sell or give away and yet still have. The value of information often expands with the number of people who share it, but the commercial value can often be lost when its initial owner loses exclusivity. The essence of patent and copyright law has been to resolve that tension and promote the greatest good for the greatest number, consistent with principles of justice and fairness. The inventor of a new algorithm or the discoverer of a new principle of electromagnetism deserves to be rewarded—partly to provide incentives for others to chase similar breakthroughs—but society as a whole also deserves to benefit from the widespread application of such new discoveries.

This inherent tension has been heightened by the world's shift onto the Internet. Longtime technology thought leader Stewart Brand is often quoted as having said in the early years of the Internet, "Information wants to be free." But what he actually said was, "On the one hand, information wants to be expensive, because it's so valuable," adding, "On the other hand, information wants to be free, because the cost of getting it out is getting lower and lower all the time. So you have these two fighting against each other."

Because digital information has become so strategic in the operations of Earth Inc., we are witnessing a global, multipronged struggle over the future of the Internet, with battlefronts scattered throughout the overlapping worlds of politics and power, commerce and industry, art and culture, science and technology:

- Between those who want information to be free and others who want to control it and exchange it for wealth or power;
- Between those who want *people* to be free and those who want to control their lives;
- Between individuals who share private information freely on social networks and others who use that information in unanticipated and sometimes harmful ways;
- Between Internet-based companies who indiscriminately collect vast amounts of information about their customers and customers who value their privacy;

- Between legacy centers of power that occupied privileged positions in the old order of information now breaking down and new centers of inchoate power seeking their own place in the new pattern struggling to emerge;
- Between activists (and "hacktivists") who value transparency and nations and corporations that value secrecy;
- Between corporations whose business models depend upon the ability to protect intellectual property contained in computers connected to the Internet and competitors who seek to steal that intellectual property by using other computers also connected to the Internet;
- Between cybercriminals intent on exploiting rich new targets in the flows of wealth and information on the Internet and law enforcement organizations whose strategy for stopping cybercrime sometimes threatens to destroy historic and hard-won boundaries between the spheres occupied by individuals and the episodic desire by their governments to invade those private spheres.

The complexity of the world's transition to the Internet is even more fraught because all of these conflicts are occurring simultaneously on the same common Internet that everyone shares. And, not surprisingly, proposed remedies for problems in one set of conflicts frequently enhance the potential for disrupting efforts to resolve problems in other sets of conflicts.

Proposals to require measures that eliminate anonymity on the Internet in order to protect cybersecurity and fight cybercrime pose a deadly threat to the ability of dissidents in authoritarian countries to propose reforms and connect with others seeking change in their governments. By the same token, the dream of reformers that the global Internet will inevitably drive global change in the direction of more freedom for individuals, regardless of where they live, strikes fear in the hearts of authoritarian rulers.

Even in free countries, activists who expose information that governments have tried to keep secret often trigger intrusive new government measures to expand the information they collect about citizens. When the Wikileaks organization, run by an Australian living in Sweden on servers based in Sweden, Iceland, and possibly other locations, publicized information stolen from the U.S. government, the subsequent crackdown enraged other hacktivists, who then broke into numerous other government and corporate websites around the world.

Because the Internet crosses national boundaries, it diminishes the ability of nation-states to manage such conflicts through laws and regulations that reflect the values in each nation (or at least the values of the governments in power). Independent groups of hacktivists have been able to break into sites controlled by the FBI, CIA, the U.S. Senate, the Pentagon, the International Monetary Fund, the official website of the Vatican, Interpol, 10 Downing Street in London, the British Ministry of Justice, and NASA (even breaking into the software of the space station while it was orbiting the Earth). When the FBI organized a secure conference call to discuss how to respond to such attacks with Scotland Yard, hackers recorded the call and put it on the web. The inmates have clearly taken over a large part of the Internet asylum when Nurse Ratched's private conversations about security are broadcast for all to hear.

The extreme difficulty in protecting cybersecurity was vividly demonstrated when EMC, a technology security company used by the National Security Agency, the Central Intelligence Agency, the Pentagon, the White House, the Department of Homeland Security, and many leading defense contractors, was penetrated by a cyberattack believed to have originated in China. EMC's security system was considered the state of the art in protecting computers connected to the Internet— which, of course, is why it was used by the organizations with the greatest need for protecting their digital data. It remains undisclosed how much sensitive information was stolen, but this attack was a sobering wake-up call.

In 2010, U.S. secretary of defense Robert Gates labeled cyberspace as the "fifth domain" for potential military conflict—alongside land, sea, air, and space. In 2012, Rear Admiral Samuel Cox, the director of intelligence at the U.S. Cyber Command (established in 2009), said that we are now witnessing "a global cyber arms race." Other experts have noted that at this stage in the development of cybersecurity technology, offense has the advantage over defense.

Securing the secrecy of important communications has always been a struggle. It was first mentioned by "the father of history," Herodotus, in his description of the "secret writing" that he said was responsible for the Greek victory in the Battle of Thermopylae, which prevented ancient Greece's conquest by Persia. A Greek living in Persia, Demaratus, witnessed the preparations for what the leader of Persia, Xerxes, intended as a surprise invasion and sent an elaborately hidden warning to Sparta. Later during the same war, a Greek leader shaved his messenger's head, wrote what he wished to convey on the messenger's scalp, and then "waited for the hair to regrow." From the use of "invisible ink" in the Middle Ages to Nazi Germany's use of the Enigma machine during World War II, cryptography in its various forms has often been recognized as crucial to the survival of nations.

The speed with which the Internet proliferated made it difficult for its original architects to remedy the lack of truly secure encryption—which they quickly recognized in the Internet's early days as a structural problem. "The system kind of got loose," said Vint Cerf.*

It is theoretically possible to develop new and more effective protections for the security of Internet data flows, and many engineers and information scientists are working to solve the problem. However, the rapidity with which Earth Inc. adapted to and coalesced around the Internet has made industry and commerce so dependent on its current architecture that any effort to change its design radically would be fraught with difficulty. And the extent to which billions of people have adapted their daily lives to the constant use of the Internet would also complicate efforts to fundamentally change its architecture.

---

* A predecessor of the Internet was demonstrated in 1969, when on October 29 the first long-distance communication between computers was sent from UCLA to SRI in Menlo Park. ARPANET was subsequently developed by the Department of Defense as a means of assuring uninterrupted communication among far-flung military units, and communication with intercontinental ballistic missile silos in the aftermath of a potential nuclear strike by the Soviet Union. However, the first description of an "Internet" based on TCP/IP appeared in a May 1974 paper by Vint Cerf and Bob Kahn, and the first three-network demonstration took place on November 22, 1977. The formal operational launch of the Internet took place on January 1, 1983. The public funding of a demonstration network linking supercomputers—the National Research and Education Network—repeated the pattern established in the 1840s, when a publicly funded demonstration of Samuel Morse's invention, the telegraph, produced the capacity to send a message from Washington to Baltimore: "What hath God wrought?" (Morse had actually received the first message seven years earlier over a distance of three miles in New Jersey—the less inspiring and less memorable sentence "A patient waiter is no loser.") The age of electronic, "instantaneous" communication was born. Five days later, the first public demonstration of the telegraph was conducted over the same two-mile line before a small assembled audience and featured a message that underscored the significance of the new invention for business: "Railroad cars just arrived, 345 passengers." On May 24, 1876, less than thirty-two years after the public demonstration of the telegraph, Alexander Graham Bell first demonstrated the ability to send voice communication electrically over wires with his message "Mr. Watson, come here; I want to see you."

McKinsey, the global management consulting firm, concluded in a recent report that four trends have converged to make cybersecurity a problem:

- Value continues to migrate online and digital data has become more pervasive;
- Corporations are now expected to be more "open" than ever before;
- Supply chains are increasingly interconnected; and
- Malevolent actors are becoming more sophisticated.

As a result, this radical transformation of the global economy has created what most experts describe as a massive cybersecurity threat to almost all companies that are using the Internet as part of their core business strategy. Particular attention has been focused on what appears to be a highly organized and persistent effort by organizations in China to steal highly sensitive information from corporations, government agencies, and organizations that have links to one or both categories.

U.S. intelligence agencies have long been assumed to conduct surveillance of foreign governments, including through cybertools to take information from computers if they have reason to believe that U.S. security is threatened. What is different about the apparent Chinese effort is that it seems to be driven not only by military and national intelligence concerns, but also by a mercantilist effort to confer advantage on Chinese businesses. "There's a big difference," says Richard Clarke, the former counterterrorism czar. "We don't hack our way into a Chinese computer company like Huawei and provide the secrets of Huawei technology to their American competitor Cisco. We don't do that."

There is no doubt that U.S. companies are being regularly and persistently attacked. Recent research published by the Aspen Institute indicates that the U.S. economy is losing more than 373,000 jobs each year—and $16 billion in lost earnings—from the theft of intellectual property. Shawn Henry, formerly a top official in the FBI's cybercrime unit, reported that one U.S. company lost a decade's worth of research and development—worth $1 billion—in a single night.

Mike McConnell, a former director of national intelligence, said recently, "In looking at computer systems of consequence—in government, Congress, at the Department of Defense, aerospace, companies with valuable trade secrets—we've not examined one yet that has not been infected by an advanced persistent threat." The U.S. Secret Service testified in 2010 that "nearly four times the amount of data collected in the archives of the Library of Congress" was stolen from the United States. The director of the FBI testified that cybersecurity will soon overtake terrorism: "The cyberthreat will be the number one threat to the country."

Another digital security company, McAfee, reported that a 2010 series of cyberattacks (called "Operation Shady RAT") resulted in the infiltration of highly secure computer systems in not only the United States, but also Taiwan, South Korea, Vietnam, Canada, Japan, Switzerland, the United Kingdom, Indonesia, Denmark, Singapore, Hong Kong, Germany, India, the International Olympic Committee, thirteen U.S. defense contractors, and a large number of other corporations—none of them in China.

But the United States—as the nation whose commerce has migrated online more than that of any other nation—is most at risk. The United States Chamber of Commerce was informed by the FBI that some of its Asia policy experts who regularly visit China had been hacked, but before the Chamber was able to secure its network, the hackers had stolen six weeks' worth of emails between the Chamber and most of the largest U.S. corporations. Long afterward, the Chamber found out that one of its office printers and one of its thermostats in a corporate apartment were still sending information over the Internet to China.

Along with printers and thermostats, billions of other devices are now connected to the Internet of Things, ranging from refrigerators, lights, furnaces, and air conditioners to cars, trucks, planes, trains, and ships to the small embedded systems inside the machinery of factories to the individual packages containing the products they produce. Some dairy farmers in Switzerland are even connecting the genitals of their cows to the Internet with a device that monitors their estrous cycles and sends a text when a cow is ready to be bred. Interspecies "sexting"?

THE PERVASIVENESS AND significance of the Internet of Things has clearly raised the possibility that cyberattacks can not only pose risks to the security of important information with commercial, intelligence, and military value, but can also have kinetic impacts. With so many Internet-connected computerized devices now controlling water and electric systems, power plants and refineries, transportation grids and other crucial systems, it is not difficult to conjure scenarios in which a coordinated attack on a nation's vital infrastructure could do real physical harm.

According to John O. Brennan, the White House official in charge of counterterrorism, "Last year alone [2011] there were nearly 200 known attempted or successful cyberintrusions of the control systems that run these facilities, a nearly fivefold increase from 2010." In the spring of 2012, Iran announced that it had been forced to sever the Internet connections of major Iranian oil terminals on the Persian Gulf, oil rigs, and the Tehran offices of the Oil Ministry because of repeated cyberattacks from an unknown source. Later that year, Saudi Arabia's state-owned oil company, Aramco, was the victim of cyberattacks that U.S. security officials said were almost certainly launched by Iran, which announced in 2011 that it had established a special military "cybercorps" after one of its nuclear enrichment facilities, in Natanz, was attacked by a computer virus. The attack on Aramco, which replaced all of the data on 75 percent of the firm's computers with an image of a burning American flag, demonstrated, in the words of former national counterterrorism czar Richard Clarke, that "you don't have to be sophisticated to do a lot of damage."

The Stuxnet computer worm, which was probably set loose by Israel and the U.S. working together, found its way—as intended—into a small Siemens industrial control system connected to the motors running the Iranian gas centrifuges that were enriching uranium as part of their nuclear program. When the Stuxnet worm confirmed that it was inside the specific piece of equipment it was looking for, it turned itself on and began to vary the speeds of the motors powering the Iranian centrifuges and desynchronize them in a way that caused them to break apart and destroy themselves. In 2010, an even more sophisticated software worm, called Flame, which analysts said "dwarfs Stuxnet" in the amount of code it contains, reportedly began infecting computers in Iran and several other nations in the Middle East and North Africa.

Although the result of the Stuxnet attack, which slowed down the Iranian effort to develop weapons-grade nuclear material, was cheered in much of the world, many experts have expressed concern that the sophisticated code involved—much of it now downloaded on the Internet—could be used for destructive attacks against Internet-connected machinery and systems in industrial countries. Some have already been inadvertently infected by Stuxnet. After a wave of cyberattacks against U.S. financial institutions in late 2012 that security officials said they believed were launched by Iran, U.S. Defense Secretary Leon Panetta publicly warned that a "cyber–Pearl Harbor" could do serious damage to U.S. infrastructure.

Because computer viruses, worms, and other threats can be resent from remote servers located in almost any country around the world, the original source of the attack is often virtually impossible to identify. Even when circumstantial evidence overwhelmingly points toward a single country—China, for example—it is difficult to identify what organization or individuals within that country are responsible for the attack, much less whether the Chinese government or a specific corporation or group was ultimately responsible. According to Scott Aken, a former counterintelligence agent and expert in cybercrime, "In most cases, companies don't realize they've been burned until years later when a foreign competitor puts out their very same product—only they're making it 30 percent cheaper."

While organizations in China have apparently been the principal offenders in this category, a large number of Western corporations have engaged in similar activities against their competitors. A division of News Corporation engaged in supermarket display advertising was found to have hacked into the private emails of its principal competitor to steal its intellectual property and then steal some of its most valuable customers. Another division of News Corp admitted to hacking into emails of individuals to gather information for news stories. And employees at yet another division have pled guilty to hacking into the telephone voicemails of thousands of individuals in the United Kingdom.

The constant reliance on Internet-connected digital devices has created a false sense of comfort that has led to the extreme vulnerability of almost all communications over the Internet. Experts generally agree that the weakest link in any security system is the role of human behavior. Independent hackers have demonstrated how easily they can hack into supposedly secure videoconferences held by venture capital companies, law firms, oil and pharmaceutical companies—even the boardroom of Goldman Sachs—because the people in charge of the videoconferencing systems forgot to, or did not know how to, use the complicated privacy settings. Many commercial targets of cybercrime have been reluctant to acknowledge the theft of important information because they have a financial incentive to keep the theft secret. Even some companies that have been explicitly warned that they are targets have failed to take action to protect themselves.

PRIVACY

Other companies are routinely collecting information about their own customers and users—often without permission. Social media sites like Facebook and search engines like Google are among the many companies whose business models are based on advertising revenue and who maximize the effectiveness of advertising by constantly collecting information on each user in order to personalize and tailor advertising to match each person's individual collection of interests.

Many Internet sites, in effect, treat their customers as their products. That is, the revenue they receive from voluminous files of information about each user is simply too valuable for them to give up. The use of Facebook's "like" button automatically "allows" the site to track users' online interests without offering them an opportunity to give their consent. In a sense, this is yet another manifestation of the underlying cyber-Faustian bargain. The revenue that is earned from the targeted advertising made possible by all of those "cookies" (small software programs placed—often surreptitiously—on a user's computer during its interaction with a website) supports the "free" distribution of voluminous amounts of valuable content on the Internet. Most Internet users seem to feel that the tradeoff is an acceptable one. After all, the advertisements they are exposed to are ones they are more likely to be interested in. The tracking technologies are, in the words of one analyst, "simply tools to improve the grip strength of the Invisible Hand."

There are generational differences in the acceptance of this tradeoff where social media sites like Facebook and Twitter are concerned. Many in my generation, for example, are often surprised at the amount of personal information shared on Facebook by those who are younger. Already, some social media users who have left school to enter the workforce have been surprised when potential employers routinely access all of their posts and sometimes discover information that one would not necessarily want a potential employer to see. More recently, some employers have demanded that job applicants provide the password to their Facebook accounts so that private sites can also be accessed. (Facebook, to its credit, has reiterated that its policy is to never give out such passwords, and they urge their users not to do so. However, in a tough job market, the pressure to expand potential employers' visibility into their online lives is obviously more acceptable to some than others.) It is also noteworthy that, after being hired, many employees have been subjected to cybersurveillance by their employers.

The extreme convenience offered by Internet websites leads many users to feel that incremental losses of privacy are a small price to pay. The very fact that I can find virtually every business in Nashville, Tennessee, where I live, and virtually every business anywhere in the United States (and in almost any other country) is almost, well, magical. This is a tangible illustration of what economists call the network effect—by which they mean that the value of any network, especially the Internet, increases exponentially as more people connect to it. Indeed, according to Metcalfe's Law, an equation proposed by one of the early pioneers of the network, Robert Metcalfe, the inherent value of any network actually increases as the *square* of the number of people who connect to it.

Similarly, the convenience provided by online navigational software—like Google's Street View—makes it easy to ignore the misgivings some have about having the picture and location of their homes displayed on the Internet. (Google's apparent collection of large amounts of information from unencrypted WiFi networks in the homes and businesses it photographed—which it says was inadvertent—is a continuing source of controversy in several countries.)

Many take comfort in the fact that hundreds of millions of others are facing the same risks. So how bad could it be? Most people are simply unaware of the nature and extent of the files being compiled on them. And for those who do become aware—and concerned—they quickly find out that there is no way that they can choose not to have their every move on the Internet tracked. The written privacy policies on websites are typically far too long, vague, and complicated to understand, and the options for changing settings that some sites offer are too complex and difficult.

There is abundant evidence that the general expectations of privacy are at wide variance with the new reality of online tracking of people connected to the Internet, and adequate legal protections have not caught up with Internet usage. In some countries, including the United States, Internet users can choose to no longer have advertising based on the tracking delivered to them. But users who try to opt out of the tracking itself are presently unable to do so. The protections that supposedly provide a "do not track" option are essentially useless, due to persistent lobbying pressure from the advertising industry. Even when people try to opt out, the tracking continues for a simple reason: there is an enormous amount of money to be made from collecting all of the information about what everyone does on the Internet. Every click is worth a minuscule fraction of a penny, but there are so many clicks that billions of dollars are at stake each year.

The Wall Street Journal has published a lengthy series of investigative articles on the way cookies report information about a user's online activities. Everyone who clicks on Dictionary.com automatically has 234 cookies installed on his or her computer or smartphone, 223 of which collect and report information about the user's online activity to advertisers and others who purchase the data.

The cumulative impact of pervasive data tracking may yet produce a backlash. The word most frequently used by users of the Internet to describe the pervasiveness of online tracking is "creepy." Although the companies that track people's use of the Internet often say that the user's name is not attached to the file that is assembled and constantly updated, experts say there is little difficulty in matching individual computer numbers with the name, address, and telephone numbers of each person.

As computer processing has grown steadily faster, cheaper, and more powerful, some companies and governments have begun using an even more invasive technology known as Deep Packet Inspection (DPI), which collects "packets" of data sent to separate routers and reassembles them to reconstitute the original messages sent, and to pick out particular words and phrases in packets in order to flag them for closer examination and reconstitution. Tim Berners-Lee, the inventor of the World Wide Web, has spoken out against the use of DPI and has described it as a grave threat to the privacy of Internet users.

In one of the most publicized examples of the exposure of private data via computer, the roommate of a gay student at Rutgers was found guilty of using a webcam to share views of his roommate engaging in intimate acts (tragically, the gay student committed suicide soon after).

Some sites, including Facebook, use facial recognition software to automatically tag people when they appear in photos on the site. Voice recognition software is also now used by many sites to identify people when they speak. These audio files are often used to enhance the ability of the software to learn the accent and diction of each user in order to improve the accuracy with which the machine interprets successive verbal communications. In order to protect the user's privacy, some companies erase the audio files after a few weeks. Others, however, keep every utterance on file forever. Similarly, many software programs and apps use location-tracking programs in order to enhance the convenience with which information can be delivered with relevance to the user's location. An estimated 25,000 U.S. citizens are also victims of "GPS stalking" each year.

But all such information—websites visited, items within each website perused, geographic location day by day and minute by minute, recordings of questions users ask, pictures of the individuals wherever and whenever they may appear on websites, purchases and credit card activity, social media posts, and voluminous archival data in accessible government databases—when combined, can constitute an encyclopedic narrative of a person's life, including details and patterns that most would not want to be compiled. Max Schrems, a twenty-five-year-old Austrian law student, used the European Union's data protection law to request all the data collected about him on Facebook, and received a CD with more than 1,200 pages of information, most of which he thought he had deleted. The case is still pending.

Even when Internet users are not connected to a social media site and have not accepted cookies from commercial websites, they sometimes suffer invasions of privacy from private hackers and cybercriminals who use techniques such as phishing—which employs enticing email messages (sometimes mimicking the names and addresses chosen from a user's contact list) in order to trick people into clicking on an attachment that contains surreptitious programs designed to steal information from the user's computer or mobile device. The new crime known as identity theft is, in part, a consequence of all the private information about individuals now accessible on the Internet.

Through the use of these and other techniques, cyberthieves have launched attacks against Sony, Citigroup, American Express, Discover Financial, Global Payments, Stratfor, AT&T, and Fidelity Investments, all of which have reported large losses as a result of cybercrime. (Sony lost $171 million.) The Ponemon Institute estimated in 2011 that the average digital data breach costs organizations more than $7.2 million, with the cost increasing each year. Yet another computer security company, Norton, calculated that the annual cost of cybercrime on a global basis is $388 billion—"more than the annual global market for marijuana, cocaine, and heroin combined." Numerous other online businesses have also been penetrated, including LinkedIn, eHarmony, and Google's Gmail. In the fall of 2012, a simultaneous cyberattack on Bank of America, JPMorgan Chase, Citigroup, U.S. Bank, Wells Fargo, and PNC prevented customers from gaining access to their accounts or using them to make payments.

The obvious and pressing need for more effective protections against cybercrime, and especially the need to protect U.S. companies against the grave cybersecurity threats they are facing from China, Russia, Iran, and elsewhere, have been married to the post-9/11 fears of another terrorist attack as a combined justification for new proposals that many fear could fundamentally alter the right of American citizens—and those in other nations that value freedom—to be protected against unreasonable searches, seizures, and surveillance by their own government.

There is reason for concern that the biggest long-term problem with cybersecurity is the use of voluminous data files and online surveillance technologies to alter the relationship between government and the governed along lines that too closely resemble the Big Brother dystopia conjured by George Orwell more than sixty years ago. The United Kingdom, where Orwell published his novel, has proposed a new law that would allow the government to store Internet and telephone communications by everyone in the country. It has already installed 60,000 security cameras throughout the nation.

Many blithely dismiss the fear that the U.S. government could ever evolve into a surveillance state with powers that threaten the freedom of its citizens. More than sixty years ago, Supreme Court Justice Felix Frankfurter wrote, "The accretion of dangerous power does not come in a day. It does come, however slowly, from the generative force of unchecked disregard of the restrictions that fence in even the most disinterested assertion of authority."

One of the founders of reason-based analysis of data to arrive at wise decisions, Francis Bacon, is credited with the succinct expression of a biblical teaching: "Knowledge is power." The prevention of too much concentrated power in the hands of too few—through the division of governmental powers into separate centers balanced against one another, including among them an independent judiciary—is one of the core principles on which all free self-governments are based. If knowledge is indeed a potent source of power, and if the executive and administrative centers of political power in governments have massive troves of information about every citizen's thoughts, movements, and activities, then the survival of liberty may well be at risk.

As the first nation founded on principles that enshrined the dignity of individuals, the United States has in many ways been the most attentive to the protection of privacy and liberty against overbearing intrusions by the central government. And many in the United States have taken comfort in the knowledge that throughout its history, America has experienced a recurring cycle: periods of crisis when the government overreached its proper boundaries and violated the liberties of individuals—followed soon after by a period of regret and atonement, during which the excesses were remedied and the proper equilibrium between government and the individual was restored.

There are several reasons, however, to worry that the period of excess and intrusion that followed the understandable reaction to the terrorist attacks of September 11, 2001, may be a discontinuity in the historical pattern. First, after 9/11, according to another former National Security Agency employee, "basically all rules were thrown out the window, and they would use any excuse to justify a waiver to spy on Americans"—including exercising their ability to eavesdrop on telephone calls as they were taking place. Former senior NSA official Thomas Drake said that the post-9/11 policy shift "began to rapidly turn the United States of America into the equivalent of a foreign nation for dragnet blanket electronic surveillance."

One former agency official estimates that since 9/11, the NSA has intercepted "between 15 and 20 trillion" communications. The "war against terror" does not seem to have an end in sight. The spread of access by individuals and nonstate organizations to weapons of mass destruction has made the fear of deadly attack a fixture of American political life. The formal state of emergency first established after September 11, 2001, was routinely extended yet again in 2012. The American Civil Liberties Union reported that a 2012 Freedom of Information inquiry showed a sharp increase in the number of Americans subjected to warrantless electronic surveillance by the Justice Department over the previous two years (even as formal requests for warrants declined). Chris Soghoian, the principal technologist at the ACLU's Speech, Privacy and Technology Project, said, "I think there's really something at a deep level creepy about the government looking through your communications records, and you never learn that they were doing it."

Second, the aggregation of power in the executive branch—at the expense of the Congress—was accelerated by the emergence of the nuclear arms race following World War II. Now, the prevailing fear of another terrorist attack serves as a seemingly unassailable justification for the creation of government surveillance capabilities that would have been shocking to most Americans even a few years ago.

History teaches, however, that unchecked powers—once granted—may well be used in abusive ways when placed in the hands of less scrupulous leaders. When Presidents Woodrow Wilson and Richard Nixon engaged in abuses of civil liberties that shocked the nation's conscience, new laws and protections were passed to protect against a recurrence of the abuses. Now, apparently, the threshold for what is required to shock the nation's conscience has been raised because of fear. The U.S. Supreme Court ruled in 2012, for example, that police have the right to conduct strip searches, including the inspection of bodily orifices, for individuals who are suspected of offenses as trivial as an unpaid parking ticket or riding a bicycle with a defective "audible bell" attached to the handlebars. George Orwell might have rejected such examples in his description of a police state's powers for fear that they would not have seemed credible to the reader. (It is important to note, however, that the same Supreme Court ruled that it was unconstitutional for police departments to secretly attach electronic GPS tracking devices to the automobiles of citizens without a court-ordered warrant.)

In another chilling example of a new common government practice that would have sparked outrage in past years, customs agents are now allowed to extract and copy all digital information contained on an American citizen's computer or other digital device when he or she reenters the country from an international trip. Private emails, search histories, personal photographs, business records, and everything else contained in the computer's files can be taken without any grounds for suspicion whatsoever. It is easy to understand how such searches are justified when the government has reason to believe that the traveler has been engaging in child pornography, for example, or has been meeting with a terrorist group in a foreign country. However, such searches are now placed in the "routine" category, and no reasonable cause for allowing the search is required. In one case, a documentary filmmaker raising questions about U.S. government policies was among those whose digital information has been searched and seized without any showing of reasonable cause.

The surveillance technologies now available—including the monitoring of virtually all digital information—have advanced to the point where much of the essential apparatus of a police state is already in place. An investigation by the American Civil Liberties Union showed that police departments in many U.S. cities now routinely obtain location-tracking data on thousands of individuals without a warrant. According to *The New York Times*, "The practice has become big business for cellphone companies, too, with a handful of carriers marketing a catalog of 'surveillance fees' to police departments." The U.S. government has also provided grants to local police departments for the installation of tracking cameras mounted on patrol cars to routinely scan and photograph the license plates of every car they pass, tagging each photograph with a day-and-time stamp and a GPS location, and adding it to the database. A *Wall Street Journal* investigation found that 37 percent of big city departments are participating in this data collection exercise, which compiles voluminous information on the whereabouts of everyone driving cars in their cities and keeps it on file. At least two private companies are also compiling similar databases by routinely photographing license plates and selling the information to repossession companies. One of them advertises it has 700 million scans thus far. The CEO of the other said he plans to sell the data to private investigators, insurers, and others interested in tracking people's locations and routines.

Especially since September 11, 2001, business has been booming for the manufacturers of surveillance hardware and software. The market for these technologies has grown over the last decade to an estimated $5 billion per year. Like the Internet itself, these technologies cross international borders with ease. U.S. companies are the principal manufacturers and suppliers of surveillance and censorship software and hardware used by authoritarian nations—including Iran, Syria, and China.

Surveillance technologies initially developed by U.S. companies for use in war zones also often make their way back into the United States. The drone technology used so widely in Iraq, Afghanistan, and Pakistan has now been adopted by some domestic police forces—with predictions that new generations of unobtrusive microdrones equipped with video cameras will become commonplace tools for law enforcement agencies. The Electronic Frontier Foundation found through a Freedom of Information Act lawsuit that as of 2012 there were already sixty-three active drone sites in the U.S. in twenty states.

Advances in microelectronics have also made hidden cameras and microphones far easier to use. Some sophisticated versions of spyware are now being used remotely to surreptitiously turn on a user's smartphone or computer microphone and camera to record conversations and take photographs and videos without a user's permission or awareness—even if the device has been turned off. Similarly, the microphones contained in the OnStar systems installed in many automobiles have also been used to monitor the conversations of some suspects. Other software programs can be surreptitiously installed to keep track of a user's keystrokes in order to reconstruct passwords and other confidential information as it is typed into a computer or device.

Significantly, the cybersecurity threats faced by U.S. corporations—alongside the threat of terrorism—are being used as a new justification for building the most intrusive and powerful data collection system that the world has ever known. In January of 2011, at the groundbreaking of this new giant, $2 billion facility in Utah, the senior official from the National Security Agency, Chris Inglis, announced the purpose of the "state of the art facility" was to "enable and protect the nation's cybersecurity." The capabilities of the facility being built there (which will become operational at the end of 2013) include the ability to monitor every telephone call, email, text message, Google search, or other electronic communication (whether encrypted or not) sent to or from any American citizen. All of these communications will be stored in perpetuity for data mining.

This system is eerily similar to a proposal put forward by the George W. Bush–Dick Cheney administration two years after the 9/11 attacks. It was called Total Information Awareness (TIA), and the suggestion caused public outrage and resulted in congressional action to cancel it. In the years since, politicians in both political parties have become fearful about challenging any intelligence-gathering proposal that is described as having a national security purpose.

In more recent years, the American people have successfully persuaded Congress to curb some government intrusions into their privacy. In 2011, the Stop Online Piracy Act and its companion Senate bill PROTECT IP Act—which were sought by entertainment and other information content companies as a means of safeguarding their intellectual property—were found to contain new government authorities to shut down websites popular with the public if they contained any copyrighted material. The resulting outrage, and the effective online campaign against these proposals, resulted in the withdrawal of both. However, the outrage generated by the prospect of losing access to online entertainment has not been matched by a similar outrage over the prospect of warrantless government surveillance of private communications among Americans.

The Cyber Intelligence Sharing and Protection Act (CISPA) is an example of a proposed U.S. law to empower the government to eavesdrop on any online communication if it has reason to suspect cybercrime. While it is easy to understand the motivation behind this proposal, the volume of Internet communication that could be deemed suspect under the broadly defined terms of the law poses a de facto exemption for government agencies from a wide variety of other laws intended to protect the privacy of Internet users.

It is yet another example of how the cyber-Faustian bargain we have made with the Internet is creating difficulty in reconciling the historic principles upon which the United States was founded with the new reality of the Global Mind. As a technology writer recently put it, "If America's ongoing experiment in democracy and economic freedom is to endure, we will need to think again about cultivating the necessary habits of the heart and resisting the allure of the ideology of technology." China and other nations dedicated to authoritarian governance are also facing a historic discontinuity because of the new reality of the Global Mind.

Every nation uses the Internet and every nation has its own ideas about the future of the Internet. The multiple overlapping conflicts accompanying the world's historic shift onto the Internet are nowhere close to being resolved. As a result, there are calls for the imposition of some form of global governance over the Internet, which has, since its inception, been governed benignly by the U.S. government (and by a quasi-independent group established by the U.S. government) according to norms and values that reflect the American tradition of free speech and robust free markets.

The fact that nations such as China, Russia, and Iran, whose values and norms are often in direct conflict with those of the U.S., are among the chief protagonists pushing to transfer authority over the global Internet to an international body is reason enough to fear the proposal and to follow the unfolding struggle with care. It is unfortunate that Brazil, India, and South Africa are following the lead of China and Russia.

Some corporations and governmental agencies are now developing "dark nets"—that is, closed networks that are not connected to the Internet—as a last resort for protecting confidential, high-value information. Some Internet companies—most significantly Facebook, which now has one billion users and prohibits anonymity—have adopted a "walled garden" approach that separates some of its information from the rest of the Internet.

In addition, some corporations that sell access to the Internet and simultaneously sell high-value content over the Internet have attempted to slow down or make more expensive similar content from competitors. Although they raise legitimate issues about allocating the cost of expanding their bandwidth, this potential conflict of interest is also an important issue for the future of the Internet. It is the reason so many have called for network neutrality laws that protect free speech and free competition.

The efforts by some corporations to control information on the Internet have led some to fear that the Internet could eventually be split apart into multiple, separate networks. However, that is unlikely to occur, because the full value generated by the Internet depends upon the fact that it is connected in one way or another to the vast majority of people, companies, and organizations in the world. For the same reason, the efforts by nations like China and Iran to isolate their citizens from disruptive forces coursing across the Internet globally are probably doomed to fail.

The world system as a whole is breaking out of an old enduring pattern that has been in place since the emergence of the system based on nation-states. No one doubts that nations will continue as the primary units of account where governance is concerned. But the dominant information system now being used by the world as a whole—the Global Mind—has an inherent unifying imperative, just as the printing press helped unify nations in the era in which they were born. And the decisions now confronting the world as a whole cannot be made by any single nation or small group of nations. For several decades, when the United States made up its mind, the world followed the U.S. lead. Now, however, along with digital information, the power to shape the world's future is being dispersed throughout the globe. As a result, the *Global* Mind is not so easy to make up.

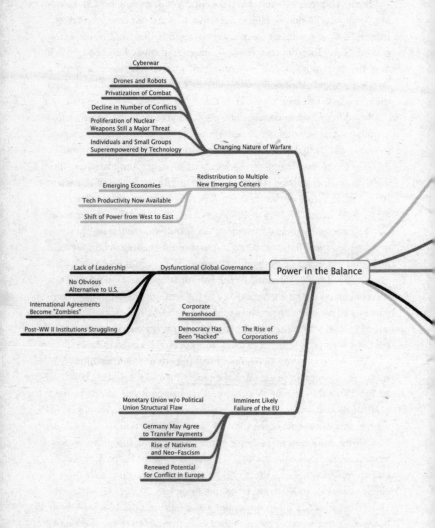

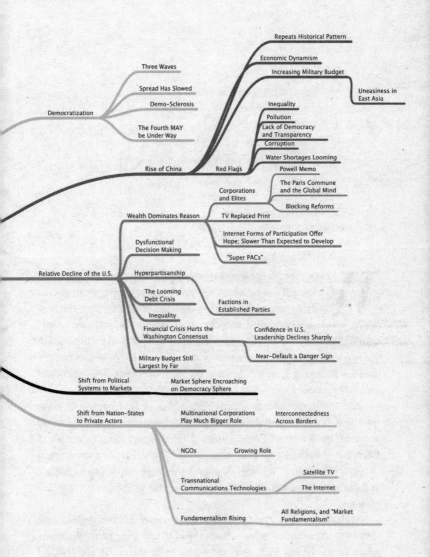

Democratization
- Three Waves
- Spread Has Slowed
- Demo–Sclerosis
- The Fourth MAY be Under Way

Rise of China
- Repeats Historical Pattern
- Economic Dynamism
- Increasing Military Budget — Uneasiness in East Asia
- Red Flags
  - Inequality
  - Pollution
  - Lack of Democracy and Transparency
  - Corruption
  - Water Shortages Looming

Relative Decline of the U.S.
- Wealth Dominates Reason
  - Corporations and Elites
    - Powell Memo
    - The Paris Commune and the Global Mind
    - Blocking Reforms
  - TV Replaced Print
  - Internet Forms of Participation Offer Hope; Slower Than Expected to Develop
  - "Super PACs"
- Dysfunctional Decision Making
- Hyperpartisanship
- The Looming Debt Crisis
  - Factions in Established Parties
- Inequality
- Financial Crisis Hurts the Washington Consensus
  - Confidence in U.S. Leadership Declines Sharply
  - Near–Default a Danger Sign
- Military Budget Still Largest by Far

Shift from Political Systems to Markets — Market Sphere Encroaching on Democracy Sphere

Shift from Nation–States to Private Actors
- Multinational Corporations Play Much Bigger Role — Interconnectedness Across Borders
- NGOs — Growing Role
- Transnational Communications Technologies
  - Satellite TV
  - The Internet
- Fundamentalism Rising — All Religions, and "Market Fundamentalism"

POWER IN THE BALANCE

ITH A TIGHTLY INTEGRATED GLOBAL ECONOMY AND A PLANET-wide digital network, we are witnessing the birth of the world's first truly global civilization. As knowledge and economic power are multiplied and dispersed far more widely and swiftly than by the Print and Industrial Revolutions, the political equilibrium of the world is undergoing a massive change on a scale not seen since the decades following Europe's linkage by sea routes to the Americas and Asia 500 years ago.

As a result, the balance of power among nations is changing dramatically. Just as the Industrial Revolution led to the dominance of the world economy by Western Europe and the United States, the emergence of Earth Inc. is shifting economic power from West to East and spreading it to the new growth economies developing throughout the world. China, in particular, is overtaking the U.S. as the center of gravity in the global economy.

More importantly, just as nation-states emerged as the dominant form of political organization in the wake of the printing press, the emergence of the Global Mind is changing many of the social and political assumptions on which the nation-state system was based. Some of the sources of power traditionally wielded primarily by nations are no longer as firmly under their exclusive control. While our individual political identities remain primarily national, and will for a long time to come, the simultaneous globalization of information and markets is transferring power once reserved for national governments to private actors—including multinational corporations, networked entrepreneurs, and billions of individuals in the global middle class.

No nation can escape these powerful waves of change by unilaterally imposing its own design. The choices most relevant to our future are now ones that confront the world as a whole. But because nation-states retain the exclusive power to negotiate policies and implement them globally, the only practical way to reclaim control of our destiny is to seek a global consensus within the community of nations to secure the implementation of policies that protect human values. And since the end of World War II—at least until recently—most of the world has looked primarily to the United States of America for leadership when facing the need for such a consensus.

Many fear, however, that the ability of the U.S. to provide leadership in the world is declining in relative terms. In 2010, China became the world's leading manufacturing nation, ending a period of U.S. leadership that had lasted for 110 years. An economic historian at Nuffield College, Oxford, Robert Allen, said this milestone marked the "closing of a 500-year cycle in economic history." When China's overall economic strength surpasses that of the United States later this decade, it will mark the first time since 1890 that any economy in the world has been larger than the American economy.

Worse, not since the 1890s has U.S. government decision making been as feeble, dysfunctional, and servile to corporate and other special interests as it is now. The gravity of the danger posed by this debasement of American democracy is still not widely understood. The subordination of reason-based analysis to the influence of wealth and power in U.S. decision making has led to catastrophically bad policy choices, sclerotic decision making, and a significant weakening of U.S. influence in the world.

Even a relative decline in the preeminence of the U.S. position in the world system has significant consequences. It remains "the indispensable nation" in reducing the potential for avoidable conflicts—keeping the sea lanes open, monitoring and countering terrorist groups, and playing a balancing role in tense regions like the Middle East and East Asia, and in regions (like Europe) that could face new tensions without strong U.S. leadership. Among its many other roles, the United States has also exercised responsibility for maintaining relative stability in the world's international monetary system and has organized responses to periodic market crises.

At the moment, though, the degradation of the U.S. political system is causing a dangerous deficit of governance in the world system and a gap between the problems that need to be addressed and the vision and cooperation necessary to address them. This is the real fulcrum in the world's balance of power today—and it is badly in need of repair. In the absence of strong U.S. leadership, the community of nations is apparently no longer able to coalesce in support of international coordination and agreements that establish the cooperative governance arrangements necessary for the solution of global problems.

Meetings of the G20 (which now commands more attention than the G8) have become little more than a series of annual opportunities for the leaders of its component nations to issue joint press releases. Their habit of wearing matching colorful shirts that represent the fashion motif of the host nation recalls the parable of the child who noticed that the emperor has no clothes. Except in this case, the clothes have no emperor.

Largely because of U.S. government decisions to follow the lead of powerful domestic corporate interests, once-hopeful multilateral negotiations—like the Doha Round of trade talks (commenced in 2001) and the Kyoto Protocol (commenced in 1997)—are now sometimes characterized as "zombies." That is, they are neither alive nor dead; they just stagger around and scare people. Similarly, the Law of the Sea Treaty is in a condition of stasis.

The global institutions established with U.S. leadership after World War II—the United Nations, the World Bank, the International Monetary Fund, and the World Trade Organization (formerly the General Agreement on Tariffs and Trade)—are now largely ineffective because of the global changes that have shaken the geopolitical assumptions upon which they were based. Chief among them was the assumption that the U.S. would provide global leadership.

So long as the United States offered the vision necessary for these institutions—and so long as most of the world trusted that U.S. leadership would move the world community in a direction that benefited all—these institutions often worked well. If any nation's goals are seen as being motivated by the pursuit of goals that are in the interest of all, its political power is greatly enhanced. By contrast, if the nation offering leadership to the world is seen as primarily promoting its own narrower interests—the commercial prospects of its corporations, for example—its capacity for leadership is diminished.

Two thirds of a century after their birth, these multilateral institutions face criticism from developing countries, environmentalists, and advocates for the poor because of what many see as "democratic deficits." Both the World Bank and the International Monetary Fund require support from 85 percent of the voting rights held by member nations. Since the United States alone has more than 15 percent of the voting rights in both organizations, it has effective veto power over their decisions. Similarly, some countries ask why France and the United Kingdom are still among only five permanent members of the U.N. Security Council when Brazil, with a GDP larger than either, and India, whose GDP is greater than both combined and will soon be the most populous country in the world, are not.

The significant loss of confidence in U.S. leadership, especially since the economic crisis of 2007–08, has accelerated the shift in the equilibrium of power in the world. Some experts predict the emergence of a new equilibrium with both the United States and China sharing power at its center; some have already preemptively labeled it the "G2."

RELATIVE OR ABSOLUTE DECLINE?

Other experts predict an unstable, and more dangerous, multipolar world. It seems most likely that the increasing integration of global markets and information flows will lead to an extended period of uncertainty before global power settles into a new more complex equilibrium that may not be defined by poles of power at all. The old division of the world into rich nations and poor nations is changing as many formerly poor nations now have faster economic growth rates than the wealthy developed nations. As the gap closes between these fast-growing developing and emerging economies on the one hand and the wealthy mature economies on the other, economic and political power are not only shifting from West to East, but are also being widely dispersed throughout the world: to São Paulo, Mumbai, Jakarta, Seoul, Taipei, Istanbul, Johannesburg, Lagos, Mexico City, Singapore, and Beijing.

Whatever new equilibrium of power emerges, its configuration will be determined by the resolution of several significant uncertainties about the future of the United States, China, and nation-states generally: First, is the United States really in a period of decline? If so, can the decline be reversed? And if not, is it merely relative to that of other nations, or is there a danger of an absolute decline? Second, is China likely to continue growing at its current rate or are there weaknesses in the foundations on which its prosperity is being built? Finally, are nation-states themselves losing relative power in the age of Earth Inc. and the Global Mind?

There is a lively dispute among scholars about whether the United States is in decline at all. The loss of U.S. geopolitical power has been a recurring theme for far longer than many Americans realize. Even before the U.S. became the most powerful nation, there were episodic warnings that American power was waning. Some argue that concerns about China overtaking the United States in forms of power other than economic output represent just another example of what happened when so many were concerned about Japan Inc. in the 1970s and 1980s—and even earlier concerns when the former Soviet Union was seen as a threat to U.S. dominance in the 1950s and 1960s.

For more than a decade following World War II, many strategic thinkers worried that the U.S. was in danger of quickly falling from the pinnacle of world power. When the USSR acquired nuclear weapons and tightened its grip on Eastern and Central Europe, these fears grew. When Sputnik was launched in 1957, making the USSR the first nation in space, the warning bells rung by declinists were heard even more loudly.

Many of the alarms currently being sounded about the decline of U.S. power are based on a comparison between our present difficulties and a misremembered sense of how completely the U.S. dominated global decision making in the second half of the twentieth century. A more realistic and textured view would take into account the fact that there was never a golden age in which U.S. designs were implemented successfully without resistance and multiple failures.

It is also worth remembering that while the U.S. share of global economic output fell from 50 percent in the late 1940s to roughly 25 percent in the early 1970s, it has remained at that same level for the last forty years. The rise of China's share of global GDP and the economic strength of other emerging and developing economies has come largely at the expense of Europe, not the United States.

The rise of the United States as the dominant global power began early in the twentieth century when it first became the world's largest economy, when President Theodore Roosevelt aggressively asserted U.S. diplomatic and military power, and when it played the crucial role in determining the outcome of World War I under President Woodrow Wilson. And of course after providing the decisive economic and military strength to defeat the Axis powers in World War II, the United States emerged as the victor in both the European and Pacific theaters and was recognized as the leading power in the world. The economies of the European nations had been devastated and exhausted by the war. Those of Japan and Germany had been destroyed. The Soviet Union, having suffered casualties 100 times greater than those of the United States, had been weakened. Whatever antithetical moral authority it might have once aspired to under Lenin had been long since destroyed by Stalin's 1939 pact with Hitler and his exceptional cruelty and brutality toward his own people.

Moving quickly, the United States provided crucial leadership to establish the postwar institutions for world order and global governance. These included the Bretton Woods Agreement, which formalized the U.S. dollar as the world's reserve currency, and a series of regional military self-defense alliances, the most important of which was NATO, the North Atlantic Treaty Organization. By using foreign aid and generous trade agreements that provided access to U.S. markets, the United States grew into an even more dominant role. And the United States promoted democratic capitalism throughout the noncommunist parts of the world.

It catalyzed the emergence of European economic and political integration by midwifing the European Coal and Steel Community (which later evolved into the Common Market and the European Union). And the visionary and generous Marshall Plan lifted the nations of Europe that had been devastated by World War II to prosperity and encouraged a commitment to democracy and regional integration. Secretary of State Cordell Hull, who was described by FDR as "the father of the United Nations," was an advocate of freer reciprocal cross-border trade in Europe and the world, arguing that "when goods cross borders, armies do not." By presiding over the reconstruction, democratization, and demilitarization of Japan, the United States also solidified its position as the dominant power in Asia.

In 1949, when the Soviet Union became the world's second nuclear power and China embraced communism after the victory of Mao Zedong, the four-decade Cold War imposed its own dynamic on the operations of the world system. The nuclear standoff between the U.S. and the USSR was accompanied by a global struggle between two ideologies with competing designs for the organization of both politics and economics.

For several decades, the structure of the world's equilibrium of power was defined by the constant tension between these two polar opposites. At one pole, the United States led an alliance of nations that included the recovering democracies of Western Europe and a reconstructed Japan, all of whom advocated the ideology of democratic capitalism. At the other pole, the Union of Soviet Socialist Republics led a captive group of nations in Central and Eastern Europe in advocating the ideology of communism. This abbreviated description belies more complex dynamics, of course, but virtually every political and military conflict in the world was shaped by this larger struggle.

When the Soviet Union was unable to compete with the economic strength of the United States (and was unable to adapt its command economy and authoritarian political culture to the early stages of the Information Revolution), it imploded. With the collapse of the Berlin Wall in 1989 and the subsequent breakup of the Soviet Union two years later (when Russia itself withdrew from the USSR), communism disappeared from the world as a serious ideological competitor.

U.S. HEGEMONY IN the world thus reached its peak, and the ideology of democratic capitalism spread so widely that one political philosopher speculated that we were seeing "the end of history"—implying that no further challenge to either democracy or capitalism was likely to emerge.

This ideological and political victory secured for the United States universal recognition as the dominant power in what appeared to be, at least for a brief period, a unipolar world. But once again, the superficial label concealed complex changes that accompanied the shift in the power equilibrium.

Well before the beginning of World War II, Soviet communism had run afoul of a basic truth about power that was clearly understood by the founders of the United States: when too much power is concentrated in the hands of one or a small group of people, it corrupts their judgment and their humanity.

American democracy, by contrast, was based on a sophisticated understanding of human nature, the superior quality of decision making to be found in what is now sometimes called the wisdom of crowds, and lessons learned from the history of the Roman Republic about the dangers posed to liberty by centralized power. Unhealthy concentrations of power were recognized to be detrimental to the survival of freedom. So power was separated into competing domains designed to check and balance one another in order to maintain a safe equipoise within which individuals could maintain their freedom to speak, worship, and assemble freely.

The ability of any nation to persuade others to follow its leadership is often greatly influenced by its moral authority. In the case of the United States, it is undeniably true that since the ratification of its Constitution and Bill of Rights in 1790–91, its founding principles have resonated in the hearts and minds of people throughout the world, no matter the country in which they live.

Since the end of the eighteenth century, there have been three waves of democracy that spread throughout the world. The first, in the aftermath of the American Revolution, produced twenty-nine democracies. When the Great Liberator, Simón Bolívar, led democratic revolutions in South America in the two decades after America's founding, he carried a picture of George Washington in his breast pocket.

This was followed by a period of decline that shrank the number to twelve by the beginning of World War II. After 1945, the second wave of democratization swelled the number of democracies to thirty-six, but once again this expansion was followed by a decline to thirty from 1962 until the mid-1970s. The third wave began in the mid-1970s and then accelerated with the collapse of communism in 1989.

The struggle within the United States over policies that promote the higher values reflected in the U.S. Constitution—individual rights, for example—has often been lost to the interests of business and calculations of realpolitik. When Western European countries began to grant independence to their overseas colonies and pull back from the spheres of influence they had established during their imperial periods, the United States partially filled the resulting power vacuums by extending aid and forming economic, political, and military relationships with many of the newly independent nations. When the United States feared that the withdrawal of France from its colonial role in Vietnam might lead to the expansion of what some mistakenly viewed as a quasi-monolithic communist sphere, this misunderstanding of Ho Chi Minh's fundamentally nationalist motivation contributed to the tragic miscalculation that resulted in the Vietnam War.

Nevertheless, in spite of its strategic mistake in Vietnam (following the earlier long and costly stalemate in the Korean War), heavy-handed military interventions in Latin America, and other difficult challenges, the U.S. consolidated its position of leadership in the world. The unprecedented growth of U.S. prosperity in the decades following World War II—along with its continued advocacy of freedom—made it an aspirational model for other countries. It is difficult to imagine that human rights and self-determination could have made as much progress throughout the world in the post–World War II era without the U.S. being in a dominant position.

More recently, the spread of democracy has slowed. Since the market crisis of 2007–08, there has been a decline in the number of democratic nations in the world and a degradation in the quality and extent of democracy in several others—including the United States. But even though the world is still in a "democratic recession," some believe that the Arab Spring and other Internet-empowered democratic movements may signal the beginning of a fourth wave of democratization, though the results are still ambiguous at best.

In any case, it is premature to predict an absolute decline in U.S power. Among positive signs that the United States may yet slow its relative decline, the U.S. university system is still far and away the best in the world. Its venture investment culture continues to make the U.S. the greatest source of innovation and creativity. Although the U.S. military budget is lower as a percentage of GDP than it has been for most of the post–World War II era, it has increased in absolute terms to the highest level since 1945. The U.S. military is still by far the most powerful, best trained (by the best officer corps), best equipped, and most lavishly financed armed force the world has ever seen. Its annual budget is equal to the combined military budgets of the next fifty militaries in the world and almost equal to the military spending of the entire rest of the world put together.

As someone who was frequently described as a pro-defense Democrat during my service in the Congress and in the White House, I have seen how valuable it has been for the United States and for the cause of freedom to maintain unquestioned military superiority. However, after more than a decade of fighting two seemingly endless wars, while simultaneously maintaining large deployments in Europe and Asia, U.S. military resources are strained to the point of breaking. And the relative decline of America's economic power and wealth is beginning to force the reconsideration of such large military budgets.

The same global trends that have dispersed productive activity throughout Earth Inc. and connected people throughout the world to the Global Mind are also dispersing technologies relevant to warfare, which used to be monopolized by nation-states. The ability to launch destructive cyberattacks, for example, is now being widely spread on the Internet.

Some of the means of waging violent warfare are being robosourced and outsourced. The use of drones and other semiautonomous robotic weapons proliferated dramatically during the wars in Iraq and Afghanistan. The U.S. Air Force now trains more pilots for unmanned vehicles than it trains pilots of manned fighter jets. (Interestingly, the drone pilots suffer post-traumatic stress disorder at the same rate as fighter pilots even though they see their targets over a television screen thousands of miles away.)

On several occasions, drones have been hacked by the forces they are targeting. In 2010, intelligence analysts found that Islamic militants in Iraq used commercially available software selling for $26 to hack into the unencrypted video signals coming from U.S. drones and watch the same video in real time that was being sent to the U.S. controllers back in the United States. In Afghanistan, insurgent forces were able to do the same thing, and at the end of 2011, Iran hacked into the control system of a U.S. stealth drone and commanded it to land on an airstrip in Kashmar, Iran.

A new generation of robotic weapons in the air, on the land, and in the sea is being rapidly developed. More than fifty countries are now experimenting with semiautonomous military robots of their own. (A new legal doctrine of "robot rights" has been developed by U.S. military lawyers to give unmanned drones and robots the legal right to unleash deadly fire when threatened, just as a fighter pilot has the right to fire at a potential attacker as soon as he is alerted to the fact that a targeting radar has "lit up" his plane.)

At the same time, some dangerous combat missions are being outsourced. During the war in Iraq, the United States shifted significant operations in the war zone to private contractors.* After the unpopular Vietnam War, the United States abandoned the draft and has since relied on a professional volunteer army—which many claim emotionally insulates the American people from some of the impact wars used to have on the general population.

THE CHINA ISSUE

Meanwhile, China's military budgets—while still only a fraction of U.S. defense spending—are increasing. Yet there are questions about the sustainability of China's present economic buildup. Many feel that it is premature to predict a future in which China becomes the dominant global power, or even occupies the center of a new power equilibrium alongside the United States, because they doubt that the social, political, and economic foundations in China are durable. In spite of the economic progress in China, experts warn that the lack of free speech, the concentrated autocratic power in Beijing, and the high levels of corruption throughout China's political and economic system raise questions about the sustainability of its recent growth rates.

For example, at the end of 2010, there were an estimated 64 million empty apartments in China. The building bubble there has been attributed to a number of causes, but for several years visitors have remarked upon the large number of subsidized high-rise apartment buildings that spring up quickly and remain unoccupied for very long periods of time. According to research by Morgan Stanley, almost 30 percent of

---

* Mercenary armies have always been present in the history of warfare, but are more prominent than ever in some long-running conflicts, such as those that have killed 400,000 people in the Democratic Republic of the Congo.

the windmills constructed by China are not connected to the electrical grid; many have been placed in remote locations with strong winds but no economical way to extend the grid to them. China's success in building its capacity to construct renewable energy systems of low cost has been of benefit to China and to the global market, but as with the many empty apartment buildings, the idle windmills serve as a warning that some trends in the Chinese economic miracle may not continue at the same pace. China's banking system suffers from the same distortions of state manipulation. Some state-owned banks are recycling their allocations of credit into black market lending at usurious and unsustainable interest rates.

There are also questions about China's social and political cohesion during what has already been a disruptive economic transition, accompanied by the largest internal migration in history and horrendous levels of pollution. Although precise statistics are hard to verify, a professor at Tsinghua University, Sun Liping, estimated that in 2010 there were "180,000 protests, riots and other mass incidents." That number reflects a fourfold increase from 2000. Numerous other reports confirm that social unrest appears to be building in response to economic inequality, intolerable environmental conditions, and opposition to property seizures and other abuses by autocratic local and regional leaders. Partly as a result of dissatisfaction and unrest—particularly among internal migrant workers—wages have been increasing significantly in the last two years.

Some scholars have cautioned against a Western bias in prematurely predicting instability in countries whose governments do not gain democratic legitimacy. In China, according to some experts, legitimacy can be and is derived from other sources besides the participatory nature of their system. Since Confucian times, legitimacy has been gained in the eyes of the governed when the policies implemented are successful and when the persons placed in positions of power are seen to have earned their power in a form of meritocracy and demonstrate sufficient wisdom to seem well chosen.

IT IS PRECISELY these sources of legitimacy that are now most at risk in the United States. The sharp decline of public trust in government at all levels—and public trust in most all large institutions—is based in large measure on the perception that they are all failing to produce successful policies and outcomes. The previous prominence of reason-based decision making in the U.S. democratic system was its greatest source of strength. The ability of the United States, with only 5 percent of the world's people, to lead the world for as long as it has is due in no small measure to the creativity, boldness, and effectiveness of its decision making in the past.

Ironically, the economic growth in China since the reforms of Deng Xiaoping, launched in 1978, were brought about not only by his embrace of a Chinese form of capitalism but also by his intellectual victory within the Chinese Central Committee in advocating reason-based analysis as the justification for abandoning stale communist economic dogma—and his political skill in portraying this dramatic shift as simply a reaffirmation of Maoist doctrine. In a speech to the All-Army Conference in the year his reforms were begun, Deng said, "Isn't it true that seeking truth from facts, proceeding from reality and integrating theory with practice form the fundamental principle of Mao Zedong Thought?"

One reason for the rise of the United States over its first two centuries to the preeminent position among nations was that American democracy demonstrated a genius for "seeking truth from facts." Over time, it produced better decisions and policies to promote its national interests than the government of any other nation. The robust debate that takes place when democratic institutions are healthy and functioning well results in more creative and visionary initiatives than any other system of government has proven capable of producing.

Unfortunately, however, the U.S. no longer has a well-functioning self-government. To use a phrase common in the computer software industry, American democracy has been hacked. The United States Congress, the avatar of the democratically elected national legislatures in the modern world, is now incapable of passing laws without permission from the corporate lobbies and other special interests that control their campaign finances.

THE LONG REACH OF CORPORATIONS

It is now common for lawyers representing corporate lobbies to sit in the actual drafting sessions where legislation is written, and to provide the precise language for new laws intended to remove obstacles to their corporate business plans—usually by weakening provisions of existing laws and regulations intended to protect the public interest against documented excesses and abuses. Many U.S. state legislatures often now routinely rubber-stamp laws that have been written in their entirety by corporate lobbies.

Having served as an elected official in the federal government for the last quarter of the twentieth century, and having observed it closely before that period and since, I have felt a sense of shock and dismay at how quickly the integrity and efficacy of American democracy has nearly collapsed. There have been other periods in American history when wealth and corporate power have dominated the operations of government, but there are reasons for concern that this may be more than a cyclical phenomenon—particularly recent court decisions that institutionalize the dominance and control of wealth and corporate power.

This crippling of democracy comes at a time of sweeping and tumultuous change in the world system, when the need for U.S. advocacy of democratic principles and human values has never been greater. The crucial decisions facing the world are unlikely to be made well, or at all, without bold and creative U.S. leadership. It is therefore especially important to restore the integrity of U.S. democracy. But in order to do so, it is necessary to accurately diagnose how it went so badly off track. The shift of power from democracy to markets and corporations has a long history.

In general, political freedom and economic freedom have reinforced one another. The new paradigm born in the era of the printing press was based on the principle that individuals had dignity, and when armed with the free flow of information could best chart their own destinies in both the political and economic realms by aggregating their collective wisdom through regular elections of representatives, and through the "invisible hand" of supply and demand.

Throughout history, capitalism has been more conducive to higher levels of political and religious freedom than any other way of organizing economic activity. But internal tensions in the compound ideology of democratic capitalism have always been present and frequently difficult to reconcile. Just as America's founders feared concentrated political power, many of them also worried about the impact on democracy of too much concentrated economic power—particularly in the form of corporations.

The longest running corporation was created in Sweden in 1347, though the legal form did not become common until the seventeenth

century, when the Netherlands and the United Kingdom allowed a proliferation of corporate charters, especially for the exploitation of trade to
and from their new overseas colonies. After a series of spectacular frauds
and other abuses, including the South Sea Company scandal (which gave
birth to the economic concept of a "bubble"), England banned corporations in 1720. (The prohibition was not lifted until 1825 when the Industrial Revolution required the capitalization of railway companies and
other new firms to exploit emerging technologies.)

The American revolutionaries were keenly aware of this history and
originally chartered corporations mostly for civic and charitable purposes,
and only for limited periods of time. Business corporations came later, in
response to the need to raise capital for industrialization.

Referring to the English experience, Thomas Jefferson wrote in a
letter to U.S. Senator George Logan of Pennsylvania in 1816, "I hope we
shall take warning from the example and crush in its birth the aristocracy
of our monied corporations which dare already to challenge our government to a trial of strength and bid defiance to the laws of our country."

Between 1781 and 1790 the number of corporations expanded by
an order of magnitude, from 33 to 328. Then in 1811, New York State
enacted the first of many statutes that allowed the proliferation of corporations without specific and narrow limitations imposed by government.

So long as the vast majority of Americans lived and worked on farms,
corporations remained relatively small and their impact on the conditions of labor and the quality of life was relatively limited. But during
the Civil War, corporate power increased considerably with the mobilization of Northern industry, huge government procurement contracts,
and the building of the railroads. In the years following the war, the corporate role in American life grew quickly, and the efforts by corporations
to take control of the decisions in Congress and state legislatures grew
as well.

The tainted election of 1876 (deadlocked on election night by disputed electoral votes in the state of Florida) was, according to historians, settled in secret negotiations in which corporate wealth and power
played the decisive role, setting the stage for a period of corrupt deal
making that eventually led the new president, Rutherford B. Hayes, to
complain that "this is a government of the people, by the people and for
the people no longer. It is a government of corporations, by corporations,
and for corporations."

As the Industrial Revolution began to reshape America, industrial accidents became commonplace. Between 1888 and 1908, 700,000 American workers were killed in industrial accidents—approximately 100 every day. In addition to providing brutal working conditions, employers also held wages as low as possible. Efforts by employees to obtain relief from these abuses by organizing strikes and seeking the passage of protective legislation provoked a fierce reaction from corporate owners. Private police forces brutalized those attempting to organize labor unions and lawyers and lobbyists flooded the U.S. Capitol and state legislatures.

When corporations began hiring lobbyists to influence the writing of laws, the initial reaction was one of disgust. In 1853, the U.S. Supreme Court voided and made unenforceable a contingency contract involving lobbying—in part because those providing the money did so in secret. The justices concluded that such lobbying was harmful to public policy because it "tends to corrupt or contaminate, by improper influences, the integrity of our . . . political institutions" and "sully the purity or mislead the judgments of those to whom the high trust of legislation is confided" with "undue influences" that have "all the injurious effects of a direct fraud on the public."

Twenty years later, the U.S. Supreme Court addressed the question once again, invalidating contingency contracts for lobbyists with these words: "If any of the great corporations of the country were to hire adventurers who make market of themselves in this way, to procure the passage of a general law with a view to the promotion of their private interests, the moral sense of every right-minded man would instinctively denounce the employer and employed as steeped in corruption, and the employment as infamous. If the instances were numerous, open and tolerated, they would be regarded as measuring the decay of the public morals and the degeneracy of the times." The state of Georgia's new constitution explicitly banned the lobbying of legislators.

Nevertheless, the "promotion of private interests" in legislation grew by leaps and bounds as larger and larger fortunes were made during the heyday of the Industrial Revolution—and as the impact of general laws on corporate opportunities grew. During the Robber Baron era of the 1880s and 1890s, according to the definitive history by Matthew Josephson, "The halls of legislation were transformed into a mart where the price of votes was haggled over, and laws, made to order, were bought and sold."

It was during this corrupt era that the U.S. Supreme Court first designated corporations as "persons" entitled to some of the protections of the Fourteenth Amendment in an 1886 decision (*Santa Clara County v. Southern Pacific Railroad Company*). The decision itself, in favor of the Southern Pacific, did not actually address the subject of corporate "personhood," but language that some historians believe was written by Justice Stephen Field was added in the "headnotes" of the case by the court reporter, who was the former president of a railway company. The chief justice had signaled before hearing the oral arguments that "the court does not wish to hear argument on the question of whether . . . the Fourteenth Amendment . . . applies to these corporations. We are all of the opinion that it does." (This backhanded precedent for the doctrine of corporate personhood was relied upon by conservative Supreme Courts in the late twentieth century for extensions of "individual rights" to corporations—and in the *Citizens United* decision in 2010.)

This pivotal case has an interesting connection to the first nerve endings of the worldwide communications networks that later became the Global Mind. The brother of Justice Field, Cyrus Field, laid the first transoceanic telegraph cable in 1858. A third Field brother, David (whose large campaign contributions to Abraham Lincoln had resulted in Stephen's appointment to the Supreme Court), happened to be in Paris with his family during the Paris Commune in 1871, and used the telegraph cable to send news of the riots, disorder, and subsequent massacre back to the United States in real time. It was the first time in history that an overseas news event was followed in the United States, as it unfolded, on a daily basis.

Though the Paris Commune had complex causes (including the bitter emotions surrounding the French defeat in the Franco-Prussian War that month and the struggle between republicans and monarchists), it became the first symbolic clash between communism and capitalism.* Karl Marx had published *Das Kapital* just four years earlier and wrote *The Civil War in France* during the two months of the Commune, saying that it would be "forever celebrated as the glorious harbinger of a new society." A half century later, at Lenin's funeral, his body was wrapped

---

* Though Marx wrote in *The Communist Manifesto* that the 1848 French revolution had been the first "class struggle."

in a torn and tattered red and white flag that had been flown by Parisians during the two months of the Commune.

But as much as the Paris Commune inspired communists, it terrified elites in the United States, among them Justice Field, who was obsessively following the daily reports from his brother and journalists in Paris. The Paris Commune received more press coverage—almost all of it hostile—than any other story that year besides government corruption. The fear provoked by the Commune was magnified by labor unrest in the U.S., particularly by many who had arrived since the 1830s from the poorer countries of Europe in search of a better life but had been victimized by the unregulated abuses in low-wage industrial jobs. Two years later, the U.S. was plunged into a depression by the bankruptcy of financier and railroad entrepreneur Jay Cooke. Wages fell even lower and unemployment climbed even higher. *The New York Times* warned, "There is a 'dangerous class' in New York, quite as much as in Paris, and they want only the opportunity or the incentive to spread abroad the anarchy and ruin of the French Commune."

According to historians, Justice Field was so radicalized by the Commune and what he feared were its implications for U.S. class warfare that he decided to make it his mission to strengthen corporations. His strategy was to use the new Fourteenth Amendment, which had been designed to confer the constitutional rights of persons on the freed slaves, as a vehicle for extending the rights of persons to corporations instead.

By the last decade of the nineteenth century, concentrated corporate power had attained such a shocking degree of control over American democracy that it triggered a populist reaction. When the Industrial Revolution resulted in the mass migration of Americans from farms to cities, and public concern grew over excesses and abuses such as child labor, long working hours, low wages, dangerous work environments, and unsafe food and medicines, reformers worked within the democracy sphere to demand new government policies and protections in the marketplace.

The Progressive movement at the turn of the twentieth century began implementing new laws to rein in corporate power, including the first broad antitrust law, the Sherman Act of 1898, though the Supreme Court sharply limited its constitutionality, as it limited the application and enforcement of virtually all Progressive legislation. In 1901, after the pro-corporate president William McKinley was assassinated only six months into his term, Theodore Roosevelt unexpectedly became president, and the following year launched an extraordinary assault on monopolies and abuses of overbearing corporate power.

Roosevelt established the Bureau of Corporations inside his new Department of Commerce and Labor. He launched an antitrust suit to break up J. P. Morgan's Northern Securities Corporation, which included 112 corporations worth a combined $571 billion (in 2012 dollars), at the beginning of the twentieth century, and was worth "twice the total assessed value of all property in thirteen states in the southern United States." This was followed by forty more antitrust suits. A seemingly inexhaustible source of presidential energy, Roosevelt also passed the Pure Food and Drug Act and protected more than 230 million acres of land, including the Grand Canyon, the Muir Woods, and the Tongass forest reserve—all while building the Panama Canal and winning the Nobel Peace Prize for resolving the Russo-Japanese War.

Roosevelt made a fateful decision at the beginning of his presidency not to run for a second full term in 1908, noting that he had served almost the full eight years that George Washington had established as the "wise custom" by serving only two terms. When Roosevelt's handpicked successor, William Howard Taft, abandoned many of TR's reforms, the march of corporate power resumed. In response, Roosevelt began to organize his Bull Moose Party campaign to replace Taft as president in the election of 1912.

In October of 1910, Roosevelt said, "Exactly as the special interests of cotton and slavery threatened our political integrity before the Civil War, so now the great special business interests too often control and corrupt the men and methods of government for their own profit." Eighteen months later, in the midst of the campaign, he said that his party was engaged in a struggle for its soul:

> The Republican party is now facing a great crisis. It is to decide
> whether it will be, as in the days of Lincoln, the party of the plain
> people, the party of progress, the party of social and industrial jus-
> tice; or whether it will be the party of privilege and of special inter-
> ests, the heir to those who were Lincoln's most bitter opponents,
> the party that represents the great interests within and without Wall
> Street which desire through their control over the servants of the

public to be kept immune from punishment when they do wrong and to be given privileges to which they are not entitled.

After Roosevelt lost that campaign to Woodrow Wilson (Taft came in third), he continued to speak out forcefully in favor of Progressive reforms and a rollback of corporate power. He said that the most important test of the country remained "the struggle of free men to gain and hold the right of self-government as against the special interests, who twist the methods of free government into machinery for defeating the popular will." He proposed that the U.S. "prohibit the use of corporate funds directly or indirectly for political purposes," and in speech after speech, argued that the Constitution "does not give the right of suffrage to any corporation." Thanks in part to his vigorous advocacy, the Progressive movement gained strength, passing a constitutional amendment to reverse the Supreme Court's prohibition against an income tax, enacting an inheritance tax, and enacting numerous regulations to rein in corporate abuses.

The many Progressive reforms continued during Woodrow Wilson's presidency, but the pendulum shifted back toward corporate dominance of democracy during the Warren Harding administration—remembered for its corruption, including the Teapot Dome scandal in which oil company executives secretly bribed Harding administration officials for access to oil on public lands.

Following three pro-corporate Republican presidents, President Franklin Roosevelt launched the second wave of reform when he took office in 1933 in the midst of the suffering caused by the Great Depression that was triggered by the stock market crash of 1929. The New Deal expanded federal power in the marketplace to a formidable scale and scope. But once again the conservative Supreme Court stopped many of the Progressive initiatives, declaring them unconstitutional. Theodore Roosevelt had declared the justices "a menace to the welfare of the nation" and FDR essentially did the same. But he went further, proposing a court-packing plan to add to the number of justices on the court in an effort to dilute the power of the pro-business majority.

Historians differ on whether Roosevelt's threat was the cause or not, but a few months later the Supreme Court reversed course and began approving the constitutionality of most New Deal proposals. To this day, some right-wing legal advocates refer to the court's switch as a "betrayal." In the twenty-first century, right-wing judicial activists are trying to return court rulings to the philosophy that existed prior to the New Deal.

In spite of FDR's initiatives, the U.S. found it difficult to escape hard times, and slipped back into depression in 1938. Then, when America mobilized to respond to the totalitarian threat from Nazi Germany and Imperial Japan, the Depression finally ended. After the U.S. emerged victorious, its remarkable economic expansion continued for more than three decades. By then, the consensus in favor of an expanded role for the federal government in addressing national problems was supported by a majority of voters across the political spectrum.

In the turbulent decade of the 1960s, however, the seeds of a corporate-led counterreform movement were planted. After the assassination of President John F. Kennedy in the fall of 1963, a variety of social reform movements swept the nation—driven in part by the restless energy and idealism of the huge postwar baby boom generation just entering young adulthood. The civil rights movement, the women's movement, the first gay rights demonstrations, the consumer rights movement, Lyndon Johnson's War on Poverty, and the escalating protests against the continuation of the ill-considered proxy war against communism in Southeast Asia all combined to produce a fearful reaction by corporate interests and conservative ideologues.

Just as the Paris Commune had radicalized Justice Stephen Field 100 years earlier, the social movements in the U.S. during the 1960s also awakened a fear of disorder, radicalized a generation of right-wing market fundamentalists, and instilled a sense of mission in soon-to-be Supreme Court Justice Lewis Powell. Powell, a Richmond lawyer then best known for representing the tobacco industry after the surgeon general's 1964 linkage of cigarettes to lung cancer, wrote a lengthy and historic 1971 memorandum for the U.S. Chamber of Commerce in which he presented a comprehensive plan for a sustained and massively funded long-term effort to change the nature of the U.S. Congress, state legislatures, and the judiciary in order to tilt the balance in favor of corporate interests. Powell was appointed to the Supreme Court by President Nixon two months later—though his plan for the Chamber of Commerce was not disclosed publicly until long after his confirmation hearings. A former president of the American College of Trial Lawyers, Powell was widely respected, even by his ideological opponents. But his aggressive expansion of corporate rights was the most consequential development during his tenure on the court.

Justice Powell wrote decisions creating the novel concept of "corporate speech," which he found to be protected by the First Amendment. This doctrine was then used by the court to invalidate numerous laws that were intended to restrain corporate power when it interfered with the public interest. In 1978, for example, Powell wrote the opinion in a 5–4 decision that for the first time struck down state laws prohibiting corporate money in an election (a citizens referendum in Massachusetts) on the grounds that the law violated the free speech of "corporate persons." Thirty-two years later, the U.S. Supreme Court relied on Powell's opinion to allow wealthy individual donors to contribute unlimited amounts to campaigns secretly, and further expanded the 1886 *Southern Pacific* precedent declaring corporations to be persons.

While it is true that corporations are made up of individuals, the absurdity of the legal theory that corporations are "persons"—as defined in the Constitution—is evident from a comparison between the essential nature and motives of corporations compared to those of flesh-and-blood human beings. Most corporations are legally chartered by the state with an ironclad mandate to focus narrowly on the financial interests of their shareholders. They are theoretically immortal and often have access to vast wealth. Twenty-five U.S.-based multinational corporations have revenues larger than many of the world's nation-states. More than half (53) of the 100 largest economies on Earth are now corporations. Exxon-Mobil, one of the largest corporations in the world, measured by revenue and profits, has a larger economic impact than the nation of Norway.

Individuals are capable of decisions that reflect factors other than their narrow financial self-interest; they are capable of feeling concern about the future their children and grandchildren will inherit—not just the money they will leave them in their wills; America's founders decided as individuals, for example, to pledge "our Lives, our Fortunes, and our Sacred Honor" to a cause deemed far greater than money. Corporate "persons," on the other hand, now often seem to have little regard for how they can help the country in which they are based; they are only concerned about how that country can help them make more money.

At an oil industry gathering in Washington, D.C., an executive from another company asked the then CEO of Exxon, Lee Raymond, to consider building additional refinery capacity inside the United States "for security" against possible shortages of gasoline. According to those present, Raymond replied, "I'm not a U.S. company and I don't make decisions based on what's good for the U.S." Raymond's statement recalls the warning by Thomas Jefferson in 1809, barely a month after leaving the White House, when he wrote to John Jay about "the selfish spirit of commerce, which knows no country, and feels no passion or principle but that of gain."

With the emergence of Earth Inc., multinational corporations have also acquired the ability to play nation-states off against one another, locating facilities in jurisdictions with lower wages and less onerous restrictions on their freedom to operate as they wish. The late chairman of the libertarian Cato Institute, William Niskanen, said, "corporations have become sufficiently powerful to pose a threat to governments," adding that this is "particularly the case with respect to multinational corporations, who will have much less dependence upon the positions of particular governments, much less loyalty in that sense." In 2001, President George W. Bush was asked by the prime minister of India, Manmohan Singh, to influence ExxonMobil's pending decision on allowing India's state-owned oil company to participate in a joint venture including the oil company and the government of Russia. Bush replied, "Nobody tells those guys what to do."

Those who advocate expanding the market sector at the expense of democratic authority believe that governments should rarely have the power to tell corporations "what to do." For the last forty years, pursuant to the Powell Plan, corporations and conservative ideologues have not only focused on the selection of Supreme Court justices favorable to their cause and sought to influence Court opinions, they have also pursued a determined effort to influence the writing of laws and the formation of policies to expand corporate power. They dramatically increased corporate advertising aimed at conditioning public opinion. They significantly expanded the number of lobbyists hired to pursue their interests in Washington, D.C., and state capitals. And they significantly increased their campaign contributions to candidates who pledged to support their agenda.

In only a decade, the number of corporate political action committees exploded from less than 90 to 1,500. The number of corporations with registered corporate lobbyists increased from 175 to 2,500. Since then, the number has continued to increase dramatically; recorded expenditures by lobbyists increased from $100 million in 1975 to $3.5 billion per year in 2010. (The U.S. Chamber of Commerce continues to top the list of lobbying expenditures, with more than $100 million per year—more than all lobbyist expenditures combined when the Powell Plan was first conceived.) One measure of how quickly attitudes toward lobbying have changed in the political culture of Washington was that in the 1970s, only 3 percent of retiring members of Congress gained employment as lobbyists; now, more than 50 percent of retiring senators and more than 40 percent of retiring House members become lobbyists.

Corporate coffers were far from the only source of funding for efforts consistent with the Powell Plan. Several wealthy conservative individuals and foundations were also radicalized by the 1960s, which Powell had described as "ideological warfare against the enterprise system and the values of Western society." When he called for an organized, well-funded response by "business to this massive assault upon its fundamental economics, upon its philosophy, upon its right to manage its own affairs, and indeed upon its integrity," many conservative business leaders rose to answer Powell's charge.

John M. Olin, for example, reacted to the armed takeover by militant black students of a campus building at Cornell University, his alma mater, by refocusing his wealthy foundation to support right-wing think tanks and a variety of right-wing efforts to change the character of American government. He embarked on a plan to not only spend the annual income from his foundation's endowment but to spend down the entire principal as quickly as possible in order to have maximum impact. Numerous other right-wing foundations also financed efforts consistent with the Powell Plan, including the Lynde and Harry Bradley Foundation and the Adolph Coors Foundation.

Perhaps the most effective part of the heavily funded conservative strategy has been their focus on populating the federal courts— particularly the U.S. Supreme Court—with ideological allies. The Powell Plan had noted specifically, "Under our constitutional system, especially with an activist-minded Supreme Court, the judiciary may be the most important instrument for social, economic and political change. . . . This is a vast area of opportunity for the Chamber . . . if, in turn, business is willing to provide the funds."

Subsequently, corporate interests became particularly active and persistent in lobbying to place judges on the bench who would be responsive to conservative legal theories that diminish individual rights, constrict the sphere of democracy, and elevate the rights and freedom of action for corporations. They have also established conservative law schools to train an entire generation of counterreformist advocates, and a network of legal foundations to influence the course of American jurisprudence. Two U.S. Supreme Court justices have even taken corporate-funded vacations at resorts where they were treated to legal instruction in seminars organized by wealthy corporate interests.

Meanwhile, the highly organized and well-funded counterreform movement also created and funded think tanks charged with producing research and policy initiatives designed to further corporate interests. In addition, they financed the creation of political movements at the local, state, and national level. By the 1980s and 1990s, this movement launched fierce battles to place opponents of robust government policies in state legislatures, in Congress, and in the White House. Ronald Reagan's defeat of Jimmy Carter was their first watershed victory, and the takeover of Congress in the mid-1990s solidified their ability to bring most Progressive reform to a halt.

In part, the policies of FDR—which had been, in the main, supported by presidents and Congresses of both parties for several decades—were victims of their own success. As tens of millions were lifted into the middle class, many lost their enthusiasm for continued government interventions, in part because they began to resist the levels of taxation necessary to support a more robust government role in the economy. Labor unions, one of the few organized forces supporting continued reform, lost members as more jobs migrated from manufacturing into services, and as outsourcing and robosourcing hollowed out the U.S. middle class. The nature and sources of America's economic strength have changed over the last several decades as manufacturing has declined. America's branch of Earth Inc. can't be driven solely by wages—investment is of course critical—but the tilt is important, and too little noted.

Slowly at first, but then with increasing momentum, the prevailing ideology of the United States—democratic capitalism—has shifted profoundly on its axis. During the decades of conflict with communism, the internal cohesion between the democratic and capitalist spheres was particularly strong. But when communism disappeared as an ideological competitor and democratic capitalism became the ideology of choice throughout most of the world, the internal tensions between the democratic sphere and the capitalist sphere reappeared. As economic globalization accelerated, the imperatives of business were relentlessly pursued by multinational corporations. With triumphalist fervor and the enormous resources made available for a sustained implementation of the Powell Plan, corporate and right-wing forces set about diminishing the role of government in American society and enhancing the power of corporations.

Market fundamentalists began to advocate the reallocation of decision-making power from democratic processes to market mechanisms. There were proposals to privatize—and corporatize—schools, prisons, public hospitals, highways, bridges, airports, water and power utilities, police, fire, and emergency services, some military operations, and other basic functions that had been performed by democratically elected governments.

By contrast, virtually any proposal that required the exertion of governmental authority—even if it was proposed, debated, designed, and decided in a free democratic process—was often described as a dangerous and despicable step toward totalitarianism. Advocates of policies shaped within the democratic sphere and implemented through the instruments of self-government sometimes found themselves accused of being agents of the discredited ideology that had been triumphantly defeated during the long struggle with communism. The very notion that something called the public interest even existed was derided and attacked as a dangerous concept.

By then, the encroachment of big money into the democratic process had convinced many Democrats as well as almost all Republicans to adopt the new ideology that supported the contraction of the democratic sphere and the expansion of the market sphere. It was during this same transition period that television supplanted newspapers as the principal source of information for the majority of voters, and the role of money in political campaigns increased, giving corporate and other special interest donors an even more unhealthy degree of power over the deliberations of the United States Congress and state legislatures.

When the decisions of the United States result not from democratic debate but are instead determined by powerful special interests, the results can be devastating to the interests of the American people. Underfunded and poorly designed U.S. social policies have produced a relative decline in the conditions of life. Compared to the other nineteen advanced industrial democracies in the Organisation for Economic Co-operation and Development (OECD), the United States has the highest inequality of incomes and the highest poverty rate; the lowest "material well-being of children" according to the United Nations' index, the highest child poverty rate and the highest infant mortality rate; the biggest prison population and the highest homicide rate; the biggest expenditures on health care and the largest percentage of its citizens unable to afford health care.

At the same time, the success by corporate interests in reducing regulatory oversight created new risks for the U.S. economy. For example, the deregulation of the financial services industry, which accompanied the massive increase in flows of trade and investment throughout the world, led directly to the credit crisis of 2007, which caused the Great Recession (which some economists are now calling "the Second Great Contraction" or "the Lesser Depression").

The international consequences of that spectacular market failure dramatically undermined global confidence in U.S. leadership of economic policy and marked the end of an extraordinary period of U.S. dominance. Nations had generally accepted the so-called Washington Consensus as the best formula for putting their economies on sound footing and building the capacity for sustainable growth. Although most of the policy recommendations contained in the consensus were broadly seen as reflecting sound economic common sense, they tended to expand the market sphere in domestic economies as they removed barriers to global trade and investment flows.

Two other factors combined with the 2007–08 economic crisis to undermine the leadership of the United States: first, the rise of China's economy, which did not follow the prescriptions of the Washington Consensus even though its success was driven by the uniquely Chinese form of capitalism; second, the catastrophic invasion of Iraq—for reasons that were later proven to be false and dishonest.

Within the United States, it is a measure of how distorted the "conversation of democracy" has become that in the aftermath of the economic catastrophe, the most significant "populist" reaction in the U.S. political system was not a progressive demand for protective regulations to prevent a recurrence of what had just happened, but instead a right-wing faux-populist demand by the Tea Party for *less* government regulation. This movement was financed and hijacked by corporate and right-wing lobbyists who took advantage of the sense of grievance and steered it toward support of an agenda that promoted corporate interests and further diminished the ability of the government to rein in abuses. Extreme partisanship by congressional Tea Party Republicans almost produced a default of the U.S. government in 2011, and threatened to again at the end of 2012.

The sudden growth of the Tea Party was also due in significant measure to its promotion by Fox News, which under the ownership of Rupert Murdoch and the leadership of a former media strategist for Richard Nixon—Roger Ailes—has exceeded the wildest dreams of the Powell Plan's emphasis on changing the nature of American television. Powell had proposed that "The national television networks should be monitored in the same way that textbooks should be kept under constant surveillance." He called for the creation of "opportunity for supporters of the American system" within the television medium.

The inability of American democracy to make difficult decisions is now threatening the nation's economic future—and with it the ability of the world system to find a pathway forward toward a sustainable future. The exceptionally bitter partisan divide in the United States is nominally between the two major political parties. However, the nature of both Democrats and Republicans has evolved in ways that sharpen the differences between them. On the surface, it appears that Republicans have moved to the right and purged their party of moderates and extinguished the species of liberal Republicans that used to be a significant minority within the party. Democrats, according to this surface analysis, have moved to the left and have largely pushed out moderates and the conservative Democrats who used to play a prominent role in the party.

Beneath the surface, however, the changes are far more complex. Both political parties have become so dependent on business lobbies for the large sums of money they must have to purchase television advertisements in order to be reelected that special interest legislation pushed by the industries most active in purchasing influence—financial services, carbon-based energy companies, pharmaceutical companies, and others—can count on large bipartisan majorities. The historic shift of the internal boundary between the overlapping capitalist and democratic spheres that make up America's reigning ideology, democratic capitalism, has resulted in increased support within both parties for measures that constrain the role of government.

This shift has now moved so far to the right that it is not unusual for Democrats to propose ideas that originated with Republicans a few years ago, only to have them summarily rejected as "socialist." The resulting impasse threatens the future of hugely popular entitlement programs, including Social Security and Medicare, and is heightening partisan divisions on questions considered basic and nonnegotiable on both sides. The tensions have grown more impassioned and bitter than at any point in American history since the decades leading up to the Civil War.

"Market fundamentalism" has acquired, in the eyes of its critics, a quasi-religious fervor reminiscent of the zeal that many Marxists displayed before communism failed—although those to whom the label applies feel that liberals and progressives pursue the ideology of "statism" with a single-minded devotion. U.S. self-government is now almost completely dysfunctional, incapable of making important decisions necessary to reclaim control of its destiny.

James Madison, one of the most articulate of America's extraordinary founders, warned in his Federalist No. 10 about the "propensity of mankind to fall into mutual animosities" and cluster into opposing groups, parties, or factions:

> The latent causes of faction are thus sown in the nature of man; and we see them everywhere brought into different degrees of activity, according to the different circumstances of civil society. A zeal for different opinions concerning religion, concerning government, and many other points, as well of speculation as of practice; an attachment to different leaders ambitiously contending for pre-eminence and power; or to persons of other descriptions whose fortunes have been interesting to the human passions, have, in turn, divided mankind into parties, inflamed them with mutual animosity, and rendered them much more disposed to vex and oppress each other than to co-operate for their common good.

Madison noted that this tendency in human nature is so strong that even "the most frivolous and fanciful distinctions have been sufficient to kindle their unfriendly passions and excite their most violent conflicts." But he went on to highlight the "most common and durable source of factions" as "the various and unequal distribution of property." The inequality in the distribution of wealth, property, and income in the

United States is now larger than at any time since 1929. The outbreak of the Occupy movement has been driven by the dawning awareness of the majority of Americans that the operations of democratic capitalism in its current form are producing unfair and intolerable results. But the weakened state of democratic decision making in the U.S., and the enhanced control over American democracy by the forces of wealth and corporate power, have paralyzed the ability of the country to make rational decisions in favor of policies that would remedy these problems.

These two trends, unfortunately, reinforce one another. The more control over democratic decision making by powerful wealthy interests, the more they are able to ensure that decisions on policy enhance their wealth and power. This classic positive feedback loop makes inequality steadily worse, even as it makes democratic solutions for inequality less accessible.

The issue of inequality has become a political, ideological, and psychological fault line. Neuroscientists and psychologists have deepened the understanding of political scientists about the true nature of the "left-right" or "liberal-conservative" divide in the politics of every country. Research shows conclusively that these differences are also "sown in the nature of man," and that in every society there is a basic temperamental divide between those who are relatively more tolerant and others who are relatively less tolerant of inequality.

The same divide separates those for whom it is relatively more or relatively less important to care for the weak and victimized, maintain respect for authority—particularly when disorder is threatening—prioritize loyalty to one's group or nation, demonstrate patriotism, and honor the sanctity of symbols and objects that represent group values. Both groups value liberty and fairness but think about them differently. Recent research indicates that these temperamental differences may be, in part, genetically based, but perhaps more importantly, the differences are reinforced by social feedback loops.

The issue of inequality also lies on the ideological fault line between democracy and capitalism. For those who prioritize capitalism, inequality is seen as an obvious and necessary condition for the incentivization of productive activity. If some receive outsized rewards in the marketplace, that is a beneficial outcome not only for those so rewarded but for the capitalist system as a whole, because it demonstrates to others what can happen if they too become more productive.

For those who prioritize democracy, the tolerance of persistent inequality is far more likely to stimulate demands for change in the underlying policies that consistently produce unequal outcomes. Inheritance taxes have become a flashpoint in American politics. Why, ask liberals, is there a social value in failing to redistribute some portion of great fortunes when a wealthy person dies? Yet for conservatives, the ability to pass on great wealth at death is just another part of the incentive to earn great wealth in the first place. And they view the imposition of what they call a "death tax" (a label coined by a conservative strategist who conducted deep research on what language would most trigger feelings of outrage) as an encroachment upon their freedom. In my own view, it is absurd to eliminate inheritance taxes; they should be raised instead. The extreme concentration of wealth is destructive to economic vitality and to the health of democracy.

Any legislative effort to address inequality with measures that require funding through taxes of any sort has also come to mark the political fault line dividing the United States into two opposing factions. The corporate-led counterreform movement that began in the 1970s adopted as one of its key tenets a cynical strategy known as "starve the beast"; while proclaiming the importance of "balancing the budget" and "reducing deficits," the movement pushed massive tax cuts as the initial step in a plan to use the resulting funding gap as an excuse to force sharp reductions in the role of government. This was part and parcel of the larger effort to diminish the democracy sphere and enhance the market sphere.

What is most troubling to advocates of American democracy is that the radically elevated role of money in politics has given the forces representing wealth and corporate power sufficient strength to advance their agenda even when a sizable majority of the American people oppose it. In effect, those who zealously advocate the expansion of the role of markets while demanding a constriction of the ability of people in democracies to enact policies that address the abuses and disruptive risks that often accompany unrestrained market activity are posing a threat to the internal logic of the nation-state itself.

America's middle class has been hollowed out by, among other causes, the emergence of Earth Inc., the increasing proportion of retired Americans, and advances in the availability of expensive health care technologies. The result is a fast growing financial crisis that is threatening the ability of the United States to provide world leadership. U.S. government indebtedness compared to GDP is threatening to spiral out of control. According to a study by the nonpartisan Congressional Budget Office, the U.S. debt-to-GDP ratio is 70 percent in 2013, and already exceeds GDP if money the government owes to itself is added to the debt.

Although a highly publicized credit downgrade by the bond rating firm Standard & Poor's in 2011 had no perceptible effect on the demand for U.S. bonds, experts have warned that a sudden loss of confidence in the dollar and in the viability of U.S. finances cannot be ruled out in the coming decade. Partly due to the weakness of the euro and a lack of trust in the Chinese yuan, or renminbi (RMB), the U.S. dollar remains the world's reserve currency. For those and other reasons, the United States is still able to borrow from the rest of the world at extremely low interest rates—as of this writing at less than 2 percent for ten-year bonds.

Yet the looming financial troubles are potentially large enough to provoke a sudden loss of confidence in the future of the dollar, and a sudden increase in the interest rates the U.S. government would be required to pay to holders of its debt. Even a one percentage point rise over projected increases in the interest rates paid on the debt would add approximately $1 trillion to interest payments over the next decade.

The strength of any nation's economy is, of course, crucial for the exertion of power in multiple ways. It undergirds the ability to finance weapons and armies, and to use foreign aid and trade concessions to build necessary alliances. It enables the building of superior infrastructures and the provision of public goods such as education, job training, public safety, pensions, enforceability of contracts, quality of the legal system, health care, and environmental protection. It also allows for the creation of a superior capacity for research and development, now crucial to gain access to the fruits of the accelerating scientific and technological revolution.

More broadly, the ability of any nation to wield power on a sustained basis—whether military, economic, political, or moral—depends upon multiple additional factors, including:

- Its ability to form intelligent policies and implement them effectively in a timely manner, which usually requires reason-based, transparent decision making and the forging of a domestic consensus in support of policies—particularly if they require a long-term commitment. The Marshall Plan, for example, would not have been possible without bipartisan support in the Congress and the willingness of the American people to commit significant resources to a visionary plan that required decades to implement.

- The cohesion of its society, which generally requires the perception of fairness in the distribution of incomes and net worth, and a social contract within which real needs are satisfactorily met and governmental power is derived from the genuine consent of the governed. The maintenance of cohesion also requires alertness to and sustained respect for the differing experiences and perspectives of minorities, and a full understanding of the benefits from the absorption of immigrants.

- The protection of property rights, the enforcement of contracts, and opportunities to invest money without an unreasonable risk of losing wealth.

- The development and enforcement of sustainable fiscal and monetary policies and bank regulations that minimize the risk of market disruptions and do not accentuate swings in the business cycle. Economic success also requires investments in infrastructure, research and development, and appropriate antitrust enforcement.

- The development of its human capital with adequate investments in education and job training, health care and mental health care, and nutrition and child care. The Information Revolution has enhanced the importance of investments in human capital, even as it requires a regular updating of appropriate strategies.

- The protection, conservation, and stewardship of natural capital with environmental protection and energy efficiency. The global climate crisis requires extensive planning for adaptation to the big changes coming, and much greater attention to the need for rapid reductions in global warming pollution.

The United States is now failing to satisfy many of these criteria. But it is not the only nation-state that is in danger of dissipating its ability to make sound decisions about the future. The larger and more significant change in the balance of power throughout the world is the relative decline in the effective power of nation-states generally. In the words of Harvard professor Joseph Nye, "the diffusion of power away from government is one of this century's great political shifts."

NATION-STATES IN TRANSITION

One of the principal reasons for the steady decline in the effective power of nation-states has been a rise in the power of multinational corporations. The redistribution of economic power and initiative to multinational corporations operating in many national jurisdictions simultaneously (even while exerting increased influence over the domestic policies of the nations in which they are based) has significantly diminished the role of nation-states.

With their ability to outsource and robosource their labor inputs, many corporations no longer have the same incentive to support improvements in national education systems and other measures that would enhance labor productivity in their home nations. And with the astonishing increase in trade and investment flows, multinational corporations are playing a far more significant role than ever before. Some political scientists have asserted that the influence of corporations on modern governance is now almost analogous to the influence of the medieval church during the era of feudalism.

The integration of the global economy has shifted power profoundly toward markets. The massive flows of capital over digital networks in Earth Inc. have made some national economies highly vulnerable to the sudden outflow of "hot money" if and when global markets reach a negative judgment about the viability of their fiscal and monetary policies. International banks and bond rating firms have become more significant players in national debates about taxing and spending. Greece is only the best known of many examples of countries no longer able to make decisions for themselves. It must first get permission from the European Union, which supports it, and international banks, which hold its debt.

The historic decline in the power, influence, and prospects for the Eurozone countries (those European nations that have joined in a monetary union) stems in large measure from a widely recognized fatal flaw in the decision by those European nations entering the monetary union to gamble that they could delay tight integration of their fiscal policies (without which a single currency is ultimately not viable) until the political momentum toward unity made that difficult step possible.

Recently released documents confirm that when the Eurozone was founded, there was widespread awareness, particularly in Germany, that Southern European countries were not even close to the fiscal conditions that would have reduced the risk of monetary integration. Yet then chancellor Helmut Kohl and other European leaders decided that the benefits of European unity were worth the gamble that cohesion could be maintained until there was sufficient Europe-wide support for tighter fiscal unity. When the financial crisis of 2007–08 exposed the fatal flaw, the global credit markets essentially called Europe's bet.

Broadly speaking, Europe now faces two options. First, it can acknowledge the failure of the Eurozone experiment and sharply contract the number of nations that remain in the Eurozone alongside Germany and France—the core of Europe's economy. This option is unattractive for several reasons: there are no legal procedures for the withdrawal of a country from the Eurozone, the transition from the euro back to a national currency—for a country like Greece, for example—promises to be exceptionally painful and expensive; and Germany would find itself once again threatened with competitive devaluations—in nations like Italy, for example—whenever the strength of the German economy significantly outpaced that of its neighbors.

The second option is to move quickly and boldly forward to a fiscal unification of the Eurozone, notwithstanding the disparities in the strength and productivity of Germany's economy compared to the nations of Southern Europe. However, the only way to maintain anything remotely approaching parity in living standards in a fiscally unified Europe would be for Germany to make transfer payments (essentially budget subsidies) to the weaker European countries for at least a generation. Yet even though this might be a long-term economic bargain for Germany, the relatively more prosperous taxpayers of the former West Germany have shouldered the burden of subsidies to those in the relatively weaker former East Germany for the two decades since reunification—at an estimated cost of $2.17 trillion—and as a result their appetite for taking on this new burden is quite low.

The inability of Europe's leaders to establish the needed fiscal integration and move more quickly toward a unified Europe has created a significant political and economic crisis that threatens to undo one of the most important U.S. successes in the aftermath of World War II. The weakening of political cohesion and economic dynamism in Western Europe (along with the long-running political paralysis and economic slowdown in Japan) have also contributed to the new difficulties faced by the United States in providing world leadership.

As with the compound ideology of democratic capitalism, the political concept of a nation-state is also made up of two ideas that overlap one another. The idea of a nation is based on the common identity of the people who live in a national territory; whether or not they share the same language (often the vast majority do), they usually share the same feeling that they are members of a national community. The state, by contrast, is an administrative, legal, and political entity that provides the infrastructure, security, and judicial basis for life within the state. When both of these concepts overlap, the result is the kind of nation that we commonly think of as the principal form by which global civilization is organized.

There is a rich historical debate about the origins of nation-states. The first large "states" emerged around 5,400 years ago when the Agricultural Revolution first produced large food surpluses in areas endowed with plant varieties that were especially suitable for cultivation: the Nile River Valley in Egypt, the Yellow River Valley in China, the Indus River Valley in India, the Tigris and Euphrates river valleys, and the Fertile Crescent (and in nearby Crete). These states also appeared in several other areas of the world, including Mexico, the Andes, and Hawaii.

The marriage of state and nation occurred much later. In a very real sense, modern nation-states were created as an outgrowth of the Print Revolution. Throughout most of human history, it was not the dominant form of organization. Empires, city-states, confederations, and tribes all coexisted in large areas of the Earth for millennia. Although there are a few examples of nation-states that existed prior to the Print Revolution, the rise of the modern nation-state as the dominant form of political organization occurred when the spread of printed books and pamphlets in a shared form of national languages stimulated the emergence of common national identities.

Prior to the Print Revolution, languages such as French, Spanish, English, and German, among others, featured a multiplicity of dialects and forms that were so distinctive that speakers of one form often had difficulty communicating with speakers of other forms. After the Print Revolution gained momentum, however, the economic imperatives of mass mechanical reproduction of texts provided a powerful push toward a common dialect of each tongue that was then adopted as a common language within each national territory. The emergence of group identities in regions where the majority of people spoke, read, and wrote in the same language created the conditions that led to the emergence of modern nation-states.

The Reformation and the Counter-Reformation unleashed passions that combined with these new national identities to trigger a long series of bloody wars that finally culminated in the Treaty of Westphalia in 1648—the treaty that formalized the construction of a new order in Europe based on the primacy of nation-states, and the principle of noninterference by any nation-state in the affairs of another.

Soon thereafter, the dissemination of news—printed in national languages and presented within a distinctively national frame of reference— further strengthened national identities. Over time, the wider availability of civic knowledge also led to the emergence of representative democracy and elected national legislatures. When the people of nations gained political authority over the making of laws and policies, the functions of the state were married to those of the nation.

During the Industrial Revolution, the introduction of transportation networks such as railroads and highways further expanded the political role of nation-states, and further consolidated national identities. At the same time, the nature and scale of industrial technologies expanded potential points of conflict between the operations of the market and the political prerogatives of the state.

The internal cohesion of modern nation-states was also strengthened by the introduction of national curricula in schools that not only reinforced the adoption of a common national dialect but also spread a common understanding of national histories and cultures—usually in ways that emphasized the most positive stories or myths in each nation's history, while often neglecting to include narratives that might diminish feelings of nationalism. (For example, Japanese textbooks that minimize its invasion and occupation of China and Korea have regularly become sources of tension in Northeast Asia.)

Transnational global technologies such as the Internet and satellite television networks are exercising influence in spheres that used to be dominated primarily by the power of nation-states. Many regional satellite television networks dispense with national frames of reference in presenting news. And the Internet, in particular, is complicating many of the strategies formerly relied upon by nation-states to build and maintain national cohesion. Just as the printing press drew adherence to single versions of national languages and solidified national identities, the Internet is making available the knowledge of every country to the people of every other country. Google Translate, the largest of many machine translation services, now operates in sixty-four different languages and provides translations from one language to another for more documents, articles, and books in one day than all of the human translators in the world provide in a full year.

And of course, the number of texts translated by computers is increasing exponentially. Seventy-five percent of the web pages translated are from English to other languages. It is often, and inaccurately, said that English is the language of the Internet. Actually there are more Chinese language users of the Internet than there are people in the United States. But the content of the Internet that is now being dispersed throughout the world is content that still mainly originates in English.

The narratives of national histories that have dominated the curricula of mandatory public education systems now have competition from alternative narratives widely available on the Internet. And they often have the persuasive ring of truth—for example, for minorities within nation-states whose historical mistreatment can no longer be as easily obscured or whitewashed.

For these and other reasons, the glue holding some nations together in spite of their ethnic, linguistic, religious and sectarian, tribal and historical differences appears to be losing some of its cohesive strength. Belgium, for example, has reallocated the power once vested in its national government to its component regional governments. Flanders and Wallonia are not technically nation-states but might as well be.

In many parts of the world, identity-driven subnational movements are becoming more impatient and, in some cases, aggressive, in seeking independence from the nations of which they are now part. Nation-states have been described as "imagined communities"; it is, after all, impossible for citizens of a nation-state to interact with all other members of the national community. It is their common identity that forms the basis of their national bonds. If those bonds no longer lay as strong a claim on their imagination, their identity bonds may attach elsewhere—often to older identities that predated the formation of the nation-state.

In many regions, the growth of fundamentalism is also connected to the weakening of the psychological bonds of identity in the nation-state. Muslim, Hindu, Christian, Jewish—even Buddhist—fundamentalism are all sources of conflict in the world today. This does not come as a surprise to historians. After all, it was the desperate need to control religious wars and sectarian violence that led to the formal codification of nation-states as the primary form of governance in the first place.

In the midst of the English Civil War, Thomas Hobbes proposed one of the first and most influential arguments for a "social contract" to prevent the "war of every man against every man" by giving a monopoly on violence to the nation-state and granting to the sovereign of that state—whether a monarch or an "assembly of men"—the sole authority "to make war and peace . . . and to command the army."

Nationalism became a potent new cause of warfare over the three centuries between the Treaty of Westphalia and the end of World War II. As the weaponry of war was industrialized—with machine guns, poison gas, tanks, and then airplanes and missiles—the destructive power unleashed led to the horrendous loss of life in the wars of the twentieth century. And the imposition of order by nation-states within their own borders sometimes created internal tensions that led their leaders to use the projection of violence against neighboring nation-states as a means of strengthening internal cohesion by demonizing "the other." Tragically, the monopoly on violence granted to the state was also sometimes brutally directed at disfavored minorities within their borders.

In the wake of World War I, a number of nation-states were formed in the imagination of the United States, the United Kingdom, and other European nations that were seeking to create stability in regions like the Middle East and Africa, where tribal, ethnic, sectarian, and other divisions threatened continued destabilizing violence. One of the premier examples of an imagined community was Yugoslavia. When the unifying ideology of communism was imposed on this amalgam of separate peoples, Yugoslavia functioned fairly well for three generations.

But when communism collapsed, the glue of its imagined nation no longer could hold it together. The great Russian poet Yevgeny Yevtushenko described what happened next with the metaphor of a prehistoric mammoth found frozen in the ice of Siberia. When the ice melted, and the mammoth's flesh thawed, ancient microbes within the flesh awakened and began decomposing the mammoth. In like fashion, the ancient antagonisms between Serbian Orthodox Christians, Croatian Catholics, and Bosnian Muslims decomposed the glue that had formed what is now referred to as the "former Yugoslavia."

Not coincidentally, the border between Serbia and Croatia had marked the border 1,500 years ago between the Western and Eastern Roman empires, while the border between Serbia and Bosnia marked the fault line between Islam and Christendom 750 years ago. After the breakup of Yugoslavia, the new leader of independent Serbia went to the disputed territory of Kosovo to mark the 600th anniversary of the great battle there in which the Serbian Empire was defeated by the Ottoman Empire; in a demagogic and warmongering speech, he revivified the ancient hatreds wrapped in memories of that long ago defeat and launched genocidal violence against both Bosnians and Croats.

The legacy of empires has continued to vex the organization of politics and power in the world long after nation-states became the dominant form of political organization. In the last three decades of the nineteenth century, European countries colonized 10 million square miles of land in Africa and Asia, 20 percent of all land in the world, putting 150 million people under their rule. (Indeed, several modern nation-states continued to govern colonial empires well into the second half of the twentieth century.) To pick one of many examples, the breakup of the Ottoman Empire in the aftermath of World War I resulted in the decision by Western powers to create new nation-states in the Middle East, some of which pushed together peoples, tribes, and cultures that had not previously been part of the same "national" community, including Iraq and Syria. It is not coincidental that both of these nations have been coming apart at the seams.

With the weakening of cohesion in nation-states, wherever peoples feel a strong and coherent identity that is separate from the one cultivated by the nation-state that contains them, there is new restlessness. From Kurdistan to Catalonia to Scotland, from Syria to Chechnya to South Sudan, from indigenous communities in the Andean nations to tribal communities in Sub-Saharan Africa, many people are shifting their primary political identities away from the nation-states in which they lived for many generations. Although the causes are varied and complex, a few nations, like Somalia, have devolved into "post-national entities."

In many parts of the world, nonstate terrorist groups and criminal organizations such as those who are now wielding power in so-called narco-states are aggressively challenging the power of nation-states. There is an overlap between these nonstate actors: nineteen of the forty-three known terrorist groups in the world are linked to the drug trade. The market for illegal narcotics is now larger than the national economies of 163 of the world's 184 nations.

It is significant that the most consequential threat to the United States in the last three decades came from a nonstate actor, Osama bin Laden's Al-Qaeda. A malignant form of Muslim fundamentalism was the primary motivation for Al-Qaeda's 9/11 attack. (According to numerous reports, bin Laden was revulsed by the presence of U.S. military deployments in Saudi Arabia, the custodian of Islam's holiest sites.)

The damage done by the attack itself—the murder of more than 3,000 people—was horrible enough, but the tragic response it provoked, the misguided invasion of Iraq, which, as everyone now acknowledges, had nothing whatsoever to do with attacking the U.S., was ultimately an even more serious blow to America's power, prestige, and standing. Hundreds of thousands died unnecessarily, $3 trillion was wasted, and the reasons given for launching the war in the first place were later revealed as cynical and deceptive.

The decision by the United States government to abandon its historic prohibitions against the torture of captives and the indefinite detention of individuals without legal process has been widely seen around the world as diminishing its moral authority. In a world divided into different civilizations, with different religious traditions and ethnic histories, moral authority is arguably an even greater source of power. Even though the ideologies of nations vary widely, the values of justice, fairness, equality, and sustainability are valued by the people of every nation, even if they often define these values in different ways.

The apparent rise of fundamentalism in its many varieties may be due, in part, to the pace of change that naturally causes many people to more tightly embrace orthodoxies of faith as a source of spiritual and cultural stability. The globalization of culture—not only through the Internet, but also through satellite television, compact discs, and other media—has also been a source of conflict between Western societies and conservative fundamentalist societies. When cultural goods from the West depict gender roles and sexual values in ways that conflict with traditional norms in fundamentalist cultures, religious leaders condemn what they view as the socially destabilizing impact.

But the impact of globalized culture goes far beyond issues of gender equity and sexuality. Cultural goods serve as powerful advertisements for the lifestyles that are depicted, and promotions for the values of the country where such goods originate. In a sense, they carry the cultural DNA of that country. As the global middle class is exposed to images of homes, automobiles, appliances, and other common features of life in industrial countries, the pressure they exert for changes in their own domestic political and economic policies often grows accordingly.

The longer-term impact may well be to break down differences. A recent study in Cairo found that there is a strong correlation between the amount of television watched and the decline of support for fundamentalism. One of the sources of the enhanced influence of Turkey in the Middle East is the popularity of its movies and television programs. The dominance of American music has enhanced the impression of the United States as a dynamic and creative society. The ability to influence the thinking of peoples through the dissemination of cultural goods such as movies, television programs, music, books, sports, and games is increasing in an interconnected world where consumption of media is rising every year.

WAR AND PEACE

The second half of the twentieth century saw a decline in the number of people killed in wars, and a decline in the number of wars in every category, international and civil—even though millions continued to die because of the pathological behavior of dictators. The decline has continued in this century, leading some to argue that humankind is maturing, humane values are spreading, and military power is less relevant in an interconnected world. It is a measure of this change that the people of the United States feel a palpable loss of national power at a time when its military budget is larger than those of the next fifty other nations combined. However, self-described foreign policy "realists" (who believe that nation-states *always* compete in an inherently anarchic international system) warn that similar predictions made in past eras proved to be false.

History provides all too many examples of unwarranted optimism about the decline of war during previous eras when a new appreciation for the benefits of peace seemed to be on the rise. The best-selling book globally in 1910 was *The Great Illusion* by Norman Angell, who argued that the increased economic integration that accompanied the Second Industrial Revolution had made war obsolete. Less than four years later, on the eve of World War I, Andrew Carnegie, the Bill Gates of his day, wrote a New Year's greeting to friends: "We send this New Year Greeting, January 1, 1914 strong in the faith that International Peace is soon to prevail, through several of the great powers agreeing to settle their disputes by arbitration under International Law, the pen thus proving mightier than the sword."

Human nature has not changed and the history of almost every nation contains sobering reminders that the use of military power has often been decisive in changing their fate. Nationalist politicians in many countries—including the United States and China—will, of course, seek to exploit fears about the future—and the fear of one another—by calling for the buildup of military strength. In the present era, some Chinese military strategists have written that a well-planned cyberattack on the United States could allow China to "gain equal footing" with the U.S. in spite of U.S. superiority in conventional and nuclear weaponry. And as has often been the case in history, fear begets fear; the buildup of a capacity for war leads those against whom it might be used to infer that there is an intent to do just that.

The fear of a surprise military attack has itself had a distorting influence on the priority given to military expenditures throughout history, and is a fear inherently difficult for the people and leaders of any nation to keep in proper perspective. That is one reason why national security depends more than ever on superior intelligence gathering and analysis in order to protect against strategic surprise and to maintain alertness to strategic opportunities.

In addition, new developments in technology have frequently changed the nature of warfare in ways that have surprised complacent nations who were focused on the technologies that were dominant in previous wars. The Maginot Line painstakingly constructed by France after World War I proved impotent in the face of new highly mobile tanks deployed by Nazi Germany. Military power now depends more than ever on the effective mastery of research and development to gain leverage from the still accelerating scientific and technological revolution, which has an enormous impact on the evolution of weaponry.

While the utility of military power may indeed be finally declining in significance in a world where the people and businesses of every nation are more closely linked than ever before, the recent decline in warfare of all kinds in the world—particularly war between nation-states—may have less to do with a sudden outbreak of empathy in mankind and may have more to do with the role played by the United States and its allies in the post–World War II era in mediating conflicts, building alliances, and sometimes intervening with a combination of limited military force and economic sanctions—as it did, for example, in the former Yugoslavia to limit the spread of violence between Serbia, Croatia, and Bosnia.

Supranational entities have also been playing an ever growing role, intervening in nations unable to halt violent conflicts and mediating the resolution of disputes. These international groups include not only U.N.-sponsored global efforts, but also, increasingly, efforts by regional supranational entities like the African Union, the Arab League, the European Union, NATO, and others. Nongovernmental organizations, faith-based charitable groups, and philanthropic foundations are playing an increasingly significant role in providing essential public goods in areas where nation-states are faltering. When sustained military operations are necessary and established supranational entities are unable to reach consensus, "coalitions of the willing" have been formed.

But in many of these interventions—particularly where NATO and coalitions of the willing were involved—the United States has played a key organizing and coordinating role, and has often provided not only the critical intelligence collection and analysis but also the decisive military force as well. If the equilibrium of power in the world continues to shift in ways that weaken the formerly dominant position of the United States, it could threaten an end to the period some historians have labeled the Pax Americana.

The recent decline in war may also be related to two developments during the long Cold War between the United States and the USSR. First of all, when these two superpowers built vast arsenals of nuclear bombs mounted on intercontinental ballistic missiles, submarines, and bombers, the quantum increase in the probable consequences of all-out war became so obviously and palpably unacceptable that both the U.S. and the USSR soberly backed away from the precipice. The escalating cost of maintaining and modernizing these arsenals also became a burden for both superpowers. (The Brookings Institution has calculated that since 1940, the U.S. has spent $5.5 trillion on its nuclear war fighting capability—more than on any other program besides Social Security.) Though the risk of such a war has been sharply reduced by arms control agreements, the partial dismantling of both arsenals, and enhanced communications and safeguards (including a recent bilateral nuclear cyber-security agreement), the risk of an escalation in tensions must still be continually managed.

Second, during the last third of the twentieth century, both the U.S. and the USSR had bitter experiences in failed efforts to use overwhelming conventional military strength against guerrilla armies using irregular warfare tactics, blending into their populations and fighting a war of attrition. The lessons learned by the superpowers were also learned by guerrilla forces. Partly as a result, the continued spread of irregular warfare tactics is now seriously undermining the nation-state monopoly on the ability to use warfare as a decisive instrument of policy.

The large excess inventories of rifles and automatic weapons manufactured during previous wars are increasingly available not only to insurgent guerrilla forces, but also to individuals, terrorist groups, and criminal organizations. When a new generation of weapons is manufactured, the older generation is not destroyed. Rather, they find their way into the hands of others, often magnifying the bloodshed in regional and civil wars. Unfortunately, the lobbying power and political influence of gun and munitions manufacturers and defense companies has contributed to this spread of weapons throughout the world. President Barack Obama reversed U.S. policy in 2009 and resumed advocacy of a treaty to limit this destructive trade, but progress is slow at best because of opposition from several countries and the dysfunctionality of global decision making.

The U.S. continues to dominate the international trade in weapons of all kinds—including long-range precision weapons and surface-to-air missiles—some of which end up being trafficked in black markets. In his final speech as president, Dwight Eisenhower warned the United States about the "military industrial complex." As the victorious commanding general in World War II, Eisenhower could hardly have been accused of being soft on national security. Although there are undeniable benefits to the United States from weapons deals, including an enhanced ability to form and maintain useful alliances, it is troubling that more than half (52.7 percent in 2010) of all of the military weapons sold to countries around the world originate in the United States.

More significantly, the dispersal of scientific and technological knowledge and expertise throughout Earth Inc. and the Global Mind has also undermined the monopoly exercised by nation-states over the .

means of inflicting mass violence. Chemical and biological agents capable of causing mass casualties are also on the list of weapons now theoretically accessible to nonstate groups.

The knowledge necessary to build weapons of mass destruction, including nuclear weapons, has already been dangerously dispersed to other nations. Instead of the two nuclear powers that faced off at the beginning of the Cold War, there are now thirty-five to forty countries with the potential to build nuclear bombs. North Korea, which has already developed a handful of nuclear weapons, and Iran, which most believe is attempting to do so, are developing longer-range missile programs that could over time result in the ability to project intercontinental power. Proliferation experts are deeply concerned that the spread of nuclear weapons to some of these countries could markedly increase the risk that terrorist groups could purchase or steal the components they need to make a bomb of their own. The former head of Pakistan's nuclear program, A. Q. Khan, developed extensive ties with Islamic militant groups. North Korea, strapped for cash as always, has already sold missile technology and many believe it is capable of selling nuclear weapons components.

National security experts are also concerned about regional cascades of nuclear proliferation in regions like the Persian Gulf and Northeast Asia. In other words, the development of a nuclear arsenal by Iran would exert pressure on Saudi Arabia and potentially other countries in the region to develop their own nuclear arsenals in order to provide deterrence. If North Korea were to gain the credible ability to threaten a nuclear attack against Japan, the pressure on Japan to develop its own arsenal would be intense in spite of Japan's historic experience and opposition to nuclear weapons.

Because leadership in the community of nations is essential, there is an urgent need to restore the integrity of democratic decision making in the United States. And there are hopeful trends, not least the awakening of reformist activism on the Internet. Throughout the world, the Internet is empowering the rapidly increasing members of the global middle class to demand the kinds of accountability and reform from their governments that middle-class citizens have historically always been more likely to demand than the poor and underprivileged. Stanford political science professor Francis Fukuyama notes that this is "most broadly accepted in countries that have reached a level of material prosperity sufficient to allow a majority of their citizens to think of themselves as middle class, which is why there tends to be a correlation between high levels of development and stable democracy."

The trends associated with the emergence of Earth Inc.—particularly robosourcing, the transfer of work from humans to intelligent interconnected machines—threaten to slow the rise of the global middle class by diminishing aggregate wages. But a recent report from the European Strategy and Policy Analysis System (ESPAS) calculates that the global middle class will double in the next twelve years from two billion to four billion people, and will reach almost five billion people by 2030.

The report adds: "By 2030, the demands and concerns of people in many different countries are likely to converge, with a major impact on national politics and international relations. This will be the result mainly of greater awareness among the world's citizenry that their aspirations and grievances are shared. This awareness is already configuring a global citizens' agenda that emphasizes fundamental freedoms, economic and social rights and, increasingly, environmental issues."

The awareness of higher living standards, higher levels of freedom and human rights, better environmental conditions, and the benefits of more responsive governments will continue to spread within the Global Mind. This new global awareness of the myriad ways in which the lives of billions can be improved is certain to exert a profound influence on the behavior of political leaders throughout the world.

Already, the spread of independence movements committed to democratic capitalism in states throughout the former Soviet Union, and the explosive spread of the Arab Spring in nations throughout the Middle East and North Africa, also serve as examples of the real possibility that such changes can occur even more quickly in a world empowered by its connections to the Global Mind.

With the ongoing emergence of the world's first truly global civilization, the future will depend upon the outcome of the struggle now beginning between the raw imperatives of Earth Inc. and the vast potential inherent in the Global Mind for the insistence by people of conscience that excesses be constrained with the imposition and enforcement of standards and principles that honor and respect human values.

Lest this sound impractical or hopelessly idealistic, there are many examples of new global norms having been established by this mechanism in the past—well prior to the enhanced potential we now have available for promoting new global norms by using the Internet. The abolition movement, the anti-Apartheid movement, the promotion of women's rights, restrictions on child labor, the anti-whaling movement, the Geneva Conventions against torture, the rapid spread of anticolonialism in the 1960s, the ban on atmospheric nuclear testing, and successive waves of the democracy movement—all gained momentum from the sharing of ideas and ideals among groups of committed individuals in multiple countries who pressured their governments to cooperate in the design of laws and treaties that led to broad-based change in much of the world.

No matter the nation in which we reside, we as human beings now face a choice: either to be swept along by the powerful currents of technological change and economic determinism into a future that may threaten our deepest values, or to build a capacity for collective decision making on a global scale that allows us to shape that future in ways that protect human dignity and reflect the aspirations of nations and peoples.

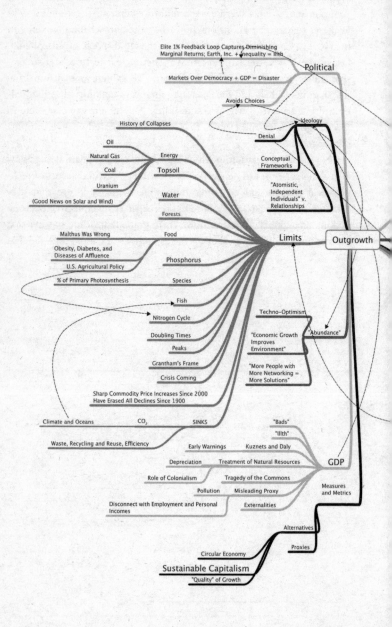

Political

Elite 1% Feedback Loop Captures Diminishing Marginal Returns; Earth, Inc. + Inequality = Illth

Markets Over Democracy + GDP = Disaster

Avoids Choices

Ideology

Denial

Conceptual Frameworks

"Atomistic, Independent Individuals" v. Relationships

History of Collapses

Oil

Natural Gas   Energy

Coal   Topsoil

Uranium

(Good News on Solar and Wind)   Water

Forests

Malthus Was Wrong   Food

Obesity, Diabetes, and
Diseases of Affluence   Phosphorus

U.S. Agricultural Policy

% of Primary Photosynthesis   Species

Fish

Nitrogen Cycle

Doubling Times

Peaks

Grantham's Frame

Crisis Coming

Sharp Commodity Price Increases Since 2000
Have Erased All Declines Since 1900

Limits   Outgrowth

Techno-Optimism

"Economic Growth
Improves
Environment"

"Abundance"

"More People with
More Networking =
More Solutions"

Climate and Oceans   CO₂   SINKS

Waste, Recycling and Reuse, Efficiency

"Bads"

"Illth"

Early Warnings   Kuznets and Daly

Depreciation   Treatment of Natural Resources   GDP

Role of Colonialism   Tragedy of the Commons

Pollution   Misleading Proxy   Measures
and Metrics

Disconnect with Employment and Personal   Externalities
Incomes

Alternatives

Circular Economy   Proxies

Sustainable Capitalism

"Quality" of Growth

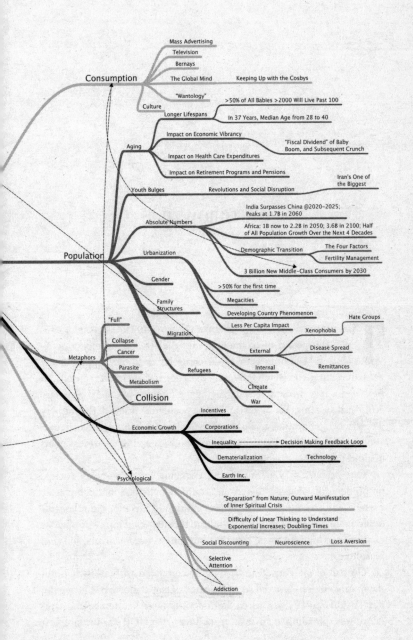

OUTGROWTH

THE RAPID GROWTH OF HUMAN CIVILIZATION—IN THE NUMBER OF PEO-ple, the power of technology, and the size of the global economy—is colliding with approaching limits to the supply of key natural resources on which billions of lives depend, including topsoil and freshwater. It is also seriously damaging the integrity of crucial planetary ecological systems. Yet "growth," in the peculiar and self-defeating way we define it, continues to be the principal and overriding objective of almost all national and global economic policies and the business plans of almost all corporations.

Our primary way of measuring economic growth—gross domestic product, or GDP—is based on absurd calculations that completely exclude any consideration of the distribution of income, the relentless depletion of essential resources, and the reckless spewing of prodigious quantities of harmful waste into the atmosphere, oceans, rivers, soil, and biosphere.

Growth of GDP used to be roughly correlated with an increase in the number of jobs and the size of average personal incomes. During the post–World War II years, when the American model of democratic capitalism was spreading, many experts believed that GDP was the simplest and most accurate measure of whether economic policy was moving in the right direction. Even then, however, the economist who had created it in 1937, Simon Kuznets, warned that it was a potentially dangerous oversimplification that could be misleading and subject to "illusion and resulting abuse" because it did not account for "the personal distribution of income" or "a variety of costs that must be recognized."

In the twenty-first century, especially since the emergence of Earth Inc., policies aimed at maximizing GDP have been driving the world toward more concentrated wealth and power, more inequality of incomes, higher long-term unemployment, more public and private debt, more social and geopolitical instability, greater market volatility, more pollution, and what biologists refer to as the Sixth Great Extinction. Some of these negative consequences are actually counted as *positive* outcomes in the functionally insane definition of growth that we still use as our compass. The heading it gives us points straight off the edge of a cliff.

The world's abject failure to acknowledge the danger to the future of civilization—and to change course—reflects the absence of coherent global leadership and the imbalance of power, as the insistent imperatives of Earth Inc. dominate decisions at the expense of participatory democracy. Even though growth of GDP no longer increases prosperity or the sense of well-being for the average person, it *is* still correlated with the incomes of elites.

The combination of Earth Inc. and the Global Mind now provides elites with an enhanced ability to manufacture consent for political decisions that serve their interests rather than the public interest—and provides corporations with an enhanced ability to manufacture wants in order to increase consumption of commodities and manufactured products. The result is rising levels of per capita consumption, with an impact that is magnified by the continuing increases in human population.

The global middle class will grow by an incredible three billion people in just the next seventeen years. And the globalization of culture on television and the Internet is linking their aspirations to living standards that no longer reflect those of their neighbors, but instead reflect standards that are more common in the wealthiest nations. That is one of the reasons why the growth in per capita consumption of food, water, meat, commodities, and manufactured goods is exceeding the rate of growth in the number of people in the world.

Earth Inc.—and its impact on ecological systems and the supply of key resources—is being driven by this combination of many more people *and* much larger per capita consumption rates. The advertiser-driven ideology that's pervasive in the Global Mind equates more consumption with more happiness. It is a false promise, of course, just like the promise that GDP growth will bring more prosperity.

The tendency to confuse increased commercial consumption with increased happiness was the subject of a letter from Thomas Jefferson to George Washington in early 1784: "All the world is becoming commercial. Was it practicable to keep our new empire separated from them we might indulge ourselves in speculating whether commerce contributes to the happiness of mankind. But we cannot separate ourselves from them. Our citizens have had too full taste of the comforts furnished by the arts and manufactures to be debarred the use of them."

Jefferson would not be surprised by the recent voluminous research into the causes of happiness, which shows that over the last half century the United States has tripled its economic output with absolutely no gain in the general public's happiness or sense of well-being. Similar results have been found in other high-consumption countries. After basic needs are met, higher incomes produce gains in happiness only up to a point, beyond which further increases in consumption do not enhance a sense of well-being.

The cumulative impact of surging per capita consumption, rapid population growth, human dominance of every ecological system, and the forcing of pervasive biological changes worldwide has created the very real possibility, according to twenty-two prominent biologists and ecologists in a 2012 study in *Nature*, that we may soon reach a dangerous "planetary scale 'tipping point.'" According to one of the coauthors, James H. Brown, "We've created this enormous bubble of population and economy. If you try to get the good data and do the arithmetic, it's just unsustainable. It's either got to be deflated gently, or it's going to burst."

In the parable of the boy who cried wolf, warnings of danger that turned out to be false bred complacency to the point where a subsequent warning of a danger that was all too real was ignored. Past warnings that humanity was about to encounter harsh limits to its ability to grow much further were often perceived as false: from Thomas Malthus's warnings about population growth at the end of the eighteenth century to *The Limits to Growth*, published in 1972 by Donella Meadows, among others.

We resist the notion that there might be limits to the rate of growth we are used to—in part because new technologies have so frequently enabled us to become far more efficient in producing more with less and to substitute a new resource for one in short supply. Some of the resources we depend upon the most, including topsoil (and some key elements, like phosphorus for fertilizers), however, have no substitutes and are being depleted.

RISING PRESSURES, CLEARER LIMITS

On every continent, the population and economy are placing new demands for more food, freshwater, energy, commodities of all kinds, and manufactured products. And worryingly, over the past ten years, multiple indicators have been showing that real physical limits are being reached.

World food prices spiked to all-time record high levels in 2008 and again in 2011. Both times, food riots and political upheavals struck several countries. Important groundwater aquifers are being depleted at unsustainable rates—especially in northern China, India, and the Western United States. Water tables are falling in countries where 50 percent of the world's people live. The unsustainable erosion of topsoil and loss of soil fertility are depressing crop yields in several important food-growing regions.

The prices of almost all commodities in the world economy have surged simultaneously in the last eleven years. After declining steadily throughout the twentieth century by an average of 70 percent—with the expected ups and downs for the Great Depression and the post–World War I depression, the two world wars, and the oil price shocks of 1973 and 1979, all of those price reductions were wiped out by price increases between 2002 and 2012—increases larger than those that accompanied either World War I or II.

Among the commodities with the fastest price increases are iron ore, copper, coal, corn, silver, sorghum, palladium, rubber, flaxseed, palm oil, soybeans, coconut oil, and nickel. An influential investor, Jeremy Grantham, warns that the growth in demand for commodities creates the danger that we may soon reach "peak everything."

The cause of these continuing price hikes is a surge of demand that reflects population increases, and even more significantly, sharply rising per capita consumption levels. This has proven particularly true in China and other emerging economies whose growth rates (since the middle of the 1990s) have been at least three times faster than those in the industrial world. China, in particular, is now consuming more than half the world's cement, nearly half of all the world's iron ore, coal, pigs, steel, and lead—and roughly 40 percent of the aluminum and copper.

Almost one quarter of the new cars being produced each year are now made in China. The largest U.S. automobile manufacturer, General Motors, now sells more automobiles in China than in its home country. In the last forty years, the world's population of cars and trucks quadrupled from 250 million to slightly over one billion in 2013. The number of cars and trucks in the world is projected to double again in the next thirty years—driving ever higher oil consumption. The production of automobiles in developing and emerging economies will overtake production in developed countries by 2015 and auto sales in the same countries will overtake those in developed economies by 2020, according to the International Energy Agency (IEA), which has also found that "All of the net growth [in the IEA's scenario, which assumes that new proposed policies to reduce emissions will be put into effect] comes from the transport sector in emerging economies."

Within the last two years, there have been some indications that consumption levels in the United States—where they are still the highest in the world—and in other developed countries may be slowing, and in some cases may have peaked. Some optimists believe that as a result, concerns about continued high growth rates may be overblown. However, even if consumption by the one billion people in the developed countries declined, it is certainly nowhere close to doing so where the other six billion of us are concerned. If the rest of the world bought cars and trucks at the same per capita rate as in the United States, the world's population of cars and trucks would be 5.5 billion. The production of global warming pollution and the consumption of oil would increase dramatically over and above today's unsustainable levels. With the increasing population and rising living standards in developing countries, the pressure on resource constraints will continue, even as robosourcing and outsourcing reduce macroeconomic demand in developed countries.

Around the same time that *The Limits to Growth* was published, peak oil production was passed in the United States. Years earlier, a respected geologist named M. King Hubbert collected voluminous data on oil production in the United States and calculated that an immutable peak would be reached shortly after 1970. Although his predictions were widely dismissed, peak production did occur exactly when he predicted it would. Exploration, drilling, and recovery technologies have since advanced significantly and U.S. oil production may soon edge back slightly above the 1970 peak, but the new supplies are far more expensive.

The balance of geopolitical power shifted slightly after the 1970 milestone. Less than a year after peak oil production in the U.S., the Organization of Petroleum Exporting Countries (OPEC) began to flex its muscles, and two years later, in the fall of 1973, the Arab members of OPEC implemented the first oil embargo. Since those tumultuous years when peak oil was reached in the United States, energy consumption worldwide has doubled, and the growth rates in China and other emerging markets portend further significant increases.

Although the use of coal is declining in the U.S., and coal-fired generating plants are being phased out in many other developed countries as well, China's coal imports have already increased 60-fold over the past decade—and will double again by 2015. The burning of coal in much of the rest of the developing world has also continued to increase significantly. According to the International Energy Agency, developing and emerging markets will account for all of the net global increase in both coal and oil consumption through the next two decades.

The prediction of *global* peak oil is fraught with controversy, largely because of uncertainty about the size of reserves yet to be discovered deep underneath the ocean floor in regions that have been difficult to access, and in unconventional sources such as the exceptionally dirty tar sands of Canada, the carbon-rich extra-heavy oil in Venezuela, and tight oil resources discovered in deep continental shale formations. Some experts are predicting that even larger new oil supplies in the U.S. will soon be produced with the same water-intensive hydraulic fracturing (commonly called fracking) techniques—combined with horizontal drilling— that have been used to exploit the newly discovered abundance of deep shale gas. Yet even if supplies are increased significantly, global demand is growing even faster—and in any case, no sane civilization would add so much additional CO_2 to the already oversaturated global atmosphere.

At current levels of growth, the global economy is now projected to require a 23.5 percent increase in the consumption of oil in less than twenty-five years—even as the marginal cost of increased supplies reaches all-time record highs, and even as the political instability in the world's largest oil-producing region threatens wars, revolutions, and the disruption of supply routes.

Global oil production from conventional wells on land actually seems to have peaked more than thirty years ago. The growth of oil production since 1982 has been of more expensive unconventional onshore sources, and particularly offshore, where production is increasingly in risky deep-water wells—like BP's Deepwater Horizon (Macondo) well in the Gulf of Mexico. Now, the same accident-prone deepwater drilling technology is being recklessly deployed in the unforgiving and environmentally fragile Arctic Ocean. And unfortunately, oil companies are also ratcheting up the political pressure to produce oil from exceptionally carbon-intensive tar sands, which would make the problem of global warming that much worse.

The projected reserves in these dirty sources, as well as in the deep reserves underneath the ocean floor, yield much more expensive oil than the world has enjoyed in the past. Even if we do not reach global peak oil in the near future, the prices we pay for oil are likely to be permanently higher than those we became accustomed to during the century and a half over which we exploited the cheaper, more easily recoverable reserves.

These higher petroleum prices have already had a big impact on food prices because modern industrial agriculture consumes prodigious quantities of diesel fuel for locomotion, and poorly contained methane for 90 percent of their fertilizer costs. According to Berkeley professor and author Michael Pollan, "it takes more than a calorie of fossil fuel energy to produce a calorie of food." No wonder the demand for both oil and food is continuing to skyrocket—particularly in fast growing emerging economies. The impact of higher food prices is significantly larger in developing countries, where low-income families frequently spend 50 to 70 percent of their income on food.

In spite of the impressive increases in food production in the last half century, and in spite of premature warnings in centuries past that humanity was reaching hard limits in its ability to provide more food for more people, many experts are nearly unanimous in pointing to multiple threats confronting the ability of the world to expand food supplies:

- Erosion of fertile topsoil at unsustainable rates; each inch of top-soil lost diminishes grain yields by 6 percent;
- Loss of soil fertility; each reduction of organic matter in soil by 50 percent reduces many crop yields by 25 percent;
- Increasing desertification of grasslands;
- Increasing competition for agricultural water from cities and in-dustry, even though agriculture is projected to require 45 percent more water by 2030;
- Slowing rate of agricultural productivity gains since the Green Rev-olution in the second half of the twentieth century—from 3.5 per-cent annually three decades ago to a little over one percent now;
- Increasing resistance of pests, weeds, and plant diseases to pesti-cides, herbicides, and other agricultural chemicals;
- Loss of a significant amount of the world's remaining plant genetic diversity; as much as three quarters of all plant genetic diversity may have already been lost;
- Increased risk of export bans by large producers facing their own domestic price hikes; according to the Council on Foreign Re-lations, data from the U.N.'s World Food Programme show that "over forty countries in 2008 imposed some form of export ban in an effort to increase domestic food security";
- Erratic and less predictable precipitation patterns associated with global warming—which leads to less frequent but larger down-pours, interrupting longer periods of deeper drought;
- Looming impact of catastrophic heat stress on important food crops that cannot survive the predicted global temperature in-crease of 6 degrees C (11 degrees F); each one degree C increase in temperature is expected by experts to produce a 10 percent decline in crop yields;
- Growing consumption of food, driven both by population growth, the growth in per capita consumption, and the increasing global preference for resource-intensive meat consumption;
- Diversion of more cropland from food crops to crops suitable for biofuel; and
- Conversion of cropland to urban and suburban sprawl.

We already know that extreme shortages of food, fertile land, and fresh-water in countries with growing populations can lead to a complete breakdown of social order and a sharp increase in violence. Studies have shown conclusively that this deadly combination was a major contributing factor in the years leading up to the 100 days of genocide in Rwanda in 1994, which at the time had one of the five highest population growth rates in the world, with 67 percent of its people under the age of twenty-four.

Jared Diamond, author of *Collapse: How Societies Choose to Fail or Succeed*, wrote, "modern Rwanda illustrates a case where Malthus's worst-case scenario does seem to have been right. . . . Severe problems of overpopulation, environmental impact, and climate change cannot persist indefinitely: sooner or later they are likely to resolve themselves, whether in the manner of Rwanda or in some other manner not of our devising, if we don't succeed in solving them by our own actions."

Several experts now worry that there is a danger that several large food-growing countries—China and India among them—will run into a wall. Were that to occur, the resulting global food shortages and price hikes could be catastrophic. In Gujarat, India, the head of the International Water Management Institute's groundwater station, Tushaar Shah, said of the looming water crisis in his region, "when the balloon bursts, untold anarchy will be the lot of rural India."

India would not be alone. For example, rapid population growth and gross overexploitation of soil, water, and other natural resources are contributing to the anarchy and increasing radicalism in Yemen. Tap water flows in the capital city of Sana'a only one day in four. Partly because of water shortages and soil erosion, the harvest of grain has declined more than 30 percent in the last four decades. Yemen is becoming, in the words of Lester Brown, president of the Earth Policy Institute, "a hydrological basket case."

THE GROWTH OF CITIES

The collective failure to clearly see the most likely future consequences of realities evolving along easily measurable trend lines also reflects a well-known human vulnerability when we try to think about the future. Neuroscientists and behavioral economists have established that we have a kind of brain glitch when it comes to making choices in the present that require an evaluation of the future. The geeky term for this glitch in our thinking is "social discounting"—which simply means that we are prone to dramatically overminimize the future effects of choices we make now.

This vulnerability is an even bigger problem for us when the particular changes we have to evaluate are part of a pattern of exponential change—the kind of change that is common in the age of Earth Inc. and the Global Mind—because we are more comfortable thinking about change as a slow, linear process. There is one exponential change in particular over the last several generations whose implications we have been slow to recognize: the change in global population.

DURING THE LAST century alone, we quadrupled the human population. By way of perspective, it took 200,000 years for our species to reach the one billion mark, yet we have added that many people in just the first thirteen years of this century. In the next thirteen years, we will add another billion, and yet another billion just fourteen years after that—for a total of nine billion souls by the middle of this century. In only thirty-seven years, our population will grow by a number equivalent to all the people in the world at the beginning of World War II. And more than 95 percent of the new additions will be in developing countries.

Moreover, 100 percent of this huge net increase in global population will take place in cities, with the largest increases in the largest cities. In all, there will be more city-dwellers in the world than the entire population of the world at the beginning of the 1990s. In fact, the population of megacities already has increased tenfold over the last forty years. In this period of hyper-urbanization, cities with less than one million people will see a reduction in their share of the world's urban population. This is a new trend that has surprised population experts, who note that it is a reversal of past urbanization patterns.

This historic transformation of human civilization from a predominantly rural pattern to a predominantly urban pattern has significant implications for the organization of society and the economy. The trend is so powerful that even with the enormous increase in overall population now under way, rural populations have leveled off and are expected to decline significantly, starting in the next decade.

Again, some perspective: for almost all of the ten millennia since the first cities were built, no more than 10 to 12 percent of people ever lived in urban areas. In the nineteenth century, the Industrial Revolution began to increase urban populations, but the percentage was still only 13 percent at the beginning of the twentieth century. By 1950, roughly a third of the world's people lived in cities, and as of 2011, for the first time, more than half of us lived in cities. Already, more than 78 percent of people in developed countries live in cities, and by 2050 that number is projected to increase to 86 percent, with 64 percent of the population in less developed countries living in cities.

Forty years ago, only two cities in the world—New York and Tokyo—had populations of 10 million people; in 2013, twenty-three cities will have more than that number. And by 2025, thirty-seven such megacities will sprawl across the Earth. The sheer geographic size of cities and their rapid expansion into surrounding rural areas that used to be primarily agricultural land is also a challenge in many nations. The sprawl is increasing even more rapidly than the population—with a projected increase of 175 percent between 2000 and 2030.

The fastest growing of the new megacities is Lagos, Nigeria—which will grow from 11 million today to just under 19 million in 2025. All five of the fastest growing cities are in developing countries. The others, in addition to Lagos, are Dhaka, Bangladesh; Shenzhen, China; Karachi, Pakistan; and Delhi, India—which is projected to have a population of almost 33 million people by 2025. The largest megacity today, Tokyo, has more than 37 million people and is expected to grow to 38.7 million people by 2025. By 2050, almost 70 percent of the world's population will be city dwellers.

One of the challenges posed by this hyper-urbanization is to the ability of municipal governments to provide adequate housing, freshwater, sanitation, and other essential needs. More than a billion people in the world live in slums today, roughly one out of every three inhabitants of cities. Without significant changes in policy and governance, the number of slum dwellers is projected to double to two billion people within the next seventeen years. The urban poor population—defined as those living on $1.25 a day or less—is growing at a rate even faster than the overall urban growth rate.

The majority of those moving to cities—especially in developing countries—do so to earn higher incomes. Even though income inequalities have been increasing within most countries, on a global basis, there has simultaneously been a historic movement of people out of poverty and into the middle class—particularly in Asia. And the vast majority of the growing global middle class will live in cities.

Already, more than 80 percent of global production takes place in cities. The per capita carbon emissions of people who live in cities are lower than that of people who live in suburbs, but in spite of the improved efficiency with which resources are used in cities, the overall per capita consumption rates in cities are significantly higher than in rural areas—primarily because incomes are higher.

In just the last thirty years, per capita consumption of meat in developing countries has doubled, while egg consumption has quintupled. The impact of skyrocketing meat consumption on topsoil, deforestation, and freshwater resources—and its production of global warming pollution and cardiovascular disease—is magnified by another factor as well: nine kilograms of plant protein are consumed in the production of one kilogram of meat protein.

HUNGER AND OBESITY

The change in diets around the world is also creating a global obesity epidemic—and in its wake a global diabetes epidemic—even as more than 900 million people in the world still suffer from chronic hunger. In the United States, where many global trends begin, the weight of the average American has increased by approximately twenty pounds in the last forty years. A recent study projects that half the adult population of the United States will be obese by 2030, with one quarter of them "severely obese."

At a time when hunger and malnutrition are continuing at still grossly unacceptable levels in poor countries around the world (and in some pockets within developed countries), few have missed the irony that simultaneously obesity is at record levels in developed countries and growing in many developing countries.

How could this be? Well, first of all, it is encouraging to note that the world community has been slowly but steadily decreasing the number of people suffering from chronic hunger.

Secondly, on a global basis, obesity has more than doubled in the last thirty years. According to the World Health Organization, almost 1.5 billion adults above the age of twenty are overweight, and more than a third of them are classified as obese. Two thirds of the world's population now live in countries where more people die from conditions related to being obese and overweight than from conditions related to being underweight.

Obesity represents a major risk factor for the world's leading cause of death—cardiovascular diseases, principally heart disease and stroke—and is the major risk factor for diabetes, which has now become the first global pandemic involving a noncommunicable disease.[*] Adults with diabetes are two to four times more likely to suffer heart disease or a stroke, and approximately two thirds of those suffering from diabetes die from either stroke or heart disease.[†]

The tragic increase in obesity among children is particularly troubling; almost 17 percent of U.S. children are obese today, as are almost 7 percent of all children in the world. One respected study indicates that 77 percent of obese children will suffer from obesity as adults. If there is any good news in the latest statistics, it is that the prevalence of obesity in the U.S. appears to be reaching a plateau, though the increases in childhood obesity ensure that the epidemic will continue to grow in the future, both in the U.S. and globally.

The causes of this surge in obesity are both simple—in that people are eating too much and exercising too little—and complex, because the manufacturing and marketing of food products has changed dramatically. Dr. David Kessler, former head of the U.S. Food and Drug Administration, has extensively documented how food manufacturers and restaurant and fast food chains carefully combine fats, sugar, and salt in precise ratios that reach the "bliss point"—which means they trigger brain systems that increase the desire to eat more, even after our stomachs are full.

---

[*] At least, it's noncommunicable by means of pathogens transferred from one person to another; research shows that it is communicable socially in families, communities, and nations in which the people one normally comes into contact with include many who are obese and overweight.

[†] Obesity is also a major risk factor for osteoarthritis and other musculoskeletal disorders, some cancers—particularly colon, breast, and endometrial—and kidney failure. Health experts estimate that the cost of treating these obesity-related diseases consumes roughly 10 to 20 percent of U.S. health care spending each year. Globally, approximately 6.4 percent of the world's adult population now has diabetes, and according to the World Health Organization that number is expected to grow to 7.8 percent in the next seventeen years, to a total of 438 million—more than 70 percent of them in low- and middle-income countries.

On a global basis, the World Health Organization has found a pattern of increased consumption of "energy-dense foods that are high in fat, salt and sugars but low in vitamins, minerals and other micronutrients."

Hyper-urbanization has separated more people from reliable sources of fresh fruit and vegetables. Quality calories in fruits and vegetables now cost ten times as much as calories per gram in sweets and foods abundant in starch. In a report for the Johns Hopkins Bloomberg School of Public Health, Arielle Traub documented the increase from 1985 to 2000 in the price of fresh fruits and vegetables by 40 percent, while prices of fats declined by 15 percent and sugared soft drinks by 25 percent. Relative price, limitation of access to healthy food, increased inactivity, and the cumulative effects of massive food advertising campaigns all contribute to the obesity epidemic.

Several studies indicate that low-income neighborhoods have less access to supermarkets with a selection of fresh fruits and vegetables and are more likely to have fast food chains and convenience stores selling Slim Jims and Big Gulps than middle- and higher-income neighborhoods. Relative income and the time and knowledge necessary for food preparation both also play a role. Once eating habits are established, they are harder to change. When the U.S. government introduced healthier foods into the school lunch program in 2012, students at many schools launched protests on social media and threw the healthier food away.

In many countries, there is an almost precise correlation between the introduction of American fast food outlets and climbing obesity rates. One of the factors that led to the surge in fast food, manufactured food, and increasing portion sizes was a historic change in U.S. agricultural policy in the 1970s, at exactly the time when obesity rates began their climb. Instead of compensating farmers to withdraw land from production, as had been the case since FDR's New Deal, the government subsidized farmers to grow as much as they possibly could. This policy change coincided with new advances in agriculture technology, including better hybrid seeds coming out of the Green Revolution. Consequently, food prices went down significantly. Dr. Carson Chow, a mathematician working at the National Institute of Diabetes and Digestive and Kidney Diseases, constructed a detailed mathematical model that strongly suggests that the changes in U.S. agricultural policy correlate precisely with the large average weight gains and increased obesity.

The advertising industry has played a major role. One fast food hamburger chain, to pick only one example, famously used in its television advertisements a skimpily clad sex symbol washing a car in a suggestive manner. The advertising budget for manufactured food items and fast food chains is already two thirds that for automobiles. And again, these interrelated trends may have started in the U.S., but have now spread around the world. The impact of obesity on the world's resources is the equivalent of adding an extra one billion people to the planet.

THE ORIGINS OF MASS MARKETING

The rising rate of consumption in the world is a relatively new phenomenon, less than a century old, and is also a trend that started in the United States. Although mass advertising began to emerge in the late nineteenth and early twentieth centuries, most historians date the true beginning of consumer culture to the 1920s, when the first mass electronic medium, radio, was introduced in the United States, along with the first national circulation magazines and the first silent films shown in theaters. Significantly, consumer credit also became more widely available during the Roaring Twenties to help buyers finance the purchase of relatively expensive new products like automobiles and radios.

Electricity, which was available in less than one percent of American households at the beginning of the twentieth century, rose to almost 70 percent of U.S. homes by the end of the 1920s. The technology of mass production with interchangeable parts and early forms of automation (all forerunners of today's Earth Inc.) began to decouple productivity from increased employment and produced a cornucopia of consumer goods that stimulated a keen interest among manufacturers and merchants in the emerging science of mass marketing. The advertising industry entered a new and distinctly different role in the marketplace.

It was at precisely this moment in history that the ideas of Sigmund Freud became popular in the United States. Freud's first trip to America was in 1909, to deliver a series of five lectures on psychoanalysis at Clark University in Worcester, Massachusetts, to an audience that included William James (whose young protégé, Walter Lippmann, was greatly influenced by Freud) and many of the other most prominent intellectuals in America. Throughout the following decade, ideas popularized by Freud—like the role of the subconscious in understanding human motivation, psychological transference, and other insights from psychoanalysis—spread, particularly on the East Coast, where the advertising industry was and is located. The American Psychoanalytic Society was founded two years after Freud's visit.

By the time the United States entered World War I in 1917, these psychological concepts had been adapted into techniques of mass persuasion that were used during the war effort. Woodrow Wilson established a Committee on Public Information. Sigmund Freud's nephew, Edward Bernays, served on the committee, alongside Walter Lippmann, only two years his elder, whose influence on Bernays almost rivaled that of his uncle Sigmund. After the war, Bernays pronounced himself astonished at the effectiveness of mass propaganda and set out to introduce the techniques into mass marketing.

Known as the "father of public relations," Bernays actually coined the phrase "public relations" in order to avoid using the word "propaganda," which had acquired a negative connotation in the U.S. due to its frequent use by Germany to describe its mass communications strategy during the war. Bernays revolutionized the field of marketing research by discarding the then standard technique of asking consumers what they liked and disliked about various products. Instead, Bernays spent time with psychoanalysts and conducted deep interviews with people designed to uncover the associations they made in their subconscious minds that might be relevant to the marketing of products and brands. Bernays's business partner, Paul Mazur, said, "We must shift America from a needs to a desires culture. . . . People must be trained to desire, to want new things, even before the old have been entirely consumed. We must shape a new mentality. Man's desires must overshadow his needs."

As Bernays later wrote, in 1928,

> the conscious and intelligent manipulation of the organized habits and opinions of the masses is an important element in democratic society. Those who manipulate this unseen mechanism of society constitute an invisible government that is the true ruling power of this country. We are governed, our minds molded, our tastes formed, our ideas suggested, largely by men we have never heard of. This is a logical result of the way in which our democratic society is organized. . . . In almost every act of our daily lives, whether in the sphere of politics or business, in our social conduct or our ethical

thinking, we are dominated by the relatively small number of persons . . . who understand the mental processes and social patterns of the masses. It is they who pull the wires which control the public mind.

In one of his early successes, Bernays tackled a problem for his client, the American Tobacco Company: how could he break down the social taboo against women smoking cigarettes? He hired a group of women to dress as suffragettes and march in formation in a parade down Fifth Avenue in New York City on Easter Sunday, 1929. When they reached the section of elevated seats reserved for the press, the faux suffragettes all pulled out cigarettes, lit them up, and proclaimed them to be "freedom torches." Decades later, the iconic cigarette advertisement aimed at women—"You've come a long way baby"—was still using Bernays's innovative but sinister association of smoking with women's rights.

In 1927, a prominent American business advisor, Edward Cowdrick, wrote that stimulating consumption had become more important than production: "the worker has come to be more important as a consumer than he is as a producer . . . not to manufacture and mine and raise enough goods, but to find enough people who will buy them—this is the vital problem of business." He described this fresh macroeconomic conventional wisdom as the "new economic gospel of consumption."

His use of the word "gospel" was not as casual as it may sound today. The struggle between capitalism and communism had taken on a new significance in the wake of Lenin's successful revolution in Russia ten years earlier and the establishment of the USSR. During the long struggle between capitalism and communism in the twentieth century, unlimited *growth* was the one assumption built into both ideologies that neither questioned.

In 1926, President Calvin Coolidge in a speech to advertisers ventured into the same sacred territory that Cowdrick had described as a new economic gospel: "Advertising ministers to the spiritual side of trade. It is a great power that has been entrusted to your keeping which charges you with the high responsibility of inspiring and ennobling the commercial world. It is all part of the greater work of regeneration and redemption of mankind."

Three years later, two months before the stock market crash of 1929,

Coolidge's successor as president, Herbert Hoover, issued the report of his Committee on Recent Economic Changes, which took note of the newly recognized power of psychology in mass marketing: "The survey has proved conclusively what has long been held theoretically to be true, that wants are almost insatiable; that one want satisfied makes way for another. The conclusion is that economically, we have a boundless field before us; that there are new wants that will make way endlessly for newer wants, as fast as they are satisfied . . . by advertising and other promotional devices, by scientific fact finding, by a carefully predeveloped consumption a measurable pull on production has been created . . . it would seem that we can go on with increasing activity."

In the 1930s, another Freudian psychoanalyst from Vienna, Ernest Dichter, immigrated to the U.S. and began working on mass marketing. Fully aware of the popularity of Freudian concepts in the advertising business, he told potential customers on Madison Avenue and Wall Street that he was not only a "psychologist from Vienna" but that he had lived on the very same street as Sigmund Freud. He promised them that he could help them "sell more and communicate better." And, like President Coolidge, he saw the importance of stimulating more mass consumption as a means of strengthening America's economy in the struggle to ensure the triumph of capitalism. "To some extent the needs and wants of people have to be continuously stirred up," Dichter said.

Inevitably, the new power of psychology-based mass electronic marketing had an enormous impact on the democracy sphere as well as the market sphere. Bernays and Lippmann had both always predicted it would. But in the desperate and dangerous interwar period in Europe, these new powers were put in the service of totalitarianism. In 1922, Joseph Stalin became general secretary of the Communist Party in the USSR and Benito Mussolini became the fascist prime minister of a coalition government in Italy. Six months earlier, Adolf Hitler had become the chairman of the National Socialist Party in Germany.

Fifteen years later, after the Nuremburg Laws and the opening of the first concentration camps, Edward Bernays was dismayed by the eyewitness report of a recent visitor to Berlin who told him that Joseph Goebbels was making intensive use of Bernays's book *Propaganda* in organizing Hitler's genocide.

In the U.S., also in 1922, Bernays's friend and former wartime propaganda colleague, Walter Lippmann, wrote:

> The manufacture of consent . . . was supposed to have died out with the appearance of democracy. But it has not died out. It has, in fact, improved enormously in technique. As a result of psychological research, coupled with the modern means of communication, the practice of democracy has turned a corner. A revolution is taking place, infinitely more significant than any shifting of economic power. . . . The knowledge of how to create consent will alter every political calculation and modify every political premise. . . . It is no longer possible, for example, to believe in the original dogma of democracy.

As noted in the previous chapter, the combination of unlimited secret campaign contributions and extremely expensive but devastatingly effective psychology-based mass electronic marketing is indeed posing a deadly threat to the continued vibrancy and good health of participatory democracy. If the current assault on the integrity of democracy is allowed to continue, Lippmann's dark prophecy may yet come to pass; if elites can use money, power, and mass persuasion to control the policies of the United States, the average person may eventually come to a point where it seems, in Lippmann's words, "no longer reasonable" to believe that America is a democracy.

In the market sphere, the amount of money spent to "manufacture wants" and stimulate increased consumption has continued to rise year by year. The appeal of Freudian-based mass marketing began to wane later in the twentieth century, but more recently the invention of more sophisticated techniques such as brain scans has reinvigorated the use of subconscious analysis in the field of neuromarketing. Mass marketing to promote increased consumption is now so pervasive that we almost consider it to be a normal part of our environment. The average person living in a city used to see an average of 2,000 commercial messages per day thirty-five years ago. According to *The New York Times*, the average city dweller now sees 5,000 commercial messages per day.

WASTE AND POLLUTION

Increased per capita consumption by a larger and larger global population is pressing against the limits of some resources. As both human population and the global economy grow in size, we are not only consuming more natural resources to make products, we are also producing larger and larger streams of waste. According to a recent report from the World Bank, the per capita production of garbage alone from urban residents in the world is now 2.6 pounds per person every day; and the total volume is projected to increase by 70 percent in a dozen years.

The cost of managing the garbage will almost double over the same period to $375 billion per year—with most of the increase in developing countries. According to the Organisation for Economic Co-operation and Development, each one percent increase in national income produces a .69 percent increase in municipal solid waste in developed countries.

And that's just the garbage. When the waste associated with energy production, the making of chemicals, manufacturing, electronic goods, agricultural waste, and waste from the paper products industries are apportioned on a per capita basis among the seven billion people who consume the results of all these processes, the actual amount of waste produced each day is more than the body weight of all seven billion people.

There is a thriving black market in the illegal disposal of waste—particularly shipments from developed countries to poor countries. In the European Union, exports of plastic waste—almost 90 percent of it to China—increased by more than 250 percent in the last decade. The news media has focused on the enormous "garbage patch" in the middle of the Pacific Ocean—made up mostly of plastic—but much larger volumes are on land in millions of waste dumps.

Although there have been some laudable efforts by many companies and cities to increase the recycling of waste, the total volumes are overwhelming the current capacity for responsible disposal practices. For example, organic waste can be used to produce valuable methane, but due to inertia and an absence of leadership, so much organic waste is simply dumped in unimproved landfills that it decomposes to produce 4 percent of all the global warming pollution each year.

The growing volumes of e-waste (waste associated with electronic products) have been the focus of increasing attention because of the presence of highly toxic materials in the waste stream. And once again, even though recycling efforts are under way, the problem is still growing faster than the solution.

Toxic chemical and biological waste poses a particular challenge. During the 1970s and 1980s, I chaired and participated in a large number of congressional hearings on the dangers of toxic chemical waste. The tough laws that were enacted in the wake of those and other hearings have since been severely weakened by chemical industry lobbying of the U.S. Congress and executive branch. A recent U.S. study by the Centers for Disease Control and Prevention identified traces of 212 chemical wastes in the average American, including pesticides, arsenic, cadmium, and flame retardants.

Flame retardants? Their presence in the tissues of Americans has an interesting backstory that provides another example of the imbalance of power in U.S. decision making and the dominance of corporate interests over the public interest. An exhaustive examination by the *Chicago Tribune* in 2012 demonstrated in detail how the cigarette industry corruptly influenced policymakers to legally compel the addition of toxic flame retardants to the foam inside most furniture in order to save lives that were being lost in thousands of fires started each year by smokers who fall asleep and drop their lit cigarettes on a couch or chair.

A far more logical and less dangerous solution—one that had been proposed since the early part of the twentieth century—would have been to require the cigarette manufacturers to *remove* the chemicals they routinely add to cigarettes to keep them burning even when they are not being puffed. But the tobacco industry did not want to be blamed for the fires, and they worried that any inconvenience for smokers might hurt sales. So they came up with a corrupt scheme to buy enough influence to require the addition of dangerous chemicals to most all furniture instead.

When the companies manufacturing the flame retardant chemicals began to understand how they benefited from this ruse, they joined in providing more money to support the tobacco companies' scheme. The same lobbyist represented state fire officials and the chemical manufacturers—and remained secretly on the payroll of the tobacco industry. Meanwhile, children continue to breathe in the dust from the decaying flame retardants and scientists continue to link their exposure with evidence of cancer, reproductive disorders, and damage to fetuses. And, by the way, the Consumer Product Safety Commission recently found that the flame retardants added to the foam in the furniture didn't work to cut down on house fires.

A few particularly dangerous chemicals, such as Bisphenol A (BPA) and phthalates (which are chemically similar to flame retardants), have been singled out for special attention by health experts, but the law enacted in 1976 to deal with such chemicals, the Toxic Substances Control Act, has never been truly implemented. An estimated 83,000 chemicals are listed in the inventory of substances that should be tested, but the Environmental Protection Agency has required testing for only 200 of them, and has restricted usage of only 5. The chemical manufacturers are allowed to withhold most of the medically relevant information about these chemicals from regulators, by claiming they are trade secrets.

THE SURGE IN the development of agricultural and industrial chemicals following World War II was based in significant part on leftover stockpiles of unused nerve gas and munitions. (The inventor of poison gas in World War I was also the inventor of synthetic nitrogen fertilizer.) These new kinds of chemical compounds introduced more toxic forms of water pollution than in the past. In prior periods, water pollution had been dominated by fecal contamination causing typhoid and cholera. Although the latter problems have been largely solved in developed countries, waterborne diseases are still among the leading causes of death in developing countries, especially in South Asia, Africa, and portions of the Middle East.

Indeed, pollution of rivers, streams, and groundwater aquifers is a serious problem contributing to water shortages in many areas of the world. The World Commission on Water for the Twenty-First Century, in which multiple United Nations agencies participate, reported in 1999 that "More than one-half of the world's major rivers are being seriously depleted and polluted." One of the reasons for this global tragedy is that neither depletion nor pollution is included in the world's prevailing system for measuring national income and productivity—GDP. As economist Herman Daly points out, "We do not subtract the cost of pollution as a bad, yet we add the value of pollution cleanup as a good. This is asymmetric accounting." As a result, decisions to clean up the environment are routinely—and inaccurately—described as hurtful to prosperity. For example, in Guangzhou, China, the vice director of the city's

planning agency felt forced to defend a decision to limit automobile traffic as a means of reducing dangerous levels of air pollution by saying, "Of
course from the government's point of view, we give up some growth,
but to achieve better health for all citizens, it's worth it."

Recently, an investigation by *The New York Times* collected hundreds
of thousands of state and federal records of water pollution under the
Freedom of Information Act that showed that approximately one out of
every ten Americans has been exposed to chemical waste or other health
threats in their drinking water.

Since 1972, the United States has pioneered clean water protections and most of the developed world has followed suit. However, the
progress in developing countries has fallen short of the 2000 Millennium
Development Goals (a blueprint for global development agreed to by
all 193 member states of the United Nations and 23 international organizations). According to the World Health Organization, "over 2 billion
people gained access to improved water sources (defined as 'likely to
provide safe water') and 1.8 billion people gained access to improved
sanitation facilities between 1990 and 2010 . . . [however] over 780 million people are still without access to improved sources of drinking water
and 2.5 billion lack improved sanitation."

If current trends continue, these numbers will remain unacceptably
high in 2015: according to the World Health Organization, "605 million
people will be without an improved drinking water source and 2.4 billion people will lack access to improved sanitation facilities." In China,
where 90 percent of the shallow groundwater contains pollution, including chemical and industrial waste, 190 million Chinese become ill each
year due to their drinking water, and tens of thousands die.

Supplies of freshwater are unevenly distributed, with more than half
of the total located in only six countries. The declining availability and
deteriorating quality of freshwater in numerous countries and regions
stands alongside the loss of topsoil as one of the two most serious limitations constraining the expansion of food production. Overconsumption
and profligate waste of freshwater—new competition for water from cities and the growing demands of Earth Inc.—are threatening to create
food crises in multiple areas of the world.

Just as urban sprawl has had an impact on the supply of agricultural
land, "energy sprawl" is also having a harsh impact on the availability

of freshwater for food crops. The unwise decision to promote the rapid growth of first generation ethanol fuels and biodiesel from palm oil has reallocated both water and land resources from food crops. And the growing craze for deep shale gas, which requires five million gallons of water per well, has put severe strain on supplies in regions that were already experiencing shortages. Many cities and counties in Texas, for example, have now been forced to choose between allocating water to agriculture and hydraulic fracturing (fracking) of gas and oil. On a global basis, the use of water for energy production is projected to grow twice as fast as energy demand.

The expansion of oil and gas fracking is adding to the injection of toxic liquid waste into areas deep underground that have been thought to be safe repositories—until recently. In the United States, an estimated 30 *trillion* gallons of toxic liquid waste have been injected into more than 680,000 wells into underground storage over the past few decades, even as the practice of fracking changes the underground geology, opening new fissures and modifying underground flow patterns. Unfortunately, some of these deep repositories have leaked waste upward into regions containing drinking water aquifers.

Groundwater resources represent approximately 30 percent of all the freshwater resources in the world, compared to one percent represented by all of the surface freshwater. In the last half century, the rate of shrinkage in groundwater aquifers has doubled. The rate at which groundwater withdrawals have been increasing has accelerated steadily over the last half century to double the rate in 1960, but in the last fifteen years (since the growth rates of China and other emerging economies have accelerated), the increases have proceeded at a much faster pace.

The introduction of new water drilling and pumping technology has also been a significant factor. In India, for example, $12 billion has been invested in new wells and pumps; more than 21 million wells have been drilled by the 100 million Indian farmers. Partly as a result, the aquifers in many communities have been completely dried up and drinking water has to be trucked in—while farmers must rely on increasingly unpredictable rainfall.

Because of the growth in population and the increase in water consumption, the surface water from many of the world's important rivers is now so overallocated that many of the rivers no longer reach the sea:

the Colorado, the Indus, the Nile, the Rio Grande, the Murray-Darling in Australia, the Yangtze and Yellow rivers in China, and the Elbe in Germany.

A SWELLING POPULATION

Although population growth rates have slowed in most of the world over the past several decades, the overall size of the population is now so large that even a slower rate of growth will add billions more people before our numbers stabilize near the end of this century at a total that is now difficult to predict but is estimated at between 10 and 15 billion. (There is also a low projection of 6.1 billion people, and a runaway projection of 27 billion people—that would occur if there are no further decreases in fertility. But the vast majority of experts assume that the most likely range is slightly above 10 billion.)

In the next dozen years, India will surpass China for at least the balance of the century as the most populous nation on Earth. In the next *two* dozen years, Africa will have more people than either China or India and by the end of the century is projected to have more people than both combined. Half of the global growth in population over the next four decades is projected to take place in Africa, which is projected to more than triple its present population, to an astonishing 3.6 billion by the end of this century. Given the dangerously low levels of soil fertility in much of Sub-Saharan Africa, shortages of freshwater, poor governance in many countries, and the projected impact of global warming, the limits to growth in Africa are likely to be a central focus of the world's attention in the balance of the twenty-first century.

The reason it is so difficult to predict peak global population, and the reason that the range of estimates varies by five billion people—as many as all the people in the world at the end of the 1980s!—is that it is inherently difficult to predict how many children the average woman will prefer to have during the next several decades. An increase in that key variable by even a half a child (demographers have long since become numb to the discomfiture of that expression) can mean the difference in several billion people in world population in the course of the next eighty-seven years. The multiple factors that have an impact on women's preferences are, in turn, also difficult to project over such a long period of time.

These new higher projections for peak global population in the latter part of the twenty-first century reflect a slower than expected decline in the average fertility rate in scores of less developed countries—the majority of them in Africa. The biggest single reason for the increased population estimates for Africa and the world is the failure of the world community to make fertility management knowledge and technology available to women who wish to use it.

Population and development specialists have learned a great deal over the last few decades about the many factors that actually drive changes in the dynamics of population growth. Voluminous research has shown conclusively that four elements of the population puzzle all fit together and act in combination to shift the pattern of population growth in any country from one equilibrium—characterized by high death rates, high birth rates, and large families—to a second equilibrium—characterized by low death rates, low birth rates, and small families.

The good news is that the global effort to slow population growth is actually a success story, albeit one that is unfolding in slow motion. Even though very large increases in our absolute numbers will continue for many decades to come, almost every nation in the world has been moving from the high equilibrium state toward the low equilibrium state. Some have changed quickly but others are lagging behind. In the U.S., the rate of growth in population has slowed to the lowest level since the Great Depression.

For several decades in the twentieth century, the prevailing view was that increases in GDP—particularly those factors associated with industrial development—were the key to falling population growth rates. This was another early illustration of the seductive convenience and illusory simplicity of GDP as a proxy measurement of generalized progress and how it can capture the attention of policymakers, even when it is only loosely connected to the real goals they are trying to reach.

Although GDP is *not* one of the four factors, economic growth is loosely correlated in many countries with the creation of social conditions that can and do have an impact on population. And conversely, in most instances, extreme poverty is certainly correlated with higher population growth rates—especially in countries with failing institutions and shortages of clean water and topsoil. All fourteen nations with those three characteristics have extremely high population growth rates; thirteen of the fourteen are in Sub-Saharan Africa.

The four relevant factors, all of which are necessary but none of which, by itself, is sufficient, are:

- First, the education of girls—the single most powerful factor. The education of boys is also important, but population statistics show clearly that the ability of girls to become literate and to obtain a good education is crucial.
- Second, the empowerment of women in society, to the point where their opinions are heard and respected, and they have the ability to participate in making decisions with their husbands or partners about family size and other issues important to their families.
- Third, the ubiquitous availability of fertility management knowledge and technology, so that women can effectively choose how many children they wish to have and the spacing of their children.
- Fourth, low infant mortality rates. As an African leader, Julius K. Nyerere, said midway through the twentieth century, "the most powerful contraceptive is the confidence by parents that their children will live."

The struggle to provide access to contraception and the knowledge of how to manage fertility has not gone as well as social scientists and population experts had hoped. The commitments by wealthy countries to finance wider access in poor countries to fertility management have not been fully met. In some developed countries where democracy is being weakened, like the United States, attacks on programs beneficial to women have been more successful in recent years. Political opposition to contraception, for example, has surprisingly reemerged in the United States in the last two years, even though the overwhelming majority of American women (including 98 percent of sexually active Catholic women) support it and even though it seemed to be a question that had been settled in the 1960s.

The religion-based opposition to contraception by a tiny minority in the U.S. has also had a harsh impact on U.S. contributions to the global effort to make fertility management available in fast growing developing countries—in part because of the disingenuous conflation of contraception and abortion. Since foreign aid is always vulnerable to budget cuts in the United States, the amount of help actually provided has fallen far short of the amount pledged. And once again, the imbalance of power and political paralysis in the U.S. has deprived the world of desperately needed leadership, which, in turn, has seriously damaged the world's ability to take action.

Partially as a result, anticipated declines in fertility rates have not been achieved—especially in Africa, where thirty-nine out of the fifty-five African countries have high levels of fertility. (There are nine high-fertility countries in Asia, six in Pacific Island nations, and four in the low-income countries of Latin America.) In thirty-four of the fifty-eight high-fertility countries in the world, population will triple in the balance of this century.

On a global basis, women now have an average of 2.5 children during their childbearing years. In Africa, however, the average is almost 4.5 children per woman. Moreover, in four African countries, the average woman is still expected to have more than six children, leading to disruptive and unsustainable population growth. Malawi, for example, which has 15 million people today, is projected to have a nearly tenfold increase in population by the end of the century to an estimated 129 million. The African nation with the largest population, Nigeria, is projected to increase from slightly more than 160 million people today to more than 730 million by 2100. That would put Nigeria's population at the level of China in the mid-1960s.

Before the improved understanding of population dynamics, many people assumed that higher death rates reduce overall population. But the impact of high death rates on high birth rates gives the lie to this former belief. The Black Death of the fourteenth century did reduce population—indeed, that is believed to be the last time population has declined. But in today's world, even the most feared diseases have not had an impact on population. The HIV/AIDS epidemic has had an impact on the overall numbers of people in a few African countries, but on a worldwide basis population grew by more in the first five months of 2011 than all of the deaths from HIV/AIDS since the disease first began rapidly spreading three decades ago.

In countries with high rates of infant mortality, the natural tendency of parents is to have more children in order to ensure that at least some of them will survive to take care of their parents in old age and to carry on the family name and tradition. In practice, when child death rates fall dramatically, birth rates generally decline a half generation later—provided the other three factors are also present. Following World

War II, revolutionary advances in health care—higher levels of sanitation, better nutrition, antibiotics, vaccines, and other achievements of modern medicine—led to significant declines in child and infant mortality in many countries around the world. This same combination of improvements in health care and nutrition has doubled life expectancy in industrial countries from the beginning of the nineteenth century—from thirty-five to seventy-seven years.

AN AGENDA FOR WOMEN AND GIRLS

The education of girls has become commonplace around the world, including in most countries that used to focus only on the education of boys. Although there is still opposition to the education of girls among groups such as Afghanistan's Taliban, most nations have long since become aware of the competitive advantages, especially in the information age, of educating all their children. Saudi Arabia used to focus on boys alone in their school systems, but according to the most recent statistics available, almost 60 percent—compared to 8 percent in 1970—of college students in Saudi Arabia were women.

The comparable figure in Qatar is 64 percent, in Tunisia and the United Arab Emirates, 60 percent; the average in Arab states is now 48 percent; in Iran 51 percent. Indeed, more women than men received college degrees in 67 of the 120 nations for which statistics are available. The world average is 51 percent. In the United States, women now receive 62 percent of associate's degrees, 58 percent of bachelor's degrees, 61 percent of master's degrees, and 51 percent of doctoral degrees.

The *empowerment* of women, on the other hand, remains a challenging goal in many traditionalist societies. None of those women college graduates in Saudi Arabia, for example, is allowed to drive a car—or vote—although their relatively progressive king has announced plans to allow women to vote beginning in 2015. Even though 93 percent of the gender gap in education of women has been closed on a global basis, less than 60 percent of the gap in economic participation and only 18 percent of the gap in political participation have been closed.

The Global Mind has accelerated demands for the empowerment of women throughout the world; women make up more than half of all social network users globally, and almost half of all Internet users. When they are exposed to the more favorable norms of gender equity in advanced countries, they naturally become impatient for change.

Women have been coming into the workplace in almost all countries in higher numbers than men, reflecting a historic change in global attitudes about women working outside the home. In fact, for the last forty years, two women have entered the workplace for every man. Women have made a particular difference in the competitiveness of rapidly growing East Asian nations, with 83 women in the workforce for every 100 men. They have had the biggest impact in several export-oriented businesses, including clothing and textiles—filling between 60 to 80 percent of the jobs.

The Economist has calculated that on a global basis, "the increased employment of women in developed economies has contributed much more to global growth than China has." In developed countries as a group, women are responsible for producing slightly less than 40 percent of GDP. However, another flaw in the way GDP is measured—one noted by Kuznets when he first introduced it in 1937—is that it does not assign any economic value to the work women (and some men) do in the home: raising children, cooking meals, keeping the household, and all the rest. If this housework in developed countries were valued at the amount that would be paid for nannies, cooks, and housekeepers, the total contribution of women to GDP would be well over 50 percent.

The movement of women into jobs outside the home has had startling social impacts. In the United States, during the three decades between the 1960s and 1990s, the percentage of married women with children younger than six years old who work outside the home skyrocketed from 12 percent to 55 percent. The percentage of all mothers with young children who chose to work outside the home rose during the same three decades from 20 to 60 percent.

These sociological changes are also among the many factors contributing to the obesity epidemic. Because many more mothers are now working outside the home, and a much higher percentage of children live in families where both parents work, more people eat fast food, other restaurant food, and manufactured meals and food products designed for minimal preparation, such as microwaving. Portion sizes have also increased along with the body mass index. It all adds up to what Kessler calls "conditioned hyper-eating."

Studies also show that children in low-income neighborhoods are often permitted and even encouraged by their parents and caregivers to watch more television than average because of greater concerns about their safety playing outside in neighborhoods that, relatively speaking, are prone to more violence. This mirrors a global trend for people of all ages, who are spending more time on electronic screens connecting to the Global Mind and are, on average, more likely to work in jobs that do not require as much physical activity as in the past. The trend toward more driving and less walking is also a factor.

THE CHANGING FAMILY

The increased participation of women in the workforce, the dramatic changes in the education of women, and changes in social values have also led to significant structural changes in the institution of the family. Divorces have increased dramatically in almost every part of the world, partly due to new legislation making them easier to obtain, and, according to experts, partly because of the increased participation of women in the workforce. Some experts also note the role of online relationships; according to several analyses, between 20 and 30 percent of all divorces in the U.S. now involve Facebook.

The age at which the average woman gets married has also increased significantly, and a larger percentage of men and women choose never to marry at all. Fifty years ago, two thirds of all Americans in their twenties were married. Now, only one quarter are. Many more couples are living together—and having children—without getting married. Forty-one percent of children in the U.S. are now born to unmarried women. Fifty years ago, only 5 percent of U.S. children were born to unmarried mothers. Today, the comparable figure among mothers under thirty is 50 percent. Among African American mothers of all ages, the percentage is now 73 percent.

In the overall rankings of countries on the basis of gender equity, the four highest are Iceland, Norway, Finland, and Sweden; the lowest rank goes to Yemen. However, the political participation of women has lagged far behind most other indicators of gender equity. On a global basis, women make up less than 20 percent of elected parliaments, with the highest percentage (42 percent) in the Nordic countries and the lowest percentage (11.4 percent) in the Arab states. The United States is barely above the global average. Only two countries in the world have a female majority in their parliaments—one of the tiniest countries, Andorra, and one of the poorest, Rwanda, which in the wake of its 1994 tragedy enacted a constitutional requirement that a minimum of 30 percent of its parliament be women. The empowerment of women in corporate governance is lower still—with only 7 percent of corporate boards in the world made up of women.

All four of the factors that bring about a reduction in population growth rates are connected to the expansion of participatory democracy and the right of women to vote. In those countries where women vote in high percentages, there is understandably more support for programs that reduce child mortality, educate girls, further empower women, and ensure high levels of access to fertility management.

In most wealthy industrial countries, birth rates have fallen so swiftly that some now have declining populations. Russia, Germany, Italy, Austria, Poland, and several other countries in Eastern and Southern Europe now have fertility rates well below the replacement rate. Japan, South Korea, China, and several countries in Southeast Asia have also fallen below the replacement rate. The U.S. birthrate fell to an all-time low in 2011.

In a few of these countries, the fertility rate has fallen so low that they are in danger of falling into what demographers call the fertility trap. That is, fewer women of childbearing age will themselves have fewer children, adding up to a sudden and sharp further decline in population. Japan's population is projected to decline from 127 million today to 100 million by midcentury, and 64 million by 2100.

Sweden and France both adopted policies some years ago to increase fertility and avoid the fertility trap; both countries spend roughly 4 percent of their national income on programs that support families and make it easier for working parents to have children if they wish: generous maternity and paternity leave, free preschool, affordable high-quality child care, excellent child and maternal health care, protections for women returning to their career paths after having children, and other benefits. Both countries are now once again nearly at their replacement rate of fertility.

By contrast, Japan and Italy have failed to provide such services and have not yet been able to slow their fertility declines. As a result, they will soon face great difficulty in financing pensions because of a dramatic change in the *ratio* of their working-age population to their retired population. Social contracts that are based on financing mechanisms that tax work to pay for retirement are far more burdensome for working people when there are many fewer of working age compared to the number who are retired.

LONGEVITY

To a greater or lesser degree, this new demographic reality is a major cause of the budget crises in most developed countries in the world today. Similarly, since health care is used more intensively by older people, the same demographic changes have contributed to developed country budget crises for health care programs—most of all in the United States, because of the greater per capita expense of U.S. health care compared to any other nation.

The relative size of the retiree population is also increasing because of a significant increase in average lifespans almost everywhere. Incredibly, more than half of the babies born in developed countries after the year 2000 are projected to live past the age of 100. In the United States, more than half of the babies born in 2007 will live to be more than 104.

This revolution in human longevity is causing dramatic adjustments throughout the world. Although statistics are hard to come by, anthropologists believe that the average human lifespan for most of the last 200,000 years was probably less than thirty years; some believe much less. After the Agricultural Revolution and the building of cities, lifespans began climbing slowly upward, but not until the middle of the nineteenth century did the average lifespan reach forty. In the last 150 years, however, average lifespans worldwide have climbed to sixty-nine—and in most industrial countries are now in the high seventies.

Improvements in sanitation, nutrition, and health care—particularly the introduction of antibiotics, vaccines, and other modern medicines—have played the most important roles in increasing lifespans. But education levels, literacy, and the distribution of information about health care have also had major impacts. Access to information online about health and wellness has also begun to play an even more significant role. Globalization and urbanization have magnified these factors in some countries, leading to even more rapid increases in longevity. China is projected to double the percentage of its population represented by those aged sixty-five and older within the next quarter century.

The larger ratio of older people in some countries represents only one illustration of how changes in societies can be driven not only by the absolute size of populations but also by changes in the distribution in different age groups. When a baby boom generation comes into the workforce, societies with an ample number of jobs to be filled can experience enormous productivity gains. Yet years later, when that same generation ages, they are sometimes less able to adapt quickly to new technology and new demands for flexibility in the workforce, as in the age of Earth Inc. If a subsequent decline in fertility leads to smaller successive generations entering the workforce to replace them, the same cohort that clamored for revolutionary change in their youth start clamoring for bigger pension checks and better health care in their old age.

China enjoyed an economic boom for the last three decades, powered by a young workforce. Yet, within the next two years, China's working age population will begin to decline, and by 2050 fully one third of Chinese will be sixty or older. Similarly, the percentage of the Indian population over sixty-five will double during the same thirty-seven years, though the percentage of the elderly will still be half that in China.

Japan had a remarkable economic boom when its workforce was predominantly young, but its slowdown over the last two decades has coincided with the aging of its population. In 2012, the Japanese bought more adult diapers than baby diapers. By midcentury its median age, the world's oldest, in 2012, at forty-three, will be fifty-six. Globally, the median is projected to increase from twenty-eight today to forty by midcentury.

Whenever there is an unusually large generation of young people compared to the rest of a society, the so-called youth bulge can also contribute to disruptive and even revolutionary pressures if the society does not have an adequate number of job opportunities—particularly for males between the ages of eighteen and twenty-five. Demographic historians believe that the relatively large proportion of young men in France more than 200 years ago contributed to the pressures that resulted in the French Revolution. The same was true during the seventeenth-century English Civil War and the majority of the revolutions in developing countries during the twentieth century. The cultural and political upheavals of the 1960s in the United States coincided with the young adulthood of the post–World War II baby boom.

In the 1990s, according to Population Action International, nations with more than 40 percent of its adults made up of those aged fifteen to twenty-nine experienced civil conflict at twice the rate of countries generally, and more than two thirds of civil conflicts since the 1970s have been in nations with youth bulges. Among the many factors causing the Arab Spring in 2011 was the disproportionate size of the young adult generation in most of the Arab countries. It is worth remembering, however, that it was a food vendor in Tunisia who set off the Arab Spring during a period of food price hikes around the world.

One of the largest youth bulges in the world today is in Iran, and although the street demonstrations and the Green Revolution have been suppressed brutally, the pressures for societal change appear to be still building. Similarly, Saudi Arabia, where dissent and demonstrations are also suppressed, faces similar demographic pressures for change, because the percentage of its population made up of young men age fifteen to twenty-nine is exceptionally high, and the number of jobs available to them is exceptionally low.

By most of these demographic measures, the United States has a more favorable outlook than many developed countries. Its median age is climbing, but will reach only 40 by midcentury. Its fertility rate is above the replacement rate, partly due to immigration and to the fertility of immigrant populations.

MIGRATIONS

In 2010, the United Nations reported that the world's migrant population had reached almost 214 million people, driving the percentage of migrants in the population of developed countries to 10 percent, an increase from 7.2 percent twenty years earlier. In the last year for which statistics are available (2009), 740 million internal migrants moved from one region to another inside countries. Cities are the primary destination for these migrants—both international migrants and those who migrate within their own countries from one region to another, almost always from rural areas to cities.

One new trend is that the number of international migrants moving from one developing country to another is now roughly equal to the number of migrants moving from a developing country to developed regions of the world. As the secretary-general of the United Nations put it, "In other words, those moving 'South-to-South' are about as numerous as those moving 'South-to-North.'"

Although migration has, of course, many beneficial effects—not least among them the enrichment of the talent pool in countries and regions to which they relocate—the number of international migrants has also been driving some dangerous trends in many countries. Xenophobia, with its associated discrimination and violence against migrants—particularly those with ethnicities, nationalities, cultures, and religions markedly different from the majority in the country they move to—has been most pronounced in regions stressed by high unemployment among natives and in countries where the percentage of international migrants has become seen as a threat to the majority's culture, traditions, and future prosperity.

In Athens, neo-Nazi vigilantes have been patrolling the streets and brutally attacking the growing number of Muslim migrants from several countries, including Afghanistan, Pakistan, and Algeria. In Moscow and some other Russian cities, neo-Nazis, skinheads, and other right-wing extremist groups are also brutalizing migrants—many of them from areas like Chechnya in the trans-Caucasus region, where there are significant Muslim populations.

Migrants now make up 20 percent or more of the people in forty-one countries around the world; three quarters of them have less than one million people. There are now thirty-eight larger countries where cross-border migrants make up 10 percent of the population or more.

India will soon complete construction of a 2,100-mile-long, 2.5-meter-high iron fence on its border with Bangladesh. As the nation most affected by the early impacts of climate change, Bangladesh has experienced a surge of internal migration from low-lying coastal areas and offshore islands in the Bay of Bengal, where four million people currently live. The overall population of Bangladesh is expected to increase from 150 million today to 242 million over the next few decades.

Bangladesh has also been the destination for a large number of international migrants from Afghanistan since the U.S. invasion. The presence among these migrants of many jihadists and Taliban members has given rise to concerns by India about an upsurge in Islamic extremism on the Bangladesh side of the border. But the continuing economic stress in Bangladesh is the principal source of pressure for trans-border migration toward India and through India to other destinations.

Even in the United States, where immigration has been a historic success story, the surge of legal and undocumented migrants in the early part of the twenty-first century created social stress. Twenty percent of all international migrants now live in the United States even though it has only 5 percent of the world's population. During the twelve-month period that ended in July 2011, the number of "non-white" babies born outnumbered Caucasian babies for the first time. The concern over illegal immigration from Mexico and other countries during the same period is cited by domestic terrorism experts as a major factor causing the surge of hate groups.

A RECENT STUDY by the Brookings Institution indicates that "minorities accounted for 92 percent of the nation's population growth in the decade that ended in 2010." The number of white children in the U.S. declined by 4.3 million as the number of Hispanic and Asian children increased by 5.5 million. Already, more than half of U.S. cities are minority majority, with the two largest groups represented by Hispanics at 26 percent and African Americans at 22 percent. Hispanics now represent the largest minority group in the United States.

U.S. domestic terrorist groups actually peaked in the 1990s just prior to the bombing of the federal office building in Oklahoma City. The number declined sharply for more than twelve years until the inauguration of Barack Obama, which appeared to trigger a renewed upsurge in 2009–12 to levels far higher than the previous peak. The Southern Poverty Law Center links the increase to the changes in America's demographic makeup: "This very real and very significant change is represented in the person of Barack Obama. We've of course seen the most remarkable growth in the radical right since 2008, precisely coinciding with Obama's first three years as president."

Ironically, net immigration from Mexico fell to zero in 2012, though immigration from several other countries has continued. Flows of Asian immigrants to the U.S. overtook Hispanics in 2009. And according to the Brookings study, "Even if immigration stopped tomorrow, we will achieve a national minority majority child population by 2050 (by around 2023 if current immigration trends continue)."

The relatively higher birth rate in the Palestinian territories, compared to the Jewish birth rate in Israel, is causing changes in the political analyses by both Palestinians and Israelis of how to evaluate potential options for resolving, or at least managing, the tensions in the region. The same differential birth rate has led to a sevenfold increase in the Arab minority population inside Israel's borders since the modern state was founded, leading to oft expressed concerns by some Israelis that demographic trends could one day force a choice between the Jewish nature of the state of Israel and the democratic principle of majority rule.

There are also often negative consequences in countries where large numbers of migrants are leaving. Chief among them is the problem caused by a brain drain when highly trained professionals—such as doctors and nurses—leave their countries of origin, in part because their skills make it much easier for them to find lucrative employment and higher standards of living in developed countries. When middle-class families migrate, there is often diminished support in their countries of origin for continued investments in public goods like education and health care. At the same time, the increasing percentage of migrant and domestic minority populations in developed countries has sometimes appeared to weaken the social contract supporting the provision of public goods—particularly public education—when phenomena such as white flight to private schools results in less support for public school budgets.

Nevertheless, many destination countries have adopted policies designed to attract higher-skilled migrants. And the need for low-wage workers in many developed countries with smaller than optimal workforces has also led to a significant expansion of temporary worker programs—particularly in the United States, Australia, and the United Kingdom. Colleges and universities have also significantly increased their recruiting of migrant students from foreign countries.

Many of the nations and regions from which migrants originate also experience some positive benefits, particularly in the form of remittances, especially from migrants leaving lower-middle-income countries. The remittances sent by migrants back to their families totaled $351 billion in 2011 and are projected to reach $441 billion in 2014.

The amount of remittances sent back to their communities of origin by internal migrants is believed to be much larger. In China, internal migrants from rural areas send an average of $545 per year back home from the cities where they work. In Bangladesh, the Coalition for the Urban Poor calculates that migrants from rural areas to the capital city of Dhaka routinely send back home as much as 60 percent of their income.

Indian migrants from the poor states of Uttar Pradesh, Bihar, and West Bengal routinely send money orders from Mumbai back to their home communities in amounts that make up the majority of money flowing into those three states.

REFUGEES

Alongside the flows of international and internal migrants are growing numbers of refugees. According to the international treaty on refugees, the definition of a refugee is someone who, unlike a migrant, leaves his or her country of origin due to the fear of violence or persecution. Almost 44 million people around the world have been forced from their nations of origin by conflict or persecution—of which 15.4 million are classified as refugees—and another 27.5 million people have been displaced by violence and persecution to new communities within their own country.

The U.N. high commissioner for refugees, António Guterres, notes that 70 percent of current refugees have been in that status for more than five years, and as a result, "it's becoming more and more difficult to find solutions for them." Twelve million among them are stateless, meaning they have no place to go home to. In the last five years, for the first time, more refugees moved to cities than to refugee camps. While equal numbers of migrants traveled to developed and developing countries, 80 percent of refugees live in poor regions of the world.

All of the large source countries for refugees are mired in violent conflicts, including Somalia, the Democratic Republic of the Congo, Myanmar, Colombia, and Sudan. The two largest source countries for refugees are Afghanistan and Iraq. The ill-fated decision by the United States in 2002 to invade Iraq—thus prolonging the conflict in Afghanistan as well, by prematurely removing troops that had encircled Osama bin Laden—has had a cascading impact on the entire region, flooding neighboring countries with refugees.

The three million Afghans displaced by the war in their home country have fled mostly to Pakistan (1.9 million) and Iran (one million). The 1.7 million refugees from Iraq have also gone mostly to neighboring countries. Indeed, according to the *World Development Report*, more than three quarters of refugees worldwide are hosted in nations neighboring their country of origin. The largest number are now living in Asia and the Pacific (2 million—most of them in South Asia), Sub-Saharan Africa

(2.2 million—403,000 of them in one country, Kenya!), and the Middle East and North Africa (another 1.9 million).

However, more than 1.6 million refugees, the vast majority of them Muslim, have found their way to Europe, further exacerbating xenophobic tensions and increasing fears of radicalization of poorly assimilated young Muslim populations living in Europe; Muslims already make up 5 percent of Europe's population. The surge of international migrants from North Africa and South Asia into Europe has also triggered a renewal of xenophobia, even in countries previously known for their commitment to tolerance. In several European countries, the combination of economic stress and the growing numbers of immigrants, particularly Muslim immigrants, has disrupted the political balance as extreme right-wing and nativist groups exploit the public's uneasiness.

THE FASTEST GROWING new category of refugees is climate refugees. Although they are not legally classified as refugees (because the definition in the Refugee Protocol requires that the source of their motivating fear be violence or persecution from other people), they are nevertheless routinely described as refugees because their migration is not voluntary. In the U.N.'s *State of the World's Refugees* report from June 2012, U.N. secretary-general Ban Ki-moon noted that the traditional causes of forced displacement, "conflict and human rights abuses," are now "increasingly intertwined with and compounded by other factors," many of them related "to the relentless advance of climate change."

Israel announced a major national plan last May on climate change that included a recommendation to build "sea fences" near its maritime borders on the Red Sea and the Mediterranean, linked with impassable barriers on its land borders, in order to protect against a predicted wave of climate refugees. "Climate change is already here and requires comprehensive preparations," said Israeli environmental protection minister Gilad Erdan. "The lack of water, warming and sea level rise, even if it will occur on a different schedule, will bring migration movements from all impoverished regions to every place where it is possible to escape this," the study noted.

One of the two leaders of the team authoring the report, Professor Arnon Soffer of the University of Haifa's Geography Department, added, "The migration wave is not a problem for the future. It is today, it is going on now. . . . It will just increase from day to day." Noting that European navies prevent most boats with migrants from reaching Europe, he said they are forced to go elsewhere, but "in India they shoot, in Nepal they shoot, in Japan they shoot." The team noted that climate refugees are expected from Africa, where approximately 800 lakes have dried up completely in the last decade, including the former largest lake in Africa, Lake Chad, which mobilized many climate refugees eastward into the Darfur region.

Persistent droughts and desertification in Somalia have also contributed to the violent conflict there. Other climate refugees attempting to migrate to Israel are expected from Jordan, the Palestinian territories, Syria, and the Nile Delta in Egypt. In addition, still more internal climate refugees are expected from the Negev, from which many Bedouins have already moved to cities in the center of Israel. Soffer added, "If we want to keep Israel a Jewish state, we will have to defend ourselves from what I call 'climate refugees,' exactly as Europe is doing now."

U.S. assistant secretary of state Kurt Campbell recently wrote that the impact of climate change on Africa and South Asia, including "the expected decline in food production and fresh drinking water, combined with the increased conflict sparked by resource scarcity," is likely to produce "a surge in the number of Muslim immigrants to the European Union (EU)," doubling Europe's Muslim population within the next twelve years, "and it will be much larger if, as we expect, the effects of climate change spur additional migration from Africa and South Asia."

A few years ago, I visited the southernmost extremity of the European Union, Spain's Canary Islands, just off the coast of West Africa. I found many conversations dominated by concerns of residents about the surge of refugees attempting to migrate by boat from Africa to their most convenient point of entry into the European Union. In some years, more than 20,000 Africans have attempted the dangerous journey across to the Canaries.

Over the next century, the global community can expect millions of climate refugees. Almost 150 million people now live in low-lying areas only one meter or less higher than the current sea level. For each additional meter of sea level rise, roughly 100 million more people will be forced to abandon the places they call home. And this number, of course, does not include refugees from desertifying dryland areas.

The dimensions of the climate crisis are described in Chapter 6, along with the difficult but cost-effective and necessary responses. What is clear now, however, is that even with global warming in its early stages, the growth of human civilization is already pressing hard against limitations that are complicating our ability to provide the essentials of life for billions of people.

ENDANGERED GROUNDWATER AND TOPSOIL

For example, where topsoil and groundwater are concerned, there is a disconnect between the frenzied rate of exploitation of both these resources on the one hand, and the extremely slow rate with which either resource can be regenerated on the other. Renewable groundwater aquifers fill back up, on average, at the rate of less than one half of one percent per year. Similarly, topsoil regenerates naturally—but at the agonizingly slow rate of approximately 2.5 centimeters every 500 years.

In just the last forty years, the overexploitation of topsoil has led to the loss of a significant amount of productivity on almost one third of the arable land on Earth. Without urgent action, the majority of the Earth's topsoil could be severely degraded or lost before the end of this century. In China, topsoil is being lost fifty-seven times faster than this natural replacement process; in Europe seventeen times faster. According to the National Academy of Sciences, it is being lost in the United States ten times faster than it can be replenished. Ethiopia is now losing almost two billion tons of topsoil every year to rain washing the erodible soils down the steep slopes of its terrain.

In the case of groundwater, the nearly total depletion of some important aquifers and the sharply dropping levels of others have now focused the attention of experts in many countries on the future of this resource. The doubling of the global withdrawal rate over the last half century—and the projection that withdrawals will continue to increase at an even faster pace—have many experts beginning to get very worried. In many areas, the withdrawals from aquifers now far exceed the rate of replenishment; many aquifers are now falling several meters per year.

IT IS AS if we are willfully blind to the basic underlying reality of our relationship to the Earth's limited resources. But this seeming blindness is reinforced by the world's principal method of accounting for natural resources, which treats their use as income rather than withdrawals from capital. This is, in the words of economist Herman Daly, "a colossal accounting error. . . . At least we should put the costs and the benefits in separate accounts for comparison."

The basic distinction between operating income and withdrawals from capital is crucial, whether one is accounting for a company or a nation. In the words of a classic accounting text, if this distinction is misunderstood and improperly made, it leads to "practical confusion between income and capital." Another seminal accounting text notes that "net income of an entity for any period is the maximum amount that can be distributed to its owners during the period and still allow the entity to have the same net worth at the end of the period as at the beginning. . . . In other words, capital must be maintained before an entity can earn income." This same principle holds true for nations and for the world as a whole. In recognition of this principle, the U.N. Statistical Commission in 2012 adopted a "system of environmental-economic accounts" as a step toward integrating environmental externalities. In 2007, the European Union launched its "beyond GDP" initiative, and is due to release an assessment by all member states of their "natural capital" in 2014.

When Simon Kuznets warned in 1937 that misuse of GDP would make us vulnerable to such accounting errors and could lead to a form of willful blindness, he noted that conflicts over resources might well exacerbate the risk inherent in the admittedly flawed design of his elaborate accounting system:

> The valuable capacity of the human mind to simplify a complex situation in a compact characterization becomes dangerous when not controlled in terms of definitely stated criteria. With quantitative measurements especially, the definiteness of the result suggests, often misleadingly, a precision and simplicity in the outlines of the object measured. Measurements of national income are subject to this type of illusion . . . especially since they deal with matters that are the center of conflict of opposing social groups where the effectiveness of an argument is often contingent on oversimplification.

In an example of the precise problem Kuznets was anticipating, today—all around the world—calculations about the impact of groundwater withdrawals are often at "the center of conflict of opposing social groups." Often, officials in regions where water supplies are shared with other regions or countries—and whose farms and businesses would be disrupted by any change in water allocations—have strong incentives to minimize the seriousness of the problem—putting off for the future a problem they would rather not deal with in the near term. It's an all too familiar challenge for anyone who works on global warming.

In just one of many examples of this particular variety of denial, when an expert from the University of Oklahoma, Luo Yiqi, visited Inner Mongolia in northern China a few years ago to study desertification, he was astonished to see fields of rice (one of the most water-intensive crops) grown with water that authorities allowed to be pumped at grossly unsustainable rates from deep aquifers. "Apparently," he noted dryly, "farmers did not get enough scientific guidance."

The regrettable decision to ignore the depreciation of natural resources, while accounting precisely for the depreciation of capital goods, may have been subtly influenced by the state of the world when this formula was created in the 1930s. We were still in the last stages of the colonial era, when limitations on supplies of natural resources seemed irrelevant; industrialized countries could simply obtain more in their colonial possessions, where the supply seemed, for all intents and purposes, limitless. Global population has tripled since the national accounts were adopted, and the dangerous illusion that Kuznets warned about is now at the heart of the world's failure to recognize the twin dangers of unsustainable depletion of both topsoil and groundwater.

Since the beginning of the Agricultural Revolution, these two strategic resources have both been essential for the production of food. The irrigation of crops emerged roughly 7,000 years ago and the Green Revolution of the twentieth century increased agriculture's dependence on irrigation—particularly in China, where 80 percent of the harvest depends on irrigation, and India, where 60 percent depends on irrigation. (The U.S. depends far less on irrigation.)

Large dams for water storage gained popularity in the late nineteenth and early twentieth centuries. There are now 45,000 large dams in the world, including on all twenty-one of the world's longest rivers. FDR's economic stimulus program in the 1930s resulted in large-scale dam construction by the Tennessee Valley Authority in my home region, and the Bonneville Power Administration in the Pacific Northwest—and of course, the majestic Hoover Dam on the Colorado River, which was the tallest in the U.S. when it was built seventy years ago.

Prior to the Industrial Revolution and the explosion of urban populations, more than 90 percent of global freshwater was used for agriculture. In more recent decades, the competition for water between agriculture, manufacturing, and fast-growing thirsty cities has led to growing disputes over water allocation—disputes that agriculture often loses. Today, more than 70 percent of the world's freshwater is used to grow food, even though 780 million people in the world still lack access to safe drinking water. As noted earlier, the world has made significant progress in reducing the number of people who lack access to improved water resources (though little progress has been made in preventing the contamination of freshwater sources—both surface and groundwater resources—from human and animal waste and other pollutants).

Some deep aquifers have long been sealed from surface water. A recently tapped aquifer in the Northeastern United States, Patapsco (under the state of Maryland) has water found to be one million years old. Similarly, water in the Nubian Aquifer (underneath the Sahara), the Great Artesian Basin (underneath northeastern Australia), and the Alberta Basin (underneath western Canada) all have water more than one million years old. But although these "fossil" aquifers are nonreplenishable, most scientists believe they are limited in their supply of water; the vast majority of aquifers are replenished slowly as rainwater filters down to them.

Until recently, the amount of information about groundwater depletion rates was spotty at best, and according to one expert, the threat to the resource is a classic case of "out of sight, out of mind." Indeed, so much water is now being withdrawn from underground aquifers that it is believed by experts to account for 20 percent of the sea level rise in recent decades (although scientists forecast that the accelerating ice loss from Greenland and Antarctica will dramatically increase sea level rise later in this century).

The highest rates of groundwater depletion are in northwest India and northeast Pakistan, the Central Valley of California, and northeastern China. One Chinese groundwater specialist found that an aquifer in northern China with water 30,000 years old was being used unsustainably to irrigate crops in dryland areas. China has embarked upon the largest water project in history—the South–North Water Transfer Project that has been under construction for decades, intended to remedy water shortages in northern China. Asia, which has 29 percent of the world's freshwater resources, is now using more than 50 percent of the world's water. According to the United Nations, "In 2000, about 57% of the world's freshwater withdrawal, and 70% of its consumption, took place in Asia, where the world's major irrigated lands are located."

Africa has only 9 percent of the world's freshwater, but is using 13 percent, and is expected by U.N. experts to have the most intensive increases in water withdrawal in the coming decades. Europe is consuming only a slightly larger percentage than its own supply. The Americas are fortunate in having more water than they use, but large regions—particularly Mexico and the Southwestern U.S.—are already experiencing severe shortages. In 2011, more than one million head of cattle were herded north from Texas to wetter, cooler pastures. Few expect them ever to return.

According to a study by the Scripps Institute, there is a "50-50 chance" that Lake Mead—the largest man-made lake in the western hemisphere, the one formed by Hoover Dam—will run completely dry before the end of this decade. In addition, according to the U.S. Department of Agriculture, the water table beneath three of the largest grain-producing states—Kansas, Texas, and Oklahoma—has dropped more than 100 feet, forcing many farmers to abandon irrigation. Reservoirs in the state of Georgia have also been running at dangerously low levels for several years.

Improving the efficiency of water use is a cost-effective option for ameliorating shortages in some areas. Many aging water distribution systems leak extraordinary amounts of water. In the U.S., for example, an important urban water line bursts every two minutes on average, twenty-four hours a day. Some portions of older urban water systems were built over 160 years ago, and since then, have been—like groundwater resources—"out of sight, out of mind." Repairing municipal water pipes is expensive, but some cities are belatedly recognizing the necessity of undertaking this task.

According to ecologist Peter Gleick, we should view efficiency as a giant wellspring that could provide vast new quantities of needed freshwater. Unfortunately, this wellspring, like many of the aquifers now being recklessly depleted, also seems to be out of sight, out of mind. The majority of agricultural irrigation practices are still extremely wasteful. Switching to scientifically precise drip irrigation techniques is cost-effective in most agricultural operations, but many farmers have been slow to make the change. Another benefit of switching to more efficient and precise methods of irrigation is that wasteful and excessive irrigation of crops increases the salinity of soils—because the irrigation water usually contains small amounts of salt that build up with continued use.

The recycling of water is growing in popularity. Some communities already require the use of greywater—used water that is not suitable for drinking but is safe for watering plants. The more controversial recycling proposals take sewage water and remove all of the contaminants, purify it, and put it into drinking water systems. There is still a great deal of consumer resistance to these plans, but some communities have successfully implemented the approach.

In regions where rainfall is becoming more concentrated in large downpours—interrupted by longer periods of drought—many experts are calling for the increased use of cisterns to capture more of the rainfall and store it for drinking water. This once common practice fell out of favor with the extension of underground water lines from reservoirs. I remember the cisterns we used to have on our family farm when I was a boy. We stopped using them when we got "city water."

THE STATE OF the world's topsoil is threatened by the same willfully blind overexploitation that has caused shortages of freshwater. In the world's prevailing system of accounting, neither water nor topsoil are assigned any value. Therefore, wasteful and destructive practices that diminish the supplies of both are invisible in the world's economic calculations. Yet topsoil, along with water, is the basis for virtually all human life on Earth. More than 99.7 percent of food consumed by human beings comes from cropland, more specifically from the six to eight inches of topsoil that cover roughly 10 percent of the Earth's surface.

On a global basis, we are effectively strip-mining this crucial resource in an unsustainable pattern, by recklessly plowing erodible soils, overgrazing grasslands, taking arable land for buildings and roads to accommodate urban and suburban sprawl, tolerating reckless deforestation, and failing to use proven land management techniques that replenish soil carbon and nitrogen.

At present, every kilogram of corn produced in the American Midwest results in the loss of more than a kilogram of topsoil. In some states, such as Iowa, the ratio is even higher: 1.5 kilograms of topsoil lost for each kilogram of grain. These rates of soil loss are not sustainable. They deplete soil carbon, thus damaging the productivity of the soil over time, and accelerate the emission of carbon dioxide into the atmosphere.

We already know how to slow and reverse soil erosion, but global leadership would be required to mobilize the community of nations in the same way that FDR mobilized the United States in the 1930s. Organic agriculture with low-till and no-till practices can sharply reduce soil loss while simultaneously increasing the fertility of the topsoil. Crop rotation, a technique that used to be widespread before industrial agriculture took over, can replenish soil carbon and nitrogen.

Another once common technique that has since been abandoned in large areas of the world is the recycling of animal manure as fertilizer for crops. Factory farming—the clustering of thousands of head of livestock in crowded feedlots and feeding them corn—has turned this natural fertilizer into highly acidic toxic waste that is harmful to crops and thus becomes an expensive liability instead of a valued asset.

A major study in 2012 by leading researchers at the University of Minnesota, Iowa State University, and the Agricultural Research Service of the U.S. Department of Agriculture showed that the use of nontoxic manure as fertilizer and a three-year crop rotation designed to replenish soil fertility reduced the need for herbicides and nitrogen fertilizer by almost 90 percent, without reducing profits. One of the researchers, Professor Matt Liebman of Iowa State, said that one of the reasons farmers do not use the approach recommended in the study is that "there's no cost assigned to environmental externalities."

For the last century, modern agriculture has been based on heavy use of synthetic nitrogen fertilizer—90 percent of the cost of which is from natural gas, from which virtually all of the nitrogen is derived. However, agricultural productivity growth has been slowing even as fertilizer use per acre has been increasing dramatically. Moreover, the heavy use of nitrogen in agriculture has caused significant water pollution problems around the world as it runs off farmland with the rain and feeds uncontrolled massive algae blooms in coastal regions of the ocean—and dead zones, areas devoid of life, which are growing in several ocean regions, including the part of the Gulf of Mexico into which the Mississippi River drains. In China the use of synthetic nitrogen fertilizer has increased by 40 percent in the last two decades even though grain production has remained relatively stable; it is this nitrogen runoff that has produced the recent spectacular algae blooms in Chinese rivers, lakes, and coastal areas.

Additional nitrogen emissions from the combustion of fossil fuels in factories, on farms, and in cars and trucks have created significant air pollution problems, particularly in the U.S., China, Southeast Asia, and parts of Latin America. More efficient and targeted use of nitrogen fertilizers, and tighter restrictions on emissions from factories and vehicles, are needed to address the problem.

While nitrogen supplies are not limited, there is a potentially serious emerging limit to the supply of another crucial component of fertilizer—phosphorus, which is a relatively rare element on the Earth. Even as conventional sources of phosphorus are running out, modern agricultural techniques have tripled the depletion of phosphorus from cropland.

A PHOSPHORUS CARTEL?

The first warning about a phosphorus shortage in a 1938 message to Congress, by President Roosevelt, led to a successful worldwide search for additional reserves—including the discovery of phosphates near Tampa, Florida, where 65 percent of U.S. production now takes place. But while the United States produces 40 percent of the world's corn and soybeans, it produces only 19 percent of the world's phosphorus, which, in the long run, is essential for agriculture to continue—and so now the search for new reserves is beginning again.

Forty percent of the world's current supply of phosphates (the most common form in which phosphorus occurs) is in Morocco, which has been called the "Saudi Arabia of phosphorus." The next largest reserves are found in China, which imposed a 135 percent tariff on exports during the 2008 food price crisis. Many experts fear that similar hoarding of phosphorus could occur if food prices continue to go up, although other experts are more sanguine about the possibility of finding new sources in unconventional locations, such as the ocean floor.

Phosphorus is essential to all life, including human life. It makes up the backbone of DNA, among other things, and fully one percent of the bodyweight of human beings is made up of phosphorus; in fact, the seven billion people on Earth discard large quantities in urine every day.

Some countries are now actively exploring the recycling of urine in order to extend the supplies of phosphorus for fertilizers.

The addition of rhizobium bacteria and mycorrhizal fungi to soils as seeds are planted can improve crop yields and also speed the recovery of soil fertility and enhance the sequestration of soil carbon. The planting of leguminous trees every thirty feet or so as buffer strips and contour hedges can replenish nitrogen in the soil and further protect against erosion. Leaving the majority of crop residue—like corn stover—on the land during and after harvesting the crop can also restore the fertility of the soil while diminishing erosion. The use of biochar (from sustainable sources) in a carefully managed way can also improve yields and soil quality. The reduction of meat as a percentage of a healthy diet can relieve pressure on the Earth's topsoil. And the expansion of small-scale organic gardens in countries with a surplus of arable land could potentially add significant volumes of fresh food to the world supply, as they did in Western countries when Victory Gardens were planted during World War II.

But perhaps the single most effective measure to protect topsoil would be to use carbon credits to provide an additional source of income for farmers who pay careful attention to safeguarding and improving the carbon content and fertility of their soils.

So long as the world ignores the value of topsoil in its constant calculations of growth and productivity, the demands placed on agriculture by the combination of growing population and growing per capita consumption of food will continue putting the future of topsoil at severe risk. At current consumption rates (which are still increasing), we need an additional 15 million hectares each year to keep up with the extra food production needed for the increasing population. Yet we are destroying and losing approximately 10 million hectares (approximately 25 million acres) every year. At present, much of the additional cropland being developed results in deforestation—often in forest areas that have very thin topsoils that are quickly depleted by wind and water after the trees are gone. In addition, the more forestland that is converted to farmland, the more biodiversity is lost.

In some respects, the global topsoil crisis is an echo of what happened in the United States in the first third of the twentieth century when the first mass market tractors—pulling the more efficient plows that had been invented three quarters of a century earlier—broke the sod of erodible grasslands in the Midwest for crops; over the next three decades the vulnerable topsoil was washed and blown away, creating the Dust Bowl of the 1930s. Less well known in the U.S. is the even larger tragedy experienced in Central Asia during the 1950s when the USSR plowed an enormous area of grassland—mostly in Kazakhstan (1954)—and created their own Dust Bowl.

Another epic land-use catastrophe occurred in Central Asia in the 1960s, when the USSR embarked on a shortsighted plan to grow thirsty cotton crops in dryland areas of Uzbekistan and Turkmenistan. So much water was diverted from two rivers—the Amu Darya and the Syr Darya—that the world's fourth largest inland sea, the Aral Sea, almost completely disappeared. I visited the Aral Sea two decades ago and saw firsthand the tragedy that resulted for the people who used to depend on it.

DUST STORMS AHEAD

My father's generation was motivated by the U.S. soil erosion crisis to adopt new land management techniques. One of the great accomplishments of FDR's New Deal was the massive program to reconvert eroded land to grassland and a nationwide effort to fight soil erosion. I still remember my father teaching me when I was a young boy how to stop gullies before they began cutting deep into the soil, and how to recognize the richest soil—it's black from all the organic carbon in the soil.

Modern-day dust storms are now once again increasing in size and frequency as drylands are being overgrazed and erodible soils are subjected to higher temperatures and stronger winds. "Drylands are on the front line of the climate change challenges for the world," said Luc Gnacadja, who heads the U.N. Convention to Combat Desertification. The U.N. Environment Programme reports that land degradation in drylands threatens the way of life for an estimated one billion people in 100 countries. Desertification is taking a toll on topsoil and destroying arable cropland—particularly in regions of Africa north and south of the Sahara, throughout the Middle East, in Central Asia, and in large areas of China, where overgrazing, poor cultivation techniques, and urban sprawl are contributing significantly to the phenomenon.

In the U.S., in July of 2011, Phoenix, Arizona, was covered with dust when, in the words of the National Weather Service, "A very large and historic dust storm moved through a large swatch of Arizona." Although these Southwestern U.S. dust storms, often called haboobs, are not new, Phoenix has had an unusually large number of them in recent years—seven in 2011 alone.

The U.S. Geological Survey and UCLA conducted a study in 2011 that predicted "accelerated rates of dust emission from wind erosion" as a result of climate change in the Southwestern United States. Climate expert Joseph Romm has recommended use of the term "dustbowlification" as a way of describing what is in store for many regions of desertifying drylands.

Lester Brown, long one of the world's leading environmental experts, points out that the two most significant desertifying areas now generating dust storms are in north-central China and in the areas of Central Africa that lie on the southern edge of the Sahara Desert. As Brown puts it, "Two huge dust bowls are forming, one across northwest China, western Mongolia, and central Asia; the other in central Africa."

According to geographer Andrew Goudie at Oxford, dust storms from the Sahara have increased tenfold during the last fifty years. The chairman of the African Union, Jean Ping, says, "The phenomenon of desertification affects 43 percent of productive lands, or 70 percent of economic activity and 40 percent of the continent's population." In large areas of Sub-Saharan Africa, soil carbon content is now lower than it was in the United States' Midwest just prior to the Dust Bowl.

In Nigeria, while human population increased fourfold over the last sixty years, the number of livestock exploded from six million to more than 100 million. Partly as a result, the northern region of Nigeria is being desertified—which is contributing to growing clashes between Muslims moving from the north into non-Muslim areas in southern Nigeria. Growth in the population of both humans and livestock is also driving competition for land in other drying areas of Africa and has led to deadly conflicts between herders and farmers (whose ethnicities and religions are also different), who have fought one another in Sudan, Mali, and elsewhere.

The same livestock population explosion is damaging the overgrazed grasslands surrounding China's Gobi Desert, where the dust storms are also increasing dramatically. While the United States and China have roughly the same amount of grazing land and roughly the same number of cattle (80–100 million), China has 284 million sheep and goats compared to less than 10 million in the United States. According to the latest statistics available, China is now losing almost 1,400 square miles of arable land to deserts every year.

The U.S. embassy in China has used satellite photos to illustrate the "desert mergers and acquisitions" in north-central China where two deserts in Inner Mongolia and Gansu Province are merging and expanding. In Xinjiang Province in northwestern China, the same thing is happening, as the Taklamakan and Kumtag deserts are also merging and expanding. More than 24,000 villages and their surrounding cropland have had to be at least partially abandoned in these northern and western regions of China. Similar tragedies are unfolding in both Iran and Afghanistan, both of which have already abandoned many villages to the encroaching desert.

While the massive sandstorms in China and Africa are capturing attention today, Lester Brown warns, "A third massive cropland expansion is now taking place in the Brazilian Amazon Basin and in the cerrado, a savannah-like region bordering the basin on its south side." These soils are highly erodible and the results are predictable: low yields, followed by soil erosion on a massive scale. The knock-on effects also include the further expansion of cattle ranching into the Amazon rainforest, adding even more risk to the integrity of that globally important ecosystem. The Amazon has already suffered from two "hundred-year droughts" in the last seven years. As the deforestation and the wildfires continue in the Amazon, many experts have expressed concern that the Amazon is in danger of being transformed over time from the greatest tropical rainforest on Earth into a massive dryland region.

With the rapidly increasing populations in Africa and the Middle East, and impending food shortages, it is remarkable that the world has paid so little attention to the desertification crisis. According to the U.N. Convention to Combat Desertification's Luc Gnacadja, the reason desertification has not become a higher priority is that 90 percent of the people affected live in developing countries. It is another example of the imbalance of power in the world—and the lack of leadership. Gnacadja added, "The top 20 centimeters of soil is all that stands between us and extinction."

The loss of arable land is particularly acute in the most populated nation of North Africa. According to the United Nations, Egypt is now losing an incredible 3.5 acres per hour of its fertile agricultural land in the Nile Delta—mainly because of new construction and urban sprawl to accommodate additional shelter for Egypt's fast growing population.

In addition, rising sea level in the Mediterranean is already pushing saltwater aquifers upward in areas near the coast, resulting in the loss of cropland to salinization. Salinization is also occurring in the rich Ganges Delta, the Mekong Delta, and in other so-called mega-deltas. A one-meter rise in sea level—less than that predicted during this century—would inundate a significant percentage of the most fertile soils in the Nile Delta—from whence 40 percent of Egypt's food production comes.

The pressure created by the increased use of water-intensive agriculture, population growth, and economic expansion is increasing tensions over the allocation of river water in several regions of the world where the management of rivers and dams affects watersheds shared by multiple countries. The potential for conflict is building in the Nile River watershed, where the largest country dependent on the Nile, Egypt, now benefits from its allocation of the majority of the Nile's water. But Ethiopia, where 85 percent of the Nile's headwaters originate but where very little of that water is now consumed, will double its population in the next thirty-seven years—and Sudan, which also depends on the Nile, is expected to increase its population 85 percent during the same period.

To the east of Egypt, the decision by Turkey to take a larger portion of the headwaters from the Tigris and Euphrates has led to growing complaints by Iraq and Syria that they are being treated unjustly. Both Iraq and Syria have been overdepleting their underground aquifers as they seek a resolution of the issue. Similarly, China's effort to take for itself a larger percentage of water from rivers that flow into Southeast Asia and India is raising tensions that will only get worse as populations in all the affected countries increase. In the United States, the growing conflicts over allocations of water in the West from the Colorado River system are being waged in court. But the underlying cause in all four of these giant watersheds is the same: there's more demand for water and less supply.

Conflicts between nations over access to freshwater have historically produced very few wars, though conflicts *within* nations over water have frequently produced social tensions and occasional violent clashes. By contrast, conflicts over land have, of course, frequently caused wars in the past.

A NEW SCRAMBLE IN AFRICA

In our new globalized economy, some nations with growing populations and shrinking resources of topsoil and water for agriculture are embarking on large-scale projects to purchase vast tracts of arable land in other countries—particularly in Africa, where an estimated one third of the world's uncultivated arable land is found. The degree of control that African governments—and the elites that run many of them—have over property rights is much greater in many parts of Africa where tribal property rights predating the colonial era are too easily ignored.

China, India, the Republic of Korea, Saudi Arabia, and other countries, along with multinational corporations and even hedge funds investing money from U.S. universities, are buying up large amounts of land in Africa to produce wheat and other crops for their own consumption and for sale in global markets. "It's a new colonialism, it's like the scramble of Africa in [the] 19th century whereby our resources were exploited to develop the Western world," said Makambo Lotorobo, an official with a Kenyan NGO, Friends of Lake Turkana.

"There is no doubt that this is not just about land, this is about water," said Philip Woodhouse from the University of Manchester. Devlin Kuyek, a researcher with GRAIN, an NGO specializing in food and agriculture issues, added, "Rich countries are eyeing Africa not just for a healthy return on capital, but also as an insurance policy."

This has led to an agricultural real estate boom in Africa. More than one third of Liberia's land, for example, has been sold to private investors. According to an analysis by the Rights and Resources Initiative, a Washington-based international coalition of NGOs, the Democratic Republic of the Congo has signed deals with foreign owners for 48.8 percent of its agricultural land; Mozambique has signed deals with foreign growers for 21.1 percent of its land. Almost 10 percent of the land in South Sudan (according to Norwegian analysts)—and 25 percent of the best acreage around its capital—was sold to investors after the country won its independence in 2011. China reached an agreement with the Democratic Republic of the Congo to grow palm oil for biofuel on 2.8 million hectares of land. There is disagreement among experts on how much of the land involved in these massive African purchases is being used for biofuels. The World Bank calculated that in 2009, 21 percent were for biofuels; the International Land Coalition calculated that 44 percent was dedicated to biofuels.

The South Korean multinational corporation Daewoo attempted to purchase almost half of the arable land in Madagascar, though public riots led to a cancellation of the contract (South Korean companies purchased 700,000 hectares in northern Sudan for wheat, according to a study by *The Guardian*, and the United Arab Emirates purchased slightly more—750,000 hectares).

In Ethiopia, where 8.2 percent of agricultural land has been signed over to foreigners, Nyikaw Ochalla, originally from the Gambella region and who is now living in the United Kingdom, told *The Guardian*, "The foreign companies are arriving in large numbers, depriving people of land they have used for centuries. There is no consultation with the indigenous population. The deals are done secretly. The only thing the local people see is people coming with lots of tractors to invade their lands. All the land round my family village of Illia has been taken over and is being cleared. People now have to work for an Indian company. Their land has been compulsorily taken and they have been given no compensation. People cannot believe what is happening. Thousands of people will be affected and people will go hungry."

The World Bank analyzed reports of international agricultural property deals between 2008 and 2009 and concluded that during that two-year period, foreign nations and corporations purchased almost 80 million hectares of land—approximately the area of the nation of Pakistan—and that two thirds of the deals were in Africa. In addition to the sheer scale of the international land purchases and long-term leases in Africa, other concerns highlighted by African and international NGOs include problems with the use of water, soil management, and the impact on local farmers whose precolonial tenure rights are often unenforceable. In Uganda, where 14.6 percent of agricultural lands have been signed over to foreign growers, 20,000 people have claimed they were unjustly evicted from their own land; the case is pending in Ugandan courts.

After examining more than thirty studies of this issue, the International Institute for Environment and Development concluded that many of the large-scale foreign investments have already failed because of miscalculations concerning difficulties in financing the projects or unrealistic business plans. Part of the underlying problem is a gross imbalance in political power, with elites in nondemocratic governments dealing with multinational corporations and foreign countries to make short-term profits at the expense of the sustainability of their nations' food production capability and often at the expense of poor farmers who are evicted from the land when ownership is transferred.

Several nations suffering from loss of topsoil, sharply declining crop yields, and shortages of freshwater have been forced to increase food imports. Saudi Arabia may have its last wheat harvest in 2013; it previously announced that it will rely entirely on wheat imports by 2016. In the 1970s, fearful that its central role in organizing the OPEC oil embargo might make it vulnerable to a counterembargo on the grain imports it relied upon heavily to feed its people, Saudi Arabia launched a crash program to subsidize (at almost $1,000 per ton) the growing of wheat irrigated with water from a deep nonrenewable aquifer underneath the Arabian Peninsula. However, years later, it belatedly realized that it was rapidly depleting the aquifer and announced cancellation of the program. "The decision to import is to preserve water," said Saudi deputy minister of agriculture for research and development Abdullah al-Obaid. Agriculture absorbs 85 to 90 percent of Saudi Arabia's water, and 80 to 85 percent of that water comes from underground aquifers. (Elsewhere in the region, Israel banned the irrigation of wheat in 2000.)

THE OCEANS

The need to meet increasing demand for freshwater and food, especially protein, has led many to look to the oceans for relief. Saudi Arabia is among many nations that have long dreamed that a logical solution to our water problems will eventually involve desalination of seawater. After all, 97.5 percent of all of the water on Earth is saltwater, and most plans to deal with the current and projected shortages of freshwater involve the use and allocation of the other 2.5 percent of Earth's water resources—70 percent of which is locked up in the ice and snow of Antarctica and Greenland.

Unfortunately, even with the best currently available technology, the amount of energy required to remove the salt and other minerals from seawater is so great that even energy-rich Saudi Arabia cannot afford it. It is more beneficial, in their view, to sell the oil they would otherwise have to burn in desalination plants and use the money to purchase the use of water-rich land in Africa. There are, of course, many desalination plants in the world—including in Saudi Arabia. However, the quantities produced are still relatively small and the expense makes wider use of desalination for the world's growing water needs financially unsustainable.

Nevertheless, there are many scientists and engineers working to invent new, more cost-effective technologies for desalination. Some believe that this challenge is yet another reason why the world should embark on a massive, large-scale global effort to accelerate the cost reductions now under way in solar energy. I have seen many intriguing business plans aimed at solving this problem, but none that yet appears to be close to financial feasibility.

As a measure of the desperation that water shortages can cause, one Saudi prince, Mohammed al-Faisal, provided funding to a French engineer, Georges Mougin, to develop a business plan for lassoing icebergs in the North Atlantic and then towing them to areas experiencing severe droughts. According to their calculations, a 30-million-ton iceberg could supply 500,000 people with freshwater for a year.

The production of food crops, of course, normally requires both freshwater *and* topsoil. Some techno-optimists, though, have touted the possibility of growing crops without topsoil in hydroponic facilities where the plants are suspended from racks and supplied with ample amounts of water, nutrients, and sunlight. Unfortunately, hydroponics is the food equivalent of desalination: it is prohibitively expensive, largely because it too is so energy-intensive.

Yet there is one source of high-quality protein that does not require topsoil—seafood. Today, more than 4.3 billion people rely on fish for approximately 15 percent of their animal protein consumption. Unfortunately, however, the demand for fish is far outstripping the supply. Consumption of fish has increased significantly because of two familiar trends: growth in population and growth in per capita consumption. Over the last half century, the average person's fish consumption globally increased from twenty-two pounds per person per year to almost thirty-eight pounds in 2012. As a result, the majority of the world's ocean fisheries have been overexploited and almost one third of fish stocks in the oceans, according to the United Nations, are in danger. Stocks of large fish—tuna, swordfish, marlin, cod, halibut, and flounder, for example—have been reduced by 90 percent since the 1960s.

Although other factors play a role—including the destruction of coral reefs and changes in ocean temperature and acidity due to global warming pollution—the overexploitation of the ocean fisheries is the principal cause of the decline. The world reached "peak fish" twenty-five years ago. According to the Secretariat of the Convention on Biological Diversity, "About 80 percent of the world marine fish stocks for which assessment information is available are fully exploited or overexploited. . . . The average maximum size of fish caught declined by 22% since 1959 globally for all assessed communities. There is also an increasing trend of stock collapses over time, with 14% of assessed stocks collapsed in 2007."

The good news is that ocean fisheries that are carefully managed *can* and *do* recover. The United States has led the way in such protections, and many of the U.S. fisheries are now improving in their health and abundance. President George W. Bush enacted an excellent system of protection for a large marine area in the Pacific Ocean northwest of the Hawaiian Islands. However, most fishing countries have not yet followed the example of the U.S. restrictions on overfishing, and global fish consumption is continuing to increase steadily.

Most of the continuing increase in fish consumption is now being supplied by farmed fish. However, there are growing concerns about the rapid expansion of aquaculture—61 percent of which will occur in China over the next seven years. Farmed fish do not have the same healthy qualities as wild fish, and often—particularly if they are imported from China or other jurisdictions that lack adequate environmental enforcement—can be tainted by pollution, antibiotics, and antifungals. In addition, most farmed fish are fed large amounts of smaller *wild* fish processed for formulated fishmeal. Salmon, for example, are fed at a ratio of five pounds of wild fish for each pound of farmed salmon produced. Consequently, the netting of enormous volumes of small fish in the oceans is now causing further disruption to the ocean food chain.

During an expedition to Antarctica in 2012, I talked with scientists who are deeply concerned about the overexploitation of the krill population in the Antarctic Ocean, largely for fishmeal and pet food. The U.S. Department of Agriculture has noted that the overexploitation of so-called industrial species that are used for fishmeal instead of direct human consumption will begin to impose limits on the production of fishmeal and fish oil for aquaculture in 2013. Over half of the fish food in agriculture is now made from plant protein, and some operators are trying to increase that percentage, but it is still difficult to provide essential nutrients economically without fishmeal.

In addition, any major expansion of plant protein dedicated to aquaculture would represent yet another diversion of arable land from the production of food that can be directly consumed by people.

The overexploitation of the oceans, like the reckless depletion of the world's resources of freshwater and topsoil, has increased the amount of attention being paid to the genetic engineering of plants and animals—to give them traits that will enable them to thrive in the new conditions we are creating in the world. Although more than 10 percent of all cropland is now planted with genetically engineered crops, the issues raised are complex, as we shall now see.

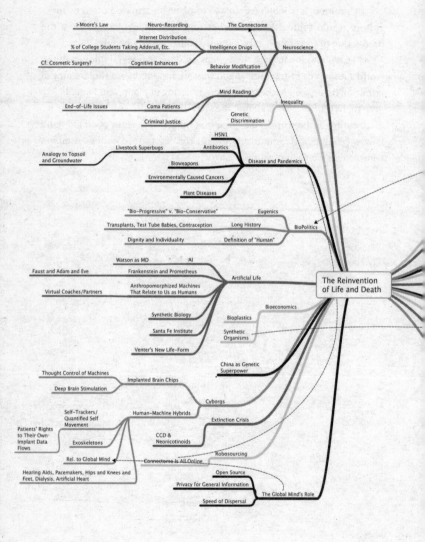

The Reinvention of Life and Death

Neuroscience
- The Connectome
  - Neuro-Recording
    - >Moore's Law
- Intelligence Drugs
  - Internet Distribution
  - % of College Students Taking Adderall, Etc.
  - Cognitive Enhancers
    - Cf. Cosmetic Surgery?
- Behavior Modification
- Mind Reading
  - Coma Patients
    - End-of-Life Issues
  - Criminal Justice

Inequality
- Genetic Discrimination

Disease and Pandemics
- Antibiotics
  - H5N1
  - Livestock Superbugs
    - Analogy to Topsoil and Groundwater
- Bioweapons
- Environmentally Caused Cancers
- Plant Diseases

BioPolitics
- Eugenics
  - "Bio-Progressive" v. "Bio-Conservative"
- Long History
  - Transplants, Test Tube Babies, Contraception
- Definition of "Human"
  - Dignity and Individuality

Artificial Life
- AI
  - Watson as MD
- Frankenstein and Prometheus
  - Faust and Adam and Eve
- Anthropomorphized Machines That Relate to Us as Humans
  - Virtual Coaches/Partners
- Synthetic Biology
- Santa Fe Institute
- Venter's New Life-Form

Bioeconomics
- Bioplastics
- Synthetic Organisms

- China as Genetic Superpower

Cyborgs
- Implanted Brain Chips
  - Thought Control of Machines
  - Deep Brain Stimulation
- Human-Machine Hybrids
  - Self-Trackers/ Quantified Self Movement
  - Exoskeletons
    - Patients' Rights to Their Own Implant Data Flows
  - Rel. to Global Mind
  - Hearing Aids, Pacemakers, Hips and Knees and Feet, Dialysis, Artificial Heart

Extinction Crisis
- CCD & Neonicotinoids

- Robosourcing

Connectome Is All Online

The Global Mind's Role
- Open Source
- Privacy for General Information
- Speed of Dispersal

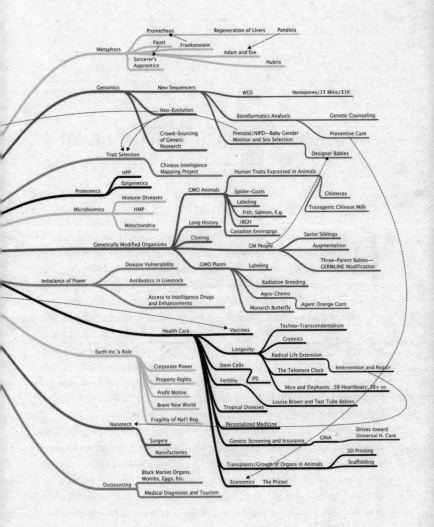

THE REINVENTION OF LIFE AND DEATH

F OR THE FIRST TIME IN HISTORY, THE DIGITIZATION OF PEOPLE IS CRE- ating a new capability to change the *being* in human being. The convergence of the Digital Revolution and the Life Sciences Rev- olution is altering not only what we know and how we communicate, not just what we do and how we do it—it is beginning to change who we are.

Already, the outsourcing and robosourcing of the genetic, biochemi- cal, and structural building blocks of life itself are leading to the emer- gence of new forms of microbes, plants, animals, and humans. We are crossing ancient boundaries: the boundary that separates one species from another, the divide between people and animals, and the distinc- tion between living things and man-made machinery.

In mythology, the lines dividing powers reserved for the gods from those allowed to people were marked by warnings; transgressions were severely punished. Yet no Zeus has forbidden us to introduce human genes into other animals; or to create hybrid creatures by mixing the genes of spiders and goats; or to surgically imbed silicon computer chips into the gray matter of human brains; or to provide a genetic menu of selectable traits for parents who wish to design their own children.

The use of science and technology in an effort to enhance human beings is taking us beyond the outer edges of the moral, ethical, and religious maps bequeathed to us by previous generations. We are now in terra incognita, where the ancient maps sometimes noted, "There Be Monsters." But those with enough courage to sail into the unknown were often richly rewarded, and in this case, the scientific community tells us with great confidence that in health care and other fields great advances await us, even though great wisdom will be needed in deciding how to proceed.

When humankind takes possession of a new and previously unimaginable power, the experience often creates a mixture of exhilaration and trepidation. In the teachings of the Abrahamic religions, the first man and the first woman were condemned to a life of toil when they seized knowledge that had been forbidden them. When Prometheus stole fire from the gods, he was condemned to eternal suffering. Every day, eagles tore into his flesh and consumed his liver, but every night his liver was regenerated so he could endure the same fate the next morning.

Ironically, scientists at Wake Forest University are now genetically engineering replacement livers in their laboratory bioreactors—and no one doubts that their groundbreaking work is anything but good. The prospects for advances in virtually all forms of health care are creating exhilaration in many fields of medical research—though it is obvious that the culture and practice of medicine, along with all of the health care professions and institutions, will soon be as disruptively reorganized as the typewriter and long-playing record businesses before it.

"PRECISION HEALTH CARE"

With exciting and nearly miraculous potential new cures for deadly diseases and debilitating conditions on the research horizon, many health care experts believe that it is inevitable that the practice of medicine will soon be radically transformed. "Personalized medicine," or, as some now refer to it, "precision medicine," is based on digital and molecular models of an individual's genes, proteins, microbial communities, and other sources of medically relevant information. Most experts believe it will almost certainly become the model for medical care.

The ability to monitor and continuously update individuals' health functions and trends will make preventive care much more effective. The new economics of health care driven by this revolution may soon make the traditional insurance model based on large risk pools obsolete because of the huge volume of fine-grained information about every individual that can now be gathered. The role of insurance companies is already being reinvented as these firms begin to adopt digital health models and mine the "big data" being created.

Pharmaceuticals, which are now aimed at large groups of individuals manifesting similar symptoms, will soon be targeted toward genetic and molecular signatures of individual patients. This revolution is already taking place in cancer treatment and in the treatment of "orphan diseases" (those that affect fewer than 200,000 people in the U.S.; the definition varies from country to country). This trend is expected to broaden as our knowledge of diseases improves.

The use of artificial intelligence—like IBM's Watson system—to assist doctors in making diagnoses and prescribing treatment options promises to reduce medical errors and enhance the skills of physicians. Just as artificial intelligence is revolutionizing the work of lawyers, it will profoundly change the work of doctors. Dr. Eric Topol, in his book *The Creative Destruction of Medicine*, writes, "This is much bigger than a change; this is the essence of creative destruction as conceptualized by [Austrian economist Joseph] Schumpeter. Not a single aspect of health and medicine today will ultimately be spared or unaffected in some way. Doctors, hospitals, the life science industry, government and its regulatory bodies: all are subject to radical transformation."

Individuals will play a different role in their own health care as well. Numerous medical teams are working with software engineers to develop more sophisticated self-tracking programs that empower individuals to be more successful in modifying unhealthy behaviors in order to manage chronic diseases. Some of these programs facilitate more regular communication between doctors and patients to discuss and interpret the continuous data flows from digital monitors that are on—and inside—the patient's body. This is part of a broader trend known as the "quantified self" movement.

Other programs and apps create social networks of individuals attempting to deal with the same health challenges—partly to take advantage of what scientists refer to as the Hawthorne effect: the simple knowledge that one's progress is being watched by others leads to an improvement in the amount of progress made. For example, some people (I do not include myself in this group) are fond of the new scales that automatically tweet their weight so that everyone who follows them will see their progress or lack thereof. There are new companies being developed based on the translation of landmark clinical trials (such as the Diabetes Prevention Program) from resource-intensive studies into social and digital media programs. Some experts believe that global access to large-scale digital programs aimed at changing destructive behaviors may soon make it possible to significantly reduce the incidence of chronic diseases like diabetes and obesity.

THE NEW ABILITIES scientists have gained to see, study, map, modify, and manipulate cells in living systems are also being applied to the human brain. These techniques have already been used to give amputees the ability to control advanced prosthetic arms and legs with their brains, as if they were using their own natural limbs—by connecting the artificial limbs to neural implants. Doctors have also empowered paralyzed monkeys to operate their arms and hands by implanting a device in the brain that is wired to the appropriate muscles. In addition, these breakthroughs offer the possibility of curing some brain diseases.

Just as the discovery of DNA led to the mapping of the human genome, the discovery of how neurons in the brain connect to and communicate with one another is leading inexorably toward the complete mapping of what brain scientists call the "connectome."* Although the data processing required is an estimated ten times greater than that required for mapping the genome, and even though several of the key technologies necessary to complete the map are still in development, brain scientists are highly confident that they will be able to complete the first "larger-scale maps of neural wiring" within the next few years.

The significance of a complete wiring diagram for the human brain can hardly be overstated. More than sixty years ago, Teilhard de Chardin predicted that "Thought might artificially perfect the thinking instrument itself."

Some doctors are using neural implants to serve as pacemakers for the brains of people who have Parkinson's disease—and provide deep brain stimulation to alleviate their symptoms. Others have used a simi-

* Olaf Sporns, a professor of computational cognitive neuroscience at Indiana University, was the first to coin the word "connectome." The National Institutes of Health now have a Human Connectome Project.

lar technique to alert people with epilepsy to the first signs of a seizure and stimulate the brain to minimize its impact. Others have long used cochlear implants connected to an external microphone to deliver sound into the brain and the auditory nerve. Interestingly, these devices must be activated in stages to give the brain a chance to adjust to them. In Boston, scientists at the Massachusetts Eye and Ear Infirmary connected a lens to a blind man's optic nerve, enabling him to perceive color and even to read large print.

Yet for all of the joy and exhilaration that accompany such miraculous advances in health care, there is also an undercurrent of apprehension for some, because the scope, magnitude, and speed of the multiple revolutions in biotechnology and the life sciences will soon require us to make almost godlike distinctions between what is likely to be good or bad for the entire future of the human species, particularly where permanently modifying the gene pool is concerned. Are we ready to make such decisions? The available evidence would suggest that the answer is not really, but we are going to make them anyway.

A COMPLEX ETHICAL CALCULUS

We know intuitively that we desperately need more wisdom than we currently have in order to responsibly wield some of these new powers. To be sure, many of the choices are easy because the obvious benefits of most new genetically based interventions make it immoral *not* to use them. The prospect of eliminating cancer, diabetes, Alzheimer's, multiple sclerosis, and other deadly and fearsome diseases ensures these new capabilities will proceed at an ever accelerating rate.

Other choices may not be as straightforward. The prospective ability to pick traits like hair and eye color, height, strength, and intelligence to create "designer babies" may be highly appealing to some parents. After all, consider what competitive parenting has already done for the test preparation industry. If some parents are seen to be giving their children a decisive advantage through the insertion of beneficial genetic traits, other parents may feel that they have to do the same.

Yet some genetic alterations will be passed on to future generations and may trigger collateral genetic changes that are not yet fully understood. Are we ready to seize control of heredity and take responsibility for actively directing the future course of evolution? As Dr. Harvey

Fineberg, president of the Institute of Medicine, put it in 2011, "We will have converted old-style evolution into neo-evolution." Are we ready to make *these* choices? Again, the answer seems to be no, yet we are going to make them anyway.

But who is the "we" who will make these choices? These incredibly powerful changes are overwhelming the present capacity of humankind for deliberative collective decision making. The atrophy of American democracy and the consequent absence of leadership in the global community have created a power vacuum at the very time when human civilization should be shaping the imperatives of this revolution in ways that protect human values. Instead of seizing the opportunity to drive down health costs and improve outcomes, the United States is decreasing its investment in biomedical research. The budget for the National Institutes of Health has declined over the past ten years, and the U.S. education system is waning in science, math, and engineering.

One of the early pioneers of in vitro fertilization, Dr. Jeffrey Steinberg, who runs the Los Angeles Fertility Institutes, said that the beginning of the age of active trait selection is now upon us. "It's time for everyone to pull their heads out of the sand," says Steinberg. One of his colleagues at the center, Marcy Darnovsky, said that the discovery in 2012 of a noninvasive process to sequence a complete fetal genome is already raising "some scenarios that are extremely troubling," adding that among the questions that may emerge from wider use of such tests is "who deserves to be born?"

Richard Hayes, executive director of the Center for Genetics and Society, expressed his concern that the debate on the ethical questions involved with fetal genomic screening and trait selection thus far has primarily involved a small expert community and that, "Average people feel overwhelmed with the technical detail. They feel disempowered." He also expressed concern that the widespread use of trait selection could lead to "an objectification of children as commodities. . . . We support the use of [preimplantation genetic diagnosis (PGD)] to allow couples at risk to have healthy children. But for non-medical, cosmetic purposes, we believe this would undermine humanity and create a techno-eugenic rat race."

Nations are competitive too. China's Beijing Genomic Institute (BGI) has installed 167 of the world's most powerful genomic sequencing machines in their Hong Kong and Shenzhen facilities that experts say will soon exceed the sequencing capacity of the entire United States.

Its initial focus is finding genes associated with higher intelligence and matching individual students with professions or occupations that make the best use of their capabilities.

According to some estimates, the Chinese government has spent well over $100 billion on life sciences research over just the last three years, and has persuaded 80,000 Chinese Ph.D.'s trained in Western countries to return to China. One Boston-based expert research team, the Monitor Group, reported in 2010 that China is "poised to become the global leader in life science discovery and innovation within the next decade." China's State Council has declared that its genetic research industry will be one of the pillars of its twenty-first-century industrial ambitions. Some researchers have reported preliminary discussions of plans to eventually sequence the genomes of almost every child in China.

Multinational corporations are also playing a powerful role, quickly exploiting the many advances in the laboratory that have profitable commercial applications. Having invaded the democracy sphere, the market sphere is now also bidding for dominance in the biosphere. Just as Earth Inc. emerged from the interconnection of billions of computers and intelligent devices able to communicate easily with one another across all national boundaries, *Life Inc.* is emerging from the ability of scientists and engineers to connect flows of genetic information among living cells across all species boundaries.

The merger between Earth Inc. and Life Inc. is well under way. Since the first patent on a gene was allowed by a Supreme Court decision in the U.S. in 1980, more than 40,000 gene patents have been issued, covering 2,000 human genes. So have tissues, including some tissues taken from patients and used for commercial purposes without their permission. (Technically, in order to receive a patent, the owner must transform, isolate, or purify the gene or tissue in some way. In practice, however, the gene or tissue itself becomes commercially controlled by the patent owner.)

There are obvious advantages to the use of the power of the profit motive and of the private sector in exploiting the new revolution in the life sciences. In 2012, the European Commission approved the first Western gene therapy drug, known as Glybera, in a treatment of a rare genetic disorder that prevents the breakdown of fat in blood. In August 2011, the U.S. Food and Drug Administration (FDA) approved a drug known as

Crizotinib for the targeted treatment of a rare type of lung cancer driven by a gene mutation.

However, the same imbalance of power that has produced levels of inequality in income is also manifested in the unequal access to the full range of innovations important to humanity flowing out of the Life Sciences Revolution. For example, one biotechnology company—Monsanto—now controls patents on the vast majority of all seeds planted in the world. A U.S. seed expert, Neil Harl of Iowa State University, said in 2010, "We now believe that Monsanto has control over as much as 90 percent of [seed genetics]."

The race to patent genes and tissues is in stark contrast to the attitude expressed by the discoverer of the polio vaccine, Jonas Salk,* when he was asked by Edward R. Murrow, "This vaccine is going to be in great demand. Everyone's going to want it. It's potentially very lucrative. Who holds the patent?" In response, Salk said, "The American people, I guess. Could you patent the sun?"

THE DIGITIZATION OF LIFE

In Salk's day, the idea of patenting life science discoveries intended for the greater good seemed odd. A few decades later, one of Salk's most distinguished peers, Norman Borlaug, implemented his Green Revolution with traditional crossbreeding and hybridization techniques at a time when the frenzy of research into the genome was still in its early stages. Toward the end of his career, Borlaug referred to the race in the U.S. to lock down ownership of patents on genetically modified plants, saying, "God help us if that were to happen, we would all starve." He opposed the dominance of the market sphere in plant genetics and told an audience in India, "We battled against patenting . . . and always stood for free exchange of germplasm." The U.S. and the European Union both recognize patents on isolated or purified genes. Recent cases in the U.S. appellate courts continue to uphold the patentability of genes.

On one level, the digitization of life is merely a twenty-first-century continuation of the story of humankind's mastery over the world. Alone

* The first effective polio vaccine was developed by Jonas Salk in 1952 and licensed for the public in 1955. A team led by Albert Sabin developed the first oral polio vaccine, which was licensed for the public in 1962.

among life-forms, we have the ability to make complex informational models of reality. Then, by learning from and manipulating the models, we gain the ability to understand and manipulate the reality. Just as the information flowing through the Global Mind is expressed in ones and zeros—the binary building blocks of the Digital Revolution—the language of DNA spoken by all living things is expressed in four letters: A, T, C, and G.

Even leaving aside its other miraculous properties, DNA's information storage capacity is incredible. In 2012, a research team at Harvard led by George Church encoded a book with more than 50,000 words into strands of DNA and then read it back with no errors. Church, a molecular biologist, said a billion copies of the book could be stored in a test tube and be retrieved for centuries, and that "a device the size of your thumb could store as much information as the whole internet."

At a deeper level, however, the discovery of how to manipulate the designs of life itself marks the beginning of an entirely new story. In the decade following the end of World War II, the double helix structure of DNA was discovered by James Watson, Francis Crick, and Rosalind Franklin. (Franklin was, historians of science now know, unfairly deprived of recognition for her seminal contributions to the scientific paper announcing the discovery in 1953. She died before the Nobel Prize in Medicine was later awarded to Watson and Crick.) In 2003, exactly fifty years later, the human genome was sequenced.

Even as the scientific community is wrestling with the challenges of all the data involved in DNA sequencing, they are beginning to sequence RNA (ribonucleic acid), which scientists are finding plays a far more sophisticated role than simply serving as a messenger system to convey the information that is translated into proteins. The proteins themselves—which among other things actually build and control the cells that make up all forms of life—are being analyzed in the Human Proteome Project, which must deal with a further large increase in the amount of data involved. Proteins take many different forms and are "folded" in patterns that affect their function and role. After they are "translated," proteins can also be chemically modified in multiple ways that extend their range of functions and control their behavior. The complexity of this analytical challenge is far beyond that involved in sequencing the genome.

"Epigenetics" involves the study of inheritable changes that do *not* involve a change in the underlying DNA. The Human Epigenome

Project has made major advances in the understanding of these changes. Several pharmaceutical products based on epigenetic breakthroughs are already helping cancer patients, and other therapeutics are being tested in human clinical trials. The decoding of the underpinnings of life, health, and disease is leading to many exciting diagnostic and therapeutic breakthroughs.

In the same way that the digital code used by computers contains both informational content and operating instructions, the intricate universal codes of biology now being deciphered and catalogued make it possible not only to understand the blueprints of life-forms, but also to change their designs and functions. By transferring genes from one species to another and by creating novel DNA strands of their own design, scientists can insert them into life-forms to transform and commandeer them to do what they want them to do. Like viruses, these DNA strands are not technically "alive" because they cannot replicate themselves. But also like viruses, they can take control of living cells and program behaviors, including the production of custom chemicals that have value in the marketplace. They can also program the replication of the DNA strands that were inserted into the life-form.

The introduction of synthetic DNA strands into living organisms has already produced beneficial advances. More than thirty years ago, one of the first breakthroughs was the synthesis of human insulin to replace less effective insulin produced from pigs and other animals. In the near future, scientists anticipate significant improvements in artificial skin and synthetic human blood. Others hope to engineer changes in cyanobacteria to produce products as diverse as fuel for vehicles and protein for human consumption.

But the spread of the technology raises questions that are troubling to bioethicists. As the head of one think tank studying this science put it, "Synthetic biology poses what may be the most profound challenge to government oversight of technology in human history, carrying with it significant economic, legal, security and ethical implications that extend far beyond the safety and capabilities of the technologies themselves. Yet by dint of economic imperative, as well as the sheer volume of scientific and commercial activity underway around the world, it is already functionally unstoppable . . . a juggernaut already beyond the reach of governance."

Because the digitization of life coincides with the emergence of the Global Mind, whenever a new piece of the larger puzzle being solved is put in place, research teams the world over instantly begin connecting it to the puzzle pieces they have been dealing with. The more genes that are sequenced, the easier and faster it is for scientists to map the network of connections between those genes and others that are known to appear in predictable patterns.

As Jun Wang, executive director of the Beijing Genomics Institute, put it, there is a "strong network effect . . . the health profile and personal genetic information of one individual will, to a certain extent, provide clues to better understand others' genomes and their medical implications. In this sense, a personal genome is not only for one, but also for all humanity."

An unprecedented collaboration in 2012 among more than 500 scientists at thirty-two different laboratories around the world resulted in a major breakthrough in the understanding of DNA bits that had been previously dismissed as having no meaningful role. They discovered that this so-called junk DNA actually contains millions of "on-off switches" arrayed in extremely complex networks that play crucial roles in controlling the function and interaction of genes. While this landmark achievement resulted in the identification of the function of 80 percent of DNA, it also humbled scientists with the realization that they are a very long way from fully understanding how genetic regulation of life really works. Job Dekker, a molecular biophysicist at the University of Massachusetts Medical School, said after the discovery that every gene is surrounded by "an ocean of regulatory elements" in a "very complicated three-dimensional structure," only one percent of which has yet been described.

The Global Mind has also facilitated the emergence of an Internet-based global marketplace in so-called biobricks—DNA strands with known properties and reliable uses—that are easily and inexpensively available to teams of synthetic biologists. Scientists at MIT, including the founder of the BioBricks Foundation, Ron Weiss, have catalyzed the creation of the Registry of Standard Biological Parts, which is serving as a global repository, or universal library, for thousands of DNA segments— segments that can be used as genetic building blocks of code free of charge. In the same way that the Internet has catalyzed the dispersal of manufacturing to hundreds of thousands of locations, it is also dispersing the basic tools and raw materials of genetic engineering to laboratories on every continent.

THE GENOME EFFECT

The convergence of the Digital Revolution and the Life Sciences Revolution is accelerating these developments at a pace that far outstrips even the speed with which computers are advancing. To illustrate how quickly this radical change is unfolding, the cost of sequencing the first human genome ten years ago was approximately $3 billion. But in 2013 detailed digital genomes of individuals are expected to be available at a cost of only $1,000 per person.

At that price, according to experts, genomes will become routinely used in medical diagnoses, in the tailoring of pharmaceuticals to an individual's genetic design, and for many other purposes. In the process, according to one genomic expert, "It will raise a host of public policy issues (privacy, security, disclosure, reimbursement, interpretation, counseling, etc.), all important topics for future discussions." In the meantime, a British company announced in 2012 that it will imminently begin selling a small disposable gene-sequencing machine for less than $900.

For the first few years, the cost reduction curve for the sequencing of individual human genomes roughly followed the 50 percent drop every eighteen to twenty-four months that has long been measured by Moore's Law. But at the end of 2007, the cost for sequencing began to drop at a significantly faster pace—in part because of the network effect, but mainly because multiple advances in the technologies involved in sequencing allowed significant increases in the length of DNA strands that can be quickly analyzed. Experts believe that these extraordinary cost reductions will continue at breakneck speed for the foreseeable future. As a result, some companies, including Life Technologies, are producing synthetic genomes on the assumption that the pace of discovery in genomics will continue to accelerate.

By contrast, the distillation of wisdom is a process that normally takes considerable time, and the molding of wisdom into accepted rules by which we can guide our choices takes more time still. For almost 4,000 years,* since the introduction by Hammurabi of the first written set of laws, we have developed legal codes by building on precedents that we have come to believe embody the distilled wisdom of past judgments made well. Yet the great convergence in science being driven by the

---

* Historians date the introduction of Hammurabi's Code to around 1780 BC.

digitization of life—with overlapping and still accelerating revolutions in genetics, epigenetics, genomics, proteomics, microbiomics, optogenetics, regenerative medicine, neuroscience, nanotechnology, materials science, cybernetics, supercomputing, bioinformatics, and other fields—is presenting us with new capabilities faster than we can discern the deeper meaning and full implications of the choices they invite us to make.

For example, the impending creation of completely new forms of artificial life capable of self-replication should, arguably, be the occasion for a full discussion and debate about not only the risks, benefits, and appropriate safeguards, but also an exploration of the deeper implications of crossing such an epochal threshold. In the prophetic words of Teilhard de Chardin in the mid-twentieth century, "We may well one day be capable of producing what the Earth, left to itself, seems no longer able to produce: a new wave of organisms, an artificially provoked neo-life."

The scientists who are working hard to achieve this breakthrough are understandably excited and enthusiastic, and the incredibly promising benefits expected to flow from their hoped-for accomplishment seem reason enough to proceed full speed ahead. As a result, it certainly seems timorous to even raise the sardonic question "What could go wrong?"

MORE THAN A little, it seems—or at least it seems totally reasonable to explore the question. Craig Venter, who had already made history by sequencing his own genome, made history again in 2010 by creating the first live bacteria made completely from synthetic DNA. Although some scientists minimized the accomplishment by pointing out that Venter had merely copied the blueprint of a known bacterium, and had used the empty shell of another as the container for his new life-form, others marked it as an important turning point.

In July 2012, Venter and his colleagues, along with a scientific team at Stanford, announced the completion of a software model containing all of the genes (525 of them—the smallest number known), cells, RNA, proteins, and metabolites (small molecules generated in cells) of an organism—a free-living microbe known as *Mycoplasma genitalium*. Venter is now working to create a unique artificial life-form in a project that is intended to discover the minimum amount of DNA information necessary for self-replication. "We are trying to understand the fundamental principles for the design of life, so that we can redesign it—in the way an intelligent designer would have done in the first place, if there had been one," Venter said. His reference to an "intelligent designer" seems intended as implicit dismissal of creationism and reflects a newly combative attitude that many scientists have understandably come to feel is appropriate in response to the aggressive attacks on evolution by many fundamentalists.

One need not believe in a deity, however, in order to entertain the possibility that the web of life has an emergent holistic integrity featuring linkages we do not yet fully understand and which we might not risk disrupting if we did. Even though our understanding of hubris originated in ancient stories about the downfall of men who took for themselves powers reserved for the gods, its deeper meaning—and the risk it often carries—is rooted in human arrogance and pride, whether or not it involves an offense against the deity. As Shakespeare wrote, "The fault, dear Brutus, is not in our stars, but in ourselves." For all of us, hubris is inherent in human nature. Its essence includes prideful overconfidence in the completeness of one's own understanding of the consequences of exercising power in a realm that may well have complexities that still extend beyond the understanding of any human.

Nor is the posture of fundamentalism unique to the religious. Reductionism—the belief that scientific understanding is usually best pursued by breaking down phenomena into their component parts and subparts—has sometimes led to a form of selective attention that can cause observers to overlook emergent phenomena that arise in complex systems, and in their interaction with other complex systems.

One of the world's most distinguished evolutionary biologists, E. O. Wilson, has been bitterly attacked by many of his peers for his proposal that Darwinian selection operates not only at the level of individual members of a species, but also at the level of "superorganisms"—by which he means that adaptations serving the interests of a species as a whole may be selected even if they do not enhance the prospects for survival of the individual creatures manifesting those adaptations. Wilson, who was but is no longer a Christian, is not proposing "intelligent design" of the sort believed in by creationists. He is, rather, asserting that there is another layer to the complexity of evolution that operates on an "emergent" level.

Francis Collins, a devout Christian who headed the U.S. government's Human Genome Project (which announced its results at the same time that Craig Venter announced his), has bemoaned the "increasing polarization between the scientific and spiritual worldviews, much of it, I think, driven by those who are threatened by the alternatives and who are unwilling to consider the possibility that there might be harmony here. . . . We have to recognize that our understanding of nature is something that grows decade by decade, century by century."

Venter, for his part, is fully confident that enough is already known to justify a large-scale project to reinvent life according to a human design. "Life evolved in a messy fashion through random changes over three billion years," he says. "We are designing it so that there are modules for different functions, such as chromosome replication and cell division, and then we can decide what metabolism we want it to have."

ARTIFICIAL LIFE

As with many of the startling new advances in the life sciences, the design and creation of artificial life-forms offers the credible promise of breakthroughs in health care, energy production, environmental remediation, and many other fields. One of the new products Venter and other scientists hope to create is synthetic viruses engineered to destroy or weaken antibiotic-resistant bacteria. These synthetic viruses—or bacteriophages—can be programmed to attack only the targeted bacteria, leaving other cells alone. These viruses utilize sophisticated strategies to not only kill the bacteria but also use the bacteria before it dies to replicate the synthetic virus so that it can go on killing other targeted bacteria until the infection subsides.

The use of new synthetic organisms for the acceleration of vaccine development is also generating great hope. These synthetic vaccines are being designed as part of the world's effort to prepare for potential new pandemics like the bird flu (H5N1) of 2007 and the so-called swine flu (H1N1) of 2009. Scientists have been particularly concerned that the H5N1 bird flu is now only a few mutations away from developing an ability to pass from one human to another through airborne transmission.

The traditional process by which vaccines are developed requires a lengthy development, production, and testing cycle of months, not days, which makes it nearly impossible for doctors to obtain adequate supplies of the vaccine after a new mutant of the virus begins spreading. Scientists are using the tools of synthetic biology to accelerate the evolution of existing flu strains in the laboratory and they hope to be able to predict which new strains are most likely to emerge. Then, by studying their blueprints, scientists hope to preemptively synthesize vaccines that will be able to stop whatever mutant of the virus subsequently appears in the real world and stockpile supplies in anticipation of the new virus's emergence. Disposable biofactories are being set up around the world to decrease the cost and time of manufacturing of vaccines. It is now possible to set up a biofactory in a remote rural village where the vaccine is needed quickly to stop the spread of a newly discovered strain of virus or bacteria.

Some experts have also predicted that synthetic biology may supplant 15 to 20 percent of the global chemical industry within the next few years, producing many chemical products more cheaply than they can be extracted from natural sources, producing pharmaceutical products, bioplastics, and other new materials. Some predict that this new approach to chemical and pharmaceutical manufacturing will—by using the 3D printing technique described in Chapter 1—revolutionize the production process by utilizing a "widely dispersed" strategy. Since most of the value lies in the information, which can easily be transmitted to unlimited locations, the actual production process by which the information is translated into production of Synthetic Biology products can be located almost anywhere.

These and other exciting prospects expected to accompany the advances in synthetic biology and the creation of artificial life-forms have led many to impatiently dismiss any concerns about unwanted consequences. This impatience is not of recent vintage. Ninety years ago, English biochemist J. B. S. Haldane wrote an influential essay that provoked a series of futurist speculations about human beings taking active control of the future course of evolution. In an effort to place in context—and essentially dismiss—the widespread uneasiness about the subject, he wrote:

> The chemical or physical inventor is always a Prometheus. There is no great invention, from fire to flying, which has not been hailed as an insult to some god. But if every physical and chemical invention is a blasphemy, every biological invention is a perversion. There is hardly one which, on first being brought to the notice of an observer from any nation which has not previously heard of their existence, would not appear to him as indecent and unnatural.

By contrast, Leon Kass, who chaired the U.S. Council on Bioethics from 2001 to 2005, has argued that the intuition or feeling that something is somehow repugnant should not be automatically dismissed as antiscientific: "In some crucial cases, however, repugnance is the emotional expression of deep wisdom, beyond reason's power completely to articulate it. . . . We intuit and we feel, immediately and without argument, the violation of things that we rightfully hold dear."

In Chapter 2, the word "creepy" was used by several observers of trends unfolding in the digital world, such as the ubiquitous tracking of voluminous amounts of information about most people who use the Internet. As others have noted, "creepy" is an imprecise word because it describes a feeling that itself lacks precision—not fear, but a vague uneasiness about something whose nature and implications are so unfamiliar that we feel the need to be alert to the possibility that something fearful or harmful might emerge. There is a comparably indeterminate "pre-fear" that many feel when contemplating some of the onrushing advances in the world of genetic engineering.

An example: a method for producing spider silk has been developed by genetic engineers who insert genes from orb-making spiders into goats which then secrete the spider silk—along with milk—from their udders. Spider silk is incredibly useful because it is both elastic and five times stronger than steel by weight. The spiders themselves cannot be farmed because of their antisocial, cannibalistic nature. But the insertion of their silk-producing genes in the goats allows not only a larger volume of spider silk to be produced, but also allows the farming of the goats.*

In any case, there is no doubt that the widespread use of synthetic biology—and particularly the use of self-replicating artificial life-forms—could potentially generate radical changes in the world, including some potential changes that arguably should be carefully monitored. There are, after all, too many examples of plants and animals purposely introduced into a new, nonnative environment that then quickly spread out of control and disrupted the ecosystem into which they were introduced.

Kudzu, a Japanese plant that was introduced into my native South-

---

* Other scientists have mimicked the molecular design of spider silk by synthesizing their own from a commercially available substance (polyurethane elastomer) treated with clay platelets only one nanometer (a billionth of a meter) thick and only 25 nanometers across, then carefully processing the mixture. This work has been funded by the Institute for Soldier Nanotechnologies at MIT because the military applications are considered of such high importance.

ern United States as a means of combating soil erosion, spread wildly and became a threat to native trees and plants. It became known as "the vine that ate the South." Do we have to worry about "microbial kudzu" if a synthetic life-form capable of self-replication is introduced into the biosphere for specific useful purposes, but then spreads rapidly in ways that are not predicted or even contemplated?

Often in the past, urgent questions raised about powerful new breakthroughs in science and technology have focused on potentially catastrophic disaster scenarios that turned out to be based more on fear than reason—when the questions that should have been pursued were about other more diffuse impacts. For example, on the eve of the Bikini Atoll test of the world's first hydrogen bomb in 1954, a few scientists raised the concern that the explosion could theoretically trigger a chain reaction in the ocean and create an unimaginable ecological Armageddon.

Their fearful speculation was dismissed by physicists who were confident that such an event was absurdly implausible. And of course it was. But other questions focused on deeper and more relevant concerns were not adequately dealt with at all. Would this thermonuclear explosion contribute significantly to the diversion of trillions of dollars into weaponry and further accelerate a dangerous nuclear arms race that threatened the survival of human civilization?

For the most part, the fears of microbial kudzu (or their microscopic mechanical counterparts—self-replicating nanobots, or so-called gray goo), are now often described as probably overblown, although the executive director of GeneWatch, an NGO watchdog organization, Helen Wallace, told *The New York Times Magazine*, "It's almost inevitable that there will be some level of escape. The question is: Will those organisms survive and reproduce? I don't think anyone knows."

But what about other questions that may seem less urgent but may be more important in the long run: if we robosource life itself, and synthesize life-forms that are more suited to our design than the pattern that has been followed by life for 3.5 billion years, how is this new capability likely to alter our relationship to nature? How is it likely to change nature? Are we comfortable forging full speed ahead without any organized effort to identify and avoid potential outcomes that we may not like?

One concern that technologists and counterterrorism specialists have highlighted is the possibility of a new generation of biological weapons. After all, some of the early developments in genetic engineering, we now know, were employed by the Soviet Union in a secret biological weapons program forty years ago. If the exciting tools of the Digital Revolution have been weaponized for cyberwar, why would we not want some safeguards to prevent the same diversion of synthetic biology into bioweapons?

The New and Emerging Science and Technology (NEST) high-level expert group of the European Commission wrote in 2005 that "The possibility of designing a new virus or bacterium *à la carte* could be used by bioterrorists to create new resistant pathogenic strains or organisms, perhaps even engineered to attack genetically specific sub-populations." In 2012, the U.S. National Science Advisory Board for Biosecurity attempted to halt the publication of two scientific research papers—one in *Nature* and the other in *Science*—that contained details on the genetic code of a mutated strain of bird flu that had been modified in an effort to determine what genetic changes could make the virus more readily transmissible among mammals.

Citing concerns that the detailed design of a virus that was only a few mutations away from a form that could be spread by human-to-human transmission, the bioterrorism officials tried to dissuade scientists from publishing the full genetic sequence that accompanied their papers. Although publication was allowed to proceed after a full security review, the U.S. government remains actively involved in monitoring genetic research that could lead to new bioweapons. Under U.S. law, the FBI screens the members of research teams working on projects considered militarily sensitive.

HUMAN CLONING

Among the few lines of experiments specifically *banned* by the U.S. government are those involving federally funded research into the cloning of human beings. As vice president, not long after the birth of the first cloned sheep, Dolly, in 1996, when it became clear that human cloning was imminently feasible, I strongly supported this interim ban pending a much fuller exploration of the implications for humanity of proceeding down that path, and called for the creation of a new National Bioethics Advisory Commission to review the ethical, moral, and legal implications of human cloning.

A few years earlier, as chairman of the Senate Subcommittee on

Science, I had pushed successfully for a commitment of 3 percent of the funding for the Human Genome Project to be allocated to extensive ethical, legal, and social implications (they are now referred to as ELSI grants), in an effort to ensure careful study of the difficult questions that were emerging more quickly than their answers. This set-aside has become the largest government-financed research program into ethics ever established. James Watson, the co-discoverer of the double helix, who by then had been named to head the Genome Project, was enthusiastically supportive of the ethics program.

The ethics of human cloning has been debated almost since the very beginning of the DNA era. The original paper published by Watson and Crick in 1953 included this sentence: "It has not escaped our notice that the specific pairing we have postulated immediately suggests a possible copying mechanism for the genetic material." As chairman of the Science Investigations Subcommittee in the U.S. House of Representatives, I conducted a series of hearings in the early 1980s about the emerging science of cloning, genetic engineering, and genetic screening. Scientists were focused at that stage on cloning animals, and fifteen years later they succeeded with Dolly. Since then, they have cloned many other livestock and other animals.

But from the start of their experiments, the scientists were clear that all of the progress they were making on the cloning of animals was directly applicable to the cloning of people—and that it was only ethical concerns that had prevented them from attempting such procedures. Since 1996, human cloning has been made illegal in almost every country in Europe, and the then director-general of the World Health Organization called the procedure "ethically unacceptable as it would violate some of the basic principles which govern medically assisted reproduction. These include respect for the dignity of the human being and the protection of the security of human genetic material."

Nevertheless, most anticipate that with the passage of time, and further development and refinement of the technique, human cloning will eventually take place—at least in circumstances where a clear medical benefit can be gained without causing a clear form of harm to the individual who is cloned or to society at large. In 2011, scientists at the New York Stem Cell Foundation Laboratory announced that they had cloned human embryos by reprogramming an adult human egg cell, engineering it to return to its embryonic stage, and then created from it a line of identical embryonic stem cells that reproduced themselves. Although the DNA of these cells is not identical to that of the patient who donated the egg cell, they are identical to one another, which facilitates the efficacy of research carried out on them.

Several countries, including Brazil, Mexico, and Canada, have banned the cloning of human embryos for research. The United States has not done so, and several Asian countries seem to have far fewer misgivings about moving forward aggressively with the science of cloning human embryos—if not actual humans. From time to time, there are reports that one or another fertility doctor working at some secret laboratory, located in a nation without a ban on human cloning, has broken this modern taboo against human cloning. Most if not all of these stories, however, have been suspected of being fraudulent. There has yet been no confirmed birth of a human clone.

In general, those in favor of proceeding with experiments in human cloning believe that the procedure is not really different from other forms of technological progress, that it is inevitable in any case, and is significantly more promising than most experiments because of the medical benefits that can be gained. They believe that the decision on whether or not to proceed with a specific clone should—like a decision on abortion—be in the hands of the parent or parents.

Those who oppose cloning of people fear that its use would undermine the dignity of individuals and run the risk of "commoditizing" human beings. Cloning does theoretically create the possibility of mass-producing many genetic replicas of the same original—a process that would be as different from natural reproduction as manufacturing is from craftsmanship.

Some base their argument on religious views of the rights and protections due to every person, though many who oppose human cloning base their views not on religion, but on a more generalized humanist assertion of individual dignity. In essence, they fear that the manipulation of humanity might undermine the definition of those individuals who have been manipulated as fully human. This concern seems to rest, however, on an assumption that human beings are reducible to their genetic makeup—a view that is normally inconsistent with the ideology of those who make the protection of individual dignity a top priority.

Both the temporary delay in the public release of details concerning how to create dangerous mutations in the H5N1 bird flu virus and the temporary ban on government-funded research into human cloning represent rare examples of thoughtful—if controversial—oversight of potentially problematic developments in order to assess their implications for humanity as a whole. Both represented examples of U.S. leadership that led to at least a temporary global consensus. In neither case was there a powerful industry seeking to push forward quickly in spite of the misgivings expressed by representatives of the public.

ANTIBIOTICS BEFORE SWINE

Unfortunately, when there *is* a strong commercial interest in influencing governments to make a decision that runs roughshod over the public interest, business lobbies are often able to have their way with government—which once again raises the question: who is the "we" that makes decisions about the future course of the Life Sciences Revolution when important human values are placed at risk? In the age of Earth Inc., Life Inc., and the Global Mind, the record of decision making includes troubling examples of obsequious deference to the interests of multinational corporations and a reckless disregard of sound science.

Consider the shameful acquiescence by the U.S. Congress in the livestock industry's continuing absurd dominance of antibiotic use in the United States. In yet another illustration of the dangerous imbalance of power in political decision making, a truly shocking 80 percent of all U.S. antibiotics are still allowed to be legally used on farms in livestock feed and for injections in spite of grave threats to human health. In 2012, the FDA began an effort to limit this use of antibiotics with a new rule that will require a prescription from veterinarians.

Since the discovery of penicillin in 1929 by Alexander Fleming, antibiotics have become one of the most important advances in the history of health care. Although Fleming said his discovery was "accidental," the legendary Irish scientist John Tyndall (who first discovered that CO_2 traps heat) reported to the Royal Society in London in 1875 that a species of *Penicillium* had destroyed some of the bacteria he was working with, and Ernest Duchesne wrote in 1897 on the destruction of bacteria by another species of *Penicillium*. Duchesne had recommended research into his discovery but entered the army and went to war immediately after publication of his paper. He died of tuberculosis before he could resume his work.

In the wake of penicillin, which was not used in a significant way until the early 1940s, many other potent antibiotics were discovered in the 1950s and 1960s. In the last few decades, though, the discoveries have slowed to a trickle. The inappropriate and irresponsible use of this limited arsenal of life-saving antibiotics is rapidly eroding their effectiveness. Pathogens that are stopped by antibiotics mutate and evolve over time in ways that circumvent the effectiveness of the antibiotic.

Consequently, doctors and other medical experts have urged almost since the first use of these miracle cures that they be used sparingly and only when they are clearly needed. After all, the more they are used, the more opportunities the pathogens have to evolve through successive generations before they stumble upon new traits that make the antibiotics impotent. Some antibiotics have already become ineffective against certain diseases. And with the slowing discovery of new antibiotics, the potency of the ones we use in our current arsenal is being weakened at a rate that is frightening to many health experts. The effectiveness of our antibiotic arsenal—like topsoil and groundwater—can be depleted quickly but regenerated only at an agonizingly slow rate.

One of the most serious new "superbugs" is multidrug-resistant tuberculosis, which, according to Dr. Margaret Chan, director-general of the World Health Organization, is extremely difficult and expensive to treat. At present, 1.34 million people die from tuberculosis each year. Of the 12 million cases in 2010, Chan estimated that 650,000 involved strains of TB that were multidrug-resistant. The prospect of a "post antibiotic world" means, Chan said, "Things as common as strep throat or a child's scratched knee, could once again kill." In response to these concerns, the FDA formed a new task force in 2012 to support development of new antibacterial drugs.

But in spite of these basic medical facts, many governments—including, shockingly, the United States government—allow the massive use of antibiotics by the livestock industry as a growth stimulant. The mechanism by which the antibiotics cause a faster growth rate in livestock is not yet fully understood, but the impact on profits is very clear and sizable. The pathogens in the guts of the livestock are evolving quickly into superbugs that are immune to the impact of antibiotics. Since the antibiotics are given in subtherapeutic doses and are not principally used for the health of the livestock anyway, the livestock companies don't particularly care. And of course, their lobbyists tendentiously dispute the science while handing out campaign contributions to office-holders.

Last year, scientists confirmed that a staphylococcus germ that was vulnerable to antibiotics jumped from humans to pigs whose daily food ration included tetracycline and methicillin. Then, the scientists confirmed that the same staph germ, after becoming resistant to the antibiotics, found a way to jump back from pigs into humans.

The particular staph germ that was studied—CC398—has spread in populations of pigs, chickens, and cattle. Careful analysis of the genetic structure of the germ proved that it was a direct ancestor of an antibiotic-susceptible germ that originated in people. It is now present, according to the American Society for Microbiology, in almost half of all meat that has been sampled in the U.S. Although it can be killed with thorough cooking of the meat, it can nevertheless infect people if it cross-contaminates kitchen utensils, countertops, or pans.

Again, the U.S. government's frequently obsequious approach to regulatory decision making when a powerful industry exerts its influence stands in stark contrast to the approach it takes when commercial interests are not yet actively engaged. In the latter case, it seems to be easier for government to sensitively apply the precautionary principle. But this controversy illustrates the former case: those who benefit from the massive and reckless use of antibiotics in the livestock industry have fought a rearguard action for decades and have thus far been successful in preventing a ban or even, until recently, a regulation limiting this insane practice.

The European Union has already banned antibiotics in livestock feed, but in a number of other countries the practice continues unimpeded. The staph germ that has jumped from people to livestock and back again is only one of many bacteria that are now becoming resistant to antibiotics because of our idiotic acceptance of the livestock industry's insistence that it is perfectly fine for them to reduce some of their costs by becoming factories for turning out killer germs against which antibiotics have no effect. In a democracy that actually functioned as it is supposed to, this would not be a close question.

Legislators have also repeatedly voted down a law that would prevent the sale of animals with mad cow disease (bovine spongiform encephalopathy, or BSE)—a neurodegenerative brain disease caused by eating beef contaminated during the slaughtering process by brain or spinal cord tissue from an animal infected by the pathogen (a misfolded protein, or prion) that causes the disease. Animals with later stages of the disease can carry the prions in other tissues as well. When an animal on the way to the slaughterhouse begins stumbling, staggering, and falling down, it is fifty times more likely to have the disease.

The struggle in repeated votes in the Congress has been over whether those specific animals manifesting those specific symptoms should be diverted from the food supply. At least three quarters of the confirmed cases of mad cow disease in people in North America were traced to animals that had manifested those symptoms just before they were slaughtered. But the political power and lobbying muscle of the livestock industry has so intimidated and enthralled elected representatives in the U.S. that lawmakers have repeatedly voted to put the public at risk in order to protect a tiny portion of the industry's profits. The Obama administration has issued a regulation that embodies the intent of the law rejected by Congress. However, because it is merely a regulation, it could be reversed by Obama's successor as president. Again, in a working democracy, this would hardly be a close question.

THE INABILITY OF Congress to free itself from the influence of special interests has implications for how the United States can make the difficult and sensitive judgments that lie ahead in the Life Sciences Revolution. If the elected representatives of the people cannot be trusted to make even obvious choices in the public interest—as in the mad cow votes or the decisions on frittering away antibiotic resistance in order to enrich the livestock industry—then where else can these choices be made? Who else can make them? And even if such decisions are made sensitively and well in one country, what is to prevent the wrong decision being made elsewhere? And if the future of human heredity is affected in perpetuity, is that an acceptable outcome?

EUGENICS

The past record of decisions made by government about genetics is far from reassuring. History sometimes resembles Greek mythology, in that like the gods, our past mistakes often mark important boundaries with warnings. The history of the eugenics movement 100 years ago provides such a warning: a profound misunderstanding of Darwinian evolution was used as the basis for misguided efforts by government to engineer the genetic makeup of populations according to racist and other unacceptable criteria.

In retrospect, the eugenics movement should have been vigorously condemned at the time—all the more so because of the stature of some of its surprising proponents. A number of otherwise thoughtful Americans came to support active efforts by their government to shape the genetic future of the U.S. population through the forcible sterilization of individuals who they feared would otherwise pass along undesirable traits to future generations.

In 1922, a "model eugenical sterilization law" (originally written in 1914) was published by Harry Laughlin, superintendent of the recently established "Eugenics Record Office" in New York State, to authorize sterilization of people regarded as

(1) Feeble-minded; (2) Insane, (including the psychopathic); (3) Criminalistic (including the delinquent and wayward); (4) Epileptic; (5) Inebriate (including drug-habitues); (6) Diseased (including the tuberculous, the syphilitic, the leprous, and others with chronic, infectious and legally segregable diseases); (7) Blind (including those with seriously impaired vision); (8) Deaf (including those with seriously impaired hearing); (9) Deformed (including the crippled); and (10) Dependent (including orphans, ne'er-do-wells, the homeless, tramps and paupers.)

Between 1907 and 1963, over 64,000 people were sterilized under laws similar to Laughlin's design. He argued that such individuals were burdensome to the state because of the expense of taking care of them. He and others also made the case that the advances in sanitation, public health, and nutrition during the previous century had led to the survival of more "undesirable" people who were reproducing at rates not possible in the past.

What makes the list of traits in Laughlin's "model law" bizarre as well as offensive is that he obviously believed they were heritable. Ironically, Laughlin was himself an epileptic; thus, under his model legislation, he would have been suitable for forced sterilization. Laughlin's malignant theories also had an impact on U.S. immigration law. His work on evaluating recent immigrants from Southern and Eastern Europe was influential in forming the highly restrictive quota system of 1924.

As pointed out by Jonathan Moreno in his book *The Body Politic*, the eugenics movement was influenced by deep confusion over what evolution really means. The phrase "survival of the fittest" did not originate with Charles Darwin, but with his cousin Sir Francis Galton, and was then popularized by Herbert Spencer—whose rival theory of evolution was based on the crackpot ideas of Jean-Baptiste Lamarck. Lamarck argued that characteristics developed by individuals after their birth were genetically passed on to their offspring in the next generation.

A similar bastardization of evolutionary theory was also promoted in the Soviet Union by Trofim Lysenko—who was responsible for preventing the teaching of mainstream genetics during the three decades of his rein in Soviet science. Geneticists who disagreed with Lysenko were secretly arrested; some were found dead in unexplained circumstances. Lysenko's warped ideology demanded that biological theory conform with Soviet agricultural needs—much as some U.S. politicians today insist that climate science be changed to conform with their desire to promote the unrestrained burning of oil and coal.

Darwin actually taught that it was not necessarily the "fittest" who survived, but rather those that were best adapted to their environments. Nevertheless, the twisted and mistaken version of Darwin's theory that was reflected in his cousin's formulation helped to give rise to the notion of Social Darwinism—which led, in turn, to misguided policy debates that in some respects continue to this day.

Some of the early progressives were seduced by this twisted version of Darwin's theory into believing that the state had an affirmative duty to do what it could to diminish the proliferation of unfavorable Lamarckian traits that they mistakenly believed were becoming more common because prior state interventions had made life easier for these "undesirables," and had enabled them to proliferate.

The same flawed assumptions led those on the political right to a different judgment: the state should pull back from all those policy interventions that had, in the name of what they felt was misguided compassion, led to the proliferation of "undesirables" in the first place. There were quite a few reactionary advocates of eugenics. At least one of them survives into the twenty-first century—the Pioneer Fund, described as a hate group by the Southern Poverty Law Center. Incidentally, its founding president was none other than Harry Laughlin.

Eugenics also found support, historians say, because of the socio-economic turmoil of the first decades of the twentieth century—rapid industrialization and urbanization, the disruption of long familiar social patterns, waves of immigration, and economic stress caused by low wages and episodic high unemployment. These factors combined with a new zeal for progressive reform to produce a wildly distorted view of what was appropriate by way of state intervention in heredity.

Although this episode in the world's history is now regarded as horribly unethical—in part because thirty years after it began, the genocidal crimes of Adolf Hitler discredited all race-based, and many genetics-based, theories that were even vaguely similar to that of Nazism. Nevertheless, some of the subtler lessons of the eugenics travesty have not yet been incorporated into the emerging debate over current proposals that some have labeled "neo-eugenics."

One of the greatest challenges facing democracies in this new era is how to ensure that policy decisions involving cutting-edge science are based on a clear and accurate understanding of the science involved. In the case of eugenics, the basic misconception traced back to Lamarck concerning what is inheritable and what is not contributed to an embarrassing and deeply immoral policy that might have been avoided if policymakers and the general public had been debating policy on the basis of accurate science.

It is worth noting that almost a century after the eugenics tragedy, approximately half of all Americans still say they do not believe in evolution. The judgments that must be made within the political system of the United States in the near future—and in other countries—are difficult enough even when based on an accurate reading of the science. When this inherent difficulty is compounded by flawed assumptions concerning the science that gives rise to the need to make these decisions, the vulnerability to mistaken judgments goes up accordingly.

As will be evident in the next chapter, the decisions faced by civilization where global warming is concerned are likewise difficult enough when they are based on an accurate reading of the science. But when policymakers base arguments on gross misrepresentations of the science, the degree of difficulty goes up considerably. When gross and willful misunderstandings of the science are intentionally created and reinforced by large carbon polluters who wish to paralyze the debate over how to reduce CO_2 emissions, they are, in my opinion, committing a nearly unforgivable crime against democracy and against the future well-being of the human species.

In a 1927 opinion by Justice Oliver Wendell Holmes Jr., the U.S. Supreme Court upheld one of the more than two dozen state eugenics laws. The case, *Buck v. Bell*, involved the forcible sterilization of a young Virginia woman who was allegedly "feeble-minded" and sexually promiscuous. Under the facts presented to the court, the young woman, Carrie Buck, had already had a child at the age of seventeen. In affirming the state's right to perform the sterilization, Holmes wrote that, "Society can prevent those who are manifestly unfit from continuing their kind. . . . Three generations of imbeciles are enough."

A half century after the Supreme Court decision, which has never been overturned, the director of the hospital where Buck had been sterilized tracked down the woman who had been forcibly sterilized, who was then in her eighties. He found that, far from being an "imbecile," Buck was lucid and of normal intelligence. Upon closer examination of the facts, it became obvious that they were not as represented in court. Young Carrie Buck was a foster child who had been raped by a nephew of one of her foster parents, who then committed her to the Virginia State Colony for Epileptics and Feebleminded in order to avoid what they feared would otherwise be a scandal.

As it happens, Carrie's mother, Emma Buck—the first of the three generations referred to by Justice Holmes—had also been committed to the same asylum under circumstances that are not entirely clear, although testimony indicated that she had syphilis and was unmarried when she gave birth to Carrie. In any case, the superintendent of the Virginia Colony, Albert Priddy, was eager to find a test case that could go to the Supreme Court and provide legal cover for the forced sterilizations that his and other institutions already had under way. He declared Buck "congenitally and incurably defective"; Buck's legal guardian picked a lawyer to defend her in the case who was extremely close to Priddy and a close friend since childhood to the lawyer for the Colony, a eugenics and sterilization advocate (and former Colony director) named Aubrey Strode.

Historian Paul Lombardo of Georgia State University, who wrote an extensively researched book on the case, wrote that the entire proceeding was "based on deceit and betrayal. . . . The fix was in." Buck's appointed defense counsel put forward no witnesses and no evidence, and conceded the description of his client as a "middle-grade moron." Harry Laughlin, who had never met Carrie Buck, her mother, or her daughter, testified to the court in a written statement that all three were part of the "shiftless, ignorant, and worthless class of anti-social whites of the South."

As for the third generation of Bucks, Carrie's daughter, Vivian, was examined at the age of a few weeks by a nurse who testified: "There is a look about it that is not quite normal." The baby girl was taken from her family and given to the family of Carrie's rapist. After making the honor roll in school, Vivian died of measles in the second grade. Incidentally, Carrie's sister, Doris, was also sterilized at the same institution (more than 4,000 sterilizations were performed there), though doctors lied to her about the operation when it was performed and told her it was for appendicitis. Like Carrie, Doris did not learn until much later in her life why she was unable to have children.

The "model legislation" put forward by Laughlin, which was the basis for the Virginia statute upheld by the Supreme Court, was soon thereafter used by the Third Reich as the basis for their sterilization of more than 350,000 people—just as the psychology-based marketing text written by Edward Bernays was used by Goebbels in designing the propaganda program surrounding the launch and prosecution of Hitler's genocide. The Nazis presented Laughlin with an honorary degree in 1936 from the University of Heidelberg for his work in the "science of racial cleansing."

Shamefully, eugenics was supported by, among others, President Woodrow Wilson, Alexander Graham Bell, Margaret Sanger, who founded the movement in favor of birth control—an idea that was, at the time, more controversial than eugenics—and by Theodore Roosevelt after he left the White House. In 1913, Roosevelt wrote in a letter,

It is really extraordinary that our people refuse to apply to human beings such elementary knowledge as every successful farmer is obliged to apply to his own stock breeding. Any group of farmers who permitted their best stock not to breed, and let all the increase come from the worst stock, would be treated as fit inmates for an

asylum. Yet we fail to understand that such conduct is rational compared to the conduct of a nation which permits unlimited breeding
from the worst stocks, physically and morally, while it encourages
or connives at the cold selfishness or the twisted sentimentality as
a result of which the men and women who ought to marry, and if
married have large families, remain celibates or have no children or
only one or two.

Sanger, for her part, disagreed with the methods of eugenics advocates,
but nevertheless wrote that they were working toward a goal she supported: "To assist the race toward the elimination of the unfit." One of
Sanger's own goals in promoting contraception, she wrote in 1919, was,
"More children from the fit, less from the unfit—that is the chief issue
of birth control."

The United States is not the only democratic nation with a troubling
history of forced sterilization. Between 1935 and 1976, Sweden forcibly
sterilized more than 60,000 people, including "mixed-race individuals,
single mothers with many children, deviants, gypsies and other 'vaga-
bonds.'" Curiously, even today Sweden is one of seventeen European
countries requiring sterilization before a transgendered person can officially change his or her gender identification on government identification documents. Parliamentarians in Sweden are debating whether
to change the law, which dates back to 1972. There is no scientific or
medical reason why the law has not already been repealed. The fear and
misunderstanding of small conservative political parties has made the
repeal of the law impossible thus far.

In Uzbekistan, forced sterilizations apparently began in 2004 and
became official state policy in 2009. Gynecologists are given a quota of
the number of women per week they are required to sterilize. "We go
from house to house convincing women to have the operation," said a
rural surgeon. "It's easy to talk a poor woman into it. It's also easy to trick
them."

In China, the issue of forced abortions has resurfaced with the allegations by escaped activist Chen Guangcheng, but the outgoing premier
Wen Jiabao has publicly called for a ban not only on forced abortion, but
also of "fetus gender identification." Nevertheless, many women who
have abortions in China are also sterilized against their will. In India,
although forcible sterilization is illegal, doctors and government officials

are paid a bonus for each person who is sterilized. These incentives apparently lead to widespread abuses, particularly in rural areas where many women are sterilized under false pretenses.

The global nature of the revolution in biotechnology and the life sciences—like the new global commercial realities that have emerged with Earth Inc.—means that any single nation's moral, ethical, and legal judgments may not have much impact on the practical decisions of other nations. Some general rules about what is acceptable, what is worthy of extra caution, and what should be prohibited have been tentatively observed, but there is no existing means for arriving at universal moral judgments about these new unfolding capabilities.

CHINA AND THE LIFE SCIENCES

As noted earlier, China appears determined to become the world's superpower in the application of genetic and life science analysis. The Beijing Genomics Institute (BGI), which is leading China's commitment to genomic analysis, has already completed the full genomes of fifty animal and plant species, including silk worms, pandas, honeybees, rice, soybeans, and others—along with more than 1,000 species of bacteria. But China's principal focus seems to be on what is arguably the most important, and certainly the most intriguing, part of the human body that can be modified by the new breakthroughs in life sciences and related fields: the human brain and the enhancement and more productive use of human intelligence.

Toward this end, in 2011 the BGI established China's National Gene Bank in Shenzhen, where it has been seeking to identify which genes are involved in determining intelligence. It is conducting a complete genomic analysis of 2,000 Chinese schoolchildren (1,000 prodigies from the nation's best schools, and 1,000 children considered of average intelligence) and matching the results with their achievements in school.

In the U.S., such a study would be extremely controversial, partly because of residual revulsion at the eugenics scandal, and partly because of a generalized wariness about linking intelligence to family heritage in any society that values egalitarian principles. In addition, many biologists, including Francis Collins, who succeeded James Watson as the leader of the Human Genome Project, have said that it is currently scientifically impossible in any case to link genetic information about a child to intelligence. However, some researchers disagree and believe that eventually genes associated with intelligence may well be identified.

Meanwhile, the speed with which advances are being made in mapping the neuronal connections of the human brain continue to move forward significantly faster than the progress measured by Moore's Law in the manufacturing of integrated circuits. Already, the connectome of a species of nematode, which has only 302 neurons, has been completed. Nevertheless, with an estimated 100 billion neurons in an adult human brain and at least 100 trillion synaptic connections, the challenge of fully mapping the human connectome is a daunting one. And even then, the work of understanding the human brain's functioning will have barely begun.

In that regard, it is worth remembering that after the completion of the first full sequencing of the human genome, scientists immediately realized that the map of genes was only their introduction to the even larger task of mapping all of the proteins that are expressed by the genes—which themselves adopt multiple geometric forms and are subject to significant biochemical modifications after they are translated by the genes.

In the same way, once the connectome is completed, brain scientists will have to turn to the role of proteins in the brain. As David Eagleman, a neuroscientist at the Baylor College of Medicine in Houston, puts it, "Neuroscience is obsessed with neurons because our best technology allows us to measure them. But each individual neuron is in fact as complicated as a city, with millions of proteins inside of it, trafficking and interacting in extraordinarily complex biochemical cascades."

Still, even at this early stage in the new Neuroscience Revolution, scientists have learned how to selectively activate specific brain systems. Exploiting advances in the new field of optogenetics, scientists first identify opsins—light-sensitive proteins from green algae (or bacteria)—and place into cells their corresponding genes, which then become optical switches for neurons. By also inserting genes that correspond to other proteins that *glow* in green light, the scientists were then able to switch the neuron on and off with blue light, and then observe its effects on other neurons with a green light. The science of optogenetics has quickly advanced to the point where researchers are able to use the switches to manipulate the behavior and feelings of mice by controlling the flow of ions (charged particles) to neurons, effectively turning them on and off at will. One of the promising applications may be the control of symptoms associated with Parkinson's disease.

Other scientists have inserted multiple genes from jellyfish and coral that produce different fluorescent colors—red, blue, yellow, and gradations in between—into many neurons in a process that then allows the identification of different categories of neurons by having each category light up in a different color. This so-called "brainbow" allows a much more detailed visual map of neuronal connections. And once again, the Global Mind has facilitated the emergence of a powerful network effect in brain research. When a new element of the brain's intricate circuitry is deciphered, the knowledge is widely dispersed to other research teams whose work in decoding other parts of the connectome is thereby accelerated.

WATCHING THE BRAIN THINK

Simultaneously, a completely different new approach to studying the brain—functional magnetic resonance imaging (fMRI)—has led to exciting new discoveries. This technique, which is based on the more familiar MRI scans of body parts, tracks blood flow in the brain to neurons when they are fired. When neurons are active, they take in blood containing the oxygen and glucose needed for energy. Since there is a slight magnetization difference between oxygenated blood and oxygen-depleted blood, the scanning machine can identify which areas of the brain are active at any given moment.

By correlating the images made by the machine with the subjective descriptions of thoughts or feelings reported by the individual whose brain is being scanned, scientists have been able to make breakthrough discoveries about where specific functions are located in the brain. This technique is now so far advanced that experienced teams can actually identify specific *thoughts* by seeing the "brain prints" associated with those thoughts. The word "hammer," for example, has a distinctive brain print that is extremely similar in almost everyone, regardless of nationality or culture.

One of the most startling examples of this new potential was reported in 2010 by neuroscientist Dr. Adrian Owen, when he was at the University of Cambridge in England. Owen performed fMRI scans on a young woman who was in a vegetative state with no discernible sign of consciousness and asked her questions while she was being scanned. He began by asking her to imagine playing tennis, and then asking her to imagine walking through her house. Scientists have established that people who think about playing tennis demonstrate activity in a particular part of the motor cortex portion of the brain, the supplementary motor area. Similarly, when people think about walking through their own home, there is a recognizable pattern of activity in the center of the brain in an area called the parahippocampal gyrus.

After observing that the woman responded to each of these questions by manifesting exactly the brain activity one would expect from someone who is conscious, the doctor then used these two questions as a way of empowering the young woman to "answer" either "yes" by thinking about playing tennis, or "no" by imagining a stroll through her house. He then patiently asked her a series of questions about her life, the answers to which were not known by anyone participating in the medical team. She answered correctly to virtually all the questions, leading Owen to conclude that she was in fact conscious. After continuing his experiments with many other patients, Owen speculated that as many as 20 percent of those believed to be in vegetative states may well be conscious with no way of connecting to others. Owen and his team are now using noninvasive electroencephalography (EEG) to continue this work.

Scientists at Dartmouth College are also using an EEG headset to interpret thoughts and connect them to an iPhone, allowing the user to select pictures that are then displayed on the iPhone's screen. Because the sensors of the EEG are attached to the outside of the head, it has more difficulty interpreting the electrical signals inside the skull, but they are making impressive progress.

A LOW-COST HEADSET developed some years ago by an Australian game company, Emotiv, translates brain signals and uses them to empower users to control objects on a computer screen. Neuroscientists believe that these lower-cost devices are measuring "muscle rhythms rather than real neural activity." Nevertheless, scientists and engineers at IBM's Emerging Technologies lab in the United Kingdom have adapted the headset to allow thought control of other electronic devices, including model cars, televisions, and switches. In Switzerland, scientists at the Ecole Polytechnique Fédérale de Lausanne (EPFL) have used a similar approach to build wheelchairs and robots controlled by thoughts. Four other companies, including Toyota, have announced they are developing a bicycle whose gears can be shifted by the rider's thoughts.

Gerwin Schalk and Anthony Ritaccio, at the Albany Medical Center, are working under a multimillion-dollar grant from the U.S. military to design and develop devices that enable soldiers to communicate telepathically. Although this seems like something out of a science fiction story, the Pentagon believes that these so-called telepathy helmets are sufficiently feasible that it is devoting more than $6 million to the project. The target date for completion of the prototype device is 2017.

"TRANSHUMANISM" AND THE "SINGULARITY"

If such a technology is perfected, it is difficult to imagine where more sophisticated later versions of it would lead. Some theorists have long predicted that the development of a practical way to translate human thoughts into digital patterns that can be deciphered by computers will inevitably lead to a broader convergence between machines and people that goes beyond cyborgs to open the door on a new era characterized by what they call "transhumanism."

According to Nick Bostrom, the leading historian of transhumanism, the term was apparently coined by Aldous Huxley's brother, Julian, a distinguished biologist, environmentalist, and humanitarian, who wrote in 1927, "The human species can, if it wishes, transcend itself—not just sporadically, an individual here in one way, an individual there in another way—but in its entirety, as humanity. We need a name for this new belief. Perhaps *transhumanism* will serve: man remaining man, but transcending himself, by realizing new possibilities of and for his human nature."

The idea that we as human beings are not an evolutionary end point, but are destined to evolve further—with our own active participation in directing the process—is an idea whose roots are found in the intellectual ferment following the publication of Darwin's *On the Origin of Species*, a ferment that continued into the twentieth century. This speculation led a few decades later to the discussion of a new proposed endpoint in human evolution—the "Singularity."

First used by Teilhard de Chardin, the term "Singularity" describes a future threshold beyond which artificial intelligence will exceed that of human beings. Vernor Vinge, a California mathematician and computer scientist, captured the idea succinctly in a paper published twenty years ago, entitled "The Coming Technological Singularity," in which he wrote, "Within thirty years, we will have the technological means to create superhuman intelligence. Shortly after, the human era will be ended."

In the current era, the idea of the Singularity has been popularized and enthusiastically promoted by Dr. Ray Kurzweil, a polymath, author, inventor, and futurist (and cofounder with Peter Diamandis of the Singularity University at the NASA Research Park in Moffett Field, California). Kurzweil envisions, among other things, the rapid development of technologies that will facilitate the smooth and complete translation of human thoughts into a form that can be comprehended by and *contained* in advanced computers. Assuming that these breakthroughs ever do take place, he believes that in the next few decades it will be possible to engineer the convergence of human intelligence—and even consciousness—with artificial intelligence. He recently wrote, "There will be no distinction, post-Singularity, between human and machine or between physical and virtual reality."

Kurzweil is seldom reluctant to advance provocative ideas simply because many other technologists view them as outlandish. Another close friend, Mitch Kapor, also a legend in the world of computing, has challenged Kurzweil to a $20,000 bet (to be paid to a foundation chosen by the winner) involving what is perhaps the most interesting long-running debate over the future capabilities of computers, the Turing Test. Named after the legendary pioneer of computer science Alan Turing, who first proposed it in 1950, the Turing Test has long served as a proxy for determining when computers will achieve human-level intelligence. If after conversing in writing with two interlocutors, a human being and a computer, a person cannot determine which is which, then the computer passes the test. Kurzweil has asserted that a computer will pass the Turing Test by the end of 2029. Kapor, who believes that human intelligence will forever be organically distinctive from machine-based intelligence, disagrees. The potential Singularity, however, poses a different challenge.

More recently, the silicon version of the Singularity has been met by a competitive challenge from some biologists who believe that genetic engineering of brains may well produce an "Organic Singularity" before the computer-based "Technological Singularity" is ever achieved. Personally, I don't look forward to either one, although my uneasiness may simply be an illustration of the difficult thinking that all of us have in store as these multiple revolutions speed ahead at an ever accelerating pace.

THE CREATION OF NEW BODY PARTS

Even though the merger between people and machines may remain in the realm of science fiction for the foreseeable future, the introduction of mechanical parts as replacements for components of the human body is moving forward quickly. Prosthetics are now being used to replace not only hips, knees, legs, and arms, but also eyes and other body parts that have not previously been replaceable with artificial substitutes. Cochlear implants, as noted, are used to restore hearing. Several research teams have been developing mechanical exoskeletons to enable paraplegics to walk and to confer additional strength on soldiers and others who need to carry heavy loads. Most bespoke in-ear hearing aids are already made with 3D printers. The speed with which 3D printing is advancing makes it inevitable that many other prosthetics will soon be printed.

In 2012, doctors and technologists in the Netherlands used a 3D printer (described in Chapter 1) to fabricate a lower jaw out of titanium powder for an elderly woman who was not a candidate for traditional reconstructive surgery. The jaw was designed in a computer with articulated joints that match a real jaw, grooves to accommodate the regrowth of veins and nerves, and precisely designed depressions for her muscles to be attached to it. And of course, it was sized to perfectly fit the woman's face.

Then, the 3D digital blueprint was fed into the 3D printer, which laid down titanium powder, one ultrathin layer at a time (thirty-three layers for each millimeter), and fused them together with a laser beam each time, in a process that took just a few hours. According to the woman's doctor, Dr. Jules Poukens of Hasselt University, she was able to use the printed jaw normally after awakening from her surgery, and one day later was able to swallow food.

The 3D printing of human organs is not yet feasible, but the emerging possibility has already generated tremendous excitement in the field of transplantation because of the current shortage of organs. However, well before the 3D printing of organs becomes feasible, scientists hope to develop the ability to generate replacement organs in the laboratory for transplantation into humans. Early versions of so-called exosomatic kidneys (and livers) are now being grown by regenerative medicine scientists at Wake Forest University. This emerging potential for people to grow their own replacement organs promises to transform the field of transplantation.

Doctors at the Karolinska Institute in Stockholm have already created and successfully transplanted a replacement windpipe by inducing the patient's own cells to regrow in a laboratory on a special plastic "scaffolding" that precisely copied the size and shape of the windpipe it replaced. A medical team in Pittsburgh has used a similar technique to grow a quadriceps muscle for a soldier who lost his original thigh muscle to an explosion in Afghanistan, by implanting into his leg a scaffold made from a pig's urinary bladder (stripped of living cells), which stimulated his stem cells to rebuild the muscle tissue as they sensed the matrix of the scaffolding being broken down by the body's immune system. Scientists at MIT are developing silicon nanowires a thousand times smaller than a human hair that can be embedded in these scaffolds and used to monitor how the regrown organs are performing.

As one of the authors of the National Organ Transplant Act in 1984, I learned in congressional hearings about the problems of finding enough organ donors to meet the growing need for transplantation. And having sponsored the ban on buying and selling organs, I remain unconvinced by the argument that this legal prohibition (which the U.S. shares with all other countries besides Iran) should be removed. The potential for abuse is already obvious in the disturbing black market trade in organs and tissues from people in poor countries for transplantation into people living in wealthy countries.

Pending the development of artificial and regenerated replacement organs, Internet-based tools, including social media, are helping to address the challenge of finding more organ donors and matching them with those who need transplants. In 2012, *The New York Times*'s Kevin Sack reported on a moving example of how sixty different people became part of "the longest chain of kidney transplants ever constructed." Recently, Facebook announced the addition of "organ donor" as one of the items to be updated on the profiles of its users.

Another 3D printing company, Bespoke Innovations of San Francisco, is using the process to print more advanced artificial limbs. Other firms are using it to make numerous medical implants. There is also a well-focused effort to develop the capacity to print vaccines and pharmaceuticals from basic chemicals on demand. Professor Lee Cronin of the University of Glasgow, who leads one of the teams focused on the 3D printing of pharmaceuticals, said recently that the process they are working on would place the molecules of common elements and compounds used to formulate pharmaceuticals into the equivalent of the cartridges that feed different color inks into a conventional 2D printer. With a manageably small group of such cartridges, Cronin said, "You can make any organic molecule."

One of the advantages, of course, is that this process would make it possible to transmit the 3D digital formula for pharmaceuticals and vaccines to widely dispersed 3D printers around the world for the manufacturing of the pharmaceuticals on site with negligible incremental costs for the tailoring of pharmaceuticals to each individual patient.

The pharmaceutical industry relied historically on large centralized manufacturing plants because its business model was based on the idea of a mass market, within which large numbers of people were provided essentially the same product. However, the digitization of human beings and molecular-based materials is producing such an extraordinarily high volume of differentiating data about both people and things that it will soon no longer make sense to lump people together and ignore medically significant information about their differences.

Our new prowess in manipulating the microscopic fabric of our world is also giving us the ability to engineer nanoscale machines for insertion into the human body—with some active devices the size of living cells that can coexist with human tissue. One team of nanotechnologists at MIT announced in 2012 that they have successfully built "nanofactories" that are theoretically capable of producing proteins while inside the human body when they are activated by shining a laser light on them from outside the body.

Specialized prosthetics for the brain are also being developed. Alongside pacemakers for hearts, comparable devices can now be inserted into brains to compensate for damage and disorders. Doctors are already beginning to implant computer chips and digital devices on the surface of the brain and, in some cases, deeper within the brain. By cutting a hole in the skull and placing a chip that is wired to a computer directly on the surface of the brain, doctors have empowered paralyzed patients with the ability to activate and direct the movement of robots with their thoughts.

In one widely seen demonstration, a paralyzed patient was able to direct a robot arm to pick up a cup of coffee, move it close to her lips, and insert the straw between her lips so she could take a sip.

Experts believe that it is only a matter of time before the increased computational power and the reduction in size of the computer chips will make it possible to dispense with the wires connecting the chip to a computer. Scientists and engineers at the University of Illinois, the University of Pennsylvania, and New York University are working to develop a new form of interface with the brain that is flexible enough to stretch in order to fit the contours of the brain's surface. According to the head of R&D at GlaxoSmithKline, Moncef Slaoui, "The sciences that underpin bioelectronics are proceeding at an amazing pace at academic centers around the world but it is all happening in separate places. The challenge is to integrate the work—in brain-computer interfaces, materials science, nanotechnology, micro-power generation—to provide therapeutic benefit."

Doctors at Tel Aviv University have equipped rats with an artificial cerebellum, which they have attached to the rat's brain stem to interpret information from the rest of the rat's body. By using this information, doctors are able to stimulate motor neurons to move the rat's limbs. Although the work is at an early stage, experts in the field believe that it is only a matter of time before artificial versions of entire brain subsystems are built. Francisco Sepulveda, at the University of Essex in the U.K., said that the complexity of the challenge is daunting but that scientists see a clear pathway to succeed. "It will likely take us several decades to get there, but my bet is that specific, well-organized brain parts such as the hippocampus or the visual cortex will have synthetic correlates before the end of the century."

Well before the development of a synthetic brain subsystem as complex as the hippocampus or visual cortex, other so-called neuroprosthetics are already being used in humans, including prosthetics for bladder control, relief of spinal pain, and the remediation of some forms of blindness and deafness. Other neuroprosthetics expected to be introduced in the near future will, according to scientists, be able to stimulate particular parts of the brain to enhance focus and concentration, that with the flip of a switch will stimulate the neural connections associated with "practice" in order to enhance the ability of a stroke victim to learn how to walk again.

"MODIFY THE KID"

As implants, prosthetics, neuroprosthetics, and other applications in cybernetics continue to improve, the discussion about their implications has broadened from their use as therapeutic, remedial, and reparative devices to include the implications of using prosthetics that *enhance* humans. For example, the brain implants described above that can help stroke victims learn more quickly how to walk again, can also be used in healthy people to enhance concentration at times of their choosing to help them learn a brand-new skill, or enhance their capacity for focus when they feel it is particularly important.

The temporary enhancement of mental performance through the use of pharmaceuticals has already begun, with an estimated 4 percent of college students routinely using attention-focusing medications like Adderall, Ritalin, and Provigil to improve their test scores on exams. Studies at some schools found rates as high as 35 percent. After an in-depth investigation of the use of these drugs in high schools, *The New York Times* reported that there was "no reliable research" on which to base a national estimate, but that a survey of more than fifteen schools with high academic standards yielded an estimate from doctors and students that the percentage of students using these substances "ranges from 15 percent to 40 percent."

The *Times* went on to report, "One consensus was clear: users were becoming more common . . . and some students who would rather not take the drugs would be compelled to join them because of the competition over class rank and colleges' interest." Some doctors who work with low-income families have started prescribing Adderall for children to help them compensate for the advantages that children from wealthy families have. One of them, Dr. Michael Anderson, of Canton, Georgia, told the *Times* he thinks of it as "evening the scales a little bit. . . . We've decided as a society that it's too expensive to modify the kid's environment. So we have to modify the kid."

A few years ago, almost 1,500 people working as research scientists at institutions in more than sixty countries responded to a survey on the use of brain-enhancing pharmaceuticals. Approximately 20 percent said that they had indeed used such drugs, with the majority saying they felt they improved their memory and ability to focus. Although inappropriate use and dangerous overuse of these substances has caused doctors to warn about risks and side effects, scientists are working on new compounds that carry the promise of actually boosting intelligence. Some predict that the use of the improved intelligence-enhancement drugs now under development may well become commonplace and carry as little stigma as cosmetic surgery does today. The U.S. Defense Advanced Research Projects Agency is experimenting with a different approach to enhance concentration and speed the learning of new skills, by using small electrical currents applied from outside the skull to the part of the brain used for object recognition in order to improve the training of snipers.

ENHANCING PERFORMANCE

At the 2012 Olympics, South Africa's Oscar Pistorius made history as the first double amputee track athlete ever to compete. Pistorius, who was born with no fibulas in his lower legs, both of which were amputated before he was one year old, learned to run on prosthetics. He competed in the 400-meter sprint, where he reached the semifinals, and the 4×400 relay, in which the South African team reached the finals.

Some of Pistorius's competitors expressed concern before the games that the flexible blades attached to his prosthetic lower legs actually gave him an unfair advantage. The retired world record holder in the 400-meter sprint, Michael Johnson, said, "Because we don't know for sure whether he gets an advantage from the prosthetics, it is unfair to the able-bodied competitors."

Because of his courage and determination, most were cheering for Pistorius to win. Still, it's clear that we are already in a time of ethical debate over whether artificial enhancements of human beings lead to unfair advantages of various kinds. When Pistorius competed two weeks later in the Paralympics, he himself lodged a protest against one of the other runners whose prosthetic blades, according to Pistorius, were too long compared to his height and gave him an unfair advantage.

In another example from athletics, the use of a hormone called erythropoietin (EPO)—which regulates the production of red blood cells—can give athletes a significant advantage by delivering more oxygen to the muscles for a longer period of time. One former winner of the Tour de France has already been stripped of his victory after he tested positive for elevated testosterone. He has admitted use of EPO, along with other illegal enhancements. More recently, seven-time Tour de France winner

Lance Armstrong was stripped of his championships and banned from cycling for life after the U.S. Anti-Doping Agency released a report detailing his use of EPO, steroids, and blood transfusions, doping by other members of his team, and a complex cover-up scheme.

The authorities in charge of the Olympics and other athletic competitions have been forced into a genetic and biochemical arms race to develop ever more sophisticated methods of detecting new enhancements that violate the rules. What if the gene that produces extra EPO is spliced into an athlete's genome? How will that be detected?

At least one former Olympic multiple gold medal winner, Eero Mäntyranta, the Finnish cross-country skier, was found years later to have a *natural* mutation that caused his body to produce more than the average EPO—and thus produce more red blood cells. Clearly, that cannot be considered a violation of Olympic rules. Mäntyranta competed in the 1960s, before the gene splicing technique was available. But if future Olympians show up with the same mutation, it may be impossible to determine whether it is natural or has been artificially spliced into their genomes. The splicing could be detected now, but scientists say that when the procedure is perfected, Olympic officials may not be able to make a ruling without genetic testing of the athlete's relatives.

In another example, scientists have now discovered ways to manipulate a protein called myostatin that regulates the building of muscles. Animals in which myostatin is blocked develop unnaturally large and strong muscles throughout their bodies. If athletes are genetically engineered to enhance their muscle development, does that constitute unfair competition? Isn't that simply a new form of doping, like the use of steroids and oxygen-rich blood injections? Yet here again, some people— including at least one young aspiring gymnast—have a rare but *natural* mutation that prevents them from producing a normal volume of myostatin, and results in supernormal musculature.

The convergence of genetic engineering and prosthetics is also likely to produce new breakthroughs. Scientists in California announced a new project in 2012 to create an artificial testicle, which they refer to as a human "sperm-making biological machine." Essentially a prosthesis, the artificial testicle would be injected every two months with sperm cells engineered from the man's own adult stem cells.

Some of the earliest applications of genetic research have been in the treatment of infertility. In fact, a great deal of the work since the beginning of the Life Sciences Revolution has focused on the beginning and the end of the human lifecycle—the reinvention of life and death.

THE CHANGING ETHICS OF FERTILITY

The birth in England of the first so-called test tube baby in 1978, Louise Brown, caused a global debate about the ethics and propriety of the procedure—a debate that in many ways established a template for the way publics react to most such breakthroughs. In the first stage, there is a measure of shock and awe, mingled with an anxious flurry of speculation as newly minted experts try to explore the implications of the breakthrough. Some bioethicists worried at the time that in vitro fertilization might somehow diminish parental love and weaken generational ties. But set against the unfocused angst and furrowed brows is the overflowing joy of the new parents whose dreams of a child have at last been realized. Soon thereafter, the furor dies down and fades away. As one U.S. bioethicist, Debra Mathews, put it, "People want children and no one wants anyone else to tell them they can't have them." Since 1978, more than five million children have been born to infertile people wanting children through the use of in vitro fertilization and related procedures.

During numerous congressional hearings on advances in life sciences research in the 1970s and 1980s, I saw this pattern repeated many times. Even earlier, in 1967, the first heart transplant by Dr. Christiaan Barnard in South Africa also caused controversy, but the joy and wonder of what was seen as a medical miracle put an end to the debate before it gained momentum. A doctor assisting in the operation, Dr. Warwick Peacock, told me that when the transplanted heart finally began to beat, Barnard exclaimed, "My God, it's working!" Later on, the first cloning of livestock and the commercialization of surrogate motherhood also caused controversies with very short half-lives.

Now, however, the torrent of scientific breakthroughs is leading to new fertility options that may generate controversies that don't fade as quickly. One new procedure involves the conception of an embryo and the use of preimplantation genetic diagnosis (PGD) to select a suitable "savior sibling" who can serve as an organ, tissue, bone marrow, or umbilical cord stem cell donor for his or her sibling. Some bioethicists have raised concerns that the instrumental purpose of such conceptions devalues the child, though others ask why this must necessarily be the case. In theory, the parents can love and value both children equally even as they pursue a medically important cure for the first with the assistance of the second. Whether truly informed consent on the part of the donor child is plausible in this scenario is another matter.

Scientists and doctors at the Department of Reproductive Medicine at Newcastle University in England outlined a procedure for creating "three-parent babies," to allow couples at high risk of passing on to their children an incurable genetic illness passed from their mother's faulty mitochondrial DNA to have a healthy child. If a third person, who does not have the genes in question, allows her genes (it must come from a female donor) to be substituted for that portion of the embryo's genome, then the baby will escape the feared genetic condition. Ninety-eight percent of the baby's DNA would come from the mother and father; only 2 percent or so would come from the gene donor. However, this genetic modification is one that will affect not only the baby, but all of its offspring, in perpetuity. As a result, the doctors have asked for a government review of the procedure to determine whether the procedure is acceptable under Britain's laws.

When choices such as these are in the hands of parents rather than the government, most people adopt a different standard for deciding how they feel about the procedure in question. The great exception is the continuing debate over the ethics of abortion. In spite of the passionate opposition to abortion among many thoughtful people, the majority in most countries seem to override whatever degree of uneasiness they have about the procedure by affirming the principle that it is a decision that should properly be left to the pregnant woman herself, at least in the earlier stages of the pregnancy.

Nevertheless, the dispersal of new genetic options to individuals is, in some countries, leading to new laws regulating what parents can and cannot do. India has outlawed genetic testing of embryos, or even blood tests, that are designed to identify the gender of the embryo. The strong preference by many Indian parents that their next child be male, particularly if they already have a daughter, has already led to the abortion of 500,000 female fetuses each year and a growing imbalance of the male to female sex ratio in the population. (Among the many cultural factors that have long been at work in producing the preference for baby boys is the high cost of the dowry that must be paid by the parents of a bride.) The 2011 provisional census in India, which showed a further steep decline in the child sex ratio, led the Indian government to launch a new campaign to better enforce the prohibition against the sex selection of children.

Most of the prenatal gender identification procedures in India utilize ultrasound machines rather than riskier procedures such as amniocentesis, and the prevalence of advertising for low-cost ultrasound clinics is a testament to the popularity of the procedure. Although sex-selective abortions are illegal in India, proposed bans on ultrasound machines have not gained support, in part because of their other medical uses. Some couples from India—and other countries—are now traveling to Thailand, where the successful "medical tourism" industry is offering preimplantation genetic diagnosis procedures to couples intent on having a baby boy. A doctor at one of these clinics said that he has never had a request for a female embryo.

Now a scientific breakthrough allows the testing of fetal DNA in blood samples taken from pregnant mothers; experts say the test is 95 percent accurate in determining gender seven weeks into the pregnancy, and becomes even more accurate as the pregnancy proceeds. One company making test kits, Consumer Genetics Inc., of Santa Clara, California, requires women to sign an agreement not to use the test results for sex selection; the company has also announced that it will not sell the kits in India or China.

In 2012, researchers at the University of Washington announced a breakthrough in the sequencing of almost the entire genome of a fetus from the combination of a blood sample from the pregnant woman and a saliva sample from the father. Although the process is still expensive (an estimated $20,000 to $50,000 for one fetal genome—last year, the cost was $200,000 per test), the cost is likely to continue falling very quickly. Soon after this breakthrough was announced, a medical research team at Stanford announced an improved procedure that does not require a genetic sample from the father and is expected to be widely available within two years for an estimated $3,000.

While so much attention has been focused on the gender screening of embryos, tremendous progress has been made on the screening for genetic markers that identify serious disorders that might be treated through early detection. Of the roughly four million babies born in the United States each year, for example, approximately 5,000 have genetic or functional disorders amenable to treatment if discovered early. Since newborn babies are routinely screened on the day of their birth for more than twenty diseases, the new ease with which genetic screening can be done on embryos is, in one sense, just an extension of the process already performed routinely immediately after birth.

The ethical implications are quite different, however, because of the possibility that knowledge of some condition or trait in the embryo could lead the parents to perform an abortion. Indeed, the termination of pregnancies involving fetuses with serious genetic defects is common around the world. A recent U.S. study, for example, found that more than 90 percent of American women who find that the fetus they are carrying has Down syndrome are terminating their pregnancies. The author of an article provocatively titled "The Future of Neo-Eugenics," Armand Leroi at Imperial College in the U.K., wrote, "The widespread acceptance of abortion as a eugenic practice suggests that there might be little resistance to more sophisticated methods of eugenic selection and, in general, this has been the case."

Scientists say that within this decade, they expect to develop the ability to screen embryos for such traits as hair and eye color, skin complexion, and a variety of other traits—including some that have been previously thought of as behavioral but which some scientists now believe have heavy genetic components. Dr. David Eagleman, a neuroscientist at the Baylor College of Medicine, notes, "If you are a carrier of a particular set of genes, the probability that you will commit a violent crime is four times as high as it would be if you lacked those genes. . . . The overwhelming majority of prisoners carry these genes; 98.1 percent of death-row inmates do."

If prospective parents found that set of genes in the embryo they were considering for implantation, would they be tempted to splice them out, or select a different embryo instead? Will we soon be debating "distributed eugenics"? As a result of these and similar developments, some bioethicists are expressing concern that what Leroi called "neo-eugenics" will soon confront us with yet another round of difficult ethical choices.

Already, in vitro fertilization clinics are now using preimplantation genetic diagnosis (PGD) to scan embryos for markers associated with hundreds of diseases before implantation. Although the United States has more regulations in the field of medical research than most countries, PGD is still completely unregulated. Consequently, it may be only a matter of time before a much wider range of criteria—including cosmetic or aesthetic factors—are presented as options for parents to select in the screening process.

One question that has already arisen is the ethics of disposing of embryos that are not selected for implantation. If they are screened out as candidates, they can be frozen and preserved for potential later implantation—and that is an option chosen by many women who undergo the in vitro fertilization procedure. However, often several embryos are implanted simultaneously in order to improve the odds that one will survive; that is the principal reason why multiple births are far more common with in vitro fertilization than in the general population.

The United Kingdom has set a legal limit on the number of embryos that doctors can implant, in order to decrease the number of multiple births and avoid the associated complications for the mothers and babies—and the additional cost to the health care system. As a result, one company, Auxogyn, is using digital imaging (in conjunction with a sophisticated algorithm), in order to monitor the developing embryos every five minutes—from the moment they are fertilized until one of them is selected for implantation. The purpose is to select the embryo that is most likely to develop in a healthy way.

As a practical matter, most realize that it is only a matter of time before the vast majority of frozen embryos are discarded—which raises the same underlying issue that motivates the movement to stop abortions: is an embryo in the earliest stages of life entitled to all of the legal protections available to individuals after they are born? Again, regardless of misgivings they may have, the majority in almost every country have reached the conclusion that even though embryos mark the first stage of human life, the practical differences between an embryo, or fetus, and an individual are nevertheless significant enough to allow the pregnant woman to control the choice on abortion. That view is consistent with a parallel view of the majority in almost every country that the government does not have the right to require a pregnant woman to have an abortion.

The furor over embryonic stem cell research grows out of a related issue. Even if it is judged appropriate for women to have the option of terminating their pregnancies—under most circumstances—is it also acceptable for the parents to give permission for "experimentation" on the embryo to which they have given the beginning of a life? Although this controversy is far from resolved, the majority of people in most countries apparently feel that the scientific and medical benefits of withdrawing stem cells from embryos are so significant that they justify such experiments. In many countries, the justification is linked to a prior determination that the embryos in question are due to be discarded in any case.

The discovery of nonembryonic stem cells (induced pluripotent, or iPS cells) by Shinya Yamanaka at Kyoto University (who was awarded the 2012 Nobel Prize in Medicine) is creating tremendous excitement about a wide range of new therapies and dramatic improvements in drug discovery and screening. In spite of this exciting discovery, however, many scientists still argue that embryonic stem cells may yet prove to have unique qualities and potential that justify their continued use. Researchers at University College London have already used stem cells to successfully restore some vision to mice with an inherited retinal disease, and believe that some forms of blindness in humans may soon be treatable with similar techniques. Other researchers at the University of Sheffield have used stem cells to rebuild nerves in the ears of gerbils and restore their hearing.

In 2011, Japanese fertility scientists at Kyoto University caused a stir when they announced that they had successfully used embryonic mouse stem cells to produce sperm when transplanted into the testicles of mice that were infertile. When the sperm was then extracted and put into mouse eggs, the fertilized eggs were transferred to the uteri of female mice, and resulted in normal offspring that could then reproduce naturally. Their work builds on an English science breakthrough in 2006 in which biologists at the University of Newcastle upon Tyne first produced functioning sperm cells that had been converted from stem cells and produced live offspring, though the offspring had genetic defects.

One reason why these studies drew such attention was that the same basic technique, as it is developed and perfected, may soon make it possible for infertile men to have biological children—and opens the possibility for gay and lesbian couples to have children that are genetically and biologically their own. Some headline writers also savored the speculation that since there is no reason why women cannot, in theory, produce their own sperm cells using this technique: "Will Men Become Obsolete?" On the lengthening list of potentially disquieting outcomes from genetic research, this possibility appears destined to linger near the bottom, though I am certainly biased in making that prediction.

LIFESPANS AND "HEALTHSPANS"

Just as scientists working on fertility have focused on the beginning of life, others have been focused on the end of life—and have been making dramatic progress in understanding the factors affecting longevity. They are developing new strategies, which they hope will achieve not only significant extensions in the average human lifespan, but also the extension of what many refer to as the "healthspan"—the number of years we live a healthy life without debilitating conditions or diseases.

Although a few scientific outliers have argued that genetic engineering could increase human lifespans by multiple centuries, the consensus among many aging specialists is that an increase of up to 25 percent is more likely to be the range of what is possible. According to most experts, evolutionary theory and numerous studies in human and animal genetics lead them to the conclusion that environmental and lifestyle factors contribute roughly three quarters to the aging process and that genetics makes a more modest contribution—somewhere between 20 and 30 percent.

One of the most famous studies of the relationship between lifestyle and longevity showed that extreme caloric restriction extends the lives of rodents dramatically, although there is debate about whether this lifestyle adjustment has the same effect on longevity in humans. More recent studies have shown that rhesus monkeys do *not* live longer with severe caloric restrictions. There is a subtle but important distinction, experts on all sides point out, between longevity and aging. Although they are obviously related, longevity measures the length of life, whereas aging is the process by which cell damage contributes over time to conditions that bring the end of life.

Some highly questionable therapies, such as the use of human growth hormone in an effort to slow or reverse unwanted manifestations of the aging process, may well have side effects that shorten longevity, such as triggering the onset of diabetes and the growth of tumors. Other hormones that have been used to combat symptoms of aging—most prominently, testosterone and estrogen—have also led to controversies about side effects that can shorten longevity for certain patients.

However, excitement was also stirred by a Harvard study in 2010 that showed that the aging process in mice could be halted and even *reversed* by the use of enzymes known as telomerases, which serve to protect the telomeres—or protective caps—on the ends of chromosomes in order to prevent them from damage. Scientists have long known that these telomeres get shorter with the aging of cells and that this shortening process can ultimately halt the renewal of the cells through replication. As a result of the Harvard study, scientists are exploring strategies for protecting the telomeres in order to retard the aging process.

Some researchers are optimistic that extensive whole genome studies of humans with very long lifespans may yet lead to the discovery of genetic factors that can be used to extend longevity in others. However, most of the dramatic extensions in the average human lifespan over the last century have come from improvements in sanitation and nutrition, and from medical breakthroughs such as the discovery of antibiotics and the development of vaccines. Further improvements in these highly successful strategies are likely to further improve average lifespans—probably, scientists speculate, at the rate of improvement we have become used to—about one extra year per decade.

In addition, the continued global efforts to fight infectious disease threats are also extending average lifespans by reducing the number of premature deaths. Much of this work is now focused on malaria, tuberculosis, HIV/AIDS, influenza, viral pneumonia, and multiple so-called "neglected tropical diseases" that are barely known in the industrialized world but afflict more than a billion people in developing tropical and subtropical countries.

THE DISEASE FRONT

There has been heartening progress in reducing the number of people who die of AIDS each year. In 2012, the number fell to 1.7 million, significantly down from its 2005 peak of 2.3 million. The principal reason for this progress is greater access to pharmaceuticals—particularly antiretroviral drugs—that extend the lifespan and improve the health of people who have the disease. Efforts to reduce the infection rate continue to be focused on preventive education, the distribution of condoms in high-risk areas, and accelerated efforts to develop a vaccine.

Malaria has also been reduced significantly over the past decade with a carefully chosen combination of strategies. Although the largest absolute declines were in Africa, according to the U.N., 90 percent of all malaria deaths still take place in Sub-Saharan Africa—most of them involving children under five. Although an ambitious effort in the 1950s to eradicate malaria did not succeed, a few of those working hard to eradicate malaria, including Bill Gates, now believe that their goal may actually be realistic within the next few decades.

The world did succeed in eliminating the terrible scourge of smallpox in 1980. And in 2011 the U.N. Food and Agriculture Organization succeeded in eliminating a second disease, rinderpest, a relative of measles that killed cattle and other animals with cloven hooves. Because it was an animal disease, rinderpest never garnered the global attention that smallpox commanded, but it was one of the deadliest and most feared threats to those whose families and communities depend on livestock.

For all of the appropriate attention being paid to infectious diseases, the leading causes of death in the world today, according to the World Health Organization, are chronic diseases that are not communicable. In the last year for which statistics are available, 2008, approximately 57 million people died in the world, and almost 60 percent of those deaths were caused by chronic diseases, principally cardiovascular disease, diabetes, cancer, and chronic respiratory diseases.

Cancer is a special challenge, in part because it is not one disease, but many. The U.S. National Cancer Institute and the National Human Genome Research Institute have been spending $100 million per year on a massive effort to create a "Cancer Genome Atlas," and in 2012 one of the first fruits of this project was published in *Nature* by more than 200 scientists who detailed genetic peculiarities in colon cancer tumors. Their study of more than 224 tumors has been regarded as a potential turning point in the development of new drugs that will take advantage of vulnerabilities they found in the tumor cells.

In addition to focusing on genomic analyses of cancer, scientists are exploring virtually every conceivable strategy for curing cancers. They are investigating new possibilities for shutting off the blood supply to cancerous cells, dismantling their defense mechanisms, and boosting the ability of natural immune cells to identify and attack the cancer cells. Many are particularly excited about new strategies that involve proteomics—the decoding of all of the proteins translated by cancer genes in the various forms of cancer and targeting epigenetic abnormalities.

Scientists explain that while the human genome is often characterized as a blueprint, it is actually more akin to a list of parts or ingredients.

The actual work of controlling cellular functions is done by proteins that carry out a "conversation" within and between cells. These conversations are crucial in understanding "systems diseases" like cancer.

One of the promising strategies for dealing with systemic disorders like cancer and chronic heart diseases is to strengthen the effectiveness of the body's natural defenses. And in some cases, new genetic therapies are showing promise in doing so. A team of scientists at University of California San Francisco Gladstone Institutes of Cardiovascular Disease has dramatically improved cardiac function in adult mice by reprogramming cells to restore the health of heart muscles.

IN MANY IF not most cases, though, the most effective strategy for combating chronic diseases is to make changes in lifestyles: reduce tobacco use, reduce exposure to carcinogens and other harmful chemicals in the environment, reduce obesity through better diet and more exercise, and—at least for salt-sensitive individuals—reduce sodium consumption in order to reduce hypertension (or high blood pressure).

Obesity—which is a major causal factor in multiple chronic diseases—was the subject of discouraging news in 2012 when the British medical journal *The Lancet* published a series of studies indicating that one of the principal factors leading to obesity, physical inactivity and sedentary lifestyles, is now spreading from North America and Western Europe to the rest of the world. Researchers analyzed statistics from the World Health Organization to demonstrate that more people now die every year from conditions linked with physical inactivity than die from smoking. The statistics indicate that one in ten deaths worldwide is now due to diseases caused by persistent inactivity.

Nevertheless, there are good reasons to hope that new strategies combining knowledge from the Life Sciences Revolution with new digital tools for monitoring disease states, health, and wellness may spread from advanced countries as cheaper smartphones are sold more widely throughout the globe. The use of intelligent digital assistants for the management of chronic diseases (and as wellness coaches) may have an extremely positive impact.

In developed nations, there are already numerous smartphone apps that assist those who wish to keep track of how many calories they consume, what kinds of food they are eating, how much exercise they are getting, how much sleep they are getting (some new headbands also keep track of how much deep sleep, or REM sleep, they are getting), and even how much progress they are making in dealing with addictions to substances such as alcohol, tobacco, and prescription drugs. Mood disorders and other psychological maladies are also addressed by self-tracking programs. During the 2012 summer Olympic Games in London, a number of athletes were persuaded by biotech companies attempting to improve their health-tracking devices to use glucose monitors and sleep monitors, and to receive genetic analyses designed to improve their individual nutritional needs.

Such monitoring is not limited to Olympians. Personal digital monitors of patients' heart rates, blood glucose, blood oxygenation, blood pressure, body temperature, respiratory rate, body fat levels, sleep patterns, medication use, exercise, and more are growing more common. Emerging developments in nanotechnology and synthetic biology also hold out the prospect of more sophisticated continuous monitoring from sensors inside the body. Nanobots are being designed to monitor changes in the bloodstream and vital organs, reporting information on a constant basis.

Some experts, including Dr. H. Gilbert Welch of Dartmouth, the author of *Overdiagnosed: Making People Sick in the Pursuit of Health*, believe that we are in danger of going too far in monitoring and data analysis of individuals who track their vital signs and more: "Constant monitoring is a recipe for all of us to be judged 'sick.' Judging ourselves sick, we seek intervention." Welch and some others believe that many of these interventions turn out to be costly and unnecessary. In 2011, for example, medical experts advised doctors to stop routinely using a new and sophisticated antigen test for prostate cancer precisely because the resulting interventions were apparently doing more harm than good.

The digitizing of human beings, with the creation of large files containing detailed information about their genetic and biochemical makeup and their behavior, will also require attention to the same privacy and information security issues discussed in Chapter 2. For the same reasons that this rich data is potentially so useful in improving the efficacy of health care and reducing medical costs, it is also seen as highly valuable to insurance companies and employers who are often eager to sever their relationships with customers and employees who represent high risks for big medical bills. Already, a high percentage of those who could benefit from genetic testing are refusing to have the information gathered for fear that they will lose their jobs and/or their health insurance.

A few years ago, the United States passed a federal law known as the Genetic Information Nondiscrimination Act, which prohibits the disclosure or improper use of genetic information. But enforcement is difficult and trust in the law's protection is low. The fact that insurance companies and employers usually pay for the majority of health care expenditures—including genetic testing—further reinforces the fear by patients and employees that their genetic information will not remain confidential. Many believe that flows of information on the Internet are vulnerable to disclosure in any case. The U.S. law governing health records, the Health Insurance Portability and Accountability Act, fails to guarantee patient access to records gathered from their own medical implants while companies seek to profit from personalized medical information.

Nevertheless, these self-tracking techniques—part of the so-called self-quantification movement—offer the possibility that behavior modification strategies that have traditionally been associated with clinics can be individualized and executed outside of an institutional setting. Expenditures for genetic testing are rising rapidly as prices for these tests continue to fall rapidly and as the wave of personalized medicine continues to move forward with increasing speed.

The United States may have the most difficulty in making the transition to precision medicine because of the imbalance of power and unhealthy corporate control of the public policy decision-making process, as described in Chapter 3. This chapter is not about the U.S. health care system, but it is interesting to note that the glaring inefficiencies, inequalities, and absurd expense of the U.S. system are illuminated by the developing trends in the life sciences. For example, many health care systems do not cover disease prevention and wellness promotion expenditures, because they are principally compensated for expensive interventions after a patient's health is already in jeopardy. The new health care reform bill enacted by President Obama required coverage of preventive care under U.S. health care plans for the first time.

As everyone knows, the U.S. spends far more per person on health care than any other country while achieving worse outcomes than many other countries that pay far less, and still, tens of millions do not have reasonable access to health care. Lacking any other option, they wait, often until their condition is so dire that they have to go to the emergency room, where the cost of intervention is highest and the chance of success is lowest. The recently enacted reforms will significantly improve some of these defects, but the underlying problems are likely to grow worse—primarily because insurance companies, pharmaceutical companies, and other health care providers retain almost complete control over the design of health care policy.

THE STORY OF INSURANCE

The business of insurance began as far back as ancient Rome and Greece, where life insurance policies were similar to what we now know as burial insurance. The first modern life insurance policies were not offered until the seventeenth century in England. The development of extensive railroad networks in the United States in the 1860s led to limited policies protecting against accidents on railroads and steamboats, and that led, in turn, to the first insurance policies protecting against sickness in the 1890s.

Then, in the early 1930s, when advances in medical care began to drive costs above what many patients could pay on their own, the first significant group health insurance policies were offered by nonprofits: Blue Cross for hospital charges and Blue Shield for doctors' fees. All patients paid the same premiums regardless of age or preexisting conditions. The success of the Blues led to the entry into the marketplace of private, for-profit health insurance companies, who began to charge different premiums to people based on their calculation of the risk involved—and refused to offer policies at all to those who represented an unacceptably high risk. Soon, Blue Cross and Blue Shield were forced by the new for-profit competition to also link premiums to risk.

When President Franklin Roosevelt was preparing his package of reforms in the New Deal, he twice took preliminary steps—in 1935 and again in 1938—to include a national health insurance plan as part of his legislative agenda. On both occasions, however, he feared the political opposition of the American Medical Association and removed the proposal from his plans lest it interfere with what he regarded as more pressing priorities in the depths of the Great Depression: unemployment compensation and Social Security. The introduction of legislation in 1939 by New York Democratic senator Robert Wagner offered a quixotic third opportunity to proceed but Roosevelt chose not to support the legislation.

During World War II, with wages (and prices) controlled by the government, private employers began to compete for employees, who were scarce due to the war, by offering health insurance coverage. Then after the war, unions began to include demands for more extensive health insurance as part of their negotiated contracts with employers.

Roosevelt's successor, Harry Truman, sought to revive the idea for national health insurance, but the opposition in Congress—once again fueled by the AMA—ensured that it died with a whimper. As a result, the hybrid system of employer-based health insurance became the primary model in the United States. Because older Americans and those with disabilities had a difficult time obtaining affordable health insurance within this system, new government programs were implemented to help both groups.

For the rest of the country, those who needed health insurance the most had a difficult time obtaining it, or paying for it when they could find it. By the time the inherent flaws and contradictions of this model were obvious, the American political system had degraded to the point that the companies with an interest in seeing this system continued had so much power that nothing could be done to change its basic structure.

With rare exceptions, the majority of legislators are no longer capable of serving the public interest because they are so dependent on campaign contributions from these corporate interests and so vulnerable to their nonstop lobbying. The general public is effectively disengaged from the debate, except to the extent that they absorb constant messaging from the same corporate interests—messages designed to condition their audience to support what the business lobbies want done.

GENETICALLY ENGINEERED FOOD

The same sclerosis of democracy is now hampering sensible adaptations to the wave of changes flowing out of the Life Sciences Revolution. For example, even though polls consistently show that approximately 90 percent of American citizens believe that genetically engineered food should be labeled, the U.S. Congress has adopted the point of view advocated by large agribusiness companies—that labeling is unnecessary and would be harmful to "confidence in the food supply."

However, most European countries already require such labeling. The recent approval of genetically engineered alfalfa in the U.S. provoked a larger outcry than many expected and the "Just Label It" campaign has become the centerpiece of a new grassroots push for labeling genetically modified (GM) food products in the United States, which plants twice as many acres in GM crops as any other country. Voters in California defeated a referendum in 2012 to require such labeling, after corporate interests spent $46 million on negative commercials, five times as much as proponents. Nevertheless, since approximately 70 percent of the processed foods in the U.S. contain at least some GM crops, this controversy will not go away.

By way of background, the genetic modification of plants and animals is, as enthusiastic advocates often emphasize, hardly new. Most of the food crops that humanity has depended upon since before the dawn of the Agricultural Revolution were genetically modified during the Stone Age by careful selective breeding—which, over many generations, modified the genetic structure of the plants and animals in question to manifest traits of value to humans. As Norman Borlaug put it, "Neolithic women accelerated genetic modifications in plants in the process of domesticating our food crop species."

By using the new technologies of gene splicing and other forms of genetic engineering, we are—according to this view—merely accelerating and making more efficient a long-established practice that has proven benefits and few if any detrimental side effects. And outside of Europe (and India) there is a consensus among most farmers, agribusinesses, and policymakers that GM crops are safe and must be an essential part of the world's strategy for coping with anticipated food shortages.

However, as the debate over genetically modified organisms (GMOs) has evolved, opponents of the practice point out that none of the genetic engineering has ever produced any increase in the intrinsic yields of the crops, and they have raised at least some ecosystem concerns that are not so easily dismissed. The opponents argue that the insertion of foreign genes into another genome is, in fact, different from selective breeding because it disrupts the normal pattern of the organism's genetic code and can cause unpredictable mutations.

The first genetically engineered crop to be commercialized was a new form of tomato known as the FLAVR SAVR, which was modified to remain firm for a longer period of time after it ripened. However, the tomato did not succeed due to high costs. And consumer resistance to tomato paste made from these tomatoes (it was clearly labeled as a GM product) caused the paste to be a failure.

Selective breeding was used to make an earlier change in the traits of commercial tomatoes in order to produce a flatter, less rounded bottom to accommodate the introduction of automation in the harvesting process. The new variety stayed on the conveyor belts without rolling off, was easier to pack into crates, and its tougher skin prevented the machines from crushing the tomatoes. They are sometimes called "square tomatoes," though they are not really square.

An even earlier modification of tomatoes, in 1930, also using selective breeding, was the one that resulted in what most tomato lovers regard as a catastrophic loss of flavor in modern tomatoes. The change was intended to enhance the mass marketing and distribution of tomatoes by ensuring that they were "all red" and ripened uniformly, without the green "shoulders" that consumers sometimes viewed as a sign that they were not yet ripe. Researchers working with the newly sequenced tomato genome discovered in 2012 that the elimination of the gene associated with green shoulders also eliminated the plant's ability to produce most of the sugars that used to give most tomatoes a delicious taste.

In spite of experiences such as these, which illustrate how changes made for the convenience and profitability of large corporations sometimes end up triggering other genetic changes that most people hate, farmers around the world—other than in the European Union—have adopted GM crops at an accelerating rate. Almost 11 percent of all the world's farmland was planted in GM crops in 2011, according to an international organization that promotes GMOs, the International Service for the Acquisition of Agri-biotech Applications. Over the last seventeen years, the number of acres planted in GM crops has increased almost 100-fold, and the almost 400 million acres planted in 2011 represented an increase of 8 percent from one year earlier.

Although the United States is by far the largest grower of GM crops, Brazil and Argentina are also heavily committed to the technology. Brazil, in particular, has adopted a fast-track approval system for GMOs and is pursuing a highly focused strategy for maximizing the use of biotechnology in agriculture. In developing countries overall, the adoption of modified crops in developing countries is growing twice as fast as in mature economies. An estimated 90 percent of the 16.7 million farmers growing genetically engineered crops in almost thirty countries were small farmers in developing markets.

Genetically modified soybeans, engineered to tolerate Monsanto's Roundup herbicide, are the largest GM crop globally. Corn is the second most widely planted GM crop, although it is the most planted in the U.S. ("Maize" is the term used for what is called corn in the U.S.; the word "corn" is often used outside the U.S. to refer to any cereal crop.) In the U.S., 95 percent of soybeans planted and 80 percent of corn are grown from patented seeds that farmers must purchase from Monsanto or one of their licensees. Cotton is the third most planted GM crop globally, and canola (known as "rapeseed" outside the United States) is the other large GM crop in the world.

Although the science of genetically engineered plants is advancing quickly, the vast majority of GM crops grown today are still from the first of three generations, or waves, of the technology. This first wave, in turn, includes GM crops that fall into three different categories:

- The introduction of genes that give corn and cotton the ability to produce their own insecticide inside the plants;
- Genes introduced into corn, cotton, canola, and soybeans that make the plants tolerant of two chemicals contained in widely used weed killers that are produced by the same company—Monsanto—that controls the GM seeds; and
- The introduction of genes designed to enhance the survivability of crops during droughts.

In general, farmers using the first wave of GM crops report initial reductions in their cost of production—partly due to temporarily lower use of insecticide—and temporarily lower losses to insects or weeds. The bulk of the economic benefits thus far have gone to cotton farmers using a strain that is engineered to produce its own insecticide (*Bacillus thuringiensis*, better known as Bt). In India the new Bt cotton made the nation a net exporter, rather than importer, of cotton and was a factor in the initial doubling of cotton yields because of temporarily lower losses to insects and weeds. However, many Indian cotton farmers have begun to protest the high cost of the GM seeds they must purchase anew each year and the high cost of the herbicides they must use in greater volumes as more weeds develop resistance. A parliamentary panel in India issued a controversial 2012 report asserting that "there is a connection between Bt cotton and farmers' suicides" and recommending that field trials of GM crops "under any garb should be discontinued forthwith."

New scientific studies—including a comprehensive report by the U.S. National Research Council in 2009—support the criticism by opponents of GM crops that the *intrinsic* yields of the crops themselves are not increased at all. To the contrary, some farmers have experienced slightly lower intrinsic yields because of unexpected collateral changes in the plants' genetic code. Selective breeding, on the other hand, was responsible for the impressive and life-saving yield increases of the Green Revolution. New research by an Israeli company, Kaiima, into a non-GMO technology known as "enhanced ploidy" (the inducement, selective breeding, and natural enhancement of a trait that confers more than two sets of chromosomes in each cell nucleus) is producing both greater yields and greater resistance to the effects of drought in a variety of food and other crops. Recent field trials run by Kaiima show more than 20 percent yield enhancement in corn and more than 40 percent enhancement in wheat.

The genetic modification of crops, by contrast, has not yet produced meaningful enhancements of survivability during drought. While some GM experimental strains do, in theory, offer the promise of increased yields during dry periods, these strains have not yet been introduced on a commercial scale, and test plots have demonstrated only slight yield improvements thus far, and only during *mild* drought conditions. Because of the growing prevalence of drought due to global warming, there is tremendous interest in drought-resistant strains, especially for maize, wheat, and other crops in developing countries. Unfortunately, however, drought resistance is turning out to be an extremely complex challenge for plant geneticists, involving a combination of many genes working together in complicated ways that are not yet well understood.

After an extensive analysis of the progress in genetically engineering drought-resistant crops, the Union of Concerned Scientists found "little evidence of progress in making crops more water efficient. We also found that the overall prospects for genetic engineering to significantly address agriculture's drought and water-use challenges are limited at best."

The second wave of GM crops involves the introduction of genes that enhance the nutrient value of the plants. It includes the engineering of higher protein content in corn (maize) that is used primarily for live-stock feed, and the engineering of a new strain of rice that produces extra vitamin A as part of a strategy to combat the deficiency in vitamin A that now affects approximately 250 million children around the world. This second wave also involves the introduction of genes that are designed to enhance the resistance of plants to particular fungi and viruses.

The third wave of GM crops, which is just beginning to be commercialized, involves the modification of plants through the introduction of genes that program the production of substances within the plants that have commercial value as inputs in other processes, including pharmaceutical inputs and biopolymers for the production of bioplastics that are biodegradable and easily recyclable. This third wave also involves an effort to introduce genes that modify plants with high cellulose and lignin in order to make them easier to process for the production of cellulosic ethanol. The so-called green plastics have exciting promise, but as with crops devoted to the production of biofuels, they raise questions about how much arable land can safely or wisely be diverted from the production of food in a world with growing population and food consumption, and shrinking assets of topsoil and water for agriculture.

Over the next two decades, seed scientists believe that they may be able to launch a fourth wave of GM crops by inserting the photo-synthesizing genes of corn (and other so-called C4 plants) that are more efficient in photosynthesizing light into energy in plants like wheat and rice (and other C3 plants). If they succeed—which is far from certain because of the unprecedented complexity of the challenge—this technique could indeed bring about significant intrinsic yield increases. For the time being however, the overall net benefits from genetically engineered crops have been limited to a temporary reduction in losses to pests and a temporary decrease in expenditures for insecticides.

In 2012, the Obama administration in the U.S. launched its National Bioeconomy Blueprint, specifically designed to stimulate the production—and procurement by the government—of such products. The European Commission adopted a similar strategy two months earlier. Some environmental groups have criticized both plans because of the growing concern about diverting cropland away from food production and the destruction of tropical forests to make way for more cropland.

The opponents of genetically modified crops argue that not only have these genetic technologies failed thus far to increase intrinsic yields, but also that the weeds and insects the GM crops are designed to control are quickly mutating to make themselves impervious to the herbicides and insecticides in question. In particular, the crops that are engineered to produce their own insecticide (*Bacillus thuringiensis*) are now so common that the constant diet of Bt being served to pests in large monocultured fields is doing the same thing to insects that the massive and constant use of antibiotics is doing to germs in the guts of livestock: it is forcing the mutation of new strains of pests that are highly resistant to the insecticide.

The same thing also appears to be happening to weeds that are constantly sprayed with herbicides to protect crops that have been genetically engineered to survive application of the herbicide (including principally Monsanto's Roundup, which is based on glyphosate, which used to kill virtually any green plant). Already, ten species of harmful weeds have evolved a resistance to these herbicides, requiring farmers to use other more toxic herbicides. Some opponents of GM crops have marshaled evidence tending to show that over time, as resistance increases among weeds and insects, the overall use of both herbicides and pesticides actually increases, though advocates of GM crops dispute their analysis.

Because so many weeds have now developed resistance to glyphosate (most commonly used in Roundup), there is a renewed market demand for more powerful—and more dangerous—herbicides. There are certainly plenty to choose from. The overall market for pesticides in the world represents approximately $40 billion in sales annually, with herbicides aimed at weeds representing $17.5 billion and both insecticides and fungicides representing about $10.5 billion each.

Dow AgroSciences has applied for regulatory approval to launch a new genetically engineered form of corn that tolerates the application of a pesticide known as 2,4-D, which was a key ingredient in Agent Orange—the deadly herbicide used by the U.S. Air Force to clear jungles and forest cover during the Vietnam War—which has been implicated in numerous health problems suffered by both Americans and Vietnamese who were exposed to it. Health experts from more than 140 NGOs have opposed the approval of what they call "Agent Orange corn," citing links between exposure to 2,4-D and "major health problems such as cancer, lowered sperm counts, liver toxicity and Parkinson's disease. Lab studies show that 2,4-D causes endocrine disruption, reproductive problems, neurotoxicity, and immunosuppression."

Insecticides that are sprayed on crops have also been implicated in damage to beneficial insects and other animals. The milkweed plants on which monarch butterflies almost exclusively depend have declined in the U.S. farm belt by almost 60 percent over the last decade, principally because of the expansion of cropland dedicated to crop varieties engineered to be tolerant of Roundup. There have been studies showing that Bt crops (the ones that produce insecticide) have had a direct harmful impact on at least one subspecies of monarchs, and on lacewings (considered highly beneficial insects), ladybird beetles, and beneficial biota in the soil. Although proponents of GM crops have minimized the importance of these effects, they deserve close scrutiny as GM crops continue to expand their role in the world's food production.

Most recently, scientists have attributed the disturbing and previously mysterious sudden collapses of bee colonies to a new group of pesticides known as neonicotinoids. Colony collapse disorder (CCD) has caused deep concern among beekeepers and others since the affliction first appeared in 2006. Although numerous theories about the cause of CCD were put forward, it was not until the spring of 2012 that several studies pinpointed the cause.

The neonicotinoids, which are neurotoxins similar in their makeup to nicotine, are widely used on corn seed, and the chemicals are then pulled from the seed into the corn plants as they grow. Commercial beekeepers, in turn, have long fed corn syrup to their bees. According to the U.S. Department of Agricultural Research Service, "Bee pollination is responsible for $15 billion in added crop value, particularly for specialty crops such as almonds and other nuts, berries, fruits, and vegetables. About one mouthful in three in the diet directly or indirectly benefits from honey bee pollination."

Bees, of course, play no role in the pollination of GM crops, because the engineered seeds must be purchased annually by farmers, and the bees' pesky habit of pollinating plants can introduce genes that do not fit into the seed company's design. According to *The Wall Street Journal*, the growers of a modified seedless mandarin threatened to sue beekeepers working with neighboring farms for allowing their bees to "trespass" into the orchards where the seedless mandarins were growing, out of worry that the seedless mandarins would be cross-pollinated with pollen from citrus varieties that have seeds. Understandably, the beekeepers protested that they couldn't control where their bees fly.

The global spread of industrial agriculture techniques has resulted in the increased reliance on monoculture, which has, in turn, accelerated the spread of resistance to herbicides and pesticides in weeds, insects, and plant diseases. In many countries, including the United States, all of the major commodity crops—corn, soybeans, cotton, and wheat— are grown from a small handful of genetic varieties. As a result, in most fields, virtually all of the plants are genetically identical. Some experts have long expressed concern that the reliance on monocultures makes agriculture highly vulnerable to pests and plant diseases that have too many opportunities to develop mutations that enable them to become more efficient at attacking the particular genetic variety that is planted in such abundance.

MUTATING PLANT DISEASES

In any case, new versions of plant diseases are causing problems for farmers all over the world. In 1999, a new mutated variety of an old fungal disease known as stem rust began attacking wheat fields in Uganda. Spores from the African fields were carried on the wind first to neighboring Kenya, then across the Red Sea to Yemen and the Arabian Peninsula, and from there to Iran. Plant scientists are concerned that it will continue spreading in Africa, Asia, and perhaps beyond. Two scientific experts on the disease, Peter Njao and Ruth Wanyera, expressed concern in 2012 that it could potentially destroy 80 percent of all known wheat varieties. Although this wheat rust was believed to be reduced to a minor threat a half century ago, the new mutation has made it deadlier than ever.

Similarly, cassava (also known as tapioca, manioc, and yucca), the third-largest plant-based source of calories for people (after rice and wheat), is consumed mostly in Africa, South America, and Asia. It developed a new mutation in East Africa in 2005, and since then, according to Claude Fauquet, who is the director of cassava research at the Donald Danforth Plant Science Center in St. Louis, "There has been explosive, pandemic-style spread. . . . The speed is just unprecedented, and the farmers are really desperate." Some experts have compared this outbreak to the potato blight in Ireland in the 1840s, which was linked in part to Ireland's heavy reliance on a monocultured potato strain from the Andes.

Sixty percent of the U.S. corn crop was destroyed in 1970 by a new variety of Southern corn leaf blight, demonstrating clearly, in the words of the Union of Concerned Scientists, "that a genetically uniform crop base is a disaster waiting to happen." The UCS notes that "U.S. agriculture rests on a narrow genetic base. At the beginning of the 1990s, only six varieties of corn accounted for 46 percent of the crop, nine varieties of wheat made up half of the wheat crop, and two types of peas made up 96 percent of the pea crop. Reflecting the global success of fast food in the age of Earth Inc., more than half the world's potato acreage is now planted with one variety of potato: the Russet Burbank favored by McDonald's."

Although most of the debate over genetically modified plants has focused on crops for food and animal feed, there has been surprisingly little discussion about the robust global work under way to genetically modify trees, including poplar and eucalyptus. Some scientists have expressed concern that the greater height of trees means that the genetically modified varieties will send their pollen into a much wider surrounding area than plants like soybeans, corn, and cotton.

China is already growing an estimated thousands of hectares of poplar trees genetically modified to make the Bt toxin in its leaves in order to protect them against insect infestations. Biotech companies are trying to introduce modified eucalyptus trees in the U.S. and Brazil. Scientists argue that in addition to pest resistance, modifications might be useful in enabling trees to survive droughts and could modify the nature of the wood in ways that will facilitate the production of biofuel.

In addition to plants and trees, genetically modified animals intended for the production of food for humans have also generated considerable controversy. Since the discovery in 1981 of a new technique that allows the transfer of genes from one species into the genome of another species, scientists have genetically engineered several forms of livestock, including cattle, pigs, chickens, sheep, goats, and rabbits. Although earlier experiments that reduced susceptibility to disease in mice generated a great deal of optimism, so far only one of the efforts to reduce livestock susceptibility has succeeded.

However, the ongoing efforts to produce GM animals have already produced, among other results, spider silk from goats (described above) and the production of a synthetic growth hormone in dairy cattle that increases their milk production. Recombinant bovine growth hormone (rBGH), which is injected into dairy cows, has been extremely controversial. Critics do not typically argue that rBGH is directly harmful to human health, but rather, that evidence suggests it causes the increase of a second hormone known as insulin-like growth factor (IGF), which is found in milk from cows treated with bovine growth hormone at levels up to ten times what is found in other milk.

Studies have shown a connection between elevated levels of IGF and a significantly higher risk of prostate cancer and some forms of breast cancer. Although other factors obviously are involved in the development of these cancers, and even though IGF is a natural substance in the human body, the concerns of opponents have been translated into a successful consumer campaign for the labeling of milk with bovine growth hormone, which has significantly decreased its use.

Chinese geneticists have introduced human genes associated with human milk proteins into the embryos of dairy cows, then implanted the embryos into surrogate cows that gave birth to the calves. When these animals began producing milk, it contained many proteins and antibodies that are found in human milk but not in milk from normal cows. Moreover, the genetically engineered animals are capable of reproducing themselves with the introduced genetic traits passed on. At present, there is a herd of 300 such animals at the State Key Laboratory of Agrobiotechnology of the China Agricultural University, producing milk that is much closer to human breast milk than cow milk. Scientists in Argentina, at the National Institute of Agribusiness Technology in Buenos Aires, claim to have improved on this process.

Scientists in the U.S. applied for regulatory approval in 2012 to introduce the first genetically engineered animal intended for direct consumption by human beings—a salmon modified with an extra growth hormone gene and a genetic switch that triggers the making of growth hormone even when the water temperature is colder than the threshold for normal production of growth hormone, resulting in a growth rate twice as fast as a normal salmon, which means it will reach market size in only sixteen months, compared to the normal thirty months.

Opponents of the "super salmon" have expressed concern about the possibility of increased levels of insulin-like growth factor—the same issue they have with milk produced from cattle injected with bovine growth hormone. And they expressed concern about these modified salmon escaping from their pens to breed with wild salmon, changing the species in an unintended way—much as the opponents of GM crops have expressed concern about the crosspollination of non-GM crops. Moreover, as noted in Chapter 4, farmed fish are fed fishmeal made from ocean fish in a pattern that typically requires three pounds of wild fish for each pound of farmed fish.

Scientists in Canada at the University of Guelph attempted to market genetically engineered pigs with a segment of mouse DNA introduced into their genome in order to reduce the amount of phosphorus in their feces. They called their creation Enviropigs because phosphorus is a source of algae blooms when dumped into rivers and creates dead zones where the rivers flow into the sea. They later abandoned their project and euthanized the pigs, in part because of opposition to what some critics have taken to calling "Frankenfood"—that is, food from genetically modified animals—but also because scientists elsewhere engineered an enzyme, phytase, which, when added to pig feed, accomplishes the same result hoped for with the ill-fated Enviropig.

In addition to the efforts to modify livestock and fish, there have also been initiatives over the last fifteen years to genetically engineer insects, including bollworms and mosquitoes. Most recently, a British biotechnology company, Oxford Insect Technologies (or Oxitec), has launched a project to modify the principal (though not the only) species of mosquito that carries dengue fever, in order to create mutant male mosquitoes engineered to produce offspring that require the antibiotic tetracycline in order to survive.

The larvae, having no access to tetracycline, die before they can take flight. The idea is that the male mosquitoes, which, unlike females, do not bite, will monopolize the females and impregnate them with doomed embryos, thereby sharply reducing the overall population. Although field trials in the Cayman Islands, Malaysia, and Juazeiro, Brazil, produced impressive results, there was vigorous public opposition when Oxitec proposed the release of large numbers of their mosquitoes in Key West, Florida, after an outbreak of dengue fever there in 2010.

Opponents of this project have expressed concern that the transgenic mosquitoes may have unpredictable and potentially disruptive effects on the ecosystem into which they are released. They argue that since laboratory tests have already shown that a small number of the offspring do in fact survive, there is an obvious potential for those that survive in the wild to spread their adaptation to the rest of the mosquito population over time.

Further studies may show that this project is a useful and worthwhile strategy for limiting the spread of dengue fever, but the focus on genetically modifying the principal mosquito that carries the disease poses a sharp contrast to the complete lack of focus on the principal cause of the rapid spread of dengue. The disruption of the Earth's climate balance and the consequent increase in average global temperatures is making areas of the world that used to be inhospitable to the mosquitoes carrying dengue part of their expanding range.

According to a 2012 Texas Tech University research study of dengue's spread, "Shifts in temperature and precipitation patterns caused by global climate change may have profound impacts on the ecology of certain infectious diseases." Noting that dengue is one of those diseases, the researchers projected that even though Mexico has been the main location of dengue fever in North America, with only occasional small outbreaks in South Texas and South Florida, it is spreading northward because of global warming.

Dengue, which now afflicts up to 100 million people each year and causes thousands of fatalities, is also known as "breakbone fever" because of the extreme joint pain that is one of its worst symptoms. Simultaneous outbreaks emerged in Asia, the Americas, and Africa in the eighteenth century but the disease was largely contained until World War II; scientists believe it was inadvertently spread by people during and after the war to other continents. In 2012, there were an estimated 37 million cases in India alone.

After it was spread by humans to the Americas, dengue's range was still limited to tropical and subtropical regions. But now, as its habitat expands, researchers predict that dengue is likely to spread throughout the Southern United States and that even northern areas of the U.S. are likely to experience outbreaks during summer months.

THIS CHAPTER BEGAN with a discussion of how we are, for the first time, changing the "being" in human being. We are also changing the other beings to which we are ecologically connected. When we disrupt the ecological system in which we have evolved and radically change the climate and environmental balance to which our civilization has been carefully configured, we should expect biological consequences larger than what we can fix with technologies like genetic engineering.

After all, human encroachment into wild areas is responsible for 40 percent of the new emerging infectious diseases that endanger humans, including HIV/AIDS, the bird flu, and the Ebola virus, all of which originated in wild animals forced out of their natural habitat by human encroachment, or brought into close proximity with livestock when farming expanded into previously wild regions. Veterinary epidemiologist Jonathan Epstein said recently, "When you disrupt the balance, you are precipitating the spillover of pathogens from wildlife to livestock to humans." Overall, 60 percent of the new infectious diseases endangering humans came originally from animals.

THE MICROBIOME

We also risk disrupting the ecological system *within* our bodies. New research shows the key role played by microbial communities within (and on) every person. Indeed, all of us have a microbiome of bacteria (and a much smaller number of viruses, yeasts, and amoebas) that outnumber the cells of our bodies by a ratio of ten to one. In other words, every individual shares his or her body with approximately 100 trillion microbes with 3 million nonhuman genes. They live and work synergistically with our bodies in an adaptive community of which we are part.

Early in 2012, 200 scientists who make up the Human Microbiome Project published the genetic sequencing of this community of bacteria and found that there are three basic enterotypes—much like blood types—that exist in all races and ethnicities, and are distributed in all populations without any link to gender, age, body mass, or any other discernible markers. All told, the team identified eight million protein-coding genes in the organisms, and said that half of them have a function that the scientists still do not understand.

One of the functions performed by this microbiome is the "tutoring" of the acquired immune system, particularly during infancy and childhood. According to Gary Huffnagle, of the University of Michigan, "The microbial gut flora is an arm of the immune system." Many scientists have long suspected that the repeated heavy use of antibiotics interferes with this tutoring process and may do damage to the process by which the adaptive immune system learns precision in discriminating between invaders and healthy cells. What all autoimmune diseases have in common is the inappropriate attack of healthy cells by the immune system, which needs to learn to distinguish invaders from cells of the body itself. "Autoimmune" means immunity against oneself.

There is mounting evidence that inappropriate and repeated use of antibiotics in young children may be impairing the development and "learning" of their immune systems—thereby contributing to the apparent rapid rise of numerous diseases of the immune system, such as type 1 diabetes, multiple sclerosis, Crohn's disease, and ulcerative colitis.

The human immune system is not fully developed at birth. Like the human brain, it develops and matures after passage through the birth canal (humans have the longest period of infancy and helplessness of any animal, allowing for rapid growth and development of the brain following birth—with the majority of the development and learning taking place in interaction with the environment). The immune system has an innate ability to activate white blood cells to destroy invading viruses or bacteria, but it also has an acquired—or adaptive—immune system that learns to remember invaders in order to fight them more effectively if they return. This acquired immune system produces antibodies that attach themselves to the invaders so that specific kinds of white blood cells can recognize the invaders and destroy them.

The essence of the problem is that antibiotics themselves do not discriminate between harmful bacteria and beneficial bacteria. By using antibiotics to wage war on disease, we are inadvertently destroying bacteria that we need in order to remain in a healthy balance. "I would like to lose the language of warfare. It does a disservice to all the bacteria that have co-evolved with us and are maintaining the health of our bodies," said Julie Segre, a senior investigator at the National Human Genome Research Institute.

One important bacterium in the human microbiome, *Helicobacter pylori* (or *H. pylori*), affects the regulation of two key hormones in the human stomach that are involved in energy balance and appetite. According to genetic studies, *H. pylori* has lived inside us in large numbers for 58,000 years. Up to 100 years ago, it was the single most common microbe in the stomachs of most human beings. As reported in an important 2011 essay in *Nature* by Martin Blaser, professor of microbiology and chairman of the Department of Medicine at NYU School of Medicine, however, studies have found that "fewer than 6 percent of children in the United States, Sweden and Germany were carrying the organism. Other factors may be at play in this disappearance, but antibiotics may be a culprit. A single course of amoxicillin or a macrolide antibiotic, most commonly used to treat middle-ear or respiratory infections in children, may also eradicate *H. pylori* in 20–50% of cases."

It is important to note that *H. pylori* has been found to play a role in both gastritis and ulcers; the Australian biologist who won the 2005 Nobel Prize in Medicine for discovering *H. pylori*, Dr. Barry Marshall, noted, "People have been killed who didn't get antibiotics to get rid of it." Still, several studies have found strong evidence that people who lack *H. pylori* "are more likely to develop asthma, hay fever or skin allergies in childhood." Its absence is also associated with increased acid reflux and esophageal cancer. Scientists in Germany and Switzerland have found that the introduction of *H. pylori* into the guts of mice serves to protect them against asthma. Among people, for reasons that are not yet fully understood, asthma has increased by approximately 160 percent throughout the world in the last two decades.

One of the hormones regulated by *H. pylori*, ghrelin, is one of the keys to appetite. Normally, the levels of ghrelin fall significantly after someone eats a meal, thus signaling to the brain that it's time to stop eating. However, in people missing *H. pylori* in their guts, the ghrelin levels do not fall after a meal—so the signal to stop eating is not sent. In the laboratory run by Martin Blaser, mice given antibiotics sufficient to kill the *H. pylori* gained significant body fat on an unchanged diet. Interestingly, while scientists have long said that they cannot explain the reason why subtherapeutic doses of antibiotics in livestock feed increase the animals' weight gain, there is now new evidence that it may be due to changes in their microbiome.

The replacement of beneficial bacteria wiped out by antibiotics has been shown to be an effective treatment for some diseases and conditions caused by harmful microbes normally kept in check by beneficial microbes. Probiotics, as they are called, are not new, but some doctors are now treating patients infected with a harmful bacterium known as *Clostridium difficile* by administering a suppository to accomplish a "fecal transplant."

Although the very idea triggers a feeling of repugnance in many, the procedure has been found to be both safe and extremely effective. Scientists at the University of Alberta, after reviewing 124 fecal transplants, found that 83 percent of patients experienced immediate improvement when the balance of their internal microbiome was restored. Other scientists are now hard at work developing probiotic remedies designed to restore specific beneficial bacteria when it is missing from a patient's microbiome.

Just as we are connected to and depend upon the 100 trillion microbes that live in and on each one of us from birth to death, we are also connected to and depend upon the life-forms all around us that live on and in the Earth itself. They provide life-giving services to us just as the microbes in and on our bodies do. Just as the artificial disruption of the microbial communities inside us can create an imbalance in the ecology of the microbiome that directly harms our health, the disruption of the Earth's ecological system—which we live inside—can also create an imbalance that threatens us.

The consequences for human beings of the large-scale disruption of the Earth's ecological system—and what we can do to prevent it—is the subject of the next chapter.

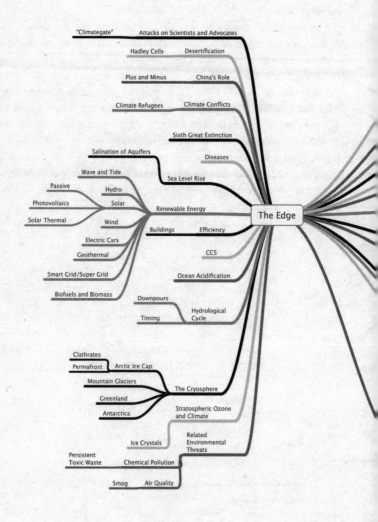

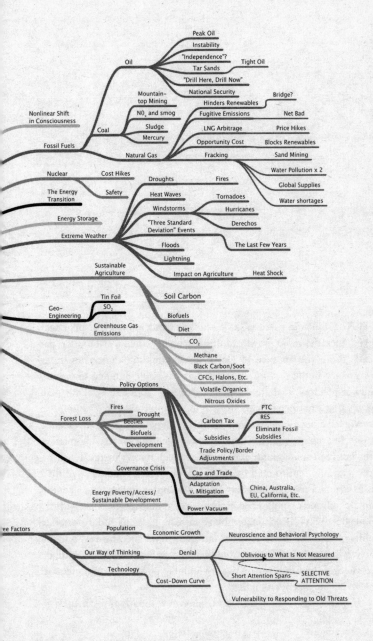

Oil
- Peak Oil
- Instability
- "Independence"?
- Tar Sands — Tight Oil
- "Drill Here, Drill Now"
- National Security

Fossil Fuels
- Coal
  - Mountain-top Mining
  - NO$_x$ and smog
  - Sludge
  - Mercury
- Natural Gas
  - Hinders Renewables — Bridge?
  - Fugitive Emissions — Net Bad
  - LNG Arbitrage — Price Hikes
  - Opportunity Cost — Blocks Renewables
  - Fracking — Sand Mining
    - Water Pollution x 2
    - Global Supplies
    - Water shortages

Nonlinear Shift in Consciousness

Nuclear — Cost Hikes
The Energy Transition — Safety
Energy Storage

Extreme Weather
- Droughts — Fires
- Heat Waves — Tornadoes
- Windstorms — Hurricanes
- "Three Standard Deviation" Events — Derechos
- Floods — The Last Few Years
- Lightning
- Impact on Agriculture — Heat Shock

Sustainable Agriculture

Geo-Engineering
- Tin Foil — Soil Carbon
- SO$_2$
- Biofuels
- Diet

Greenhouse Gas Emissions
- CO$_2$
- Methane
- Black Carbon/Soot
- CFCs, Halons, Etc.
- Volatile Organics
- Nitrous Oxides

Policy Options
- Carbon Tax — PTC / RES
- Subsidies — Eliminate Fossil Subsidies
- Trade Policy/Border Adjustments
- Cap and Trade
- Adaptation v. Mitigation — China, Australia, EU, California, Etc.
- Power Vacuum

Forest Loss
- Fires
- Drought
- Beetles
- Biofuels
- Development

Governance Crisis

Energy Poverty/Access/ Sustainable Development

ee Factors
- Population — Economic Growth
- Our Way of Thinking — Denial
- Technology — Cost-Down Curve

Neuroscience and Behavioral Psychology
Oblivious to What Is Not Measured
Short Attention Spans — SELECTIVE ATTENTION
Vulnerability to Responding to Old Threats

THE EDGE

THE EMERGENT POWER AND ACCELERATING MOMENTUM OF EARTH Inc., the rapid growth of destructive resource-consumption patterns, the absence of global leadership, and the dysfunctional governance in the community of nations have all combined to produce flows of pollution that are seriously damaging the integrity of the planetary climate balance that is essential to the survival of civilization.

We have been slow to recognize the extreme danger we are creating, in part because of the suddenness with which the underlying relationship between humankind and the ecological system of the Earth has been radically transformed by the relatively recent confluence of three basic factors. First, our numbers have quadrupled in less than a century and are still increasing; second, our way of thinking—both individually and collectively—is dominated by short-term horizons and distorted by habits of thought inherited from our prehistoric ancestors, who had to survive threats very different from the ones we face today; and third, the technologies now in common use are far more powerful than those available even a few generations ago.

In particular, our continued burning of carbon-rich fossil fuels for

85 percent of the energy that powers Earth Inc. spews 90 million extra tons of heat-trapping global warming pollution every twenty-four hours into the extraordinarily thin shell of atmosphere surrounding our planet, as if it is an open sewer. That means we are adding the equivalent by weight of more than 5,000 Deepwater Horizon Gulf oil spills *every day* to the dangerous concentrations that started accumulating with the Industrial Revolution at a rate that picked up speed dramatically throughout the last half century and is still accelerating.

As a result, human civilization is colliding with the natural world and causing grave harm to important natural systems on which our continued thriving as a species depends. There are multiple manifestations of this collision: the prospective extinction of 20 to 50 percent of all the living species on Earth within this century; the assault on the largest and most important forests in the world; the acidification of the oceans, depletion of important fish species, and imminent loss of coral reefs; the buildup of long-lived toxic chemical wastes that pose a persistent threat to people and other forms of life; the depletion of topsoil and groundwater resources at unsustainable rates; and more.

But the single most important and threatening manifestation of this collision is the climate crisis. Because the atmosphere surrounding our planet is so thin, it is highly vulnerable to the drastic change in its chemical composition brought about when we recklessly and constantly pollute it with such prodigious volumes of gaseous chemical waste. This growing blanket of pollution is smothering the atmosphere's ability to mediate the radiative balance between the Earth and the sun, trapping more extra heat energy each day in the lower atmosphere than would be released by 400,000 Hiroshima atomic bombs. In the process, we are profoundly altering the water cycle of the Earth, destroying crucial ecological balances, and compounding all of the other injuries we are inflicting on nature, including the plants and animals upon which we depend.

The good news is that we do have the capacity to begin solving the climate crisis—if we awaken to the reality of our circumstances and decide that saving the future of human civilization is a priority. That means recognizing not only the danger but also the opportunity inherent in this crisis. It means abandoning the illusion that there may be some clever technological quick fix for a planetary emergency that requires a multipronged global strategy to convert our energy systems—particularly electricity generation—manufacturing, agriculture, forestry, building technologies, transportation, mining, and other sectors of the world's economy to a low-carbon, highly efficient pattern.

And yes, when you lay out the complexity and magnitude of the response needed, it can sound daunting. But there have been recent stunning improvements in the technologies enabling us to succeed. They're increasing in efficiency and being deployed much faster than predicted. The scale of renewable energy markets has led to much sharper cost reductions than anyone predicted. The price of electricity from solar and wind has dropped so rapidly that in some areas of the world, both are already competitive with the average grid price for electricity. Globally, renewables will be the second-largest source of power generation by 2015.

REAL ALTERNATIVES

The more energy we produce from solar and wind, the cheaper it gets; the more energy we get from oil and coal, the more expensive it gets. And of course, the "fuel" for solar and wind is effectively limitless. For example, more potentially usable energy is received by the Earth from sunlight each and every hour than would be needed for all of the world's energy consumption in a full year. The potential for wind energy also exceeds the world's total energy demand several times over.

In the summer of 2012, there were periods when Germany received more than half its electricity from renewable energy sources. Some skilled investment experts are now projecting that on a global basis, even a conservative estimate of continued cost reductions for photovoltaic (PV) electricity will lead to a meteoric rise in its market share for new generating capacity over the next few years—to the point where almost half of the entire world's additional electricity generation will come from PV by midway through the next decade.

In 2010, for the first time in history, global investments in renewable energy exceeded those in fossil fuels ($187 billion, compared to $157 billion). The same year, solar photovoltaic installations in the United States rose 102 percent over those installed just one year earlier. Also, during the previous decade, 166 proposed new U.S. coal-fired generating plants were canceled, in large part due to public opposition.

Architects and builders are incorporating new designs and technologies that reduce energy consumption and the operating cost of buildings. This is particularly important because approximately 30 percent of all CO_2 emissions come from buildings, and of all buildings needed by 2050, two thirds have yet to be built. According to an EPA report, "On average, 30 percent of the energy consumed in commercial buildings is wasted. Energy efficiency is the single largest way to eliminate this waste, reduce emissions, and save money."

Hundreds of millions of people have already made changes in their purchasing decisions in order to seek out lower-carbon goods and services. In response, many businesses and industries have demonstrated leadership in accelerating carbon reductions and shifting to profitable strategies based on sustainability and a switch to renewable energy. Energy efficiency improvements are being implemented on a large scale. In the aggregate, however, greenhouse gas emissions will continue their steep rise until and unless government policies are enacted that accelerate the transition to a low-carbon world.

In order to move forward with this transition much faster, at a pace that is necessary to begin solving the climate crisis, we must first build a global political consensus—starting with a consensus in the United States—strong enough to support the policy changes that will solve the crisis: we have to put an adequate market price on the emissions of global warming pollution with a carbon tax, a steadily declining limit on emissions, and market mechanisms that promote maximum efficiency in the allocation of expenditures to achieve overall reductions.

Leaders in civic society must also place a political and social price on the dishonest distribution of false information about this existential crisis by cynical global warming deniers, many of whom know better but are trying to preserve destructive yet highly profitable business models by sowing confusion, false doubt, and political discord to delay the recognition of reality and prevent the congealing of a consensus.

Ultimately, here is the choice we face: we can either make the solution to the climate crisis the central organizing principle of global civilization—or the hostile conditions we are creating will grow rapidly worse, thickening the smothering blanket of global warming pollution surrounding our planet and destroy the viability of civilization as we know it.

For all of recorded history, we have configured the patterns of our lives and the design of our civilization to fit precisely into a relatively narrow envelope of familiar variations in temperatures, winds and rains, shorelines, river flows, frost lines, and snowfalls. We have built our communities in the places we call home—near reliable sources of the water we drink and the productive fields that give us food—in a world whose natural contours have varied little for more than 300 generations.

Since the glaciers retreated at the end of the last Ice Age, not long before the first cities were built and the invention of writing preserved the memory of man, we have taken for granted the enduring and relatively stable pattern of jet streams and ocean currents, warm spells and cold snaps, rainy seasons and dry seasons, spring planting and fall harvesting, tadpoles and butterflies, and the other enduring natural phenomena that have characterized our world for almost ten millennia. Just as the proverbial fish doesn't know it is in water—because it knows nothing *but* water—we have never known anything other than the planetary conditions that have given rise to the flourishing of humankind.

All of those who preceded us added their contributions to the elaborate legacy of the human enterprise bequeathed to us in our time. And each generation in turn has been sustained by gifts from nature itself: the pollination of crops and wild plants by insects and other animals, the natural purification of water by soils, and numerous other ecological benefits that modern economists call "ecosystem services."

All of this and more we take for granted. All of this and more we are putting at risk. Very large human-caused changes in the long predictable climate pattern we have always known could so radically reorder the nature of nature that it is difficult for us to imagine the challenges our species would confront. When a fish is taken out of the water, it cannot survive. By the same token, if we completely disrupt the conditions on which our civilization is based—not just for a few years, but for many thousands of years—it too would be unlikely to survive in anything resembling its current form.

SECURITY AND STABILITY

One of the many consequences of huge disruptions in the climate pattern we have always known would be a much higher risk of political instability. In fact, this risk is one of the principal reasons why military and national security experts in the United States have long expressed more concern about global warming than most elected officials. In many regions of the world, governance is already under tremendous stress with several failed states—Somalia, Yemen, and Zimbabwe, for example— creating difficult challenges for their regional neighbors. The increased stress that would accompany large alterations in climate patterns could push many other countries to the breaking point.

After a war game run by the National Defense University in the U.S. to simulate the geopolitical consequences of a mass migration of climate refugees from low-lying areas of Bangladesh, the head of the Bangladesh Institute of Peace and Security Studies, Major General A. N. M. Muniruzzaman, said, "By 2050, millions of displaced people will overwhelm not just our limited land and resources but our government, our institutions and our borders."

The few exceptions to the relative climate stability we have always enjoyed prove the rule. A recent study by David Zhang and others of the relationship between relatively small climate fluctuations in the past and civil conflict, published in the *Proceedings of the National Academy of Sciences*, reported, "Climate-driven economic downturn was the direct cause of large-scale human crises in pre-industrial Europe and the northern hemisphere." Indeed, our histories record the disruptive effects of comparatively small variations in the prevailing conditions in which we have thrived:

- The medieval warm period was connected to the disappearance of the Mayan civilization in Central America and the temporary colonization of southern Greenland by farmers from Scandinavia;
- During the Little Ice Age, Eskimos wrapped in fur paddled their kayaks to Scotland; farther south, millions died in a European famine centered in France;
- The huge downpours in fourteenth-century China triggered a chain of events leading to the Black Death that wiped out one quarter of the population of Europe;
- The unusually large eruption of the Tambora volcano in Indonesia in 1815 filled the Earth's atmosphere with particulates and led to the "year without a summer" in 1816 that caused widespread crop failures around the world, a wave of revolutions in Europe, and mass migrations in many regions by people searching for food and warmth.

All of these events were rare extremes that nevertheless fell within the natural boundaries of variations consistent with the same overarching climate pattern we have always known. And as terrible as the resulting catastrophes were, they were mostly temporary and relatively short-lived. By contrast, the much larger climate disruptions we are now causing threaten to create a planetary emergency lasting for time periods beyond the scope of human imagination. An estimated 25 percent of the CO_2 we put into the atmosphere this year will still be contributing to higher temperatures at least 10,000 years from now. If we force the melting of giant ice sheets in Antarctica and Greenland, they are not likely to return on a timescale that has any relevance whatsoever to our species.

Nine of the ten hottest years ever recorded since accurate measurements began in the 1880s have occurred in the last ten years. And the extra heat energy is already disrupting millions of lives. Extreme and destructive weather events that used to occur infrequently are becoming both more common and more destructive. Sometimes described as "once in a thousand year events," many bring with them enormous economic and human losses. And they are predicted to get much more common and much, much worse.

Among the recent examples: the epic flooding in Pakistan that displaced 20 million people, further destabilizing a nuclear-armed country; unprecedented heat waves in Europe in 2003 that killed 70,000 people, and in Russia in 2010 that led to 55,000 deaths, massive fires, and crop damage that pushed global food prices to record levels; the flooding of northeastern Australia in 2011 covering an area the size of France and Germany combined; the huge droughts in southern China and southwestern North America in 2011; the even deeper drought in over half of the U.S. in 2012; Superstorm Sandy in 2012, which devastated portions of New Jersey and New York City; multiple historic windstorms and downpours in many regions of the world.

The global water cycle—in which evaporation from the oceans falls as precipitation on the land and flows back to the oceans through streams that become rivers—is being radically intensified and accelerated by global warming. The warmer oceans allow significantly *more* water vapor to evaporate into the sky. More important still is the fact that warmer air *holds* more water vapor. If you take a cold shower, the mirror above your sink won't steam up, but if you take a hot shower it may. With so much more water in the atmosphere, there is also more energy fueling the size and destructive power of the storms.

Scientists have already measured an extra 4 percent of water vapor in the atmosphere above the oceans, and even though 4 percent doesn't sound like much, it has a large effect on the hydrological cycle. Because storms often reach out up to 2,000 kilometers, they gather water vapor from a large area of the sky and funnel it inward into the regions where storm conditions trigger a downpour.

By analogy, if you pull the drain in a bathtub filled with water, the water rushing down the drain does not come just from the part of the tub directly over the drain, it comes from the whole tub. In the same way, the great basins of water vapor in the sky are funneled to the "drains" opened above the land by rainstorms and snowstorms. When these basins are filled with much more water vapor than in the past, the downpours are more intense. The bigger downpours lead to bigger floods. The floods rush across the land, eroding the soil. And less of the water seeps down through the soil to recharge the underground aquifers.

Climate change is also driving desertification by altering atmospheric circulation patterns and drying out the land and vegetation. The same extra heat that evaporates more water vapor from the oceans also speeds up the evaporation of soil moisture—leading to longer, deeper, and more widespread droughts. Since the refilling of the atmospheric "basins" of moisture still takes a lot of time, many areas of the world are experiencing longer periods without rain in between the intense downpours. These longer periods of hotter temperatures in between precipitation events lead to more widespread and even deeper droughts. Once it is devoid of vegetation, the surface begins to absorb more heat. When the soil moisture is gone, the ground is baked, local temperatures rise higher still, and the topsoil becomes more vulnerable to wind erosion.

The parching and desiccation of the most highly productive agricultural breadbaskets of the world portend a food crisis in the future that could have humanitarian and political consequences too horrific to imagine. A top official with the International Maize and Wheat Improvement Center in Mexico, Marianne Bänziger, said, "There's just such a tremendous disconnect, with people not understanding the highly dangerous situation we are in."

The consequences for food production and water availability are already extremely harsh. In 2012, largely because of climate-related events that reduced crop yields, the world experienced a record one-month price increase for food, with additional record price hikes predicted for 2013. More than 65 percent of the U.S. suffered from drought conditions in 2012. In addition to the impacts on industrial agriculture in North America, Russia, Ukraine, Australia, and Argentina, subsistence agriculture has been hit hard in many tropical and subtropical countries by large alterations in the timing, duration, and magnitude of precipitation patterns due to global warming's disruption of the hydrological cycle. As a rice farmer in northeastern India, Ram Khatri Yadav, told Justin Gillis of *The New York Times*, "It will not rain in the rainy season, but it will rain in the non-rainy season. The cold season is also shrinking."

Along with the impacts discussed in Chapter 4—including the depletion of topsoil and groundwater and the competition that farmers face for land and water from fast-growing cities, industry, and biofuels production—the rising temperatures threaten many food crops with catastrophic yield reductions from heat stress alone. Stanford researcher David Lobell, who recently completed a study of the impact of temperature increases on crop yields with Columbia researcher Wolfram Schlenker, said recently, "I think there's been an under-recognition of just how sensitive crops are to heat, and how fast heat exposure is increasing."

In the last three years, new scientific research has overturned the long-held view by agricultural experts that, in the absence of drought, food crops would be relatively unharmed by rising temperatures. Many had thought that the higher CO_2 levels might fertilize plant growth by enough to counterbalance any yield decreases due to heat stress. But unfortunately, intensive research designed to confirm that hypothesis now shows that food crop yields are likely to decline much more rapidly with higher temperatures than previously believed, and that the CO_2 fertilization effect is much smaller than predicted. Moreover, weeds appear to benefit from extra CO_2 much more than food crops.

As temperatures continue to increase, corn (maize)—the most widely grown crop in the world—appears to be the most vulnerable to heat stress. Corn yields start to decrease at a range of temperatures the Earth is already experiencing regularly in summer months. Every day during the growing season (roughly from the beginning of March to the end of August) that temperatures climb above a threshold of 84 degrees F (29 degrees C), corn yields drop by 0.7 percent.

As temperatures grow hotter than 84 degrees F, the yield declines plummet further with every degree added. If temperatures in the United States are allowed to rise as much as is now projected as a result of global warming, by the end of this century corn yields could fall by as much as a third from heat stress alone, with the impact of worsening droughts and the disruption of precipitation patterns taking a larger toll still. Soybeans have a higher threshold for heat stress than corn (86 F/30 degrees C), but the same accelerated drops in yields begin when temperatures reach and exceed that level.

The warm season is longer; spring is arriving about a week earlier (and fall about a week later) in both the northern and southern hemispheres. Moreover, the decreasing size of mountain snowpacks and glaciers is adding to the worsening shortages of water for agriculture in several important regions, bringing bigger spring floods earlier in the year and depriving these regions of water during the hot summer months when it is most needed. And while the focus is normally on daytime high temperatures, nighttime temperatures are at least as important. Both the computer models and consistent observations confirm that global warming increases nighttime temperatures more than daytime temperatures.

According to some studies, each degree increase in nighttime temperatures corresponds with a linear decrease in wheat yields. A large global review of the impact of climate change on crop yields between 1980 and 2010 showed that worldwide wheat production fell due to climate-related factors by 5.5 percent. A researcher at the International Rice Institute in the Philippines, Shaobing Peng, published findings in the *Proceedings of the National Academy of Sciences* showing that yields of rice declined by 10 percent with each one degree Celsius increase in nighttime temperatures during the dry part of the growing season, even though there were no significant drops in yield associated with increasing maximum temperatures during the daytime.

Crop diseases and pests are also increasing with global warming. Higher temperatures are leading to a dramatic expansion in the range of insects harmful to food crops, sending them farther north in the northern hemisphere and farther south in the southern hemisphere, and into higher altitudes. A team of crop scientists publishing in *Environmental Research Letters* wrote, "These range expansions could have substantial economic impacts through increased seed and insecticide costs, decreased yields, and the downstream effects of changes in crop yield variability."

Other scientists have determined that higher levels of CO_2 also stimulate insect populations. Evan DeLucia, a plant biologist working with a team of entomologists at the University of Illinois, tested the impact of higher carbon dioxide levels on soybeans and found that aphids and Japanese beetles flocked to the soybeans grown in higher CO_2 environments, ate more of the plants, lived longer, and produced more eggs. "That means crop losses may go up in the future," DeLucia said.

Other scientists on DeLucia's team found that higher carbon dioxide levels caused soybeans to deactivate genes that are crucial to the production of chemicals that help to defend them against insects by blocking enzymes in the stomachs of beetles that digest soybean plants, and by deactivating other genes used by soybeans to lure the natural enemies of the beetles. As a result, according to team member Clare Casteel, the soybeans grown in higher levels of CO_2 "appear to be helpless against herbivores."

Higher temperatures are having the same effect in boosting pest populations in most areas of the world. One of the leaders of an Asian international agricultural research group, Pramod K. Agrawal, said, "Warmer conditions and longer dry seasons linked to climate change could prove to be the perfect catalyst for outbreaks of pests and diseases. They are already formidable enemies affecting food crops." A team of Indian scientists noted that, because insects are cold-blooded, "Temperature is probably the single most important environmental factor influencing insect behavior, distribution, development, survival, and reproduction. . . . It has been estimated that with a 2 degree C temperature increase, insects might experience one to five additional lifecycles per season."

Scientists at the International Center for Tropical Agriculture, for example, found that the cassava crop in Southeast Asia—worth an estimated $1.5 billion each year—is seriously threatened by pests and plant diseases that expand with warmer temperatures. According to cassava entomologist Tony Bellotti, "The cassava pest situation in Asia is pretty serious as it is. But according to our studies, rising temperatures could make things a whole lot worse." Bellotti adds, "One outbreak of an invasive species is bad enough, but our results show that climate change could trigger multiple, combined outbreaks across Southeast Asia, Southern China and the cassava-growing areas of Southern India."

Microbes that cause human diseases—and the species that carry them—are also expanding their range. In the highly populated temperate zones of the world, the prevailing climate conditions in which civilization developed were unfavorable to the survival of many disease-causing organisms. But now that warmer climate bands are moving pole-ward, some of these pathogens are moving with them.

According to a study in *Science* by Princeton University researcher Andrew Dobson and others, global warming is causing the spread of bacteria, viruses, and fungi that cause human diseases into areas that were formerly hostile to them. "Climate change is disrupting natural ecosystems in a way that is making life better for infectious diseases," said Dobson. "The accumulation of evidence has us extremely worried." Another coauthor of the study, Richard S. Ostfeld, said, "We're alarmed because in reviewing the research on a variety of different organisms, we are seeing strikingly similar patterns of increases in disease spread or incidence with climate warming."

Although the prevalence of international travel has increased dramatically and some disease-carrying insects have been unwittingly transported from the mid-latitudes to other regions, the shifting climatic conditions are contributing to the spread of diseases like dengue fever, West Nile virus, and others. The Union of Concerned Scientists wrote that, "Climate change affects the occurrence and spread of disease by impacting the population size and range of hosts and pathogens, the length of the transmission season, and the timing and intensity of outbreaks."

They also noted, "Extreme weather events such as heavy rainfall or droughts often trigger disease outbreaks, especially in poorer regions where treatment and prevention measures may be inadequate. Mosquitoes in particular are highly sensitive to temperature." Improvements in public health systems are crucial to control the spread of these migrating diseases, but many lower-income countries are pressed to find the resources needed for hiring and training more doctors, nurses, and epidemiologists. They also warned that in many of the areas to which these pathogens and their hosts spread with warmer temperatures, "The affected populations will have little or no immunity, so that epidemics could be characterized by high levels of sickness and death."

In the summer of 2012, the United States experienced the worst outbreak of West Nile virus since it first arrived on the Eastern Shore of Maryland in 1999 and spread rapidly to all fifty states in only four years, during a period of unusually warm temperatures. Dallas, Texas, was the first to declare a public health emergency and began aerial spraying of the city for the first time since 1966. As concern peaked, public safety officials issued an appeal for people to stop calling 911 when they were bitten by mosquitoes. The disease eventually spread by the end of 2012 to forty-eight of the fifty states, killing at least 234 people.

The late Paul Epstein, a professor at Harvard Medical School and a close friend, wrote in 2001 about the relationship between West Nile virus and the climate crisis. More recently, he said, "We have good evidence that the conditions that amplify the lifecycle of the disease are mild winters coupled with prolonged droughts and heat waves—the long-term extreme weather phenomena associated with climate change."

According to Christie Wilcox with *Scientific American*:

They have been predicting the effects of climate change on West Nile for over a decade. If they're right, the US is only headed for worse epidemics. . . . Studies have found that mosquitos pick up the virus more readily in higher temperatures. Higher temperatures also increase the likelihood of transmission, so the hotter it is outside, the more likely a mosquito that bites an infected bird will carry the virus and the more likely it will pass it along to an unwitting human host. In the United States, epicenters of transmission have been linked closely to above-average summer temperatures. In particular, the strain of West Nile in the US spreads better during heat waves, and the spread of West Nile westward was correlated with unseasonable warmth. High temperatures are also to blame for the virus jumping from one species of mosquito to a much more urban-loving one, leading to outbreaks across the US. . . . Record-breaking incidences of West Nile are strongly linked to global climate patterns and the direct effects of carbon dioxide emissions.

In 2010, the world experienced the hottest year since records have been kept, and ended the hottest decade ever measured. Last year, 2012, broke even more high temperature records. October 2012 was the 332nd month in a row when global temperatures were above the twentieth-century average. The worst drought since the Dust Bowl of the 1930s ravaged crops and dried up water supplies in many communities. Many farmers have already been forced to adjust to the drying of soil. The lack of water has caused a buildup of toxins in corn and other crops unable to process nitrogen fertilizer.

WORLD FEVER

In order to pinpoint the difference between global warming and natural variability, Dr. James Hansen, the single most influential climate expert in the scientific community, produced with two of his colleagues, Makiko Sato and Reto Ruedy, a groundbreaking statistical analysis of extreme temperatures all over the world from the years 1951 through 2010 that compared the more normal baseline period of 1951 through 1980 to more recent decades and especially the last several years, when the impacts of global warming have been more prominently manifested, 1981 through 2010.

By breaking down the surface temperatures of almost the entire world into blocks of 150 square miles each, Hansen was able to calculate the frequency of extremely high temperatures (and all other temperatures) during the last sixty years. The results—which do not rely on climate models, climate science, or any theories of causation—demonstrate clearly that there has been up to a 100-fold increase in extreme high temperatures in recent years compared to earlier decades. The statistical analysis shows that in the last several years, extreme temperatures have been occurring regularly on approximately 10 percent of the Earth's surface, while during the earlier decades such events occurred on only 0.1 to 0.2 percent of the Earth's surface.

Hansen's chosen metaphor to explain the difference consists of two dice, each with the requisite six sides. The first die, which shows the range of temperatures over the years between 1951 and 1980, has two sides representing "normal" seasons, two other sides representing "warmer than normal" seasons, and the final two sides representing "cooler than normal" seasons. That used to be the "normal" distribution of temperatures. The second die, however, showing the range of temperatures in more recent years, has only one side representing a normal season and only one side representing a cooler than normal season, but three sides representing warmer than normal seasons and the remaining side now representing *extremely hot* seasons—seasons that are way outside the boundary of the statistical range that used to prevail.

In the language of statisticians, a standard deviation quantifies how far the range, or spread, of a particular set of phenomena differs from the average spread. Extreme—in this case, either unusually hot or unusually cold—seasons naturally occur far less frequently than average or near-average seasons. Because seasons with extreme temperatures used to be so much less frequent, they nevertheless often surprised us, even though they fell within the normally expected range. Seasons that are *three* standard deviations from the average were exceedingly rare, but still did occur from time to time as part of the normal range.

The average temperature is warmer overall even though extremely cold events still continue to occur, though rarely. In other words, the entire distribution of temperatures has moved to much warmer values, and the bell curve of distribution has widened and flattened slightly, so that there is much more temperature variability than used to occur. But the most significant finding is that the frequency of extremely hot temperatures has gone up dramatically.

Hansen infers that the cause is global warming—and indeed, these results turn out to be perfectly consistent with what global warming science has long predicted. (In voluminous other studies, Hansen and climate scientists around the world have proven causality to a degree judged "unequivocal" and "indisputable" by virtually all of the world's scientific community.) But the results themselves are based on observations of real temperatures in the real world. They cannot be argued with, and the implications are powerfully clear.

As the old saying has it in Tennessee, if you see a turtle on top of a fence post, it is highly likely that it didn't get there by itself.* And now we are seeing turtles on every tenth fence post in every field in the world. They didn't get there on their own. It is now abundantly obvious that all the extreme temperatures and the extreme weather events associated with them are like turtles on a fence post. They didn't happen without human interference in the climate.

In 2012, new World Bank president Jim Yong Kim released a study showing temperatures will likely rise by 4 degrees C (7.2 degrees F) without bolder steps to reduce CO_2, and that there is "no certainty that adaptation to a 4 degree world is possible." Gerald Meehl, of the National Center for Atmospheric Research, uses a different metaphor to explain what is happening: if a baseball player who takes steroids hits a home run, it's possible that he might have hit the home run even without the steroids. But the fact that he took the illegal performance-enhancing drug makes it much more likely

---

* In the old days before pesticides, farmers understood that turtles, birds, and bats were their friends. To protect the turtles from the plow, farm boys and girls would walk the fields in many areas prior to plowing to rescue turtles. They would put them on fence posts, and after the tilling was done the turtles would be released, generally at sunset.

that he will hit a home run in his next at bat. Within Meehl's metaphor, the 90 million tons of global warming pollution that we are putting into the atmosphere every twenty-four hours are like steroids for the climate. An innovative 2012 study of the previous decade's climate predictions showed that the "worst case" future projections are the ones most likely to occur.

The increases in the global average temperature and the greater frequency of extremely high temperatures that Hansen and others are documenting are also melting all of the ice-covered regions of the Earth. Only thirty years ago, the Arctic Ocean was almost completely covered by ice in summer as well as winter. Remember? Some called it the North Polar Ice Cap. Tell your grandchildren how it used to separate Eurasia from North America and the Atlantic from the Pacific all year round. Last year's record low in the volume and the area it covered marked an acceleration of a melting pattern that has led to a 49 percent loss in three decades and could, in the view of many ice scientists, produce a 100 percent loss in as little as a decade.

Some shipping companies are excited that the fabled Northern Sea Route is now open for several months a year. A Chinese ship, the *Snow Dragon*, traversed the North Pole to Iceland and back in the summer of 2012. A high-speed fiber optic cable is now being installed to link the Tokyo stock markets with their counterparts in New York City so that computer-driven trades can be executed more quickly. Fishing fleets are preparing to exploit the rich biological resources of the Arctic Ocean, which until now have been protected by the ice. Navies from some countries are discussing the movement of military assets into the region, though discussions have also begun on the possibility of agreements to foster the peaceful resolution of issues involving the safety, sovereignty, and development of the Arctic Ocean as it becomes ice-free in summer.

Several oil companies are thrilled at the prospect of new drilling opportunities and some are already moving their rigs into place. But the consequences of an accidental wellhead blowout at the bottom of the Arctic Ocean similar to BP's disaster in 2010 would be far more catastrophic and far more difficult to deal with than in the Gulf of Mexico, or in any of the other numerous deepwater locations where wellhead blowouts have produced large oil spills. The relatively new and unperfected technology used for deepwater drilling involves more risk than conventional drilling because the pressures at the ocean's bottom are so great. Drilling for oil at the bottom of the Arctic Ocean, and running the risk of a large spill in a pristine ecosystem where repair and rescue operations are all but impossible for much of the year, is an absurdly reckless endeavor. The CEO of the French multinational oil company Total broke ranks with his industry in 2012 and expressed his view that drilling for oil in the Arctic Ocean posed unacceptable ecological risks and should not be carried out.

The ecology of the Arctic Ocean is already experiencing significant changes. Scientists were shocked in 2012 at the discovery of the largest algae bloom ever recorded on Earth extending from open areas of the Arctic Ocean underneath the remaining ice cover—a phenomenon that has never been seen before and was considered impossible. The researchers explained that the most likely cause of this new occurrence was that the remaining ice is now thin enough, and has so many pools of water dotting its surface, that enough sunlight was penetrating to the ocean below to provide energy for algal growth.

The consequences of melting the North Polar Ice Cap will include large impacts on weather patterns extending far southward into the heavily populated temperate zones. The dramatically increased heat absorption in an Arctic Ocean that is ice-free in summer will have consequences for the location and pattern of the northern jet stream and storm track through the fall and winter seasons, modifying ocean currents and weather patterns throughout the northern hemisphere, and perhaps beyond. Moreover, if the world's long familiar pattern of wind and ocean currents is pushed into a completely new design, the old one may never reemerge.

The land area surrounding the Arctic Ocean is also heating up, thawing frozen tundra that contains enormous amounts of carbon embodied in dead plants. They warm up and rot as the tundra thaws. Microbes turn the carbon into CO_2 or methane, depending on the amount of soil moisture. Huge deposits of methane are also contained within frozen ice crystal formations called clathrates in the tundra, at the bottom of the many shallow frozen lakes and ponds surrounding the Arctic, and in some parts of the seabed underneath the Arctic Ocean. The bubbling methane carries heat energy upward, melting the underside of the ice— which then increases the heat absorption by the water when the sun's rays are no longer reflected off the ice.

Scientists are struggling to quantify the amount of CO_2 and methane that could be released, but the area involved is so vast that their work is extremely difficult. Already, however, they have found outgassing under way that exceeded what they expected at this early stage of global warming.

Moreover, scientists discovered in 2012 that there are likely to be enormous deposits of methane underneath the Antarctic ice sheet, in amounts that may be as large as the methane presently trapped in Arctic tundra and coastal sediments. Since the clathrates are kept in place by cold temperatures and high pressures, the thinning of the Antarctic ice sheets could, scientists fear, reduce pressures underneath the ice enough to trigger the release of methane.

The changes under way in Antarctica and Greenland are the focus of intense study by scientists who are trying to calculate how much sea levels will rise, and at what rate. Both ice sheets are being destabilized and are losing mass at an increasing rate, which is leading to a much faster sea level rise than was predicted just a decade ago.

Throughout the history of urban civilization, the seas have been slowly and gently rising, as the warmer temperatures of the interglacial period have caused thermal expansion of the ocean's volume, and the melting of some terrestrial ice. But with the rapid accumulation of CO_2 and other greenhouse gases in the atmosphere during the last half century, global warming has accelerated and so has the melting of ice almost everywhere on the planet.

Predictions of the rate of sea level rise have been notoriously difficult, in part because many scientists use models calibrated using data derived from their studies of retreating glaciers at the end of the last Ice Age when conditions were very different from the ones we are now confronting. New real-time satellite measurements of ice mass in Greenland and Antarctica will soon improve scientific understanding of this process, but these measurements have been made for only a few years and more time is required to build confidence in what they are telling the scientific community. Recent observations in both west Antarctica and Greenland, however, already confirm a rapid and accelerating loss of ice. After a highly unusual melting event that affected 97 percent of Greenland's surface in July 2012, Bob Corell, chairman of the Arctic Climate Impact Assessment, said, "It shocked the hell out of us."

James Hansen, for one, surmises that we are witnessing an exponential process of ice mass loss, and that, as a result, the most relevant statistic is the doubling time of the observed loss. Based on his preliminary analysis of the data, Hansen believes it is likely that we will see a "multi-meter" sea level rise in this century. Others note that the last time temperatures on Earth were consistently as high as they are now, sea level was twenty to thirty feet higher than the present—although it took millennia for the seas to rise that much.

Because so many countries were settled by migrants, and in some cases colonialists, arriving by ship—and because trade and supply routes rely so heavily on oceangoing vessels—a disproportionate percentage of the world's largest cities are located near the sea. In fact, 50 percent of the world's population lives within fifteen miles of the coast, and according to the U.S. National Academy of Sciences, "Coastal populations around the world are also growing at a phenomenal pace. Already, nearly two-thirds of the world's population—almost 3.6 billion people—live on or within 100 miles of a coastline. Estimates are that in three decades, 6 billion people—that is, nearly 75 percent of the world's population—will live along coasts. In much of the developing world, coastal populations are exploding."

Those in low-lying areas are therefore especially vulnerable to the increases in sea level produced mainly by the melting and breakup of large masses of ice in Antarctica and Greenland. A recent study by Deborah Balk and her colleagues at the CUNY Institute for Demographic Research showed that approximately 634 million people live in low-elevation coastal zones and that the ten nations with the most people in threatened areas are: China, India, Bangladesh, Vietnam, Indonesia, Japan, Egypt, the United States, Thailand, and the Philippines. Moreover, two thirds of the world's cities with more than five million people are at least partly in vulnerable low-elevation areas.

Some of the populations who live on low-lying islands in the Pacific and Indian oceans and in coastal deltas are already beginning to relocate. Large island populations are also at risk in the Philippines and Indonesia. The number of climate refugees is expected to grow and could potentially involve more than 200 million people in this century, especially because of those who will have to move away from the mega-deltas of South Asia, Southeast Asia, China, and Egypt. Refugees from coastal areas of Bangladesh have already crowded into the capital city of Dhaka, and many have moved farther north across the border into northeastern India, where their arrival has contributed to the worsening of preexisting tensions based on religious and complex tribal conflicts. In 2012, these conflicts generated contagious fear that was spread by text messaging and email into cities throughout India.

All these regions and others are also threatened by climate-related flooding during storm surges as stronger cyclones (known as hurricanes in the U.S.) gain energy from warmer seas. Even small vertical increases are magnified by storm surges that carry the ocean inland. And with stronger storms, these surges are already having a bigger impact. In 2011, for example, New York City was put on emergency alert as a hurricane threatened to flood its subway system. In 2012, Superstorm Sandy did. London has long since built barriers between the ocean and the Thames River that can be closed to protect the city against such surges—at least for a while; the city is already discussing plans for further steps.

As noted in Chapter 4, the surge in population growth in the balance of this century will be completely in urban areas. The cities with the highest population at risk from rising seas are, in order: Calcutta, Mumbai, Dhaka, Guangzhou, Ho Chi Minh City, Shanghai, Bangkok, Rangoon, Miami, and Hai Phong. The cities with the most exposed assets vulnerable to sea level rise are: Miami, Guangzhou, New York/Newark, Calcutta, Shanghai, Mumbai, Tianjin, Tokyo, Hong Kong, and Bangkok.

In addition, as the chief scientific advisor in the United Kingdom, Sir John Beddington, recently noted, many climate refugees have migrated to low-lying coastal cities vulnerable to increased climate-related flooding and rising seas. They are unknowingly relocating into areas from which they may once again become climate refugees.

Contrary to most popular thinking, the rate of sea level rise is not uniform around the world, because some of the tectonic plates on which the landmasses rest are still slowly "rebounding" from the last Ice Age.* Scandinavia and eastern Canada, for example, were pushed down by the weight of the last glaciation and are still moving slowly upward long after the ice retreated. Conversely, areas at the opposite ends of the same tectonic plates—the coastal nations of Western Europe and the mid-Atlantic states of the U.S., for example—are slowly sinking, in a kind of seesaw effect. Cities like Venice, Italy, and Galveston, Texas, are also sinking—for a mixture of complicated reasons.

Because warmer oceans expand when their molecules push apart

---

* Additionally, climate alterations caused by changes in the gravitational pull from ice sheets have measurable effects on relative sea level rise in some areas.

from one another (thermal expansion of the oceans has contributed significantly to the relatively small increases in sea level we have experienced thus far), areas of the ocean with large accumulations of warmer water are experiencing more rapid sea level increases—the coast of the U.S. between South Carolina and Rhode Island, for example. But all the increases in sea level thus far are nothing compared to what scientists warn is in store for the entire world as Antarctica and Greenland are affected by the sharp increases in global temperatures now in store.

Many agricultural areas in low-lying coastal regions and areas adjacent to river deltas are already suffering impacts from rising seas because of saltwater invasion of the freshwater aquifers on which their farms depend. In 2012, the combination of sea level rise and sharply diminished flows in the Mississippi River, due to the drought in the U.S., led to saltwater intrusion into drinking water wells and aquifers in southern Mississippi.

The characteristics of the seawater itself are also being profoundly altered by global warming. Approximately 30 percent of human-caused CO_2 emissions end up in the ocean, where they dissolve into a weak acid, building up in such enormous volumes that it has nevertheless already made the world's oceans more acidic than at any time in the last 55 million years, which was during one of the five previous great extinction events in the history of the Earth. And the *rate* of acidification is faster than at any time in the last 300 million years.

One of the immediate concerns is that the higher levels of acidity are reducing the concentration of carbonate ions that are essential to species that make shells and coral reefs. All such structures are made from various forms of calcium carbonate, which the coral polyps and shell-making creatures scavenge from seawater. But the increasing acidity of the ocean interferes with the solidifying of these hard structures. The director of the U.S. National Oceanic and Atmospheric Administration, Jane Lubchenco, calls ocean acidification global warming's "evil twin."

The warmer ocean temperatures—also caused by man-made global warming—are especially stressful to the specialized algae that form the brightly colored skin of coral reefs and live in an intricate symbiosis with the coral polyps. When water temperatures rise too high, these specialized algae—known as zooxanthellae (also called zoox)—leave the skin of the coral, rendering it transparent and revealing the white bony skeleton underneath. These events are known as coral bleaching. Reefs can and do recover from bleaching events, but several events in the space of a few years can and do kill the reefs.

Coral reefs are particularly important because, according to experts, approximately one quarter of all ocean species spend at least part of their lifecycles in, on, and around reefs. Shockingly, scientists warn that the world is in danger of killing almost all of the coral reefs in the ocean within a generation. Between 1977 and 2001, 80 percent of the coral reefs in the Caribbean were lost. All of the rest, experts say, are threatened with destruction before the middle of the century. And the same fate threatens reefs in every ocean, including the largest of all, the Great Barrier Reef off the eastern coast of Australia. In 2012, the Australian Institute of Marine Science announced that half of the Great Barrier Reef corals had died in just the previous twenty-seven years.

The most visible and familiar reefs are warm-water reefs at relatively shallow depths. However, there may be an equal or even larger number of deeper, cold-water reefs. Because of their depth, they have been less studied and documented, but scientists say that since colder water absorbs more CO_2 than warmer water (just as a cold container of soda stays more carbonated than a warm one), many of the cold-water reefs may be in even greater danger. Some scientists hold out hope that coral reefs might yet survive, but many of their colleagues are now convinced that virtually all corals are likely to be killed off by the combination of higher ocean acidity, higher temperatures, pollution, and overfishing of species important to reef health.

The growing absorption of CO_2 in the oceans also interferes with the reproduction of some species. And among the shell-making creatures at risk are tiny zooplankton with very thin shells that play an important role at the bottom of the ocean food chain. Although much research remains to be done, many scientists are concerned about what has been happening to this crucial link that lies at the base of the ocean food chain.

Some areas of the ocean, including some off the coast of Southern California that have been sampled, are actually *corrosive*. In coastal areas of Oregon, newly corrosive seawater is killing commercially valuable shellfish. Experts have noted that even if human-caused CO_2 emissions were somehow ended in the near term, it would take tens of thousands of years before the chemistry of the oceans returned to a state comparable to that which existed prior to the last century.

Global warming and CO_2-caused acidification are exacerbating declines in fisheries and marine biodiversity that have already been caused by other human activities, such as overfishing. According to the United Nations, almost a third of all fish species are presently overexploited. Overfishing, described in Chapter 4, has already led to the dangerous depletion of up to 90 percent of large fish like tuna, marlin, and cod.

Some fishing techniques such as dynamite fishing (which still takes place in some developing countries with coral reefs) and bottom trawling (the northeast Atlantic has been particularly damaged by this practice) do extra damage to the ocean ecosystems important to the survival of sea life. Although there have been some notable success stories in some ocean fisheries, the overall picture is still extremely troubling. The combination of many factors poses a synergistic threat to the continued health of the oceans.

Along with coral reefs, critical ocean habitats like mangrove forests in many coastal areas and so-called sea grass meadows are also at risk. In addition, the number of dead zones growing in the oceans near the mouths of major river systems is doubling every decade. The heavy concentrations of nitrogen and phosphorus contained in agricultural runoff water and wastewater feed algae growth and when the algae are consumed by bacteria, the large areas of the ocean are completely depleted of oxygen, leading to the dead zones.

Ironically, the historic North American drought of 2012 reduced the flow of water from the Mississippi into the Gulf of Mexico so much—and the nitrogen, phosphorus, and other chemicals normally carried with the water—that the large dead zone spreading from the mouth of the Mississippi began to temporarily clear up.

A conference of ocean experts meeting at Oxford University in the summer of 2011 reported their conclusions as a group: "This examination of synergistic threats leads to the conclusion that we have underestimated the overall risks and that the whole of marine degradation is greater than the sum of its parts, and that degradation is now happening at a faster rate than predicted. . . . When we added it all up, it was clear that we are in a situation that could lead to major extinctions of organisms in the oceans. . . . It is clear that the traditional economic and consumer values that formerly served society well, when coupled with current rates of population increase, are not sustainable."

MITIGATION VERSUS ADAPTATION

For at least three decades, there has been a debate in the international community about the relative importance of reducing greenhouse gas emissions to *mitigate* the climate crisis compared to strategies for *adapting* to the climate crisis. Some of those who try to minimize the significance of global warming and oppose most of the policies that would mitigate it often speak of adaptation as a substitute for mitigation.

They promote the idea that since humankind has adapted to every environmental niche on the planet, there is no reason to believe that we shouldn't merely accept the consequences of global warming and get busy adapting to them. For example, the CEO of ExxonMobil, Rex Tillerson, recently said in an exchange provoked by longtime activist David Fenton, "We have spent our entire existence adapting, OK? So we will adapt to this."

FOR MY OWN part, I used to argue many years ago that resources and effort put into adaptation would divert attention from the all-out push that is necessary to mitigate global warming and quickly build the political will to sharply reduce emissions of global warming pollution. I was wrong— not wrong that deniers would propose adaptation as an alternative to mitigation, but wrong in not immediately grasping the moral imperative of pursuing both policies simultaneously, in spite of the difficulty that poses.

There are two powerful truths that must inform this global discussion about adaptation and mitigation: first, the consequences that are already occurring, let alone those that are already built into the climate system, are particularly devastating to low-income developing countries. Infrastructure repair budgets have already skyrocketed in countries where roads, bridges, and utility systems have been severely damaged by extreme downpours and resulting floods and mud slides. Others have been devastated by the climate-related droughts.

And the disruptions of subsistence agriculture by both the floods and the droughts have led to skyrocketing expenditures for food imports in many developing countries. Also, as noted earlier, some low-lying nations are also already struggling to relocate refugees from coastal areas affected by the early stages of sea level rise, while other nations are struggling to integrate arriving refugee groups into already fast-growing populations.

Since these and other developments will not only continue but worsen, the world does indeed have a moral duty and practical economic necessity to assist these nations with adaptation. Disturbingly, the world has yet to fully realize the effects of the global warming pollution already in the atmosphere. Even if we drastically reduce our emissions today, another degree Fahrenheit of warming is already "in the pipeline" and will manifest itself in the coming years. In other words, so many harmful changes are already built into the climate system by the enormous increase in the emissions, and particularly the increased *concentration*, of greenhouse gases in the atmosphere that adaptation is absolutely essential—even as we continue building the global political consensus needed to prevent the worst consequences from occurring. We have no choice but to pursue both sets of policies simultaneously.

But the second truth that must inform this debate is still by all odds the most powerful imperative: unless we quickly start reducing global warming pollution, the consequences will be so devastating that adaptation will ultimately prove to be impossible in most regions of the world. For example, higher greenhouse gas emissions are already beginning to cause large-scale changes in atmospheric circulation patterns and are predicted to bring almost unimaginably deep and prolonged drought conditions to a wide swath of highly populated and agriculturally productive regions, including all of Southern and south-central Europe, the Balkans, Turkey, the southern cone of Africa, much of Patagonia, the populated southeastern portion of Australia, the American Southwest and a large portion of the upper Midwest, most of Mexico and Central America, Venezuela and much of the northern Amazon Basin, and significant portions of Central Asia and China.

The scientific reasoning behind this devastating scenario requires some explanation. The basic nature of the global climate system, when viewed holistically, is that it serves as an engine for redistributing heat energy: from the equator toward the poles, between the oceans and the land, and from the lower atmosphere to the upper atmosphere and back again. The large increase in heat energy trapped in the lower atmosphere means—to state the obvious—that the atmospheric system is becoming more energetic.

In the northern hemisphere, this climate engine transfers heat energy from south to north in the Gulf Stream—which is the best known component of the so-called ocean conveyor belt, a Möbius Strip–like loop that connects all of the world's oceans. Other components include deep currents that travel along the bottom of the ocean, redistributing cold water from the poles back to the equator, where they return to the ocean surface. The largest of these are the Antarctic circumpolar current, which travels around the Antarctic continent and feeds the shallower Humboldt current, which flows from the Southern Ocean northward along the west coast of South America and upwells—laden with nutrients—to nourish the rich concentration of sea life off the coast of Peru; and, less well known, the deep cold current that travels north to south from an area of the North Atlantic in the vicinity of southern Greenland, *underneath* the Gulf Stream, back to the tropical Atlantic waters.

Energy is also redistributed by cyclones, by thunderstorms, and by multiyear patterns such as the alternating El Niño/La Niña phenomenon (known to scientists as the ENSO, or El Niño/Southern Oscillation). Moreover, all of these energy transfers are affected by the Coriolis effect, which is driven by the spinning of the Earth on its axis, from west to east.

THE HADLEY CELLS

Until recently, relatively less attention has been paid to the relationship between global warming and the atmospheric patterns that move energy vertically up and down in the atmosphere. The so-called Hadley cells spanning the tropics and subtropics are enormous barrel-shaped loops of wind currents that circle the planet on both sides of the equator, like giant pipelines through which the trade winds flow from east to west.

Warm and moist wind currents rise from the ground vertically into the sky in both of these cells at the edge of each respective loop that is adjacent to the equator. When their ascent reaches the top of the troposphere (the top of the lower atmosphere, approximately ten miles high in the tropics), each loop turns poleward—which means northward in the northern hemisphere cell and southward in the other. By the time these currents reach the top of the sky, much of the moisture they carried upward has fallen back to the ground as rain in the tropics.

At the apex of its ascent, each of these air currents starts flowing poleward along the top of the troposphere and travels about 2,000 miles (approximately 30 degrees of latitude), until it has discharged most of its heat. Then it descends vertically as a cooler and much drier downdraft. When each loop reaches the surface again, it turns back toward the equator, recharging itself with heat and moisture as it travels across the surface of the Earth. As it returns to the equator, it completes its loop and repeats the cycle by rising vertically once again, laden once more with heat and water vapor.

As a result of the dry downdrafts of the Hadley cells, the areas of the Earth 30 degrees north and 30 degrees south of the equator are highly vulnerable to desertification. Most of the driest regions of the Earth, including the largest of the planet's deserts, the Sahara, are located under these dry downdrafts. (Other factors contributing to the location of deserts include the "rain shadows" of mountain ranges—the areas downwind from mountain peaks—because the prevailing winds rise when they hit the windward side of the mountains and lose their moisture before descending as dry downdrafts on the leeward side. In addition, the location of deserts is influenced by what geographers call continentality—which means that the areas in the middle of large continents typically get much less moisture because they are farther away from the oceans.) But on a global basis, the most powerful desertifying factor is the downdraft of the Hadley cells.

The problem—which climate scientists have long predicted with computer models and are now observing in the real world—is that the massive warming of the atmosphere is changing the locations of these great global downdrafts, moving them farther away from the equator and toward the poles, thus widening the subtropics and intensifying their aridity. Indeed, in the northern hemisphere, the downdraft has already moved northward by as much as 3 degrees latitude—approximately 210 miles—although measurements are still imprecise. The downdraft of the Hadley cell south of the equator has also moved poleward.

There are several theories for why global warming is causing a shift in the Hadley cells, none of which are as yet confirmed. The solar heating of the lower atmosphere in the tropics and subtropics is much greater than anywhere else on the planet for obvious reasons: the sunlight strikes the Earth at a more direct angle all year round. On a percentage basis, surface temperatures are rising faster in the higher latitudes because the melting of ice and snow is dramatically changing the reflectivity of the surface,* thereby increasing the absorption of heat energy. This means, among other things, that the difference in average temperatures between the tropics and the polar regions is diminishing over time—which also has consequences for the climate balance.

However, the much larger amounts of overall heat energy absorbed in the mid-latitudes is still much greater, and causes the warmer (and thus less dense) air in the tropics to rise higher. As a result, the extra heat raises the top of the troposphere, where the wind currents deflect at a right angle from their vertical trajectory and begin traveling poleward.

The widening of the Hadley cells moves the downstroke of its circular path farther north in the northern hemisphere and farther south in the southern hemisphere. As with many of the realities connected to global warming, while this one sounds technical and can seem abstract, the real consequences for real people, animals, and plants are extremely severe.

For the areas now subjected to this downdraft, it's a bit like being under a giant hairdryer in the sky. The results include not just more frequent and more severe droughts, but *consistent* drought patterns likely leading to desertification in many of the countries in the line of fire. Moreover, most of the areas affected, like Southern Europe, Australia, Southern Africa, the American Southwest, and Mexico—are already on the edge of persistent water shortages anyway.

The word "desert," by the way, is derived from the relationship of people to the land involved: deserts are *deserted* by people. Consider the significance of Greece, Italy, and the Fertile Crescent—the cradles of Western civilization—turned into deserts by human alteration of the same natural climate feature that created the Sahara Desert beginning 7,300 years ago.

The jet stream that controls the location of storm tracks in most of North America and Eurasia is also being affected by the impact of global warming on atmospheric circulation patterns and the unusually chaotic weather patterns in these latitudes in recent years. There are actually two jet streams in both hemispheres—a subtropical jet stream flowing from east to west along the poleward margin of the barrel loop of the

---

* Another reason is that at low latitudes, a much greater fraction of the trapped energy goes into evaporation (evaporative cooling) than into heating the air.

Hadley cells (the trade winds), and the so-called polar jet stream—which flows from west to east on the poleward side of a second set of barrel loop atmospheric currents known as the Ferrel cells.

The location of the northern polar jet stream (which North Americans and north Eurasians typically call *the* jet stream) is determined in part by the wall of cold air extending southward from the Arctic Circle. But in recent years, the melting of the Arctic ice cap has led to so much extra heat absorbed there that the northern boundary of the jet stream flowing across North America and Eurasia appears to have been profoundly and radically dislocated—changing storm tracks, pulling cold Arctic air southward in winter, and disrupting precipitation patterns.

All of these energy transfer mechanisms—the wind and ocean currents, storms and cyclones, and atmospheric cells—define the shape and design of the Earth's climate pattern that has remained relatively stable and constant since shortly before the Agricultural Revolution began. Yet global warming is changing all of the energy balances that have given definition to this climate envelope, and is both intensifying and changing the locations of the weather phenomena we are used to.

Some of these balances are being changed to such a degree that scientists worry that they could be pushed far enough out of the pattern we have always known that they could flip into a very new pattern that would produce weather phenomena with intensity, distribution, and timing that are completely unfamiliar to us and inconsistent with the assumptions upon which we have built our civilization.

By way of illustration, take a leather belt and hold one end in either hand; push your hands together until a loop forms sticking upward. As you move your hands and change the inflection of your wrists, the shape of the belt loop will vary but it will remain in the same basic shape. But if you inflect your wrists a little more, it will suddenly flip into a new basic pattern with the loop pointing downward instead of upward. The variations in climate that we have always known, large as they are, are like the variations in the belt loop pointing upward. There would still be similar variations if the loop pointed downward, but if we push the boundary conditions of the loops to a point that causes it to adopt an entirely new pattern, the consequences for our climate would be extreme indeed.

We have already been confronted by unwelcome surprises in our experimentation with changing the chemical composition of the Earth's atmosphere. The sudden appearance of a continent-sized stratospheric ozone hole above Antarctica in the 1980s raised the specter of a deadly threat to many forms of life on Earth, because it allowed powerful ultraviolet radiation normally blocked by the stratospheric ozone layer to reach the surface. And unless the progressive destruction of the stratospheric ozone layer had been halted, scientists say it would have spread to the stratosphere above highly populated areas.

Even though the Antarctic ozone hole lasted each year for only approximately two months, it had already begun to produce a slight thinning of ozone in the stratosphere surrounding the entire planet. Scientists warned at the time that if the concentrations of chemicals causing ozone destruction continued to build, this dangerous thinning process would accelerate, and an even more dangerous ozone hole above the Arctic might form on a more regular basis.

Luckily, almost immediately after this frightening discovery, President Ronald Reagan and Prime Minister Margaret Thatcher helped to organize a global conference in 1987 to negotiate and quickly approve a treaty (the Montreal Protocol) that required the phasing out of the group of industrial chemicals—including the best-known, chlorofluorocarbons (CFCs)—that two scientists, Sherwood Roland and Mario Molina, had proven conclusively in 1974 were interacting with the unique atmospheric conditions in the cold stratosphere above Antarctica to produce this progressive destruction of the protective ozone layer that shields humans and other life-forms from deadly ultraviolet radiation.

EVEN THOUGH THE Montreal Protocol has been a historic success, it is important to understand the precise mechanism through which these chemicals led to the stratospheric ozone hole in the first place—because of new threats to the ozone layer from global warming. To begin with, there is a third and final set of barrel loop atmospheric cells at both the North Pole and the South Pole, called polar cells, within which the winds form a vortex around each pole.

The south polar vortex is much stronger and more coherent, especially in the austral winter, because Antarctica is land surrounded by ocean—whereas the Arctic is ocean surrounded by land—and while the Arctic Ocean is covered, at least in winter, by a thin layer of ice only several feet thick, Antarctica is covered year-round by two kilometers of ice. That also makes it the continent with the highest average altitude, which means it is closer to the top of the sky and radiates the reflected sunlight back into space more powerfully. Consequently, the air above Antarctica is much colder than anywhere else on Earth, which produces an unusually high concentration of ice crystals in the stratosphere there.

The tight vortex formed by the Antarctic circumpolar wind currents during winter holds the CFCs and ice crystals in place above the continent, almost like a bowl. And it is on the surface of these ice crystals that the CFCs react with stratospheric ozone. One other crucial ingredient must be present before the chemical reaction that destroys the ozone starts taking place: a little bit of sunlight.

At the end of the southern hemisphere winter, around the middle of September, when the first rays of sunlight strike the ice crystals held in this "bowl," the chemical reaction is ignited. Then it quickly spreads, destroying virtually all of the stratospheric ozone inside the bowl. As the atmosphere absorbs more heat, the vortex formed by the wind currents weakens and the bowl breaks up, signaling the end of the ozone hole for that year. Some large blobs of ozone-free air sometimes move northward, like the blobs in an old lava lamp from the 1960s—exposing populated areas in the southern hemisphere like Australia and Patagonia to high levels of ultraviolet radiation when air with low concentrations of ozone is no longer able to provide a screen for those at the surface.

Stratospheric ozone depletion and global warming have always been considered almost completely separate phenomena, but in 2012 scientists discovered that global warming is producing an unexpected and unwelcome threat to the stratospheric ozone layer—this time above highly populated areas in the temperate zone of the northern hemisphere.

Just as the extra heat energy absorbed in the tropics is causing the updraft of the Hadley cells to nudge the top of the troposphere higher, the extra heat energy being absorbed in the temperate zone of the northern hemisphere is causing more powerful thunderstorms to punch through the top of the troposphere, injecting water vapor into the stratosphere, where it freezes into a new and dangerous concentration of ice crystals—thus creating the conditions for triggering stratospheric ozone loss by providing the surfaces on which the CFCs still in the atmosphere can come into contact with stratospheric ozone and sunlight to destroy the protective ozone layer. This new phenomenon has begun to appear at a time when the stratosphere is also getting colder, in inverse proportion to the warming of the lower atmosphere. Long predicted by climate models, stratospheric cooling is a result of the Earth's atmosphere attempting to maintain its energy "balance." Much more work will need to be performed before this troubling surprise is fully understood, but it already illustrates the recklessness of this "planetary experiment" that humanity has under way. We are not only playing with fire, but ice as well. As Robert Frost wrote, "Some say the world will end in fire; some say in ice." Either one, he added, "would suffice."

THE RISKIEST OF EXPERIMENTS

The idea that we are engaged in an unplanned experiment with the planet was first articulated by Roger Revelle, who was my teacher and mentor on global warming. In 1957, Revelle wrote with his coauthor, Hans Suess, that, "Human beings are now carrying out a large scale geophysical experiment." They also noted, "The increase of atmospheric CO_2 from this cause [combustion of fossil fuels] is at present small but may become significant during future decades if industrial fuel combustion continues to rise exponentially."

The word "experiment" is worth a little reflection. There are ethical prohibitions against human experimentation that puts lives at risk or seriously damages those who are subjects of the experimentation. Since there are millions of lives put at risk by the "unplanned experiment" that is radically changing the Earth's atmosphere and threatening the future of human civilization, surely the same ethical principle should apply.

Climate science began more than 150 years ago when the legendary Irish scientist John Tyndall discovered that carbon dioxide traps heat. The actual mechanism by which this occurs is more complicated than the popular metaphor of a "greenhouse effect"; the bonds holding together the atoms of the CO_2 molecule absorb and radiate energy at infrared wavelengths, impeding the flow of energy from the surface outward toward space much like a blanket.

But the consequences are the same—the CO_2 in the atmosphere, like the glass in a greenhouse, retains heat that comes in from the sun. Tyndall's historic finding occurred the same year, 1859, as the drilling of the first oil well by Colonel Edwin Drake in Pennsylvania.

Thirty-seven years later, in 1896, the Swedish chemist Svante Arrhenius cited Tyndall in a landmark paper in which he addressed the following question: "Is the mean temperature of the ground in any way influenced by the presence of heat-absorbing gasses in the atmosphere?" Arrhenius performed more than 10,000 calculations by hand in order to arrive at his conclusion that a doubling of CO_2 concentrations in the atmosphere would raise global average temperatures by several degrees Celsius.

In the second half of the twentieth century, in the midst of the postwar burst of industrialization, research into global warming picked up considerably. The International Geophysical Year of 1957–58 led to the establishment by Roger Revelle and Charles David Keeling of a historic project to begin the long-term systematic measurement of CO_2 concentrations in the global atmosphere. The results were astonishing. After only a few years of measurements it became obvious that the concentration was increasing steadily by a significant amount, a result confirmed in the following years by installation of observation stations all over the world.

Because most of the landmass and deciduous vegetation is in the northern hemisphere, the CO_2 concentration shows an annual cycle of CO_2 intake and outgassing by the terrestrial biosphere, which is so much larger north of the equator than south. As a result, the CO_2 concentration in the northern hemisphere goes up in winter (when uptake of CO_2 by leaves and plants is low) and down in summer (when the trees and grasses are once again pulling CO_2 from the air).

But the observations also showed clearly that the overall concentration of CO_2 throughout this yearly seasonal cycle was being shifted steadily upward. After the first seven years of the iconic measurements contained in what is now known as the Keeling Curve, the low point in the annual cycle was already higher than the high point when the measurements began. Fifty-six years later, these measurements still continue every day—from the top of Mauna Loa; at the South Pole; in American Samoa; in Trinidad Head, California; and in Barrow, Alaska. In addition, there are sixty other "distributed cooperative" sets of measurements, including aircraft profiles, ship transects, balloons, and trains. The project is now overseen, by the way, by an outstanding scientist, Ralph Keeling, who happens to be Dave's son. He is also now monitoring the small but steady reduction in the concentration of oxygen in the atmosphere—not a cause for concern in itself, but yet another validation of the underlying climate science, which has long predicted this result, and an effective cross-check on the accuracy of the CO_2 measurements.

Ten years after Revelle and Keeling began measuring CO_2 in the atmosphere, I had the privilege of becoming Revelle's student in college and was deeply impressed by the clarity with which he described this phenomenon and the prescience with which he projected what would happen in the future if the exponential increase in fossil fuel combustion and consequent CO_2 emissions continued.

A decade after leaving college, I began holding hearings about global warming in Congress, and in 1987–88, I first ran for president in order to focus more attention on the need to solve the climate crisis. In June of 1988, NASA scientist Jim Hansen testified that the evidence of human-caused global warming had become statistically significant in observations of rising global temperatures. Six months later, in December, the United Nations established a global scientific body—the Intergovernmental Panel on Climate Change (IPCC)—to provide authoritative summaries of the evidence being found by scientific studies around the world.

Today, a quarter century after the IPCC began its work, the international scientific consensus confirming the dominance of human activities in causing global warming is as strong as any consensus ever formed in science. The threat is real, is linked primarily to human activities, is serious, and requires an urgent response in the form of sharply reduced greenhouse gas emissions. Every national academy of science and every major scientific society in the world supports the consensus view.

In a joint statement in 2009, the national academies of the G8 nations and five other nations declared, "The need for urgent action to address climate change is now indisputable." According to a peer-reviewed study published in the *Proceedings of the National Academy of Sciences* in the U.S., "97–98 percent of the climate researchers most actively publishing in the field support the tenets of ACC (anthropogenic climate change) outlined by the IPCC."

It is also significant that virtually all of the projections made by scientists in recent decades about the effects of global warming have been exceeded by the actual impacts as they later unfolded in the real world. As many have noted, scientists in general and the scientific process in particular are inherently cautious in coming to a conclusion, even, you might say, conservative. Not conservative in the political sense of the word, but conservative in their methodology and approach. This tradition and long-established culture of caution is reinforced by the peer-review process, which demands convincing proof of any claims that are published. The same culture discourages statements about even seemingly obvious implications that may reflect common sense but cannot be adequately proven to the degree required for publication in a peer-reviewed journal.

Nevertheless, in spite of this conservative culture, the global scientific community has loudly and publicly warned policymakers that we must act quickly to avert a planetary calamity. Yet even with the mounting toll from climate-related disasters and the obvious warming of the Earth that is now viscerally apparent to almost everyone, there have been very few significant policy changes designed to confront this existential threat.

With the future of human civilization hanging in the balance, both democracy and capitalism are badly failing to serve the deepest interests of humankind. Both are unwieldy and both are in a state of disrepair. But if the flaws in our current version of democracy and capitalism are addressed, if the barnacles of corruption, corporate control, and domination by elites can be scraped away, both of these systems will be invaluable in turning world civilization in the right direction before it is too late. Yet this difficult policy transition will require leadership and political courage that is presently in short supply, particularly in the United States.

In order to understand why so many political leaders are failing to address this existential crisis, it is important to explore the way public perceptions of global warming have been manipulated by global warming deniers, and how the psychology of the issue has made that manipulation easier than it should be. Powerful corporations with an interest in delaying action have lavished money on a cynical and dishonest public campaign to manipulate public opinion by sowing false doubts about the reality of climate crisis. They are taking advantage of the natural desire that all of us have to seize upon any indication that global warming isn't real after all and the scientists have somehow made a big mistake.

Many have described the climate crisis as "the issue from hell," partly because its complexity, scale, and timeframe all make public discussion of the crisis, its causes, and its solutions more difficult. Because its consequences are distributed globally, it masquerades as an abstraction. Because the solutions involve taking a new path into the future, improving long-familiar technologies, and modifying long-standing patterns, it triggers our natural reluctance to change. And because the worst damages stretch into the future, while our attention spans are naturally short, it makes us vulnerable to the illusion that we have plenty of time before we have to start solving it.

"Denial" is a psychological tendency to which all of us are vulnerable. One of the first to explore how this phenomenon works was Elisabeth Kübler-Ross, who taught, according to the organization she founded, that "Denial can be conscious or unconscious refusal to accept facts, information, or the reality of the situation. Denial is a defense mechanism and some people can become locked in this stage." The modern psychiatric definition of this condition is: "An unconscious defense mechanism characterized by refusal to acknowledge painful realities, thoughts, or feelings."

Certainly the prospect of a catastrophic threat to the future of all global civilization qualifies as "an unpleasant thought." And the natural tendency for all of us is to hope that the scientific consensus on global warming is not an accurate depiction of the real danger that we face. Those who become locked into this psychological strategy typically respond to the stronger and stronger evidence of global warming with stronger and stronger denunciations of the entire concept, and stronger attacks on those who insist that we must take action.

We have learned a lot about human nature over the last century. We now know, for example, that the "rational person" assumed by Enlightenment thinkers—and the definition of human behavior implicit in the work of Adam Smith and other classical economists, which some now refer to as "*Homo economicus*"—is really not who we are. Quite to the contrary, we are heirs to the behavioral legacy shaped during our long period of development as a species. Along with our capacity for reason, we are also hardwired to be more attentive and responsive to short-term and visceral factors than longer-term threats that require the use of our capacity for reason.

Two social scientists—Jane Risen at the University of Chicago, and Clayton Critcher at the University of California, Berkeley—asked two groups of people the same series of questions about global warming, with the only difference being the temperature in each room. Those who responded in a room that was ten degrees warmer gave answers indicating significantly larger support for doing something to counter global warming than the group in the cooler room. The differences showed up among both liberals and conservatives. In a second study, two groups were asked for their opinions about drought, and those given salty pretzels to eat had a markedly different outlook than the group that wasn't as thirsty.

At a time when the world is undergoing the dramatic changes driven by the factors covered in this book—globalization and the emergence of Earth Inc., the Digital, Internet, and computing revolutions, the Life Sciences and biotechnology revolutions, the historic transformation of the balance of political and economic power in the world, and the commitment to a form of "growth" that ignores human values and threatens to deplete key resources vital to our future—the climate crisis easily gets pushed down the list of political priorities in most nations.

The flawed definition of growth described in Chapter 4 is at the center of the catastrophic miscalculation of the costs and benefits of continuing to rely on carbon-based fuels. The stocks of publicly traded carbon fuel companies, for example, are valued on the basis of many factors, especially the value of the reserves they control. In arriving at the worth of these underground deposits, the companies assume that they will be produced and sold at market rates for burning. Yet any reasonable person familiar with the global scientific consensus on the climate crisis knows that these reserves *cannot* all be burned. The very idea is insane. Yet none of the environmental consequences of burning them is reflected in their market valuation.

In addition to denial and our misplaced blind reliance on a deeply flawed economic compass, there is another ingrained tendency to which all of us are prone: we want to believe that ultimately all is right with the world, or at least that part of the world in which each of us lives. Social psychologists call this the system justification theory, which holds that everyone wants to think well of themselves, the groups they identify with, and the social order in which they live their lives. Because of the scale of the changes necessary to confront global warming, any proposal to embark on this necessary journey can easily be portrayed as a challenge to the status quo and trigger our tendency to defend it by automatically rejecting any potential alternative to the status quo.

When there is an existential threat that requires quick mass mobilization—the attack on Pearl Harbor in 1941, for example—the natural reluctance to break out of our comfortable patterns is overridden by a sense of emergency. But most such examples are rooted in the same group conflict scenarios that characterized the long period in which we as human beings developed. There is no precedent (except the ozone hole) for a fast global response to an urgent global threat—especially when the response called for poses a big challenge to business as usual.

President Reagan, when confronting the need for nuclear arms control, expressed the same thought on many occasions, including once in a speech to the United Nations General Assembly: "In our obsession with antagonisms of the moment, we often forget how much unites all the members of humanity. Perhaps we need some outside, universal threat to make us recognize this common bond. I occasionally think how quickly our differences worldwide would vanish if we were facing an alien threat from outside this world." Some members of my political party ridiculed Reagan's formulation during his presidency, but I always thought that it embodied an important insight.

THE POLITICS OF DIVISION

We do, of course, face a common threat to all humanity where the climate crisis is concerned. But it is not from aliens; it is from *us*. So our capacity to respond by uniting to overcome the threat can be undermined by "antagonisms of the moment." America's founders recognized the importance of this ingrained trait in human nature. More than two centuries later scientists tell us that the tendency to form opposing factions is deeply rooted in the history of our species.

As E. O. Wilson recently wrote, "Everyone, no exception, must have a tribe, an alliance with which to jockey for power and territory, to demonize the enemy, to organize rallies and raise flags. And so it has ever been. . . . Human nature has not changed. Modern groups are psychologically equivalent to the tribes of ancient history. As such, these groups are directly descended from the bands of primitive humans and prehumans."

That is one of the underlying reasons that the denial of global warming has somehow become a "cultural" issue, in the sense that many who reject the scientific evidence feel a group kinship—almost a "tribal identity"—with others who are also locked into denial. In the U.S., the extreme conservative ideology that has come to dominate the Republican Party is based in part on a mutual commitment to passionately fight against a variety of different reform proposals opposed by members of a disparate coalition.

It could be called the Three Musketeers Principle: all for one and one for all. Those primarily interested in opposing any form of gun regulation agree to support the position of oil and coal companies opposed to any efforts to reduce global warming pollution. Antiabortion activists agree to support large banks in their opposition to new financial regulations. As Kurt Vonnegut said, "So it goes."

Over the last four decades, the largest carbon polluters have become charter members of the antireform counterrevolution described in Chapter 3 that was organized in the 1970s under the auspices of the U.S. Chamber of Commerce—out of fear that the tumultuous protest movements of the 1960s (against the Vietnam War, for civil rights, women's rights, gay rights, disability rights, the consumer movement, the passage of Medicare and programs to assist the poor, and so on) were threatening to spin out of control in ways that would disadvantage powerful corporations and elites. In their view, these movements threatened to undermine capitalism itself.

One of the enduring consequences of this counterreform movement was the establishment of a large network of think tanks, foundations, institutes, law schools, and activist organizations that turn out an endless stream of mostly contrived "reports," "studies," lawsuits, testimony before congressional and regulatory panels, op-eds, and books that all promote the philosophy and agenda of the new corporate Musketeers:

- Government is bad, cannot be trusted, should instead be feared, and must be starved of resources so that it is capable of interfering as little as possible with the plans of corporations and the interests of elites;
- Hardship is good for poor people because it's the only thing that will give them an incentive to become more productive; hardship also makes them more willing to accept lower wages and fewer benefits;
- Rich people, on the other hand, should be taxed as little as possible in order to encourage them to make even more money— which is the only tried-and-true way to produce more growth in the economy, even if there is too little demand because consumers don't have enough money to buy more goods and services;
- More inequality is a good thing, because it simultaneously inspires poor people to more ambition and rich people to more investing,

even if the evidence shows that the highest-income groups are primarily interested in wealth preservation when the economy is weak; and

- The environment can take care of itself nicely, no matter how much pollution we dump into it. Anyone who believes otherwise is motivated by a barely concealed love for socialism and an abiding determination to thwart business.

To one degree or another, of course, there is a natural incentive to build broad coalitions among differing interests in most political parties. I certainly experienced such pressures as a member of the Democratic Party when I served in Congress. Yet there is something different about the lockstep discipline in the new U.S. right-wing coalition—a discipline that is enforced by extremely wealthy contributors who are primarily interested in policies that increase their already unhealthy share of America's aggregate income.

In today's world, the challenge of global warming has, unfortunately, led to an almost tribal division between those who accept the overwhelming scientific consensus—and the evidence of their own senses—and those who are bound and determined to reject it. The ferocity of their opposition is treated as a kind of badge signifying their membership in the second group and antagonism toward the first.

The organized deniers know that in order to maintain their control of the coalition opposed to policies reducing greenhouse gas emissions, they do not have to prove that man-made global warming is not real—though many of them do assert as much over and over again. All they really need to do is create enough doubt to convince the public that "the jury is still out." This strategic goal was explicitly spelled out in an internal document from a business coalition dominated by large carbon polluters.

Leaked to the press in 1991, the document stated that the group's strategic goal was to "reposition global warming as theory not fact." A charitable interpretation would be that these companies had long felt besieged by what they perceived as hyperbolic claims on the part of environmental activists seeking more regulation of various forms of pollution, and that they developed a habit of reflexively countering any claim of impending harm by going all-out to undermine the credibility of the claims and of those making them.

However, in light of the decades of extensive documentation making this deadly threat crystal clear, and in light of the national academies of science around the world proclaiming that the evidence is now indisputable, it is no longer easy to be charitable in assessing what these wealthy, powerful, and self-interested deniers are doing. They reject the spirit of reasonable dialogue. They reject and vilify the integrity of the scientific process. Nothing has worked to hold them to their obligation to the greater good. Some, it is true, have examined both the evidence and their conscience and have changed. But those who have done so are still in the tiny minority. The deniers' assault on the future of our world continues.

There is, after all, no longer *any* reasonable doubt whatsoever that man-made emissions of CO_2 and the other global warming pollutants are seriously damaging the planetary ecological system that is crucial to the future survival of human civilization. Many of the extreme weather disasters that have already claimed so many lives and caused so much suffering are now being directly linked to global warming. The damage that is being done to hundreds of millions in the present generation makes it impossible, in my view, to ignore the moral consequences of what is being done.

Most legal systems in the world make it a criminal offense, as well as a civil offense, for anyone to knowingly misrepresent material facts for the purpose of self-enrichment at the expense of others who rely on the false representations and suffer harm or damage as a result. If the misrepresentation is merely negligent, it can still be a legal offense. If the false statements are *reckless* and if the harm suffered by those induced to rely on the false statements is grave, the offense is more serious still. The most common legal standard for determining whether or not the person (or corporation) misrepresenting the material facts did so "knowingly" is not "beyond a reasonable doubt," but rather the "preponderance of the evidence."

The large public multinational fossil fuel companies have an estimated $7 trillion in assets that are at risk if the global scientific consensus is accepted by publics and governments around the world. That is the reason that several of them have been misrepresenting to the public—and to investors—the material facts about the grave harm to the future of human civilization that results from the continued burning of their principal assets in such a reckless manner. The value of similar and larger reserves owned by sovereign states, when combined with the assets owned by private and public companies, adds up to a total of $27 trillion. That is why Saudi Arabia, until recently at least, has been so vehement in its efforts to block any international agreement to limit global warming pollution. In 2012, a member of the royal family, Prince Turki al-Faisal, called for Saudi Arabia to convert its domestic energy use to 100 percent renewables in order to preserve its oil reserves for sale to the rest of the world.

"SUBPRIME CARBON ASSETS"

The oil, coal, and gas assets carried on the books of fossil fuel companies is valued at market rates based on the assumption that they will eventually be sold to customers who will burn them and dump the gaseous global warming pollution that results into the Earth's atmosphere. In the past, I have referred to these reserves as "subprime carbon assets," in order to draw an analogy to subprime mortgages, which the market and most banking experts also believed had extremely high value. Actually, however, these subprime mortgages had an illusory value that was based on the absurd assumption that people who obviously couldn't pay them back somehow would. They were often referred to in the industry as "low documentation loans," or more simply as "liar loans."

I remember vividly when I signed my first home mortgage as a young man. I sat across the desk from Walter Glenn Birdwell Jr., the man in charge of Citizens Bank in Carthage, Tennessee. Before giving me the mortgage, Mr. Birdwell required me to provide written answers to a long series of questions about my income and net worth. Even though neither was very high, he gained enough confidence that I would be able to make the monthly payments. He then required me to make what was for me at the time a considerable down payment.

By contrast, the subprime mortgages were given to people who had no earthly way of paying them back—a fact that would have been immediately clear if any of them had been required to answer questions from Mr. Birdwell. Nor were these homebuyers asked to make any down payment. So, if a reasonable person could easily determine that the mortgages were unlikely to be paid back, and that it was only a matter of time before the homebuyers defaulted, why would the banks nevertheless enter into such transactions?

The answer is that in the age of Earth Inc. and the Global Mind, the banks originating these flawed mortgages were able to use powerful computers to combine many thousands of such mortgages—in the aggregate, 7.5 million of them in the U.S. alone—slice them and dice them into financially engineered derivatives products too complex for most of us to comprehend, and then sell them into the global marketplace. In other words, the ridiculous assumption was that the risk inherent in providing a mortgage to someone who couldn't pay it back could be magically eliminated if a great many such mortgages were all packaged together and sold into the global marketplace.

When this assumption was tested during the slowdown of the global economy in 2007–08, it suddenly collapsed and the bankers had an unpleasant encounter with reality. The unpleasantness didn't linger for them, however, because they were able to use the overwhelming political power they had purchased with campaign contributions and lobbying activities—with a little help from officials that had gone through the revolving door connecting governments and banks—to be bailed out by the taxpayers, who had to borrow the money for the purpose. The net result was a credit crisis and a global Great Recession, which economists may yet relabel a depression.

Subprime carbon assets have a similarly inflated value in the marketplace, undergirded by an assumption even more absurd than the ridiculous idea that it was perfectly okay to give mortgages to millions of people who couldn't ever pay them back. In this case, the assumption is that it is perfectly all right to burn every last drop of oil in the oil companies' reserves and destroy the future of civilization. It's not all right.

Yet the market value to the oil, coal, and natural gas companies of this particular absurd assumption is extremely high. Ultimately, that is the reason they have been willing to devote billions of dollars to defend it—by organizing a massive and highly sophisticated campaign of deception designed to convince people—and policymakers—that it may very well be fine to burn as much carbon fuel as we can.

These carbon polluters have also deceived coal miners and other employees in the fossil energy industry into ignoring the reality of the change that is inevitable. In a courageous and eloquent speech on the Senate floor in 2012, Senator Jay Rockefeller, from the most coal-dependent state in the U.S., West Virginia, said, "My fear is that concerns are also being fueled by the narrow view of others with divergent motivations—one that denies the inevitability of change in the energy industry, and unfairly leaves coal miners in the dust. The reality is that many who run the coal industry today would rather attack false enemies and deny real problems than find solutions."

The dominance of wealth and corporate influence in decision making has so cowed most politicians that they are scared to even discuss this existential threat in any meaningful way. There are more than a few honorable exceptions, but on issues that engage the interests of Earth Inc., Earth Inc. is fully in control of global policy. The carbon fuel companies hired four anti-climate lobbyists for every single member of the U.S. Senate and House of Representatives in their fight to defeat climate legislation. They have become one of the largest sources of campaign contributions to candidates in both parties—though significantly more goes to Republicans.

Many of these companies have provided large amounts of money over the last two decades to "liars for hire" who turn out a seemingly endless stream of misleading, peripheral, irrelevant, false, and unscientific claims:

- Global warming is a hoax perpetrated by scientists who are scheming to receive more government research funding and by activists who want to impose socialism or worse.
- Global warming isn't occurring; it stopped several years ago.
- If it is occurring, it is not caused by global warming pollution, but is instead the result of a natural cycle.
- The Earth's climate system is so resilient that it can, in any event, absorb unlimited quantities of global warming pollution with no harmful consequences.
- If global warming does occur, it will actually be good for us.
- Even if it's not good for some people, we certainly have the ability to adapt to it with little hardship.
- The ice caps on Jupiter are also melting, therefore it is logical to assume that some poorly understood phenomenon endemic to our solar system is the true cause (never mind that Jupiter doesn't *have* ice caps).
- Global warming is being caused by sunspots (never mind that temperatures have continued to go upward during the long "cool phase" of the sunspot cycle now coming to an end).
- Global warming is caused by volcanoes (never mind that human-

caused CO_2 emissions are 135 to 200 times greater than volcanic emissions, which are in any case part of a natural process that is, in the long term, carbon neutral).

- Computer models are unreliable (never mind that more than a dozen separate and independent temperature records from the real world completely confirm what the computer models have long predicted).
- Clouds will cancel out global warming (never mind the growing evidence that the net feedback from clouds is likely to make global warming even worse, not better).

There are more than 100 other bogus arguments, or red herrings, that are pushed relentlessly in the media, by lobbyists, and by captive politicians beholden to the carbon polluters. The only thing the deniers are absolutely certain about is that 90 million tons per day of global warming pollution is certainly *not* causing global warming—even if the entire global scientific community says the opposite. There are, to be sure, some opponents of the scientific consensus who genuinely believe that the science is wrong. Some of them have backgrounds and personal stories that predispose them to fight on for a variety of reasons. But they are the exceptions, and their complete lack of any credible supporting evidence would quickly marginalize them except for the fact that climate science denial has become a cottage industry generously supported by carbon polluters.

To undermine the public's confidence in the integrity of science, the carbon companies and their agents and allies constantly insinuate that climate scientists are lying about the facts they have uncovered, and/or are secretly part of a political effort to expand the role of government. The political assault against climate scientists has been designed not only to demonize them, but also to intimidate them—which has added to the naturally cautious approach that scientists habitually adopt.

One right-wing state attorney general in the United States took legal action against a climate scientist simply because his findings were inconvenient for coal companies. Right-wing legal foundations and think tanks have repeatedly sued climate scientists and vilified them in public statements. Right-wing members of Congress have repeatedly sought to slash climate research funding. To mention only one of the many consequences, the ability of the U.S. to even monitor climate change adequately is being severely damaged with multiple launches of essential monitoring satellites being delayed or canceled—just at the time when the data is most needed.

On the eve of the global negotiating session on climate in December of 2009 in Copenhagen, the entire climate science community was assaulted by what appears to have been a well-planned hacking of their private, internal emails among one another. The cherry-picking of misleading phrases taken out of context led to the trumpeting by the right-wing media of charges that the climate science community was lying to the public and to their governments. An extensive investigation determined that the hacking came from outside the targeted research center but did not identify the perpetrator. Meanwhile, four separate independent investigations all completely cleared the climate scientists of any wrongdoing.

THE DENIAL MACHINE

The ability of the public to see through the lies and deceptions of the carbon polluters and their allies has been hampered because the traditional role of the news media has changed significantly in the past few decades—especially in the United States. Many newspapers are going bankrupt and most others are under severe economic stress that reduces their ability to fulfill their historic role of ensuring that the foundation of a democracy is a "well-informed citizenry."

As noted in Chapter 3, the rising prominence of the Internet is a source of hope, but for the time being television is still far and away the dominant medium of information. And yet the news divisions of television networks are now required to focus on ways to contribute more profit to the corporate bottom line. As a result, they have been forced to blur the distinction between news and entertainment. Since ratings are the key to profitability, the kinds of news stories that are given priority have changed.

Virtually every news and political commentary program on television is sponsored in part by oil, coal, and gas companies—not just during campaign seasons, but all the time, year in and year out—with messages designed to soothe and reassure the audience that everything is fine, the global environment is not threatened, and the carbon companies are working diligently to further develop renewable energy sources.

The fear of discussing global warming has influenced almost all mainstream television news networks in the U.S. The denier coalition unleashes vitriol at almost anyone who dares to bring up the subject of global warming and, as a result, many news companies have been intimidated into silence. Even the acclaimed BBC nature program *The Frozen Planet* was edited before the Discovery Network showed it in the United States to remove the discussion of global warming. Since one of the overarching themes of the series was the melting of ice all over the planet, it was absurd to remove the discussion of global warming, which is of course the principal cause of the ice melting. As activist Bill McKibben wrote, "It was like showing a documentary on lung cancer and leaving out the part about the cigarettes."

During the hot summers of 2011 and 2012, the evening newscasts often resembled a nature hike through the Book of Revelation. But each time, the droughts and fires and windstorms and floods were covered as lead stories, the explanation was often something like, "a high pressure area" or "La Niña."

On the few occasions when global warming is discussed, the coverage is distorted by the tendency of the news media to insist on including a contrarian point of view to falsely "balance" every statement by a climate scientist about global warming—as if there was a legitimate difference of opinion. This problem has been worsened by the shrinking budgets for investigative reporting.

For someone who grew up believing in the integrity of the American democratic process—and who *still* believes that its integrity can be redeemed and restored—it is profoundly troubling that special interests have been able to capture control of decision making and policy formation in the nation that Abraham Lincoln eloquently described as "the last best hope of earth." But the fight is far from over. Its epicenter is in the United States, simply because the U.S. remains the only nation capable of rallying the world to save our future. As Edmund Burke said, "The only thing necessary for evil to prevail is for good men to do nothing." That is what it now comes down to: will good men and women do nothing, or will they respond to the emergency that is now at hand?

In the last few years, the frequency and magnitude of extreme weather events connected to the climate crisis have begun to have a significant impact on public attitudes toward global warming. Even in the U.S., where the denier propaganda campaign is still in full force, public support for actions to reduce greenhouse gas emissions has gone up significantly. Proposals to do more have been supported by a majority for many years, although the intensity of the majority's feeling has been too low to overcome the efforts of the carbon polluters to paralyze political action. More recently, however, support for action has been building steadily.

At the beginning of President Barack Obama's administration in 2009, hopes were high that U.S. policy on global warming would change—and for a time, it did. His stimulus bill put a major emphasis on green provisions, including measures to accelerate the research and development, production, and use of renewable energy systems in the United States. His appointment of the extremely able Lisa Jackson as administrator of the Environmental Protection Agency set the stage for a series of breakthrough rules and initiatives that have contributed to the reduction of CO_2 emissions and the cleaning of pollutants from the environment.

The EPA rules requiring a reduction of CO_2 emissions from new power plants and automobiles were courageous, and the EPA's ruling that mercury emissions from coal plants must be sharply reduced has contributed to the decisions by many utilities to cancel planned construction of new coal-fired generating plants. The success by Jackson, her cabinet colleague, transportation secretary Ray LaHood, and White House adviser Carol Browner in reaching an agreement with U.S. carmakers to require significant improvements in auto mileage—eventually almost doubling the current average to 54.5 miles per gallon—was described by one environmentalist, Dan Becker, who runs the Safe Climate Campaign for the Center for Auto Safety, as "The biggest single step that any nation has taken to cut global warming pollution."

But several things happened over the last few years to make the political challenge more difficult than Obama expected. First, the economic crisis and Great Recession he inherited made the administration reluctant to confront a longer-term challenge when the economic distress of the present was so pressing. The effects of the recession lingered because of its unusual depth, the massive deleveraging (repayment of debt) it triggered, the collapse of the housing market, and the inadequate size of the fiscal stimulus that injected some—but not enough—demand back into the economy.

Second, China surprised the world with its massive commitment to dominate the production and export of windmills and solar panels, heavily subsidized with government-backed cheap credit and low-wage labor—which allowed them to flood the global market with equipment priced well below the cost of production in the United States and other developed countries.

Third, even though his climate legislation passed the House of Representatives while it was still under his party's control, the obsolete and dysfunctional rules of the U.S. Senate empowered a minority to kill it in that chamber. Senators in both parties said privately that passage of the climate plan might have been within reach but that it seemed to them that President Obama was not prepared to make the all-out effort that would have been necessary to build a coalition in support of the plan. Earlier, he had chosen to make health care reform his number one priority, and the badly broken U.S. political system produced a legislative gridlock on his health plan that lasted until the midterm campaign season began, leaving no time for even Senate discussion of the climate change issue.

By then, Obama and his political team in the White House had apparently long since made a sober assessment of the political risks involved in states where the power of the fossil fuel industries would punish him for committing himself to the passage of this plan. So instead, when his opponents in Congress took up the cry "drill, baby, drill," the president proposed the expansion of oil drilling—even in the Arctic Ocean—and opened up more public land to coal mining. For these and other reasons, the positive impacts of the energy and climate proposals with which he began his presidency were nearly overwhelmed by his sharp turn toward a policy that he described as an "all of the above" approach—one that has contributed to the increased reliance on carbon-rich fossil fuels.

Fourth, the discovery of enormous reserves of deep shale gas depressed electricity prices as more coal-fired generating plants switched to cheaper gas—thus pushing the price of kilowatt hours below the level needed for wind and solar to be competitive at their present early stage of development. Shale gas has flooded the market since the discovery and perfection of a new drilling technology that combines horizontal drilling and hydraulic fracturing (fracking). Although most of the debate about fracking has involved its use in the production of shale gas, it is used in the production of oil as well, opening previously inaccessible supplies and increasing the yield of oil from fields previously nearly depleted.

THE IMPACT OF FRACKING

Experts have cautioned that the world can expect a steady increase in the price of shale gas as liquefied natural gas (LNG) exports transfer the gas from low-priced markets like the United States to much higher-priced markets in Asia and Europe, with the average cost of shale gas going up significantly in the process. Nevertheless, the size of the new reserves opened up with fracking have at least temporarily overturned the pricing structure of energy markets. And the resulting enthusiasm for the exploitation of these reserves has obscured several crucial questions and controversies that should, and over time will, inspire caution about shale gas.

To begin with, the fracking process results in the leakage of enormous quantities of methane (the principal component of natural gas), which is more than *seventy-two* times as potent as CO_2 in trapping heat in the atmosphere over a twenty-year time frame. After about a decade, methane breaks down into CO_2 and water vapor, but its warming impact, molecule for molecule, is still much larger than that of CO_2 over shorter timescales.

The global warming potency of methane has led to proposals for a global effort to focus on sharp reductions in methane emissions as an emergency short-term measure to buy time for the implementation of the more difficult strategies necessary to reduce CO_2 emissions. Similarly, others have proposed a near-term focus on sharply reducing black carbon emissions, or soot, which trap incoming heat from the sun and which settle on the surface of ice and snow to increase heat absorption and magnify melting. Taken together, these two actions could significantly reduce warming potential by 2050. Given how long the world has waited to get started on controlling emissions, we need both and more.

There are huge leakages of methane in the fracking process before the equipment is put in place to capture the gas at the surface. After the underground formation is fracked by the injection of high-pressure liquids, there is a "flowback." That is, when the fracking water, chemicals, and sand used to do the fracking flow back to the surface and out of the well, this material contains large amounts of methane, which is either vented into the atmosphere or burned. Although many of the largest drilling operators take steps to prevent this leakage, the majority of smaller "wildcat" drillers do not. Additional methane is typically leaked into the atmosphere during the processing, storage, and distribution of gas. The total volume of methane leakage is so large that multiple studies—including a recent lifecycle analysis by Nathan Myhrvold, formerly of Microsoft and co-founder of Intellectual Ventures, and Ken Caldeira, a climate scientist at the Carnegie Institution's Department of Global Ecology—have now found that virtually all of the benefit natural gas might have because of its lower carbon content compared to coal is negated.

In its ongoing operations, the fracking process also requires the continuing injection of huge amounts of water mixed with sand and toxic chemicals into the shale where the gas is confined. The requirement of an average of five million gallons of water for each well is already causing conflict in regions suffering from droughts and water shortages. In many communities, particularly in arid areas of the American West, the competition for scarce water resources was acute even before the spread of the thirsty fracking process. In parts of Texas, fracking wells are being drilled in communities where water supply limitations are already constraining usage for drinking water and agriculture.

The fracking process sometimes also inadvertently contaminates precious underground aquifers. Although the gas-bearing rock is typically much deeper than the aquifers supplying drinking water, the upward migration of liquids underground is not well understood and is difficult to predict or control. Many of the deposits where fracking is taking place are found in oil and gas fields that are dotted with old abandoned shafts drilled decades ago in the search for reserves that could be produced through conventional means. These old wells can serve as chimneys for the upward migration of both methane and drilling fluids.

Some have speculated that abandoned drill holes and other poorly understood features of the underground geology may be responsible for the fact that numerous existing water wells located far above the ongoing horizontal drilling have already been poisoned by fracking fluids. The U.S. Environmental Protection Agency has found that the fluids used to drill for gas in Wyoming are the likely cause of pollution in the aquifer above an area that was fracked there. Reports of similar pollution from fracking in other areas have been made but the EPA has been hampered in its investigations because of an unusual law passed in 2005 at the behest of then vice president Dick Cheney, which provides a special exemption for fracking activities from U.S. government oversight under the Fresh Drinking Water Act and the Clean Water Act.

The industry disputes most of these reports, and believes that in any case the pollution of some water wells is a small price to pay; the CEO of ExxonMobil, Rex Tillerson, for example, said recently, "The consequences of a misstep in a well, while large to the immediate people that live around that well, in the great scheme of things are pretty small." Nevertheless, political resistance from landowners has been growing in several regions.

Once the fracking fluid has been used, it must be disposed of as toxic wastewater. Often, it is reinjected deep underground in a manner that has caused multiple small (usually harmless) earthquakes and, on some occasions, is alleged to have infiltrated water aquifers. Indeed, the disposal of used fracking fluids is a more common source of complaints than the initial injections that begin the fracking process. In other locations, this used fracking fluid has been stored in large open-air holding ponds that sometimes overflow following heavy rainfalls. It has also at times been spread on roads, ostensibly for dust control.

Advocates of shale gas argue that there are safety measures that can mitigate many of these problems, although most claim disingenuously that the industry will adopt them voluntarily, in spite of the expense involved. By contrast, the oil and gas industry veteran who pioneered the fracking process, George P. Mitchell of Houston, Texas, has publicly called for more government regulation. "They should have very strict controls. The Department of Energy should do it," Mitchell told *Forbes* magazine. "If they don't do it right, there could be trouble. . . . It's tough to control these independents. If they do something wrong and dangerous, they should punish them," he added.

But even if new safety regulations worked as planned and even if the leakage of methane is tightly controlled, the burning of natural gas still results in an enormous volume of CO_2 emissions. The fact that these emissions can in theory be brought down to a level that represents only half of the emissions from coal has been used by some advocates of shale gas as a new twist on the old question: is the glass half full or is it half empty? They make the seductive case that switching to gas means we can bring emissions halfway down in the sectors that now rely on coal. But here is the rub: the atmosphere itself is already *full*. The concentrations of global warming pollution are already at dangerous levels.

GETTING REAL

As a result, solving the climate crisis requires reducing emissions not by a little, but by a lot. We have to begin reducing net additions of greenhouse gases by at least 80 to 90 percent—not 50 percent—in order to ensure that overall concentrations do not exceed a potential tipping point before starting to decline. Continuing to add additional amounts of greenhouse gases at a rate that far exceeds the slow rate at which CO_2 is drawn out of the atmosphere by the oceans and the biosphere would push far into the future any possibility of reducing the overall concentration levels. Reliance on gas to "bridge" the time needed to convert to renewables can help, but a longer commitment would, in fact, be tantamount to sur-rendering in the struggle to ensure that civilization survives.

In some ways, this challenge is similar to what is happening with the depletion of groundwater and topsoil. The natural replenishment of those resources takes place on a timescale far slower than the rate at which they are being depleted by human activities. The natural rate of CO_2 removal from the atmosphere takes place far more slowly than the rate at which we are adding to the overall concentrations. In all three cases, human activities are causing changes far faster than nature can adjust to them.

The underlying problem is that the new power and momentum of Earth Inc. is colliding violently with and overwhelming the environ-mental balance of the Earth. The overconsumption of limited resources and the production of unlimited pollution are both inconsistent with the continued functioning of the Earth's ecological system in a manner that supports the survival of human civilization. As noted earlier, the CO_2 con-tained in the "proven reserves" of oil, coal, and gas already on the books of carbon fuel companies and sovereign states exceeds by many times the amount we could safely add to the atmosphere—and the unconventional reserves now starting to be drawn on are potentially even larger.

The shale gas boom in the United States has led to a frenzy of explo-ration for shale gas in China, Europe, Africa, and elsewhere, raising the specter of a long-term global commitment to gas at the longer-term ex-pense of renewables. Nevertheless, production of this resource outside the U.S. has thus far been limited. In China, where geologists believe that the supply may be two and a half times the size of U.S. shale gas reserves, the underground geology requires technologies that are different from those being used in the U.S., which complicates the option of simply transferring the U.S. horizontal drilling and hydraulic fracturing technologies to China. Also, as in the Western United States, the profligate use of water in fracking may impose a limitation on use of the process—particularly in northern and northwestern China, where water shortages are endemic.

Even so, momentum is building in the global economy toward the full exploitation and production of shale gas. Some analysts make a persuasive case that if "fugitive emissions" are tightly controlled, the substitution of gas for coal might produce a temporary but still significant net reduction in the emissions of greenhouse gasses. In 2012, in what most analysts described as a surprising development, U.S. CO_2 emissions dropped to their lowest level in twenty years—in part because of the economic slowdown, because of a mild fall and winter, because of more renewable energy use and increases in efficiency, but also because of the switch from coal to natural gas by electric utilities.

Years ago I was among those who recommended the greater use of conventional natural gas as a bridge fuel to phase out coal use more quickly while solar and wind technologies were produced at sufficient scales to bring their price down even more. However, it is increasingly clear that the net effect of shale gas on the environment may ultimately be inconsistent with its use as a bridge fuel. Global society as a whole would find it difficult to make the enormous investments necessary to switch from coal to gas, and then turn right around and make equally significant investments to substitute renewable technologies for gas. It strains credulity. In other words, it may be a bridge to nowhere.

Not only have the new supplies of shale gas temporarily depressed energy prices to the point where renewable energy technologies have more trouble competing, if the studies showing that there is no net greenhouse gas benefit to switching to shale gas are correct, this might lead to the worst of all possible worlds: huge investments in shale gas diverting money from renewable energy, and a worsening of the climate crisis in the meantime. The only virtue of shale gas is that it is leading to a faster phase-out of coal, at least in the United States.

Coal has the highest carbon content of any fuel and emits the most CO_2 for each unit of energy it produces. It causes local and regional air pollution, including emissions of nitrous oxide (the leading cause of smog), sulfur dioxide (the continuing cause of acid rain), and toxic pollutants like arsenic and lead. The burning of coal also leaves huge quantities of toxic sludge—the second largest industrial waste stream in the United States—that is typically pumped to huge lagoons like the one that burst a holding wall and flooded portions of Harriman, Tennessee, in my home state four years ago.

Of particular importance, coal burning is the principal source of human-caused mercury in the environment, an extremely toxic pollutant that causes neurological damage, negatively impacting cognitive skills, the ability to focus, memory, and fine motor skills, among other effects. In the United States nearly all fish and shellfish include at least some amount of methyl-mercury that originated in coal-burning power plants. It is primarily for this reason that many fish and shellfish are considered dangerous in the diets of pregnant women, women who may become pregnant, nursing mothers, and young children. (Since the eating of fish is beneficial for brain development, pregnant women are advised to seek out fish that are low in mercury content and not avoid fish altogether.)

But the worst harm from coal burning is its dominant role in causing global warming. Although public opposition in the U.S. has contributed to cancellation of 166 new coal plants that had been planned, coal use is still growing rapidly in the world as a whole. An estimated 1,200 new coal plants are now planned in 59 countries. Under current plans, the global use of coal is expected to increase by another 65 percent in the next two decades, replacing oil as the single largest source of energy worldwide.

Coal is considered cheap, primarily because the absurdly distorted accounting system we use for measuring its cost arbitrarily excludes any consideration of all of the harm caused by burning it. Some engineers are working on improvements to a long-known process for converting underground coal reserves into gas that could be brought to the surface as fuel. But even if this technology were to be perfected, the CO_2 emissions would continue destroying the Earth's ecosystem.

Oil, the second largest source of global warming pollution, contains 70 to 75 percent of the carbon in coal for each unit of energy produced. Moreover, most of the projected new supplies of oil—in the form of shale oil, deep ocean drilling, and tar sands (not only in Canada, but also in Venezuela, Russia, and elsewhere)—are considerably more expensive to produce and carry even harsher impacts for the environment.

Conventional oil is burdened with other problems that coal does not have. Most of the easily recoverable oil in the world is found in regions such as the Persian Gulf that are politically and socially unstable. Several wars have already been initiated in the Middle East for reasons that include competition for access to oil supplies. And with Iran's determined effort to develop nuclear weapons, and ongoing political unrest in multiple countries in the region, the strategic threat of losing access to these oil supplies makes the price of oil highly volatile.

Although most of the discussions about reductions of CO_2 emissions have focused on industrial, utility, and vehicle emissions, it is also important to reduce CO_2 emissions and enhance CO_2 sequestration in the agriculture and forestry sectors, which together make up the second largest source of emissions. As the Keeling Curve demonstrates, the amount of CO_2 contained in vegetation, particularly trees, is enormous. It is roughly equal to three quarters of the amount in the atmosphere.

The largest tropical forest, the Amazon, has been under assault from developers, loggers, cattle ranchers, and subsistence farmers for decades, and even though the government of former president Luiz Inácio Lula da Silva took effective measures to slow down the destruction of the Amazon, his successor has made policy changes that are reversing some of the progress, though the rate of deforestation fell in 2012. In the last decade, the Amazon region was hit hard in 2005 and again in 2010 by "once-in-a-century" droughts (or rather, by what used to be once-in-a-century droughts before human modification of the climate). This led some forest researchers to renew their concern about a controversial computer model projection that has predicted the possibility of a dramatic "dieback" of the Amazon by mid-century if temperatures continue rising.

An increasing amount of the world's CO_2 emissions are coming from the cutting, drying, and intentional burning of peat forests and peat lands—especially in Indonesia and Malaysia—in order to establish palm oil plantations. According to the United Nations Environment Programme, peatlands contain more than one third of all the global soil carbon. Although both governments have given lip service to efforts to rein in this destructive practice, endemic corruption has undermined their stated goals. Extremely poor governance practices are among the chief causes of deforestation almost everywhere it is occurring—partly because 80 percent of global forest cover is in publicly owned forests.

Tropical forests are also under assault in Central and south-central Africa—particularly in Sudan and Zambia, and the Southeast Asian archipelago—including areas in Papua New Guinea, Indonesia, Borneo, and the Philippines. In many tropical countries, the increased demand for meat in the world's diet has contributed greatly to the clearing of forests for ranching—especially cattle ranching. As noted in Chapter 4, the growing meat intensity of diets around the world has an especially large impact on land use because each pound of animal protein requires the consumption of more than seven pounds of plant protein.

The enormous northern boreal forests in Russia, Canada, Alaska, Norway, Sweden, and Finland (and parts of China, Korea, and Japan) are also at great risk. Recent reestimates of the amount of carbon stored in these forests—not only in the trees, but also in the deep soils, which include many carbon-rich peatlands—calculate that as much as 22 percent of all carbon stored on and in the Earth's surface is in these boreal forests.

In Russia's boreal forest—by far the largest continuous expanse of trees on the planet—the larch trees that used to predominate are disappearing and are being replaced by spruce and fir. When the needles of the larch fall in the winter, unlike those of the spruce and fir, the sunlight passing through the barren limbs is reflected by the snow cover back into space, keeping the ground frozen. By contrast, when the conifer needles stay on the trees and absorb the heat energy from the sunlight, temperatures at ground level increase, thus accelerating the melting of the snow and the thawing of the tundra. The intricate symbiosis between the larch and the tundra is thereby disrupted, causing both to disappear. Millions of similar symbiotic relationships in nature are also being disrupted.

Although some Canadian provinces have impressive policies requiring sustainable forestry and limiting the damage from logging operations, Russia does not. And in both Russia and North America, the forests are being ravaged by the impact of global warming on droughts, fires, and insects. Beetles have expanded their range as average temperatures have increased, and have multiplied quickly as the number of cold snaps that used to hold them back has diminished. In many areas they are now reproducing three generations per summer rather than one. In the last decade, more than 27 million acres (110,000 square kilometers) of forests in the Western U.S. and Canada have been devastated by what the United Nations biodiversity experts described as "an unprecedented outbreak of the mountain pine beetle."

In mountainous areas, the earlier melting of snowpacks is depriving trees of needed water supplies during the hot summer months, which further increases their vulnerability to drought. One expert studying these issues, Robert L. Crabtree, told *The New York Times* recently, "A lot of ecologists like me are starting to think all these agents, like insects and fires, are just the proximate cause, and the real culprit is water stress caused by climate change."

The drought conditions weaken the trees and make them more vulnerable to beetles. And the increasing numbers of forest fires, scientists have long since established, are going up in direct proportion to the rising temperatures. There is no doubt that changes in forest management practices over the last several decades have contributed to the risk, frequency, and size of many forest fires. But the myriad impacts of global warming on fires far exceeds the impact of management practices.

The scale of the losses in the areas being deforested is completely unprecedented, according to experts, and as a result, enormous quantities of CO_2 are being released to the atmosphere. Like the Arctic tundra, the great forests of the world contain large amounts of CO_2, in the trees and plants themselves, in the soil beneath them, and in the forest litter that covers it. The great northern boreal forest of Canada and Alaska may have already become a net contributor to CO_2 levels in the atmosphere, rather than a net "sink," withdrawing CO_2 as the trees grow.

If adequate nutrients are available, the extra CO_2 in the atmosphere has the potential to stimulate some additional tree growth, though most experts point out that other limiting factors such as water availability and increased threats from insects and fire are overwhelming this potential. However, in spite of these devastating losses in forestland, the *net* loss of forests has slowed in recent years, primarily due to the planting of new forests and due to the natural regrowth of trees on abandoned agricultural land. According to the United Nations, most of the regrowth has been in temperate zones, including in forested areas of eastern North America, Europe, the Caucasus, and Central Asia. According to one study, successfully cutting the rate of deforestation in half by 2030 would save the world $3.7 trillion in environmental costs.

China has led the world in new tree planting; in fact, over the last several years, China has planted 40 percent as many trees as the rest of the world put together. Since 1981, all citizens of China older than age eleven (and younger than sixty) have been formally required to plant at least three trees per year. To date, China has planted approximately 100 million acres of new trees. Following China, the countries with the largest net gains in trees include the U.S., India, Vietnam, and Spain. Unfortunately, many of these new forests include only a single tree species, which results in a sharp decline in the biodiversity of animals and plants supported by the monoculture forest, compared to the rich variety supported by a healthy, multispecies primary forest.

For all of the needed attention paid to the sequestration of carbon in trees and vegetation, the amount of carbon sequestered in the first few feet of soil (mainly on the 10.57 percent of the Earth's land surface covered by arable land) is almost twice as much as all the carbon in the vegetation and the atmosphere combined. Indeed, well before the Industrial Revolution and the adoption of coal and oil as the world's principal energy sources, the release of CO_2 from plowing and land degradation contributed significantly to the excess of CO_2 in the air. By some estimates, approximately 60 percent of the carbon that used to be stored in soils, trees, and other vegetation has been released to the atmosphere by land clearing for agriculture and urbanization since 1800.

Modern industrial agricultural techniques—which rely on plowing, monoculture planting, and heavy use of synthetic nitrogen fertilizers—continue to release CO_2 into the atmosphere by depleting the organic carbon contained in healthy soils. The plowing facilitates wind and water erosion of topsoils; the reliance on monocultures, instead of mixed planting and crop rotation, prevents the natural restoration of soil health; and the use of synthetic nitrogen fertilizers has an effect not dissimilar from steroids: they boost the growth of the plants at the expense of the health of the soil and interfere with the normal sequestration of organic carbon in soils.

The diversion of cropland to biofuel plantations also results in a net increase in CO_2, while encouraging the destruction of yet more forestland, either directly, as in the case of the peat forests—or indirectly, by pushing subsistence farmers to clear more forests to replace the land they used to plant. As I have previously acknowledged publicly, I made a mistake supporting first generation ethanol programs while serving in the U.S. government, because I believed at the time that the net CO_2 reductions would be significant as biofuels replaced petroleum products. The calculations done since then have proven that assumption to be wrong. I and others also failed to anticipate the rapid growth of biofuels and the enormous scale they have now reached worldwide.

THE EXTINCTIONS OF SPECIES

The destruction of forests—particularly tropical forests that are rich in biodiversity—is also one of the principal factors, alongside global warming, that is driving what most biologists consider the worst consequence of the global environmental crisis: a spasm of extinction that has the potential to cause the loss of 20 to 50 percent of all living species on Earth within this century.

So much heat is already being trapped by global warming pollution that average world temperatures are increasing much more rapidly than the pace to which many animals and plants can adapt. Amphibians appear to be at greatest risk during this early stage, with multiple species of frogs, toads, salamanders, and others going extinct at a rapid rate all over the world. Approximately one third of all amphibian species are at high risk of extinction and 50 percent are declining. Experts have found that in addition to climate change and habitat loss, many amphibians have been hit by a spreading fungal disease, which may also be linked to global warming. Coral species, as noted earlier, are also facing a rapidly increasing risk of extinction.

According to experts, the other factors driving this global extinction event include, in addition to global warming and deforestation, the destruction of other key habitats like wetlands and coral reefs, human-caused toxic pollution, invasive species, and the overexploitation of some species by humans. Many wildlife species in Africa are particularly threatened by poaching and the encroachment of human activities into their territories, particularly the conversion of wild areas into agriculture.

There have been five previous extinction events in the last 450 million years. Although some of them are still not well understood, the most recent, 65 million years ago (when the age of the dinosaurs ended) was caused by a large asteroid crashing into the Earth near Yucatan. Unlike the previous five extinction events, all of which had natural causes, the one today is, in the words of the distinguished biologist E. O. Wilson, "precipitated entirely by man."

Many species of plants and animals are being forced to migrate to higher latitudes—north in the northern hemisphere and south in the southern hemisphere (one large study found that plants and animals are moving on average 3.8 miles per decade toward the poles)—and to higher altitudes (at least where there are higher areas to migrate *to*). One study of a century of animal surveys at Yosemite National Park found that half of the mountain species had moved, on average, more than 500 meters higher.

Some, when they reach the poles and the mountaintops and can go no farther, are being pushed off the planet and into extinction. Others, because they cannot move to new habitats as quickly as the climate is changing, are also being driven toward extinction. A recent Duke University study for the National Science Foundation found that more than half of the tree species in the eastern United States are at risk because they cannot adapt to climate change quickly enough.

Almost 25 percent of all plant species, according to scientists, are facing a rising risk of extinction. Agricultural scientists are especially concerned about the extinction of wild varieties of food crop plants. There are twelve so-called Vavilovian Centers of Diversity, named after Nikolai Vavilov, the great Russian scientist whose colleagues died of starvation during the siege of Leningrad protecting the seeds he had gathered from all over the world. One of them left a letter along with the enormous untouched collection of seeds, saying, "When all the world is in the flames of war, we will keep this collection for the future of the people." Vavilov himself died in prison after his criticism of Trofim Lysenko led to his persecution, arrest, conviction, and death sentence.

The ancient homes of food crops are sources of abundant genetic diversity that serve as treasure troves for geneticists looking for traits that can assist in the survival and adaptation of food crops to new pests and changing environmental conditions. But many of these have already gone extinct and others are threatened by a variety of factors, including development, monoculture, row cropping, war, and other threats.

The United Nations Convention on Biological Diversity notes, among other examples, that the number of local rice varieties being cultivated in China has declined from 46,000 in the 1950s to only 1,000 a few years ago. Seed banks like the one Vavilov first established are now cataloguing and storing many seed varieties. Norway has taken the lead with a secure storage vault hollowed out of solid rock in Svalbard, north of the Arctic Circle, as a precautionary measure for the future of mankind.

THE LOSS OF living species with whom we share the Earth and the wide-spread destruction of landscapes and habitats that hundreds of generations have called "home" should, along with the manifold other consequences of the climate crisis, lead all of us to awaken to the moral obligation we have to our own children and grandchildren. Many of those who have recognized the gravity of this crisis have not only made changes in their own lives but have begun to urge their governments to make the big policy changes that are essential to securing the human future.

THE PATH FORWARD

Generally speaking, there are four groups of policy options that can be used to drive solutions to the climate crisis. First and most important, we should use tax policy to discourage CO_2 emissions and drive the speedier adoption of alternative technologies. Most experts consider a large and steadily rising CO_2 tax to be the most effective way to use market forces to drive a large-scale shift toward a low-carbon economy.

Economists have long understood that taxes do more than raise revenue for the governments that impose them; to some extent, at least, they also discourage and reduce the economic activities that are taxed. By using taxes to adjust the overall level of cost attributed to the production of CO_2 and other greenhouse gases, governments can send a powerful signal to the market that, in the best case, unleashes the creativity of entrepreneurs and CEOs in searching for the most cost-effective ways of reducing global warming pollution. That is the reason I have advocated the use of CO_2 taxes for thirty-five years as the policy most likely to be successful. And implementing the tax in a way that escalates over time would provide the long-term signal to industry and the public that is needed to plan effective changes over coming decades.

Taxes, of course, are always and everywhere unpopular with those who pay them. Therefore, the enactment of this policy requires strong and determined leadership and, to the extent possible, bipartisanship. In recognition of those simple but significant political facts of life, I have always recommended that CO_2 taxes be coupled with reductions in other taxes by an equal amount.

Unfortunately, most people are far more willing to believe that government will indeed impose a new tax, but far less willing to believe that it will give that revenue back in another form. The forty-year campaign in the U.S. by the conservative counterreform alliance led by corporate interests and business elites has been effective in demonizing government at all levels and pursuing a "starve the beast" strategy that focuses on shrill opposition to any tax of any kind—unless the tax in question falls on low-income wage earners.

Other versions of this proposal have coupled the CO_2 tax with a rebate plan, to send a check to each taxpayer. Under this approach, sometimes labeled the "fee and dividend" approach or "feebate," those who were more successful in reducing their CO_2 emissions would actually *make* money, or use it to pay for more efficient or renewable energy technologies. Yet another version, which was introduced in the U.S. Congress in 2012 but never voted upon, would return two thirds of the revenue raised by a carbon tax to the taxpayers but would have applied one third to a reduction in the budget deficit. Unfortunately, the ingrained opposition to any new taxes—even if they are revenue neutral—has thus far made it difficult to build support for the single most effective strategy for solving the climate crisis, a CO_2 tax.

A second set of policy options involves the use of subsidies. To begin with, we should immediately remove existing subsidies that encourage fossil fuel consumption. In the United States, for example, approximately $4 billion each year—mainly in the form of special tax subsidies—go to carbon fuel companies. In India, to pick another example, the dirtiest liquid fuel, kerosene, is heavily subsidized.

Instead, governments should provide robust subsidies for the development of renewable energy technologies, at least until they reach the scale of production that will bring sufficient cost reductions to enable them to be competitive with unsubsidized fossil fuels. This policy would be even more effective in combination with a CO_2 tax, which would appropriately include in the price of fossil fuels some of the enormous costs they impose on society.

Limited government subsidies have already been successful in promoting more rapid adoption of renewable energy technologies. In fact, the cost reductions associated with the increasing scale of production have now put some renewable technologies much closer to a price that makes them competitive with coal and oil. Both solar and wind technologies are only a few years away from reaching that threshold. Yet the large carbon polluters and their allies have been working hard to eliminate subsidies for renewable energy before these clean technologies can become competitive with dirty energy—which is ironic, given that the global subsidies for the burning of fossil fuels, described above, greatly exceed the subsidies for renewable sources of energy, even though the latter are often miscalculated and misstated by opponents, who lump them in with subsidies for nuclear energy, so-called clean coal technologies, and other nonrenewable options.

The third policy option is an indirect subsidy for renewable energy in the form of a mandate requiring utilities to achieve a certain percentage of electricity production from renewable sources. This mechanism has already worked in numerous nations and regions, though many in the utility sector oppose such measures. Several U.S. states—including, most prominently, California—have successfully implemented this approach, and it is a major factor in the increased renewable energy installations in the United States. Germany has been perhaps the most successful nation in the world in using this policy option to stimulate the rapid adoption of both solar and wind technologies.

On a global basis, the combination of government subsidies for the speedier development of renewable energy technologies and the requirements that some utilities use them to produce a higher percentage of the electricity they generate has contributed to dramatic advances far beyond what most predicted. In 2002, a leading energy consulting firm projected that one gigawatt of solar electricity would be produced worldwide by 2010; that goal has been exceeded by seventeen times. The World Bank projected in 1996 that China would install 500 megawatts of solar energy by 2020. China installed double that amount by 2010.

The past projections of increased wind energy have also turned out to be overly pessimistic. The U.S. Department of Energy projected in 1999 that the U.S. wind capacity would reach ten gigawatts by 2010. Instead, that goal was met in 2006 and has now been exceeded four times over. In 2000, the U.S. Energy Information Agency projected that worldwide wind capacity would reach thirty gigawatts by 2010. Instead, that goal was exceeded by a factor of seven. The same agency projected that China would install two gigawatts of wind by 2010; that goal was exceeded 22-fold and is expected to be exceeded 75-fold by 2020.

As Dave Roberts of the environmental magazine *Grist* has pointed out, the world has previously witnessed predictions for the adoption of new technology that "weren't just off, they were *way* off." Industry and investor predictions at the beginning of the mobile telephone revolution, for example, wildly underestimated how quickly that new technology would spread. After the Arab-OPEC oil embargos in the 1970s, projections for the adoption of energy efficiency measures were also way off. What both of these prior examples have in common with renewable energy technologies is that all three are "widely dispersed" technologies that experienced unpredicted exponential growth because of a virtuous cycle, within which the increasing scale of production drove sharply lower costs—which in turn drove even faster growth.

The most frequently cited precedent for this phenomenon is the computer chip industry. As noted earlier, Moore's Law—which accurately predicted the relentless 50 percent cost reduction for computer chips every eighteen to twenty-four months—is not a law of nature, but instead a law of investment. In the early days of the computer revolution sixty years ago, chip manufacturers came to two conclusions: first, the potential market for computer chips was enormous and fast-growing—almost limitless; second, the technology development path was highly sensitive to innovation.

These dual realizations caused the leading chip manufacturers to devote enormous sums to research and development in order to protect their prospective market share against competitors. Over time, a collective consensus emerged that so long as they could continue reducing their costs on the pathway described by Moore's Law, they would be likely to retain or grow their market share. In other words, Moore's Law was transformed from a description of the past into a self-fulfilling prophecy about the future. Policies designed to create the rational expectation of steadily growing markets for renewable energy technologies can steepen a similar self-sustaining cost reduction curve for renewable energy.

The fourth policy option is widely known as cap and trade. This proposal is also designed to mobilize market forces as an ally in achieving CO_2 reductions. In spite of the relentless attacks on the mechanism, cap and trade remains favored by many policy experts as the best approach for securing a global agreement. Although I strongly favor a CO_2 tax, one of its disadvantages is that it is difficult to imagine coordinating national tax policies in many countries around the world with widely differing tax systems and differing compliance records. By contrast, a global cap and trade system would be inherently easier to harmonize among countries around the world with widely varying tax systems.

Cap and trade is based on an extremely successful policy innovated by former president George H. W. Bush to reduce emissions of sulfur dioxide (SO_2) in order to mitigate the acid precipitation in states downwind to the north and east of the Midwestern coal plants. The policy was embraced by Republicans as an alternative to government regulations mandating reductions in each plant.

The theory was that a slowly declining limit on emissions, when coupled with an ability to buy and sell emission "permits," would maximize reductions by giving a market incentive to those companies that were most efficient in limiting emissions, while simultaneously allowing a little more time for those companies having difficulty. The results were astoundingly successful. Emissions dropped much faster than predicted at a cost that was only a fraction of what was predicted. Consequently, advocates of CO_2 reductions felt that this mechanism could serve as a bipartisan compromise that would effectively reduce global warming pollution.

Unfortunately, as soon as cap and trade was presented as a bipartisan compromise, many conservatives who had originally supported the idea turned against it and began calling it "cap and tax." Thus have fossil fuel companies and their ideological allies paralyzed the policymaking process both at the global level and in the United States.

For many years, the effort to achieve a global consensus on action to solve the climate crisis was bedeviled by the international fault line between rich and poor nations, with poor countries insisting that the priority they placed on quickly replicating the economic development that had already occurred in wealthy countries meant that they could not afford to participate in a global effort to reduce global warming pollution. Proposed treaties routinely placed the first obligations on wealthy countries alone, leaving any requirements on developing nations to future rounds of negotiation.

After all, the need for more energy to power sustainable economic development in poor countries is acute. An estimated 1.3 billion people in the world still have no access whatsoever to electricity, and in spite of historic reductions in global poverty, the per capita income levels in many energy-poor countries are so low that it is easy to understand why they have resisted any constraints on potential increases in CO_2 emissions at a time when the wealthier countries have made such profligate use of fossil energy during their own past periods of economic takeoff and development.

Much has changed, however. The reality of the climate crisis has become much more apparent in developing nations as they experience harsh impacts and struggle to find the resources for disaster recovery and adaptation that are more readily available in developed countries. As a result, many developing countries have now changed their tune and are actively pushing the world community to take action on climate, even if it means that they too must shoulder part of the burden for responding. The World Bank estimates that more than three quarters of the costs from climate disruption will be borne by developing countries, most of which lack the resources and capacities to respond on their own.

Expenditures for the installation of renewable energy sources in the developing world now exceed those in rich countries. According to David Wheeler at the Center for Global Development, developing countries now are responsible for two thirds of the new renewable energy capacity since 2002 in the world, and overall have more than half of the installed global renewable energy capacity.

Even the richest countries are now being forced to recognize the economic toll of climate-related disasters. In the U.S.—still the richest country in the world—political controversies over the rising costs of disaster relief have resulted in cutbacks to emergency recovery programs that have hampered the ability of many communities to get back on their feet after climate calamities. But 2011–12 was a wakeup call.

In 2011, the U.S. had eight climate-related disasters, each costing over $1 billion. Tropical Storm Irene, which mostly missed New York City, nevertheless caused more than $15 billion in damage. Texas experienced the worst drought and highest temperatures in its history, and wildfires in 240 of its 242 counties. Thousands of daily all-time-high temperature records were broken or tied. Tornadoes, which climate researchers are still unwilling to link to global warming (partly because the records of past tornadoes are incomplete and imprecise), ravaged Tuscaloosa, Alabama, Joplin, Missouri, and many other communities; seven of them caused more than $1 billion in damage. In 2012, more than half of the counties in the U.S. suffered from drought. Hurricane Sandy cost at least $71 billion.

One of the principal objections to cap and trade in the United States has been based on the fear that developing countries would not be subject to the proposal and that U.S. industries would therefore be at a competitive disadvantage. In the last two decades, the emergence of Earth Inc. has inspired fear among factory workers in the U.S. and other developed nations that their jobs were being taken away and redistributed to factory workers in poorer countries where labor was cheap and advanced technologies were becoming available. Consequently, any perceived additional competitive advantage for developing countries became politically toxic in much of the industrial world.

That is one of many reasons why there is support for proposals to integrate CO_2 reductions into the World Trade Organization's definition of what is permitted by way of "border adjustments" to add the cost of CO_2 reductions to the price of imported goods from a country that does not require them to a country that does. In 2009, the World Trade Organization and the United Nations Environment Programme jointly published a report supporting such border adjustments.

I have long been a vocal advocate of reciprocal free trade even though that position did not endear me to my own political party. And I continue to strongly believe in free and fair international trade. But a fair set of rules is one that is designed to create and maintain a level playing field, and, in my view, CO_2 reductions certainly qualify as one of the factors that should be included in border adjustments.

When I was vice president, I joined with others in negotiating a global treaty in Kyoto, Japan, to adopt the cap and trade mechanism as the basis for the world's effort to reduce CO_2 emissions. The Kyoto Protocol was adopted by 191 countries and by the European Union as a whole, and in spite of the U.S. refusal to participate, and in spite of implementation problems, has been a success in most of the nations, provinces, and regions that are striving to meet its commitments.

Even though some nations using carbon credit trading have manipulated and abused the system, and even though problems emerged in the early days of the European system, Europe has taken action to address the problems and most nations with well-designed systems are on course to sharp emissions reductions. One policy analyst with the Potsdam Institute for Climate Impact Research, Bill Hare, said, "I can't see any other way to do it. Other policies are not easier to negotiate. The carbon market may be complex, but we live in a complex world."

Unfortunately, the decision by the United States not to join the

Kyoto Protocol and the failure to gain commitments from China and other "developing countries" (China in those years was still labeled a developing country) meant that the two largest emitters of global warming pollution were not included. If the U.S. *had* joined, the momentum for global participation and compliance would have been overwhelming and developing countries would have faced unrelenting pressure to join in the treaty's second phase, as anticipated.

Yet even though the U.S. political system is still paralyzed at the federal level, governments of many other nations are beginning to adopt new policies in recognition of the dangers we face and the opportunities to be seized. In addition to the European Union, Switzerland, New Zealand, Japan, one Canadian province, and twenty U.S. states will imminently begin cap and trade systems. Most significantly, California began implementing its system in 2012.

Australia, the largest coal exporter in the world, has adopted a plan that includes both a CO_2 tax and a cap and trade system that has been linked to the European Union's system. South Korea is in the process of setting up its own system and fourteen other countries have announced formally that they are planning to launch cap and trade systems: Brazil, Chile, Colombia, Costa Rica, India, Indonesia, Jordan, Mexico, Morocco, South Africa, Thailand, Turkey, Ukraine, and Vietnam.

Wolfgang Sterk of the Wuppertal Institute in Germany says, "The carbon market is not dead. . . . If a national system emerges in China, depending on the design and scope, it may become the biggest in the world, and allowances in that system would then give a global price signal."

China is implementing a cap and trade system in five cities (Beijing, Tianjin, Shanghai, Chongqing, and Shenzhen) and two provinces (Guangdong and Hubei). These pilots are intended to be up and running in 2013 in order to provide a learning experience that will be used to implement a nationwide cap and trade system by 2015.

As with some of the other commitments made by the Chinese government, some experts remain skeptical that they will follow through on this plan, but observers report that progress has already been made in most of the pilots that were designated. Together, the areas in the pilot program represent almost 20 percent of the Chinese population and almost 30 percent of its economic output.

China's commitment to sustainability and renewable energy has at once helped and hurt the world's ability to solve the climate crisis. By limiting imports while using subsidies to drive the cost of renewable energy technologies below the level at which Western companies can compete, China has served its own interest in dominating what everyone expects to be a key industry of the twenty-first century, but has damaged the rest of the world's ability to reap the benefits of fair competition in quickly advancing the state of these technologies.

In 2011, the United States filed a formal complaint against China for allegedly providing unfair subsidies to its wind and solar manufacturers. As of 2012, the U.S. imposed tariffs of approximately 30 percent on Chinese-imported solar panels, and the European Union began its consideration of a similar complaint. Nevertheless, in spite of these problems, the low prices that resulted from China's commitment and subsidies helped drive the scale of production to higher levels than anyone had anticipated, thus producing sharper cost reductions than anticipated.

China's impressive commitment to move forward aggressively with the deployment of wind and solar has inspired many other nations around the world, but its continuing enormous investment in new coal-fired generating plants has caused it to overtake the United States as the largest global warming polluter on the planet. Everyone realizes the importance to China of continuing its development of business and industry in order to continue reducing the levels of abject poverty in its country, but protests inside China against dirty energy projects are growing in several regions.

In the last ten years, China's energy consumption has gone up more than 150 percent, surpassing that of the U.S. And, unlike the United States, China still gets approximately 70 percent of its energy from coal. Its coal consumption has increased 200 percent over the same decade, to a level three times that of U.S. coal consumption. China is both the largest importer of coal in the world (followed by Japan, South Korea, and India) and the largest producer of coal, by far—producing half of the world's coal, two and a half times more than the U.S. (which is the second-largest producer of coal). Indeed, the amount by which China's coal consumption *increased* from 2007 to 2012 amounts to additional demand that is equivalent to all of the U.S. annual consumption. Beijing has proposed a cap on coal production and use to be implemented in 2015, though many experts are skeptical about their ability to stay within the cap.

Even though its appetite for oil pales in comparison to its consumption of coal, the amount of oil China used doubled during the 1990s, doubled again in the first decade of this century, and is now second only to that of the United States. For the first time, in 2010, Saudi Arabia's oil exports to China exceeded those to the U.S. In 2012, China's domestic oil reserves appeared to have peaked. And even though they are aggressively developing offshore oilfields, the Chinese already import half the oil they use, and the U.S. Energy Information Agency predicts that China will import three quarters of its oil within the next two decades.

Security experts have noted that this trend has implications for Chinese foreign policy in areas like the disputed reserves in the South China Sea and its forward-leaning engagement with oil-rich countries in the Middle East and Africa. Many observers found it ironic that after the United States invaded Iraq—at least in part to ensure the security of Persian Gulf oil supplies—the Chinese became the largest investor in Iraq's oilfields.

On a per capita basis, energy consumption in China is only a fraction of that in the U.S. and other more developed countries, though its per capita CO_2 emissions are approaching those of Europe. Since the reforms of Deng Xiaoping were implemented more than thirty years ago, China has converted much of its economy from agriculture to industry and the transition has been even more energy-intensive because of subsidies to fossil fuels—which reduce energy efficiency in every country that uses them. In fact, electricity rates, petroleum product prices, and natural gas prices are all fixed by the government at below market levels, though there is active debate in Beijing about letting all energy prices float further upward to global market levels. Overall, China is lagging behind other leading global economies in crucial areas of energy efficiency.

In spite of its energy challenges and its massive CO_2 emissions, China has implemented an extremely impressive set of policies to stimulate the production and use of renewable energy technologies. In its latest Five Year Plan, China announced that it will invest almost $500 billion in clean energy. The Chinese make use of "feed-in tariffs," a complex subsidy plan that worked extremely well in Germany. China also uses a full range of other policies, including tax subsidies and the imposition of renewable energy percentage targets on utilities.

In addition to capping the use of coal, it has also established a number of hard targets for the reduction of CO_2 emissions per unit of economic growth. A former vice minister of environmental protection, Pan Yue, said in 2005 that China's economic "miracle will end soon, because the environment can no longer keep pace."

In the last decade, there has been tension between goals set by the national government and implementation strategies pursued by regional governments, which are typically intertwined with industrial energy users. As a measure of the national government's seriousness in enforcing the CO_2 reduction and energy intensity reduction targets, Beijing sent officials to these regions in 2011 to impose forced closings of factories and even blackouts in order to ensure that the goals were met. More recently, the central government has linked promotions of local and regional officials to their success in achieving these goals.

In the renewable energy sector, China has dominated global production of windmills and solar panels, as noted above, but has made less progress in the installation of solar panels than it has in installing windmills—partly because it exports 95 percent of the solar panels it produces, many of them to the United States. In some recent years, 50 percent of all the windmills installed globally were in China, though almost a third of its windmills either are not connected to the electricity grid or are connected to lines that cannot handle the electricity flow.

The central government is also directing an ambitious plan to build the most sophisticated and extensive "super grid" in the world in order to remedy this problem. Beijing has announced that it will spend $269 billion over the next few years on construction of 200,000 kilometers of high-voltage transmission lines, which one industry trade publication noted is "almost the equivalent of rebuilding the United States' 257,500-kilometer transmission network from scratch."

As many countries have realized, high-capacity, high-efficiency electricity grids are essential in order to use intermittent sources of electricity like those produced by windmills and solar panels, and to transmit renewable electricity from the areas of highest potential production to the cities where it is used. As the percentage of electricity from the sources increases, the importance of smart grids and super grids will increase.

Plans are proceeding to link the high-sun areas of North Africa and the Middle East to large electricity consumers in Europe. Similar plans are on the drawing boards in North America, where high-sun areas of the Southwestern U.S. and northern Mexico can easily provide all of the electricity needed in both countries. And in both India and Australia, plans are under way to link high-sun and -wind regions with high-electricity-consuming regions.

There is, in any case, a powerful need to upgrade the reliability, carrying capacity, and advanced features of the electricity distribution grid in rich and poor countries alike. In the U.S., for example, interruptions in electrical service and unplanned blackouts, combined with inefficiencies in distribution and transmission, impose an estimated annual cost of more than $200 billion per year. In India, the largest blackout in history—by far—occurred in 2012 when more than 600 million people lost power due to problems in managing electricity flows through the antiquated grid system.

In addition to the development of super grids and smart grids—which can empower end-users of electricity with tools to become far more efficient in their ability to reduce energy consumption and save money—there is a pressing need for more efficient ways to *store* energy. A great deal of investment has gone into the research and development of new batteries that can be distributed throughout the electrical grid and in homes and businesses in order to reduce the need for wasteful over-capacity in electrical generation that is needed during the peak hours of use. These batteries can also provide valuable electricity storage when used in electric cars that, like most cars, spend the vast majority of their time in garages or parking spaces.

Toward that end, automakers around the world are launching fleets of electric vehicles in anticipation of a shift toward renewable electricity and away from expensive and risky petroleum supplies. At least some manufacturers in almost every industry are also converting to strategies that emphasize lower energy and material consumption. Energy efficiency expert Amory Lovins, of the Rocky Mountain Institute, has thoroughly documented the impressive movement by many companies to take advantage of these opportunities.

In addition to solar and wind, wave and tidal energy are both being explored—in Portugal, Scotland, and the United States, for example—and although the contribution from these sources is still minuscule, many believe that they may have great potential in the future. Nevertheless, the Intergovernmental Panel on Climate Change, in a special report on renewable energy sources in 2011, said that wave and tidal power are "unlikely to significantly contribute to global energy supply before 2020."

Geothermal energy has made a significant contribution in nations like Iceland, New Zealand, and the Philippines, where there is an abundance of easily exploitable geothermal energy. The vast potential for geothermal energy derived from much deeper geological regions has been unexpectedly difficult to develop, but here again, entrepreneurs in many countries are working hard to perfect this technology.

Although the potential for hydroelectric energy has been almost fully exploited in major areas of the world, there are undeveloped resources in Russia, Central Asia, and Africa that have great potential, though critics also warn about serious ecological risks in particular locations.

The use of biomass is expanding, and in some countries is beginning to play a significant role. In addition to traditional uses of manure and other forms of biomass for cooking, modern biomass techniques are being used to burn wood from renewable forests in far more efficient processes to produce heat and electricity. As with biofuels, the net impact of biomass use, when analyzed on a lifecycle basis, depends a great deal on the careful calculation of all of the energy inputs, the impact on land use and biodiversity, and the time periods required to recycle the carbon through the regrowth of the plants and trees.

There is also a global movement to produce methane and syngas from landfills containing large amounts of organic waste, and to produce biogas from large concentrations of animal waste gathered in animal feedlot operations. China, for example, has a major focus on biogas—requiring the installation of biogas digesters at all large cattle, pig, and chicken farms to derive the gas from animal waste, though enforcement of this mandate has been lagging. The U.S., which has a voluntary program, and other countries should follow their lead.

FALSE SOLUTIONS

There are two strategies for responding to global warming that are unlikely to work, even though each one has enthusiastic supporters. The first is carbon capture and sequestration (CCS). I have long supported research and development of CCS technologies, but have been skeptical that they will play more than a minor role. It is always possible that there will be an unexpected technological breakthrough that greatly reduces the cost of capturing CO_2 emissions and either storing them safely underground or transforming them in some manner into building materials or other forms that make them useful and safe. My friend Richard Branson has established a generous prize for the removal of CO_2 from the atmosphere, and invited NASA scientist and global warming expert Jim Hansen and me to be judges in the competition.

Barring breakthroughs, however, the cost of the CCS technologies presently available—both in money and energy—is so high that utilities and others are unlikely to use it. A utility operating a coal-fired generating plant and selling electricity to its customers would have to divert approximately 35 percent of all the electricity it produces just to provide power for the capture, compression, and storage of the CO_2 that would otherwise be released into the atmosphere. While that might be interpreted as a bargain if it saved civilization's future, the utility could not afford to do it and still stay in business. And the volumes of CO_2 emissions involved are so enormous that taxpayers do not have much appetite for shouldering the expense.

While safe and secure underground storage areas do exist, the process of locating them and then painstakingly investigating their characteristics in order to ensure that the CO_2 will not leak to the surface and into the air is quite significant. There has been notable public opposition to the siting of such underground storage facilities near populated areas. The consensus among those scientists and engineers who are experts in this subject is that the longer the CO_2 is stored, the safer it becomes—because it begins to be absorbed into the geological formation itself. Nevertheless, the overall expense of CCS has prevented its adoption by large carbon polluters.

Both the United States and China announced large government-financed demonstration projects for CCS, though the Chinese project—known as GreenGen—is behind schedule, and the U.S. project—called FutureGen—is mired in the endemic political paralysis that characterizes the present state of democracy in the United States. Norway, the United Kingdom, Canada, and Australia are among the other countries pursuing CCS. However, one of the world's leading experts on CCS, Howard Herzog of the Massachusetts Institute of Technology, has said for years that the real key to making this technology profitable and viable is to put a price on carbon.

The second technology that is sometimes described as a silver bullet that could eliminate most CO_2 emissions, at least from the electricity-generating sector, is one with a long and fraught history—nuclear power.

The present generation of 800 to 1,200 megawatt pressurized light water reactors is, unfortunately, probably a technological dead end. For a variety of reasons, the cost of reactors has been increasing significantly and steadily for decades. In the aftermath of the triple tragedy in Fukushima, Japan, the prospects for nuclear energy have further declined.

The safety record, while much improved, is still one that has been producing public opposition. France, which used to have a global reputation as the most advanced and efficient nation in nuclear power, has had difficulties with its new generation of reactors. South Korea, on the other hand, has been moving forward with a design that many experts believe is promising. Several new reactors are under construction around the world, but as our low-carbon energy options are evaluated, nuclear energy is severely hampered by both cost and perceived safety issues. There is still a distinct possibility that the research and development of a new generation of smaller and hopefully safer reactors may yet play a significant role in the world's energy future. We should know by 2030.

In spite of their problems, both CCS and nuclear power have had enduring appeal, partly because they are technological solutions that offer the possibility that a single strategy might lead to a relatively quick fix. Indeed, psychologists tell us that one of the other glitches in our common way of thinking about big problems is what they call "single-action bias," a deeply ingrained preference for single solutions to problems, however complex the problems may be.

This same common flaw in our way of thinking helps to explain the otherwise inexplicable support for a number of completely bizarre proposals that are collectively known as geoengineering. Some engineers and scientists argued several years ago that we should float billions of tiny strips of tinfoil in orbit around the Earth to reflect more incoming sunlight and thereby cool down the global temperature. The public record does not indicate whether they were wearing tinfoil hats when they launched their idea. An earlier proposal in the same vein featured a giant space parasol, also intended to block incoming sunlight. It would have had to be 1,000 miles in diameter and would have required a moon base for its construction and launch. Others have suggested that we attempt to accomplish the same result by injecting massive quantities of sulfur dioxide into the upper atmosphere in order to block sunlight.

The fact that any reputable scientist would lend his or her name to such proposals is certainly a measure of the desperation that those who understand the climate crisis feel about the abject failure of the world's political leadership to begin reducing the rate of emissions of global warming pollution. But given the unanticipated consequences of the planetary experiment we already have under way—pumping 90 million tons of heat-trapping pollution into the atmosphere every twenty-four hours—it would, in my opinion, be utterly insane to launch a second planetary experiment in the faint hope that it might temporarily cancel out some of the consequences of the first experiment without doing even more harm in the process.

Among the other consequences of the SO_2 proposal that was pointed out in a 2012 scientific study is this startling change: the sky we have gazed at since the beginning of humankind's existence on Earth would no longer be blue—or at least no longer be *as* blue. Does that matter? Perhaps we could explain to our grandchildren why there were so many references to "blue skies" in the history of the cultures on Earth. Maybe they would understand that it was necessary to sacrifice the blueness of the sky in order to accommodate the political agenda of oil, coal, and gas companies. The levels of pollution above cities have already changed the color of the night sky from black to reddish black.

No one has any idea what such proposals would mean for the photosynthesis of food crops and other plants; light needed for life would be partially blocked in order to create more "thermal space" to be occupied by steadily increasing emissions from the burning of fossil fuels. The effectiveness of photovoltaic conversion of sunlight into electricity—one of the most promising renewable energy technologies—might also be damaged. And none of these exotic proposals would do anything whatsoever to halt the acidification of the oceans.

In addition, if we failed to reduce CO_2 emissions, the sulfur dioxide injections or orbiting tinfoil strips would have to be increased steadily, year by year. Nor does anyone have the faintest idea of what these wackadoodle proposals would do to climate patterns, precipitation, storm tracks, and all of the other phenomena that are already being disrupted. Have we gone stark raving mad?

No, we haven't gone mad. It's just that our way of communicating about global challenges and debating reasonable solutions has been subjected to an unhealthy degree of distortion and control by wealthy corporate interests who are themselves desperate to prevent serious consideration of reducing global warming pollution.

Technically, there are a range of *benign* geoengineering proposals that may well offer marginal benefits without imposing reckless risks. Painting roofs white, for example, or planting millions of roof gardens are both examples of riskless changes to the reflective characteristics of the Earth's surface that could bounce more of the incoming sunlight back into space before the heat energy it carries is absorbed in the lower atmosphere. In a variation on this theme, Peru is painting rocks white high in the Andes in a desperate effort to slow the melting of glaciers and snowpacks on which they rely for drinking water and irrigation.

If we continue to delay the launching of a serious multipronged global effort to reduce the emissions of heat-trapping greenhouse gas pollution, we will find ourselves pushed toward increasingly desperate measures to mitigate the growing impacts of global warming. We will try to muddle through, argue and fight with one another, pursue our self-interest at the expense of others, often deceiving them and ourselves in the process. That is the course that we are on now.

But when the survival of what we hold most dear is clearly at risk, then we must act. In all of human history, there have been rare moments when we have risen to transcend our past and charted a new course to safeguard our deepest values. At one such challenging moment in history, Abraham Lincoln said, "The occasion is piled high with difficulty, and we must rise with the occasion. As our case is new, so we must think anew, and act anew. We must disenthrall ourselves, and then we shall save our country."

This time, our world is at stake. Not the planet itself; it would, of course, survive nicely without human civilization, albeit in an altered state. Rather, what is at stake is the set of environmental conditions and the health of the natural systems on which our civilization depends. And the fact that this crisis is global in nature is part of the unique challenge we face.

Only twice before in all of human history has the future of our entire global civilization been at risk. Once, at the dawn of *Homo sapiens'* time on Earth 100,000 years ago, anthropologists tell us that our numbers were reduced to less than 10,000 people, yet somehow we prevailed. The second occasion was when the United States and the former Soviet Union came all too close to unleashing massive nuclear arsenals against one another, killing hundreds of millions and risking a nuclear winter with potentially apocalyptic consequences. And again, somehow we prevailed.

This time, the threat to our future is one that would not arrive in a matter of minutes with bright flashes and deafening sounds. It would be drawn out, and generations yet to come would live all their lives with the painful knowledge that once upon a time the Earth was hospitable to humans. It sustained and nourished us with cool breezes and abundant food and water. It inspired and renewed us with its majestic beauty.

When memories of that Earth faded, the story would still be told: in the early decades of the twenty-first century, a generation gifted by those that came before them with the greatest prosperity and most advanced technologies the Earth had ever known broke faith with the future. They thought of themselves and enjoyed the bounty they had received, but cared not for what came after them. Would they forgive us? Or would they curse us with the dying breaths of each generation to come?

If, on the other hand, we do find a way to rise to this occasion, we will have the rare privilege of meeting and overcoming a challenge that is worthy of the best in us. We have the tools we need. Some of them, it is true, need repair. Others need to be improved and perfected for the task ahead. All that we lack is the will to prevail, but political will can be renewed and strengthened by acknowledging the truth of our circumstances and accepting our obligation to safeguard the future for the next generation and all who will follow them.

What we most need is a shift in our way of thinking and a rejection of the toxic illusions that have been so assiduously promoted and continually reinforced by opponents of actions, principally large carbon polluters and their allies. In some ways, this struggle to save the future will be played out in a contest between Earth Inc. and the Global Mind. The interconnection of people all over the world by means of the Internet has created the potential for an unprecedented global effort to communicate clearly among ourselves about the challenge that now confronts us and the solutions that are now available.

On the other hand, the increasing interconnections among businesses and industries all over the world has generated powerful commercial momentum that is highly resistant to any effort by governments to rein in its more destructive tendencies. Earth Inc. is now the dominant source of influence over governments. Fortunately, there are a great many examples of the emergence of a global conscience on the Internet that has exerted powerful pressure to correct injustices and moral failures such as child labor, abusive working conditions, false imprisonment, sex slavery, persecution of vulnerable minorities, and destruction of the environment, among other causes.

In some countries, this new emergent capacity for the development of a collective global conscience has also contributed greatly to policies aimed at solving the climate crisis. The number of grassroots, Internet-based NGOs devoted to safeguarding the ecological system of the Earth has been growing. The remaining question that is crucial to our future is whether the requisite force of truth necessary to bring about a shift in consciousness powerful enough to change the current course of civilization will emerge in time.

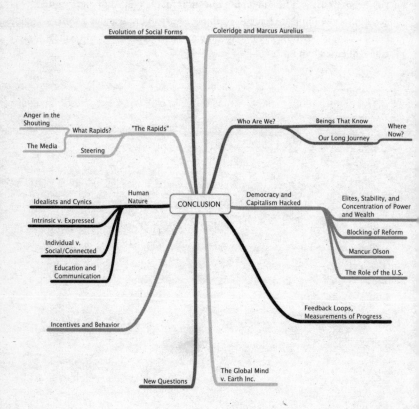

CONCLUSION

"So often do the spirits
Of great events stride on before the events,
And in today already walks tomorrow."

—SAMUEL TAYLOR COLERIDGE

THE PERSONAL JOURNEY I HAVE TAKEN IN WRITING THIS BOOK BEGAN with that single question which demanded an answer more thoughtful than the one I first offered. My search for a better answer has led me to new questions that also demand answers—especially from political, business, civic, and faith leaders around the world.

To begin with, *who are we?* The initial answer, once again, is readily available: we are *Homo sapiens.* "Beings that *know.*" The usual suspects. We have been on a very long journey already—from forests to savannahs to farms to megacities; from two to thousands to millions to billions; from stones to plows to assembly lines to nanobots; from syllables to encyclopedias to airwaves to the Global Mind; from families to tribes to communities to nations.

But that is the way we *have* been. Where our journey takes us next will depend upon what kind of beings we humans choose to be. To put it another way, our decision about the way we choose to live will determine whether the journey takes *us*, or whether *we* take the journey.

The currents of change are so powerful that some have long since taken their oars out of the water, having decided that it is better to surrender, enjoy the ride, and hope for the best—even as those currents sweep us along faster and faster toward the rapids ahead that are roaring so deafeningly we can hardly hear ourselves think.

"Rapids?" they shout above the din. "What rapids? Don't be ridiculous; there *are* no rapids. Everything is fine!" There is anger in the shouting, and some who are intimidated by the anger learn never to mention the topic that triggers it. They are browbeaten into keeping the peace by avoiding any mention of the forbidden subject.

For now at least, that is how some in the news media behave. They are terrified to even utter certain words—like "climate," for example— lest they trigger rage from those who don't want to hear about the destructive changes gaining momentum. The result is an almost pathological silence concerning the most important challenges we face, and a dangerous collective disregard for the future consequences of our present actions. But ultimately, that is not really who we are.

Anyone who spends time thinking about the possibilities of a better future must first make an assumption about human nature. Idealists who want and hope for the best sometimes make the mistake of thinking that intrinsic human nature can change, and will improve according to their hopes. Cynics enjoy catching that mistake and pointing out that human nature doesn't change at all.

My own way of thinking about human nature is neither idealistic nor cynical. I believe there is a difference between *intrinsic* human nature— which I agree does not change, and the aspects of human nature we routinely *express*, which can and do change. The 35,000-year-old paintings in the caves at Chauvet, in France, and the figurines made by our ancient ancestors in Eurasia and Africa, clearly reflect a consciousness and sensibility not very different—perhaps not intrinsically different at all—from our own. But in other ways, we are very different indeed.

We are each individuals, but as all of our major faith traditions teach us, we are all connected to one another. And science teaches us that human nature is inherently social. The social groups to which we belong have their own form of evolution. Some behaviors and norms survive from one generation to the next and others are discouraged. Habits and customs become rituals and rules, which evolve over time into cultures, social systems, laws and institutions, and which exercise a profound influence over which aspects of human nature we express.

Consider what we have learned about the human genome: even

though 99.9 percent of them are identical in every human being, our 23,000 genes—and millions of proteins—contain a universe of possibilities. Some genes are expressed while others remain inchoate, vestigial. Sometimes, capacities that evolved in the distant past are awakened for new purposes when our circumstances change. Consider also what neuroscientists have learned about the human brain: neuron trees grow dense and vibrant when they are used; others atrophy when they are not.

Some have long believed that the most important strategy for empowering the "better angels of our nature" is education. And while I certainly agree that high-quality, universal education is not only desirable but essential, it is not sufficient. Some of the worst atrocities in human history have been organized and perpetrated by well-educated villains.

Ignorance and misunderstanding are certainly enemies of genuine progress, just as knowledge, integrity, and character are crucial to our success. But the evolution of our collective behaviors, and the emergence of a genuine understanding of how deeply our connected fates are intertwined with the health of the ecological system of the Earth, will depend upon the choices we make about the structure of the systems we use. The way we measure what we do and the results of our actions, the way we communicate with one another, and the incentives and disincentives we build into our political, economic, and social systems all have a powerful influence on the future.

Behaviors that bring rewards become more common. Those that don't diminish. The elements of our nature that are activated by rewarded behaviors gain strength. Social groups establish values that reflect both the behaviors they wish to reward and those they want to discourage. These values become embedded in tribes, communities, nations, economic systems, institutions, and cultures.

I fall back on the example that inspired me and has inspired people throughout the world for more than two centuries: the enduring genius of the U.S. Constitution stemmed from its authors' clear-eyed, dead-on understanding of human nature—even though it was limited to white males—and their design of structural safeguards that discouraged the impulse to egotistical power-seeking and incentives that rewarded the impulse to resolve their differences through collective reasoning that maximized the likelihood of creative compromises based on the pursuit of the greater good.

The separation of powers and checks and balances woven into the design of the Constitution embodied a sophisticated understanding of how to discourage some behaviors inherent in human nature and encourage positive ones instead. Others have tried to structure economic systems with incentives that unleash creativity and dynamism, encourage behaviors deemed of value, and discourage other behaviors that are destructive to the common good.

Over time, we have come to recognize that the way we measure economic value also exerts a kind of evolutionary force on behavior—and that the things not measured at all are ignored as if they have no value, either positive or negative. When we change the measurements of value, the nature of the incentives, and the structure of the systems we use for making political, economic, and social decisions, we inevitably encourage the expression of some aspects of human nature and discourage others. So while intrinsic human nature may not change, the *expressions* of human nature—the aspects of our nature manifested in our behaviors and choices—can and do change readily in response to the incentives we establish as a basis for civilization. And they shape our future.

If we signal to business, for example, that unlimited pollution will incur no cost or penalty, it is of little use to then decry them as immoral when they respond predictably to the incentives we give them. When we signal to our politicians that victory in elections is best assured by spending most of their time asking for large sums of money from people and corporations that have special agendas for the shaping of public policies after the election, we incentivize politicians to express in their behavior negative aspects of human nature familiar to all of us—because they are intrinsic to all of us—even though most of us suppress them and understand fully why we should be discouraging the soft bribery and betrayal of the public trust that predictably results.

More serious problems arise when those who benefit from these distorted incentives and dysfunctional rules manage to gain sufficient political power to prevent reforms that would encourage the aspects of human nature that we want to see manifested in political and economic decision making.

Long periods of stability, which most of us naturally prefer, can enhance the vulnerability of any political or economic system to exploitation by those who have learned to distort its rules and incentives. Decades ago, the late University of Maryland political economist Mancur Olson published an extensive analysis of how elites in any society come to accumulate a steadily larger share of wealth and influence, and then use it to block reforms of the incentives and rules that work to their advantage.

Consider the vulnerability of monocultured crops to the steady evolution of pests who learn to circumvent the natural defenses of the plants to eat their fill. Consider the vulnerability of computer systems to hackers when the passwords and other safeguards remain unchanged for an extended period of time. The intrinsic nature of the pests doesn't change. Their learned behaviors—and the genes they express—do.

Democracy and capitalism have both been hacked. The results are palpably obvious in the suffocating control of policy decisions by elites, the ever increasing inequalities of income and growing concentrations of wealth, and the paralysis of any efforts at reform. And the public's ability to express their revulsion in more constructive ways, rather than surrendering to cynicism, is dampened by the structure of our dominant means of mass communication, television, which serves mainly to promote consumption of products and entertain the public, while offering no means for interactive dialogue and collaborative decision making.

Fortunately, the awakening of the Global Mind is disrupting established patterns—creating exciting new opportunities for emergent centers of influence not controlled by elites and the potential for reforms in established dysfunctional behaviors. Yet the emergence of Earth Inc. has magnified the power and reach of our economic engines, even as it has hardened the incentives, measurements of value, and rules of behavior that reward unsustainable exploitation of limited resources, the destruction of ecosystems crucial to the survival of civilization, unlimited flows of pollution, and the disregard of human and social values.

The outcome of the struggle to shape humanity's future that is now beginning will be determined by a contest between the Global Mind and Earth Inc. In a million theaters of battle, the reform of rules and incentives in markets, political systems, institutions, and societies will succeed or fail depending upon how quickly individuals and groups committed to a sustainable future gain sufficient strength, skill, and resolve by connecting with one another to express and achieve their hopes and dreams for a better world.

Here are the most important questions to be answered and battles to be won:

Can Americans reestablish the healthy functioning of the U.S. political and economic system to the point where it can once again provide visionary leadership to the community of nations? It may well be that an alternative form of global leadership will emerge in the Global Mind, but that is uncertain for now and is likely to take time that we don't have.

It is theoretically possible, though extremely unlikely, that some other nation will rise to this challenge. It is also possible that the tectonic changes that have reorganized the equilibrium of power in the world, shifting it from West to East and redistributing it throughout the world, will make it difficult for the U.S. to once again provide the strength and quality of leadership it offered during the second half of the twentieth century. The world's loss of confidence in the United States following the catastrophic political, military, and economic mistakes of the early twenty-first century accelerated this shift in power, but was not its fundamental cause.

Still, the best chance for success in shaping a positive future and avoiding catastrophe is the reestablishment of a transcendent capacity for global leadership by the United States. And for those who have difficulty believing that the promise of American democracy can be redeemed, remember that the promise America offers the world has been resurrected in the past during some very dark days. Its revolution was almost still-born. It nearly tore itself in two during the Civil War. The domineering crimes of the robber barons exceeded the excesses of today's ambitious titans. Destitution during the Great Depression, the devastating blow at Pearl Harbor while Hitler rampaged through Europe, and the brush with Armageddon during the Cuban Missile Crisis were all followed by renewals of the American spirit and a flourishing of the values at the heart of the American Dream. So America can certainly be renewed again, and its potential for world leadership can be restored.

Will it be? The answer to that question will have a profound effect on the future of humankind.

How quickly can institutions be adapted to the Internet? Even though the potential for the reestablishment of reason-based decision making through collaborative processes empowered by the Global Mind are exciting and promising, long-established institutions are notoriously resistant to change. The speed with which business models have been disintermediated and new models have emerged offers reason for hope.

But attention and focus are diluted on the Internet. The variety of experiences available, the ubiquity of entertainment, and the difficulty

in aggregating a critical mass of those committed to change all complicate the use of the Internet as a tool for institutional reform. The addition of three billion people to the global middle class by the middle of
this century, however, may be accompanied by new and more forceful
demands for democratic reforms of the kind that have so often emerged
with the growth of a prosperous and well-educated middle class in so
many nations.

Will there be sufficient safeguards and constraints placed on the impulse of governments to use the Internet as a means of gathering information about individuals and using it to establish unhealthy forms
of centralized control? Will the impulse of nations to engage in conflict
produce more destructive forms of cyberwar and mercantilist nationalism? As the severity of our challenges becomes ever clearer, I am hopeful, even confident, that enough concerned committed individuals and
groups will join together in time and self-organize creatively to become
a force for reform.

Will China's economic juggernaut continue, and if so, will its emergent commitment to safeguard the environment overtake its mercantilist
imperative? Will its success in lifting standards of living and diminishing
poverty lead to political reforms that produce a transition to democratic
governance?

Will the progressive substitution of intelligent machines for human
labor result in increased structural unemployment, or will we find ways
to create new jobs and adequately compensate those filling them? There
is no shortage of work to be done, but the dominance of corporations
and the encroachment of the market sphere into the democracy sphere
have taken a toll on the initiative and will necessary to structure new
employment opportunities in the creation of public goods in fields like
education, environmental remediation, health and mental health, family
services, community building, and many other challenges that must be
met.

Will the emergent potential for altering the fabric of life and the
genetic design of human beings be accompanied by the emergence of
wisdom sufficient for the far-reaching decisions that will soon confront
us, or will these technologies be widely dispersed without adequate consideration of the full spectrum of consequences they could entail?

Will the social compacts in developed nations survive the simultaneous effects of demographic changes that are placing heavier per capita

burdens on those in the workforce even as jobs and incomes are lost to the combination of robosourcing and outsourcing? Will new models for restoring income support and health care to the growing population of older people be created to replace the twentieth-century model?

Will the world community adequately support fertility management in developing countries with high population growth rates, continue to empower women, and improve child survival rates? The answer to these questions will determine the level of global population and the degree of stress humanity places on the natural systems of the planet. Will the unique plight of Africa be recognized and adequately addressed?

Will we provide the incentives to quickly decarbonize the global economy and sharply reduce global warming pollution in time to stabilize and then reduce the global warming pollution that is so threatening to the climate stability on which the thriving of our civilization depends?

These are hard questions that imply hard choices. Human civilization—indeed, the human *species*—is already in the early stages of the six emergent changes described in this book. They are beginning to transform our planet, our civilization, and the way we work and live our lives. Some of them are degrading self-governance, the fabric of life, the species with which we share the Earth, and the physical, mental, and spiritual nature of humanity.

The complexity of these changes, the unprecedented speed with which they are occurring, their simultaneity and the fact that they are converging, each with the others, have all contributed to a crisis of confidence in our ability as a civilization to think clearly about where they are taking us, much less to change their trajectory or slow their momentum.

But if we face these choices with courage, the right answers are pretty clear. They're controversial, to be sure. And making the right choices will be hard. Yet we do have to make them. We do have to decide. Soon. If we were to decide *not* to reclaim control over our destiny, the rest of our journey would become very hard indeed.

These currents of change are strong, and they are indeed sweeping us into a future that is very different from what we have known before. What we have to do—in the context of this metaphor—is deceptively simple: steer! That means fixing the prevailing flaws and distortions in capitalism and self-governance. It means controlling the corrosive corruption of money in politics, breaking the suffocating rule of special interests, and restoring the healthy functioning of collective decision making in representative democracy to promote the public interest. It means reforming markets and making capitalism sustainable by aligning incentives with our long-term interest. It means, for example, taxing carbon pollution and reducing taxes on work—raising revenue from what we burn, not what we earn.

More than 1,800 years ago, the last of Rome's "Five Good Emperors," Marcus Aurelius Antoninus, wrote, "Never let the future disturb you. You will meet it, if you have to, with the same weapons of reason which today arm you against the present." His advice is still sound, though soon after his reign the Roman Empire began the long process of dissolution that culminated in its overthrow 300 years later.

SO WHAT DO WE DO NOW?

Arming ourselves with the "weapons of reason" is necessary but insufficient. The emergence of the Global Mind presents us with an opportunity to strengthen reason-based decision making, but the economic and political systems within which we implement even the wisest decisions are badly in need of repair. Confidence in both market capitalism and representative democracy has fallen because both are obviously in need of reform. Fixing both of these macro-tools should be at the top of the agenda for all of us who want to help shape humanity's future.

Our first priority should be to restore our ability to communicate clearly and candidly with one another in a broadly accessible forum about the difficult choices we have to make. That means building vibrant and open "public squares" on the Internet for the discussion of the best solutions to emerging challenges and the best strategies for seizing opportunities. It also means protecting the public forum from dominance by elites and special interests with agendas that are inconsistent with the public interest.

It is especially important to accelerate the transition of democratic institutions to the Internet. The open access individuals once enjoyed to the formerly dominant print-based public forum fostered the spread of democracy and elevated the role of reason and fact-based public discourse. But the massive shift in the last third of the twentieth century from print to television as the primary medium of communication stifled democratic discourse and gave preferential access to those with wealth and power. This shift eclipsed the role of reason, diminished the importance of collective searches for the best available evidence, and elevated the role of money in politics—particularly in the United States—thereby distorting our search for truth and degrading our ability to reason together.

The same is true for the news media. The one-way, advertiser-dominated, conglomerate-controlled television medium has been suffocating the free flow of ideas necessary for genuine self-determination. In 2012, for example, it was nothing short of bizarre when the United States held its quadrennial presidential election in the midst of epic climate-related disasters—including a widespread drought affecting more than 65 percent of the nation, historic fires spreading across the West, and an epic hybrid hurricane and nor'easter that shut down large portions of New York City for the second time in two years—with not a single question about the climate crisis from any member of the news media in any of the campaign debates.

The profit-driven blurring of the line between entertainment and news, the growing influence of large advertisers on the content of news programs, and the cynical distortion of news narratives by political operatives posing as news executives have all degraded the ability of the Fourth Estate to maintain sufficient integrity and independent judgment to adequately perform their essential role in democracy.

The Internet offers a welcome opportunity to reverse this degradation of democracy and reestablish a basis for healthy self-governance once again. Although there is as yet no standard business model that yields sufficient profit to support high-quality investigative journalism on the Internet, the expansion of bandwidth to accommodate more and higher-quality video on the Internet may soon make profitable business models viable. In addition, the use of hybrid public/private models for the support of excellence in Internet-based journalism should be vigorously pursued.

The loss of privacy and data security on the Internet must be quickly addressed. The emergent "stalker economy," based on the compilation of large digital files on individuals who engage in e-commerce, is exploitive and unacceptable. Similarly, the growing potential for the misuse by governments of even larger digital files on the personal lives of their citizens—including the routine interception of private communications—

poses a serious threat to liberty and must be stopped. Those concerned about the quality of freedom in the digital age must make new legal protections for privacy a priority.

The new digital tools that provide growing access to the Global Mind should be exploited in the rapid development of personalized approaches to health care, what is now being called "precision medicine," and of self-tracking tools to reduce the cost and increase the efficacy of these personalized approaches to medicine. The same Internet-empowered precision should be applied to the speedy development of a "circular economy," characterized by much higher levels of recycling, reuse, and efficiency in the use of energy and materials.

Capitalism, like democracy, must also be reformed. The priority for those who agree that it is crucial to restore the usefulness of capitalism as a tool for reclaiming control of our destiny should be to insist upon full, complete, and accurate measurements of value. So-called externalities that are currently ignored in standard business accounting must be fully integrated into market calculations. For example, it is simply no longer acceptable to pretend that large streams of harmful pollution do not exist where profit and loss statements are concerned.

Global warming pollution, in particular, should carry a price. Placing a tax on CO_2 is the place to start. The revenue raised could be returned to taxpayers, or offset by equal reductions in other taxes—on payrolls, for example. Placing a steadily declining limit on emissions and allowing the trading of emission rights within those limits is an alternative that would also work. For those nations worried about the competitive consequences of acting in the absence of global agreement, the rules of the World Trade Organization allow the imposition of border adjustments on goods from countries that do not put a tax on carbon pollution.

The principles of sustainability—which are designed, above all, to ensure that we make intelligent choices to improve our circumstances in the present without degrading our prospects in the future—should be fully integrated into capitalism. The ubiquitous incentives built into capitalism—which embody the power of capitalism to unleash human ingenuity and productivity—should be carefully designed to ensure that they are aligned with the goals that are being pursued. Compensation systems, for example, should be carefully scrutinized by investors, managers, boards of directors, consumers, regulators, and all stakeholders in every enterprise—no matter its size.

Our current reliance on gross domestic product (GDP) as the compass by which we guide our economic policy choices must be reevaluated. The design of GDP—and the business accounting systems derived from it—is deeply flawed and cannot be safely used as a guide for economic policy decisions. For example, natural resources should be subject to depreciation and the distribution of personal income should be included in our evaluation of whether economic policies are producing success or failure. Capitalism requires acceptance of inequality, of course, but "hyper" levels of inequality—such as those now being produced—are destructive to both capitalism and democracy.

The value of public goods should also be fully recognized—not systematically denigrated and attacked on ideological grounds. In an age when robosourcing and outsourcing are systematically eliminating private employment opportunities at a rapid pace, the restoration of healthy levels of macroeconomic demand is essential for sustainable growth. The creation of more public goods—in health care, education, and environmental protection, for example—is one of the ways to provide more employment opportunities and sustain economic vibrancy in the age of Earth Inc.

Sustainability should also guide the redesign of agriculture, forestry, and fishing. The reckless depletion of topsoil, groundwater reserves, the productivity of our forests and oceans, and genetic biodiversity must be halted and reversed.

In order to stabilize human population growth, we must prioritize the education of girls, the empowerment of women, the provision of ubiquitous access to the knowledge and techniques of fertility management, and the continued raising of child survival rates. The world now enjoys a durable consensus on the efficacy of these four strategies—used in combination—to bring about the transition to smaller families, lower death rates, lower birth rates, and stabilized population levels. Wealthy countries must support these efforts in their own self-interest. Africa should receive particular attention because of its high fertility rate and threatened resource base.

Two other demographic realities should also command priority attention: The continued urbanization of the world's population should be seen as an opportunity to integrate sustainability into the design and construction of low-carbon, low-energy buildings, the use of sustainable architecture and design to make urban spaces more efficient and productive, and the redesign of urban transportation systems to minimize both energy use and pollution flows. And second, the aging of populations in the advanced economies—and in some emerging markets, like China— should be seen as an opportunity for the redesign of health strategies and income support programs in order to take into account the higher dependency ratios that threaten the viability of using payroll taxes as the principal source of funding for these programs.

With respect to the revolution in the life sciences, we should place priority on the development of safeguards against unwise permanent alterations in the human gene pool. Now that we have become the principal agents of evolution, it is crucially important to recognize that the pursuit of short-term goals through human modification can be dangerously inconsistent with the long-term best interests of the human species. As yet, however, we have not developed adequate criteria—much less decision-making protocols—for use in guiding such decisions. We must do so quickly.

Similarly, the dominance of the profit motive and corporate power in decisions about the genetic modification of animals and plants— particularly those that end up in the food supply—are beginning to create unwise risks. Commonsense procedures to analyze these risks according to standards that are based on the protection of the long-term public interest are urgently needed.

The continued advance of technological development will bring many blessings, but human values must be preserved as we evaluate the deployment and use of powerful new technologies. Some advances warrant caution and careful oversight: the proliferation of nanomaterials, synthetic life-forms, and surveillance drones are examples of new technologies rife with promise and potential, but in need of review and safeguards.

There are already several reckless practices that should be immediately stopped: the sale of deadly weapons to groups throughout the world; the use of antibiotics as a livestock growth stimulant; drilling for oil in the vulnerable Arctic Ocean; the dominance of stock market trading by supercomputers with algorithms optimized for high-speed, high-frequency trades that create volatility and risk of market disruptions; and utterly insane proposals for blocking sunlight from reaching the Earth as a strategy to offset the trapping of heat by ever-mounting levels of global warming pollution. All of these represent examples of muddled and dangerous thinking. All should be seen as test cases for whether or not we have the will, determination, and stamina to create a future worthy of the next generations.

Finally, the world community desperately needs leadership that is based on the deepest human values. Though this book is addressed to readers in the world at large, it is intended to carry a special and urgent message to the citizens of the United States of America, which remains the only nation capable of providing the kind of global leadership needed.

For that reason, and for the pride that Americans ought to feel in what the United States has represented to humanity for more than two centuries, it is crucial to halt the degradation and decline of America's commitment to a future in which human dignity is cherished and human values are protected and advanced. Two priority goals for those who wish to take action are limiting the role of money in politics and reforming outdated and obfuscatory legislative rules that allow a small minority to halt legislative action in the U.S. Senate.

Human civilization has reached a fork in the road we have long traveled. One of two paths must be chosen. Both lead us into the unknown. But one leads toward the destruction of the climate balance on which we depend, the depletion of irreplaceable resources that sustain us, the degradation of uniquely human values, and the possibility that civilization as we know it would come to an end. The other leads to the future.

ACKNOWLEDGMENTS

I AM GRATEFUL FOR THE SUPPORT, ENCOURAGEMENT, AND LOVE OF MY partner, Elizabeth Keadle, during the writing of this book, for her advice while reading and listening to successive drafts of every chapter, and for her particular insights into the life sciences chapter. Special thanks also to my brother-in-law, Frank Hunger, whose constant wise counsel and lifelong friendship have been so important to me throughout this project; and to my entire family for their encouragement and support.

This book would not have been possible without my extremely able research team, Brad Hall and Alex Lamballe, whose dedication, diligence, loyalty, and skill are exceptional in every way. I also want to thank their families for their understanding and support during the long hours of work that often spilled over into the weekends and holidays that were a significant part of the time I have devoted to writing this book over the last two years. Their character, good humor, stamina, and grit are impressive and greatly appreciated. For the initial stage of the research, Adam Abelkop was also of invaluable help, and I am especially grateful for his willingness to postpone his doctoral program to be a part of this project. After Adam's sabbatical was over, Dan Myers, on my staff in Nashville, pitched in often, always with a commitment to excellent research.

As I wrote in the Introduction, this book had its origin eight years ago when I began to focus on the drivers of global change and started collecting ideas and research. I regarded the initial detailed outline as primarily a personal exploration of an unusually compelling question, and I was gratified that it also turned out to have practical value as an input to the investing road map that my partners and I used at Generation Investment Management in launching a new initiative in "sustainable investing." I am particularly grateful to my Generation co-founder, David Blood, and all of my other Generation partners for the conversations over the years that have enriched my understanding of so many of these issues.

As I continued to elaborate on the outline, I began to think it might have value to a larger audience, but it was not until Jon Meacham decided to join Random House as a senior editor that I actually set out to write this book. Upon reading that news, I called my agent, Andrew Wylie (to whom I also once again express my gratitude here), and told him why I thought Jon was the perfect editor for this book. The three of us met in New York to discuss the idea, and a week later the project was launched. Upon its completion, I can say without exaggeration that I could not have written it without Jon, who has become a close friend and neighbor in Nashville. His wisdom, insights, and guidance have, unsurprisingly, been truly extraordinary. Thanks also to Gina Centrello, Susan Kamil, Tom Perry, Beck Stvan, Ben Steinberg, London King, Sally Marvin, Steve Messina, Benjamin Dreyer, Erika Greber, Dennis Ambrose, and the entire editorial, production, and marketing team at Random House.

Graham Allison, my close friend and a mentor for forty-five years, organized a two-day scoping exercise at the Belfour Center of the John F. Kennedy School of Government at Harvard after the initial stage of research two years ago. I am extremely grateful to Graham and the extraordinary group of other thinkers who generously took the time (and in many cases traveled long distances) to spend two days in Cambridge in intensive and stimulating discussions of the issues covered in the outline: Rodney Brooks, David Christian, Leon Fuerth, Danny Hillis, Mitch Kapor, Freada Kapor Klein, Ray Kurzweil, Joseph Nye, Dan Schrag, and Fred Spier.

I am also indebted to the distinguished group of expert reviewers who took the time to read parts or all of the first draft of the manuscript. Their assistance in correcting mistakes, suggesting additional material, providing nuance, and aiding my understanding of subjects about which they have forgotten more than I will ever learn is deeply appreciated: Graham Allison, Rosina Bierbaum, Vint Cerf, Bob Corell, Herman Daly, Jared Diamond, Harvey Fineberg, Dargan Frierson, Danny Hillis, Rattan Lal, Mike MacCracken, Dan Schrag, Beth Seidenberg, Laura Tyson, and E. O. Wilson.

In addition, numerous experts generously spent time in lengthy conversations during the research process, including Ragui Assaad, Judy Baker, Thomas Buettner, Andrew Cherlin, Katherine Curtis, Richard Hodes, Paul Kaplowitz, David Owen, Hans Rosling, Saskia Sassen, Annemarie Schneider, Joni Seager, and Audrey Singer.

I sometimes read statements by authors in the acknowledgments sections of their books to the effect that those to whom they are grateful for help bear no responsibility for any mistakes that remain. That sentiment certainly applies to this book.

I also want to thank Maggie Fox, CEO of the Climate Reality Project; Joel Hyatt, my co-founder and CEO of Current TV; and John Doerr, managing partner at Kleiner Perkins Caufield & Byers, along with David Blood at Generation and my colleagues at all four organizations, not only for their support and encouragement, but also for their patience in sometimes adjusting the schedule for calls and meetings to accommodate the time I have taken to work on this book, especially over the past two years.

(Disclosure: in addition to Generation Investment Management, there are 9 other firms, among the 120 mentioned in the text, in which I have a direct or indirect investment: Apple, Auxogyn, Citizens Bank, Coursera, Facebook, Google, JPMorgan Chase, Kaiima, and Twitter.)

Special thanks to Matt Taylor for loaning me a set of very cool gigantic whiteboards for the duration of this project.

Finally, Beth Alpert, chief of staff in my personal office in Nashville, was in overall charge of coordinating the team that helped to produce this book, even as she continued managing my other ongoing activities. Every member of my staff contributed time and effort to making this book possible: Joey Schlichter, Claudia Huskey, Lisa Berg, Betsy McManus, Jill Martin, Kristy Jeffers, Jessica Cox, and, during the early phases of the work, Kalee Kreider, Patrick Hamilton, and Alex Thorpe. And Bill Simmons went way beyond the call of duty in preparing terrific meals during the innumerable working sessions in Nashville throughout this long process. Thank you all!

BIBLIOGRAPHY

BOOKS

Acemoglu, Daron, and James A. Robinson. *Why Nations Fail: The Origins of Power, Prosperity, and Poverty*. New York: Crown Business, 2012.

Anderson, Benedict. *Imagined Communities: Reflections on the Origin and Spread of Nationalism*. New York: Verso, 2006.

Bakan, Joel. *The Corporation: The Pathological Pursuit of Profit and Power*. New York: Free Press, 2004.

Barker, Graeme. *The Agricultural Revolution in Prehistory: Why Did Foragers Become Farmers?* New York: Oxford University Press, 2009.

Beatty, Jack. *The Age of Betrayal: The Triumph of Money in America, 1865–1900*. New York: Vintage Books, 2008.

Brock, David. *The Republican Noise Machine: Right-Wing Media and How It Corrupts Democracy*. New York: Random House, 2005.

Brown, Lester. *Plan B 4.0: Mobilizing to Save Civilization*. New York: Norton, 2009.

———. *Full Planet, Empty Plates: The New Geopolitics of Food Scarcity*. New York: Norton, 2012.

———. *World on the Edge: How to Prevent Environmental and Economic Collapse*. New York: Norton, 2011.

Brzezinski, Zbigniew. *Strategic Vision: America and the Crisis of Global Power*. New York: Basic Books, 2012.

Buchanan, Allen. *Better than Human: The Promise and Perils of Enhancing Ourselves*. New York: Oxford University Press, 2011.

Carr, Nicholas. *The Shallows: What the Internet Is Doing to Our Brains*. New York: Norton, 2010.

Church, George, and Ed Regis. *Regenesis: How Synthetic Biology Will Reinvent Nature and Ourselves*. New York: Basic Books, 2012.

Coll, Steve. *Private Empire: ExxonMobil and American Power*. New York: Penguin Press, 2012.

Coyle, Diane. *The Weightless World: Strategies for Managing the Digital Economy*. Oxford: Capstone, 1997.

Diamond, Jared. *Collapse: How Societies Choose to Fail or Succeed*. New York: Viking, 2005.

———. *Guns, Germs, and Steel: The Fates of Human Societies*. New York: Norton, 1998.

Dobson, Wendy. *Gravity Shift: How Asia's New Economic Powerhouses Will Shape the Twenty-First Century*. Toronto: University of Toronto Press, 2009.

Edsall, Thomas Byrne. *The Age of Austerity: How Scarcity Will Remake American Politics*. New York: Doubleday, 2012.

Ford, Martin. *Lights in the Tunnel: Automation, Accelerating Technology and the Economy of the Future*. N.p.: Acculant, 2009.

Franklin, Daniel, and John Andrews, eds. *Megachange: The World in 2050*. Hoboken, NJ: Wiley, 2012.

Freeman, Walter J. *How Brains Make Up Their Minds*. New York: Columbia University Press, 2000.

Fukuyama, Francis. *The End of History and the Last Man*. New York: Harper Perennial, 1993.

———. *Our Posthuman Future: Consequences of the Biotechnology Revolution*. New York: Farrar, Straus & Giroux, 2002.

Gazzaniga, Michael. *Human: The Science Behind What Makes Us Unique*. New York: HarperCollins, 2008.

Goldstein, Joshua S. *Winning the War on War: The Decline of Armed Conflict Worldwide*. New York: Dutton/Penguin, 2011.

Gore, Al. *The Assault on Reason*. New York: Penguin Press, 2007.

———. *Earth in the Balance: Ecology and the Human Spirit*. Boston: Houghton Mifflin, 1992.

———. *An Inconvenient Truth: The Planetary Emergency of Global Warming and What We Can Do About It*. Emmaus, Pa.: Rodale, 2006.

———. *Our Choice: A Plan to Solve the Climate Crisis*. Emmaus, Pa.: Rodale, 2009.

Hacker, Joseph S., and Paul Pierson. *Winner-Take-All Politics: How Washington Made the Rich Richer—and Turned Its Back on the Middle Class*. New York: Simon & Schuster, 2011.

Haidt, Jonathan. *The Religious Mind: Why Good People Are Divided by Politics and Religion*. New York: Pantheon Books, 2012.

Hansen, James. *Storms of My Grandchildren: The Truth About the Coming Climate Catastrophe and Our Last Chance to Save Humanity*. New York: Bloomsbury USA, 2009.

James, Harold. *The Creation and Destruction of Value: The Globalization Cycle*. Cambridge, MA: Harvard University Press, 2009.

Johnson, Steven. *Emergence: The Connected Lives of Ants, Brains, Cities and Software*. New York: Scribner, 2001.

Jones, Steven E. *Against Technology: From the Luddites to Neo-Luddism*. New York: Routledge, 2006.

Kagan, Robert. *The World America Made*. New York: Knopf, 2012.

Kaku, Michio. *Physics of the Future: How Science Will Shape Human Destiny and Our Daily Lives by the Year 2100*. New York: Doubleday, 2011.

———. *Visions: How Science Will Revolutionize the 21st Century*. New York: Anchor Books, 1997.

Kaplan, Robert D. *The Revenge of Geography: What the Map Tells Us About Coming Conflicts and the Battle Against Fate*. New York: Random House, 2012.

Kelly, Kevin, *What Technology Wants*. New York: Viking, 2010.

Klare, Michael T. *The Race for What's Left: The Global Scramble for the World's Last Resources*. New York: Metropolitan Books, 2012.

Korten, David C. *When Corporations Rule the World*. Bloomfield, CT: Kumarian Press, 1995.

Kupchan, Charles A. *No One's World: The West, the Rising Rest, and the Coming Global Turn*. New York: Oxford University Press, 2012.

Kurzweil, Ray. *The Age of Spiritual Machines: When Computers Exceed Human Intelligence*. New York: Penguin, 1999.

———. *The Singularity Is Near: When Humans Transcend Biology*. New York: Penguin, 2006.

Lanier, Jaron. *You Are Not a Gadget: A Manifesto*. New York: Knopf, 2010.

Lessig, Lawrence. *Republic, Lost: How Money Corrupts Congress—and a Plan to Stop It*. New York: Twelve, 2011.

Lovins, Amory. *Reinventing Fire: Bold Business Solutions for the New Energy Era*. White River Junction, VT: Chelsea Green, 2011.

Luce, Edward. *Time to Start Thinking: America in the Age of Descent*. New York: Atlantic Monthly Press, 2012.

McKibben, Bill. *Eaarth: Making a Life on a Tough New Planet*. New York: Times Books, 2010.

———. *The Global Warming Reader*. New York: Penguin Books, 2012.

McLuhan, Marshall. *The Gutenberg Galaxy: The Making of Typographic Man*. Toronto: University of Toronto Press, 1962.

———. *Understanding Media: The Extensions of Man*. Cambridge, MA: MIT Press, 1994.

Meyer, Christopher, and Stan Davis. *It's Alive: The Coming Convergence of Information, Biology and Business*. New York: Crown Business, 2003.

Moreno, Jonathan D. *The Body Politic: The Battle Over Science in America*. New York: Bellevue Literary Press, 2011.

Morowitz, Harold J. *The Emergence of Everything: How the World Became Complex*. New York: Oxford University Press, 2002.

Moyo, Dambisa. *Winner Take All: China's Race for Resources and What It Means for the World*. New York: Basic Books, 2012.

Naisbitt, John. *Megatrends: Ten New Directions Transforming Our Lives*. New York: Warner Books, 1982.

Nye, Joseph S., Jr. *The Future of Power*. New York: PublicAffairs, 2011.

Olson, Mancur. *The Rise and Decline of Nations: Economic Growth, Stagflation, and Social Rigidities*. New Haven, CT: Yale University Press, 1982.

Otto, Shawn Lawrence. *Fool Me Twice: Fighting the Assault on Science in America*. New York: Rodale, 2011.

Owen, David. *Green Metropolis: Why Living Smaller, Living Closer, and Driving Less Are the Keys to Sustainability*. New York: Riverhead Books, 2009.

Pagel, Mark. *Wired for Culture: Origins of the Human Social Mind*. New York: Norton, 2012.

Polak, Fred. *The Image of the Future*. Amsterdam: Elsevier Scientific, 1973.

Postman, Neil. *Amusing Ourselves to Death*. New York: Viking, 1985.

Reich, Robert. *Aftershock: The Next Economy and America's Future*. New York: Knopf, 2010.

Rifkin, Jeremy. *The Empathic Civilization: The Race to Global Consciousness in a World in Crisis*. New York: Penguin, 2009.

———. *The End of Work: The Decline of the Global Labor Force and the Dawn of the Post-Market Era*. New York: Putnam, 1995.

———. *The Third Industrial Revolution: How Lateral Power Is Transforming Energy, the Economy, and the World*. New York: Palgrave Macmillan, 2011.

Rothkopf, David. *Power, Inc.: The Epic Rivalry Between Big Business and Government—and the Reckoning That Lies Ahead*. New York: Farrar, Straus & Giroux, 2012.

Salk, Jonas. *The Survival of the Wisest*. New York: Harper & Row, 1973.

Sandel, Michael J. *What Money Can't Buy: The Moral Limits of Markets*. New York: Farrar, Straus & Giroux, 2012.

Schor, Juliet B. *The Overworked American: The Unexpected Decline of Leisure*. New York: Basic Books, 1991.

———. *True Wealth: How and Why Millions of Americans Are Creating a Time-Rich, Ecologically Light, Small-Scale, High-Satisfaction Economy*. New York: Penguin Books, 2011.

Seager, Joni. *The Penguin Atlas of Women in the World*. New York: Penguin Books, 2009.

Seung, Sebastian. *Connectome: How the Brain's Wiring Makes Us Who We Are*. Boston: Houghton Mifflin Harcourt, 2012.

Singer, P. W. *Wired for War: The Robotics Revolution and Conflict in the 21st Century*. New York: Penguin Press, 2009.

Singh, Simon. *The Code Book: The Science of Secrecy from Ancient Egypt to Quantum Cryptography*. New York: Doubleday, 1999.

Spence, Michael. *The Next Convergence: The Future of Economic Growth in a Multispeed World*. New York: Farrar, Straus & Giroux, 2011.

Speth, James Gustave. *America the Possible: Manifesto for a New Economy*. New Haven, CT: Yale University Press, 2012.

Stiglitz, Joseph E. *The Price of Inequality: How Today's Divided Society Endangers Our Future*. New York: Norton, 2012.

Teilhard de Chardin, Pierre. *The Future of Man*. New York: Harper & Row, 1964.

———. *The Phenomenon of Man*. New York: Harper, 1959.

Toffler, Alvin. *Future Shock*. New York: Random House, 1970.

Topol, Eric. *The Creative Destruction of Medicine: How the Digital Revolution Will Create Better Health Care*. New York: Basic Books, 2012.

Turkle, Sherry. *Alone Together: Why We Expect More from Technology and Less from Each Other*. New York: Basic Books, 2011.

Vollmann, William T. *Uncentering the Earth: Copernicus and the Revolutions of the Heavenly Spheres*. New York: Norton, 2006.

Washington, Harriet A. *Deadly Monopolies: The Shocking Corporate Takeover of Life Itself—and the Consequences for Your Health and Our Medical Future*. New York: Doubleday, 2011.

Weart, Spencer. *The Discovery of Global Warming*. Cambridge, MA: Harvard University Press, 2003.

Wells, H. G. *World Brain*. London: Ayer, 1938.

Wilson, E. O. *The Social Conquest of Earth*. New York: Liveright, 2012.

Wolfe, Nathan. *The Viral Storm: The Dawn of a New Pandemic Age*. New York: Times Books, 2012.

ARTICLES

Alterman, Jon. "The Revolution Will Not Be Televised." Middle East Notes and Comment, Center for Strategic and International Studies, March 2011.

Archer, David, and Victor Brovkin. "The Millennial Atmospheric Lifetime of Anthropogenic CO_2." *Climatic Change* 90 (2008): 283–97.

Barnosky, Anthony, et al. "Has the Earth's Sixth Mass Extinction Already Arrived?" *Nature*, March 2011.

Bartlett, Bruce. "'Starve the Beast': Origins and Development of a Budgetary Metaphor." *Independent Review*, Summer 2007.

Bergsten, C. Fred. "Two's Company." *Foreign Affairs*, September/October 2009.

Bisson, Peter, Elizabeth Stephenson, and S. Patrick Viguerie. "The Global Grid." *McKinsey Quarterly*, July 26, 2011.

Blaser, Martin. "Antibiotic Overuse: Stop the Killing of Beneficial Bacteria." *Nature*, August 25, 2011.

Bohannon, John. "Searching for the Google Effect on People's Memory." *Science*, July 15, 2011.

Bostrom, Nick. "A History of Transhumanist Thought." *Journal of Evolution and Technology* 14 (April 2005).

Bowden, Mark. "The Measured Man." *Atlantic*, July/August 2012.

Bowley, Graham. "The New Speed of Money, Reshaping Markets." *New York Times*, January 2, 2011.

Bradford, James. "The NSA Is Building the Country's Biggest Spy Center (Watch What You Say)." *Wired*, March 15, 2012.

Carmody, Tim. "Google Co-founder: China, Apple, Facebook Threaten the 'Open Web.'" *Wired*, April 16, 2012.

Caruso, Denise. "Synthetic Biology: An Overview and Recommendations for Anticipating and Addressing Emerging Risks." *Science Progress*, November 12, 2008, http://scienceprogress.org/2008/11/synthetic-biology/.

Caryl, Christian. "Predators and Robots at War." *New York Review of Books*, August 30, 2011.

Cookson, Clive. "Synthetic Life." *Financial Times*, July 27, 2012.

Council on Foreign Relations. "The New North American Energy Paradigm: Reshaping the Future." June 27, 2012.

Cudahy, Brian J. "The Containership Revolution: Malcolm McLean's 1956 Innovation Goes Global." *Transportation Research News*, Transportation Research Board of the National Academies, no. 246 (September/October 2006).

Day, Peter. "Will 3D Printing Revolutionise Manufacturing?" BBC, July 27, 2011.

Diamond, Jared. "What Makes Countries Rich or Poor?" *New York Review of Books*, June 7, 2012.

Diamond, Larry. "A Fourth Wave or False Start?" *Foreign Affairs*, May 22, 2011.

———. "Liberation Technology." *Journal of Democracy* 21, no. 3 (July 2010).

Dunbar, R.I.M. (1993). "Coevolution of Neocortical Size, Group Size and Language in Humans." *Behavioral and Brain Sciences* 16, no. 4 (1993): 681–735.

Economist. "The Dating Game." December 27, 2011.

———. "Hello America." August 16, 2010.

———. "How Luther Went Viral." December 17, 2011.

———. "No Easy Fix." February 24, 2011.

———. "The Printed World." February 10, 2011.

———. "The Third Industrial Revolution." April 21, 2012.

———. "Unbottled Gini." January 20, 2011.

Etling, Bruce, Robert Faris, and John Palfrey. "Political Change in the Digital Age: The Fragility and Promise of Online Organizing." *SAIS Review* 30, no. 2 (2010).

Evans, Dave. "The Internet of Things." Cisco Blog, July 15, 2011.

Farrell, Henry, and Cosma Shalizi. "Cognitive Democracy." *Three-Toed Sloth*, May 23, 2012.

Feldstein, Martin. "China's Biggest Problems Are Political, Not Economic." *Wall Street Journal*, August 2, 2012.

Fernandez-Cornejo, J., and M. Caswell. "The First Decade of Genetically Engineered Crops in the United States." U.S. Department of Agriculture, Economic Research Service, 2006.

Financial Times. "Job-Devouring Technology Confronts US Workers." December 15, 2011.

Fineberg, Harvey. "Are We Ready for Neo-Evolution?" TED Talks, 2011.

Fishman, Ted. "As Populations Age, a Chance for Younger Nations." *New York Times Magazine*, October 17, 2010.

Fortey, Richard A. "Charles Lyell and Deep Time." *Geoscientist* 21, no. 9 (October 2011).

Fox, Justin. "What the Founding Fathers Really Thought About Corporations." *Harvard Business Review*, April 1, 2010.

Freeman, David. "The Perfected Self." *Atlantic*, June 2012.

Generation Investment Management. "Sustainable Capitalism." February 15, 2012. http://www.generationim.com/media/pdf-generation-sustainable-capitalism-v1.pdf.

Gillis, Justin. "Are We Nearing a Planetary Boundary." *New York Times*, June 6, 2012.

———. "A Warming Planet Struggles to Feed Itself." *New York Times*, June 6, 2011.

Gladwell, Malcolm. "Small Change: Why the Revolution Will Not Be Tweeted." *New Yorker*, October 4, 2010.

———. "The Tweaker." *New Yorker*, November 14, 2011.

Grantham, Jeremy. "Time to Wake Up: Days of Abundant Resources and Falling Prices Are Over Forever." *GMO Quarterly Letter*, April 2011.

Gross, Michael Joseph. "Enter the Cyber-Dragon." *Vanity Fair*, September 2011.

———. "World War 3.0." *Vanity Fair*, May 2012.

Haidt, Jonathan. "Born This Way? Nature, Nurture, Narratives, and the Making of Our Political Personalities." *Reason*, May 2012.

Hansen, James, et al. "Perception of Climate Change." *Proceedings of the National Academy of Sciences*, August 2012.

Harb, Zahera. "Arab Revolutions and the Social Media Effect." M/C Journal [Media/ Culture Journal] 14, no. 2 (2011).

Hillis, Danny. "Understanding Cancer Through Proteomics." TEDMED 2010, October 2010.

Huntington, Samuel P. "The U.S.—Decline or Renewal?" *Foreign Affairs*, Winter 1988/1989.

Ikenson, Daniel J. "Made on Earth: How Global Economic Integration Renders Trade Policy Obsolete." Cato Trade Policy Analysis No. 42, December 2, 2009.

International Monetary Fund. World Economic Outlook. September 2011.

Joffe, Josef. "Declinism's Fifth Wave." *American Interest*, January/February 2012.

Johnson, Toni. "Food Price Volatility and Insecurity." Council on Foreign Relations, August 9, 2011.

Kagan, Robert. "Not Fade Away." *New Republic*, January 11, 2012.

Kaufman, Edward E., Jr., and Carl M. Levin. "Preventing the Next Flash Crash." *New York Times*, May 6, 2011.

Keim, Brandon. "Nanosecond Trading Could Make Markets Go Haywire." *Wired*, February 16, 2012.

Kennedy, Pagan. "The Cyborg in Us All." *New York Times Magazine*, September 18, 2011.

Kleiner, Keith. "Designer Babies—Like It or Not, Here They Come." Singularity Hub, February 25, 2009.

Kristof, Nicolas D. "America's 'Primal Scream.'" *New York Times*, October 15, 2011.

Krugman, Paul. "We Are the 99.9%." *New York Times*, November 24, 2011.

Kuznetsov, V. G. "Importance of Charles Lyell's Works for the Formation of Scientific Geological Ideology." *Lithology and Mineral Resources* 46, no. 2 (2011): 186–97.

Lavelle, Marianne. "The Climate Change Lobby Explosion." Center for Public Integrity, February 24, 2009.

Levinson, Marc. "Container Shipping and the Economy." *Transportation Research News*, Transportation Research Board of the National Academies, no. 246 (September/October 2006).

Lewis, Mark. "The History of the Future." *Forbes*, October 15, 2007.

MacKenzie, Donald. "How to Make Money in Microseconds." *London Review of Books*, May 19, 2011.

MacKinnon, Rebecca. "Internet Freedom Starts at Home." *Foreign Policy*, April 3, 2012.

Macklem, Peter T. "Emergent Phenomena and the Secrets of Life." *Journal of Applied Physiology* 104 (2008): 1844–46.

Madrigal, Alexis. "I'm Being Followed: How Google—and 104 Other Companies—Are Tracking Me on the Web." *Atlantic*, February 29, 2012.

Markoff, John. "Armies of Expensive Lawyers, Replaced by Cheaper Software." *New York Times*, March 5, 2011.

———. "Cost of Gene Sequencing Falls, Raising Hopes for Medical Advances." *New York Times*, March 8, 2012.

———. "Google Cars Drive Themselves, in Traffic." *New York Times*, October 10, 2010.

McKibben, Bill. "Global Warming's Terrifying New Math." *Rolling Stone*, July 2012.

Milojević, Ivana. "A Selective History of Futures Thinking." Ph.D. diss., University of Queensland, 2002.

Mooney, Chris. "The Science of Why We Don't Believe Science." *Mother Jones*, June 2011.

Moore, Stephen, and Julian L. Simon. "The Greatest Century That Ever Was: 25 Miraculous Trends of the Past 100 Years." Cato Policy Analysis No. 364, December 15, 1999.

New York Times. "Dow Falls 1,000, Then Rebounds, Shaking Market." May 7, 2010.

Nisbet, Robert. "The Idea of Progress." *Literature of Liberty: A Review of Contemporary Liberal Thought* 2, no. 1 (1979).

Noah, Timothy. "Introducing the Great Divergence." *Slate*, September 3, 2010.

———. "Think Cranks." *New Republic*, March 30, 2012.

Nye, Joseph S. "Cyber War and Peace." Project Syndicate, April 10, 2012.

Organisation for Economic Co-operation and Development. "Divided We Stand: Why Inequality Keeps Rising." December 2011.

Peters, Glen, et al. "Rapid Growth in CO_2 Emissions After the 2008–2009 Global Financial Crisis." *Nature Climate Change*, 2011.

Purdum, Todd. "One Nation, Under Arms." *Vanity Fair*, January 2012.

Rosen, Jeffrey. "POTUS v. SCOTUS." *New York*, March 17, 2010.

Salvaris, Mike. "The Idea of Progress in History: Future Directions in Measuring Australia's Progress." Australian Bureau of Statistics, 2010.

Sargent, John F., Jr. "Nanotechnology: A Policy Primer." Congressional Research Service, April 13, 2012.

Speth, James Gustave. "America the Possible: A Manifesto, Part I." *Orion*, March/April 2012.

Steiner, Christopher. "Wall Street's Speed War." *Forbes*, September 27, 2010.

Steinhart, Eric. "Teilhard de Chardin and Transhumanism." *Journal of Evolution and Technology* 20, no. 1 (December 2008).

Stern, Nicholas. "The Economics of Climate Change: The Stern Review." *Population and Development Review* 32 (December 2006).

Stiglitz, Joseph E. "Of the 1%, by the 1%, for the 1%." *Vanity Fair*, May 2011.

Trenberth, Kevin. "Changes in Precipitation with Climate Change." *Climate Research* 47 (2010).

Trivett, Vincent. "25 US Mega Corporations: Where They Rank If They Were Countries." *Business Insider*, June 27, 2011.

Vance, Ashlee. "3-D Printing Spurs a Manufacturing Revolution." *New York Times*, September 14, 2010.

Walt, Steven M. "The End of the American Era." *National Interest*, October 25, 2011.

Wilford, John Noble. "Who Began Writing? Many Theories, Few Answers." *New York Times*, April 6, 1999.

Wilson, Daniel H. "Bionic Brains and Beyond." *Wall Street Journal*, June 1, 2012.

Wilson, E. O. "Why Humans, Like Ants, Need a Tribe." *Daily Beast*, April 1, 2012.

Worstall, Tim. "Six Waltons Have More Wealth than the Bottom 30% of Americans." *Forbes*, December 14, 2011.

Zhang, David, and Harry Lee. "The Causality Analysis of Climate Change and Large-Scale Human Crisis." *Proceedings of the National Academy of Sciences* 108 (March 2011): 17296–301.

Zimmer, Carl. "Tending the Body's Microbial Garden." *New York Times*, June 18, 2012.

NOTES

INTRODUCTION

xvi **most important issues they expected to emerge**
Peter Lindström, *The Future Agenda as Seen by the Committees and Subcommittees of the United States House of Representatives: A Workbook for Participatory Democracy* (Washington, DC: Congressional Clearinghouse on the Future and the Congressional Institute for the Future, 1982).

xvi **born in Russia a few months before the 1917 Revolution**
The Nobel Prize in Chemistry, 1977: Ilya Prigogine, "Autobiography," http://www.nobelprize.org/nobel_prizes/chemistry/laureates/1977/prigogine-autobio.html#.

xvi **educated in Belgium**
Ibid.

xvi **responsible for irreversibility in nature**
Peter T. Macklem, "Emergent Phenomena and the Secrets of Life," *Journal of Applied Physiology* 104 (2008): 1844–46; Ray Kurzweil, *The Age of Spiritual Machines: When Computers Exceed Human Intelligence* (New York: Penguin, 1999).

xvi **donut with clearly defined boundaries**
Macklem, "Emergent Phenomena and the Secrets of Life."

xvii **at a higher level of complexity**
Ibid.

xvii **stabilized underground for millions of years**
Farrington Daniels, "A Limitless Resource: Solar Energy," *New York Times*, March 18, 1956.

xviii **"which made so manifest the 'arrow of time'"**
Prigogine, "Autobiography."

xviii **"they are understood in different ways in different societies"**
Ivana Milojević, "A Selective History of Futures Thinking," from "Futures of Education: Feminist and Post-Western Critiques and Visions" (Ph.D. diss., University of Queensland, 2002).

xviii **trying to divine the future with the help of oracles or mediums**
Ibid.

xviii **animals sacrificed to the gods**
Ibid.

xviii **by studying the movements of fish**
Ibid.

xviii **marks on the Earth**
Ibid.

xix **each molecule of the gaseous cylinders onto which**
Tracy V. Wilson, "How Holograms Work," HowStuffWorks, http://science
.howstuffworks.com/hologram.htm.

xix **astrologers of ancient Babylon used a double clock**
Fred Polak, *The Image of the Future* (Amsterdam: Elsevier Scientific, 1973), p. 5.

xix **will still linger there—still trapping heat**
Daniel Schrag, personal interview.

xx **"need to enlarge knowledge, through the development of sciences and arts"**
Mike Salvaris, "The Idea of Progress in History: Future Directions in Measuring
Australia's Progress," Australian Bureau of Statistics, 2010.

xx **"earthly to heavenly things, and from the visible to the invisible"**
Robert Nisbet, "The Idea of Progress," *Literature of Liberty: A Review of Contemporary Liberal Thought* 2, no. 1 (1979).

xx **China as a guide for those who wish to progress**
Peter Hubral, "The Tao: Modern Pathway to Ancient Wisdom," *Philosopher* 98,
no. 1 (2010), http://www.the-philosopher.co.uk/taowisdom.htm; Abu al-Hasan Ali
ibn al-Husayn al-Mas'udi, "How Do We Come Upon New Ideas?," *First Break* 29
(March 2011).

xx **"true scientific progress and hence actual achievement"**
Salvaris, "The Idea of Progress in History."

xx **contributed to a fascination with the physical as well as the philosophical legacies**
Polak, *The Image of the Future*, pp. 82–95.

xxi **In the seventeenth century, the father of microbiology**
Jonathan Janson, "Antonie van Leeuwenhoek (1632–1723)," Essential Vermeer,
http://www.essentialvermeer.com/dutch-painters/dutch_art/leeuwenhoek.html.

xxi **been invented in Holland less than a century earlier**
Nobel Media, "Microscopes: Time Line," http://www.nobelprize.org/educational/
physics/microscopes/timeline/index.html.

xxi **through them discovered cells and bacteria**
Ibid.

xxi **camera obscura, made possible by the new understanding of optics**
In addition to their friendship and possible artistic collaboration, Van Leeuwenhoek was also the executor of Vermeer's will. Jonathan Janson, "Vermeer and the
Camera Obscura," Essential Vermeer, http://www.essentialvermeer.com/camera
_obscura/co_one.html; Philip Steadman, "Vermeer and the Camera Obscura,"
BBC History, February 17, 2011, http://www.bbc.co.uk/history/british/empire
_seapower/vermeer_camera_01.shtml.

xxi **"the steady step of amelioration, and will in time, I trust, disappear from the earth"**
Thomas Jefferson, "To William Ludlow," September 6, 1824, *The Portable Thomas
Jefferson* (New York: Penguin, 1977), p. 583.

xxii but Lyell amply proved that the Earth was not thousands
Richard A. Fortey, "Charles Lyell and Deep Time," *Geoscientist* 21, no. 9 (October 2011); V. G. Kuznetsov, "Importance of Charles Lyell's Works for the Formation of Scientific Geological Ideology," *Lithology and Mineral Resources* 46, no. 2 (2011): 186–97; Mark Lewis, "The History of the Future," *Forbes*, October 15, 2007.

xxii 4.5 billion, we now know
"History of Life on Earth," BBC Nature, http://www.bbc.co.uk/nature/history_of _the_earth.

xxii Darwin took Lyell's books with him during his voyage on the *Beagle*
Fortey, "Charles Lyell and Deep Time"; Kuznetsov, "Importance of Charles Lyell's Works for the Formation of Scientific Geological Ideology."

xxiii Aristotle wrote that the end of a thing defines its essential nature
Aristotle, *Eudemian Ethics*, Book 2, Section 1219a.

xxiv More than a decade before writing *Faust*
Scott Horton, "The Sorcerer's Apprentice," *Harper's*, December 2007; Cyrus Hamlin, "Faust in Performance: Peter Stein's Production of Goethe's Faust, Parts 1 & 2," *Theatre* 32 (2002).

xxiv emergent wisdom and creativity that is on a completely different plane
Henry Farrell and Cosma Shalizi, "Cognitive Democracy," *Three-Toed Sloth*, May 23, 2012, http://masi.cscs.lsa.umich.edu/~crshalizi/weblog/917.html.

xxvii Alfred North Whitehead called the obsession with measurements "the fallacy of misplaced concreteness"
Polak, *The Image of the Future*, p. 196.

xxx 4.5 billion years ago would be at the far left end
"History of Life on Earth," BBC Nature.

xxx emergence of life 3.8 billion years ago
Ibid.

xxx multicellular life 2.8 billion years ago
Ibid.

xxx land 475 million years ago
Ibid.

xxx first vertebrates more than 400 million years ago
Ibid.

xxx primates 65 million years ago
Blythe A. Williams, Richard F. Kay, and E. Christopher Kirk, "New Perspectives on Anthropoid Origins," *Proceedings of the National Academy of Sciences*, March 8, 2010.

xxx would appear 7.5 billion years from now
David Appell, "The Sun Will Eventually Engulf Earth—Maybe," *Scientific American*, September 8, 2008.

CHAPTER 1: EARTH INC.

5 efficacy, utility, and power with each passing year
Martin Ford, *Lights in the Tunnel: Automation, Accelerating Technology and the Economy of the Future* (N.p.: Acculant, 2009).

7 one million new robots within two years
"Foxconn to Replace Workers with 1 Million Robots in 3 Years," *Xinhuanet*, July 30, 2011, http://news.xinhuanet.com/english2010/china/2011-07/30/c_131018764.htm.

8 New companies have emerged to connect online workers with jobs
Quentin Hardy, "The Global Arbitrage of Online Work," *New York Times*, October 10, 2012.

8 Narrative Science, a robot reporting company
Joe Fassler, "Can the Computers at Narrative Science Replace Paid Writers?," *Atlantic*, April 12, 2012.

9 Latin America is the rare exception
Jonathan Watts, "Latin America's Income Inequality Falling, Says World Bank," *Guardian*, November 13, 2012.

9 income inequality reached a twenty-year high
Natasha Lennard, "Global Inequality Highest in 20 Years," *Salon*, November 1, 2012, http://www.salon.com/2012/11/01/global_inequality_highest_in_20_years/.

9 risen in the United States from 35 to 45
"Unbottled Gini," *Economist*, January 20, 2011; *CIA World Factbook*, https://www.cia.gov/library/publications/the-world-factbook/fields/2172.html, accessed January 20, 2012.

9 China from 30 to the low 40s
Data Set, University of Texas Inequality Project, Estimated Household Income Inequality Data Set (EHII). This is a global data set, derived from the econometric relationship between UTIP-UNIDO, other conditioning variables, and the World Bank's Deininger & Squire data set (http://utip.gov.utexas.edu/data.html); World Data Bank, http://data.worldbank.org/indicator/SI.POV.GINI.

9 Russia from the mid 20s to the low 40s
Ibid.

9 United Kingdom from 30 to 36
Data Set, University of Texas Inequality Project, Estimated Household Income Inequality Data Set; "Growing Income Inequality in OECD Countries: What Drives It and How Can Policy Tackle It?," OECD Forum on Tackling Inequality, May 2, 2011, http://www.oecd.org/dataoecd/32/20/47723414.pdf.

9 make compared to six times just two decades ago
"India Income Inequality Doubles in 20 Years, Says OECD," BBC, December 7, 2011, http://www.bbc.co.uk/news/world-asia-india-16064321.

9 investment income at the lowest tax rate of all—15 percent
Joseph E. Stiglitz, "Of the 1%, by the 1%, for the 1%," *Vanity Fair,* May 2011.

9 capital gains income goes to the top one thousandth of one percent
Paul Krugman, "We Are the 99.9%," *New York Times*, November 24, 2011.

9 now has more inequality than either Egypt or Tunisia
Nicholas D. Kristof, "America's 'Primal Scream,'" *New York Times*, October 15, 2011.

10 have more wealth than the people in the bottom 90 percent
Ibid.

10 the 150 million Americans in the bottom 50 percent
Ibid.

10 have more wealth than the bottom 30 percent of Americans
Tim Worstall, "Six Waltons Have More Wealth Than the Bottom 30% of Americans," *Forbes*, December 14, 2011.

10 up from 12 percent just a quarter century ago
Stiglitz, "Of the 1%, by the 1%, for the 1%."

10 the top 0.1 percent increased over the same period by 400 percent
Krugman, "We Are the 99.9%."

10 from 5 to 40 percent of the GDP in developed countries
UnctadStat, Statistical Database for the United Nations Conference on Trade and Development, http://unctadstat.unctad.org/ReportFolders/reportFolders.aspx.

10 capital flows are expected to continue increasing three times faster than GDP
International Monetary Fund, World Economic Outlook, September 2011, http://www.imf.org/external/pubs/ft/weo/2011/02/weodata/WEOSep2011alla.xls; Peter Bisson, Elizabeth Stephenson, and S. Patrick Viguerie, "The Global Grid," *McKinsey Quarterly*, July 26, 2011.

10 from 5 to 30 percent of GDP from 1980 to 2011
UnctadStat, Statistical Database for the United Nations Conference on Trade and Development.

10 United States, paying them wages that are 20 percent higher
Daniel J. Ikenson, "Made on Earth: How Global Economic Integration Renders Trade Policy Obsolete," Cato Trade Policy Analysis No. 42, December 2, 2009, http://www.cato.org/pubs/tpa/tpa-042.pdf.

10 for more than five million U.S. citizens
Ibid.

10 China of processed polysilicon and advanced manufacturing equipment
Steven Mufson, "China's Growing Share of Solar Market Comes at a Price," *Washington Post*, December 16, 2011.

11 largest economy in the world within this decade
Brett Arends, "IMF Bombshell: Age of America Nears End," *Market Watch*, April 25, 2011, http://www.marketwatch.com/story/imf-bombshell-age-of-america-about-to-end-2011-04-25.

11 It now has twice the number of Internet users
"The Dating Game," *Economist*, December 27, 2011, http://www.economist.com/blogs/dailychart/2010/12/save_date; "Survey: China has 513 million Internet users," CBS News, January 15, 2012, http://www.cbsnews.com/8301-205_162-57359546/survey-china-has-513-million-internet-users/; Internet World Stats, "Top 20 Countries with the Highest Number of Internet Users," August 7, 2011, http://www.internetworldstats.com/top20.htm.

11 productivity growth has been higher than in any decade since the 1960s
"Job-Devouring Technology Confronts US Workers," *Financial Times*, December 15, 2011.

11 rates of increase while unemployment has barely declined
"U.S. Tax Haul Trails Profit Surge," *Wall Street Journal*, January 4, 2012.

11 spending on private sector jobs increased by only 2 percent
Catherine Rampell, "Companies Spend on Equipment, Not Workers," *New York Times*, June 10, 2011.

11 new industrial robots in North America increased 41 percent
Robotic Industries Association, "North American Robot Orders Jump 41% in First Half of 2011," July 29, 2011, http://www.robotics.org/content-detail.cfm/Industrial-Robotics-News/North-American-Robot-Orders-Jump-41-in-First-Half-of-2011/content_id/2922.

11 GDP of advanced economies for the first time in the modern era
"Special Report: Developing World to Overtake Advanced Economies in 2013," *Euromonitor*, February 19, 2009, http://blog.euromonitor.com/2009/02/special-report-developing-world-to-overtake-advanced-economies-in-2013-.html.

12 growing much faster than the developed countries
Mark Mobius, "Emerging Markets May See More Capital Flow, Away from Assets and Currencies of Countries Burdened with High Debt," *Economic Times*, September 27, 2011.

12 Some analysts doubt the sustainability of these growth rates
Ruchir Sharma, "Broken BRICs: Why the Rest Stopped Rising," *Foreign Affairs*, November/December 2012.

12 wealthy investors to encourage them to build more factories in the West
Don Lee, "U.S. Jobs Continue to Flow Overseas," *Los Angeles Times*, October 6, 2010.

12 resulted in the loss of 27 million jobs worldwide
United Nations Department of Economic and Social Affairs, *The Report on the World Social Situation: The Global Social Crisis*, 2011, http://social.un.org/index/LinkClick.aspx?fileticket=cO3JAiiX-NE%3D&tabid=1562.

13 notional value twenty-three times larger than the entire global GDP
"Why Derivatives Caused Financial Crisis," Seeking Alpha, April 12, 2010, http://seekingalpha.com/article/198197-why-derivatives-caused-financial-crisis.

13 daily trades in all of the world's stock markets put together
Ibid.

13 thirteen times larger than the combined value of every stock and every bond on Earth
Ibid.

13 it represented more than 60 percent of all trades
Nathaniel Popper, "High-Speed Trading No Longer Hurtling Forward," *New York Times*, October 14, 2012.

13 ability to complete a transaction in 124 microseconds
Donald MacKenzie, "How to Make Money in Microseconds," *London Review of Books*, May 19, 2011.

13 according to some experts will further increase
Ibid.

13 "financial market of which we have virtually no sound theoretical understanding"
Brandon Keim, "Nanosecond Trading Could Make Markets Go Haywire," *Wired*, February 16, 2012.

13 "mystery algorithm"
John Melloy, "Mysterious Algorithm Was 4% of Trading Activity Last Week," CNBC, October 8, 2012, http://www.cnbc.com/id/49333454/Mysterious_Algorithm_Was_4_of_Trading_Activity_Last_Week.

13 enabling them to make a fortune by shorting French bonds
Christopher Steiner, "Wall Street's Speed War," *Forbes*, September 27, 2010.

13 make a similar fortune by shorting bonds from the Confederacy
Ibid.

14 to transmit information over the 825 miles
Ibid.

14 what many economists call the financialization of the economy
Ibid.

14 1980 to more than 8 percent at present
Thomas Philippon, "The Future of the Financial Industry," Stern on Finance blog, October 16, 2008, http://sternfinance.blogspot.com/2008/10/future-of -financial-industry-thomas.html.

15 6 percent are based on credit derivatives
"America's Big Bank $244 Trillion Derivatives Market Exposed," Seeking Alpha, September 15, 2011, http://seekingalpha.com/article/293830-america-s-big-bank -244-trillion-derivatives-market-exposed.

15 value of actual commodities
Ibid.

15 fourteen times the value of all the actual barrels of oil traded on that same day
Roderick Bruce, "Making Markets: Oil Derivatives: In the Beginning," Energyrisk.com, p. 31, July 2009, http://db.riskwaters.com/data/energyrisk/ EnergyRisk/Energyrisk_0709/markets.pdf.

15 because banks hold collateral equal to a large percentage
But see Mazen Labban, "Oil in Parallax: Scarcity, Markets, and the Financializa- tion of Accumulation," *Geoforum* 41 (2010): 546 ("Although financial derivatives allowed investors and traders to manage risk and hedge against the volatility of fi- nancial markets, the 'aggregate impact' of trading in derivatives was to increase risk and contribute to the volatility of the market"), citing Adam Tickell, "Unstable Futures: Controlling and Creating Risks in International Money," *Global Capital- ism Versus Democracy*, edited by Leo Panitch and Colin Leys (New York: Monthly Review Press, 1999), pp. 248–77; Adam Tickell, "Dangerous Derivatives: Control- ling and Creating Risks in International Money," *Geoforum* 31 (2000): 87–99.

15 implicitly reflected in the collective behavior found in the market (it isn't)
Peter J. Boettke, "Where Did Economics Go Wrong? Modern Economics as a Flight from Reality," *Critical Review* 11, no. 1 (1997): 11–64; Al Gore and David Blood, "A Manifesto for Sustainable Capitalism," *Wall Street Journal*, December 14, 2011.

15 Joseph Stiglitz says that high-speed trading produces only "fake liquidity"
Personal conversation with Joseph Stiglitz.

15 combined total of all of the reserves in the central banks
Morris Miller, "Global Governance to Address the Crises of Debt, Poverty and Environment," background paper prepared for the 42nd Pugwash Conference, Berlin, Germany, September 1992, http://www.management.uottawa.ca/miller/ governa.htm.

16 one another's simultaneous operations rather than underlying market realities
Donald MacKenzie, "How to Make Money in Microseconds," *London Review of Books*, May 19, 2011.

16 all in the time span of sixteen minutes—for no apparent reason
"Dow Falls 1,000, Then Rebounds, Shaking Market," *New York Times*, May 7, 2010.

16 **"P&G plunged to $39.37 from more than $60 within minutes"**
Ibid.

16 **"it was almost like 'the Twilight Zone'"**
Ibid.

16 **algorithmic echo chamber that caused prices to suddenly crash**
Graham Bowley, "The New Speed of Money, Reshaping Markets," *New York Times*, January 2, 2011; Felix Salmon and Jon Stokes, "Algorithms Take Control of Wall Street," *Wired*, December 27, 2010.

16 **offers to buy or sell must remain open for *one second***
Personal conversation with Joseph Stiglitz.

16 **would bring the global economy to its knees**
Ibid.

17 **corrupted and captive ratings agencies, then sold around the world**
"'Robo-Signing' of Mortgages Still a Problem," Associated Press, July 18, 2011, http://www.cbsnews.com/stories/2011/07/18/national/main20080533.shtml.

17 **a practice that's been popularly labeled "robosigning"**
Alan Zibel, Matthias Rieker, and Nick Timiraos, "Banks Near 'Robo-Signing' Settlement," *Wall Street Journal*, January 19, 2012.

17 **increasing since 2000 at an average of 65 percent per year**
Mark Jickling and Rena S. Miller, "Derivatives Regulation in the 111th Congress," Congressional Research Service Report for Congress, March 3, 2011, Table I, http://assets.opencrs.com/rpts/R40646_20110303.pdf.

17 **and campaign contributions to prevent them from being regulated**
"Why Derivatives Caused Financial Crisis," Seeking Alpha, April 12, 2010, http://seekingalpha.com/article/198197-why-derivatives-caused-financial-crisis.

18 **are continuing to grow at a rate half again faster than global production**
Organisation for Economic Co-operation and Development, "Divided We Stand."

18 **others that linked Europe to the New World and to Asia**
Ronald Findlay and Kevin H. O'Rourke, "Commodity Market Integration, 1500–2000," in *Globalization in Historical Perspective*, edited by Michael D. Bordo, Alan M. Taylor, and Jeffrey G. Williamson (Chicago: University of Chicago Press, 2003).

18 **Middle East, trade flows that were largely controlled by Venice and Egypt**
Ibid.

19 **Europe and Africa—revolutionized the old pattern**
Ibid.

19 **nineteenth century prior to the First Opium War, which began in 1839**
"Hello America," *Economist*, August 16, 2010 (citing Angus Maddison).

19 **Then, when the East gained more access to the new technologies**
Derek Thompson, "The Economic History of the Last 2,000 Years in 1 Little Graph," *Atlantic*, June 19, 2012.

20 **"make macro inventions highly productive and remunerative"**
Malcolm Gladwell, "The Tweaker," *New Yorker*, November 14, 2011.

22 **"conflict between that interest and any other, that other should yield"**
Wayne D. Rasmussen, U.S. Department of Agriculture National Agricultural Li-

brary, "Lincoln's Agricultural Legacy," January 30, 2012, http://www.nal.usda.gov/lincolns-agricultural-legacy.

22 1789 to a little under 60 percent
U.S. Department of Agriculture, "A History of American Agriculture: Farmers & the Land," Agriculture in the Classroom, http://www.agclassroom.org/gan/timeline/farmers_land.htm.

22 to establish colleges of agriculture and the mechanical arts
Rasmussen, "Lincoln's Agricultural Legacy."

22 Every state did so
U.S. Department of Agriculture, "A History of American Agriculture."

23 every one of the 3,000 counties in the United States
Representative Butler Derrick, *Congressional Record* 140, no. 138 (September 28, 1994).

23 global production of eggs has increased by 350 percent
United Nations Food and Agriculture Organization, *World Livestock 2011*, http://www.fao.org/docrep/014/i2373e/i2373e.pdf.

23 with 70 million tons annually—four times the production of the United States
Ibid.

23 has increased over the same period by more than 3,200 percent
Ibid.

23 very day that the first space satellite, Sputnik, was launched by the Soviet Union
Brian J. Cudahy, "The Containership Revolution: Malcolm McLean's 1956 Innovation Goes Global," *Transportation Research News,* Transportation Research Board of the National Academies, no. 246 (September/October 2006): 5–9, http://onlinepubs.trb.org/onlinepubs/trnews/trnews246.pdf.

24 will carry goods from one country to another
Ibid.; Marc Levinson, "Container Shipping and the Economy," *Transportation Research News,* Transportation Research Board of the National Academies, no. 246 (September/October 2006): 10, http://onlinepubs.trb.org/onlinepubs/trnews/trnews246.pdf.

24 now in surplus supply (much as food grains were a few decades ago)
"Plunging Prices Set to Trigger Tech Boom," *Financial Times,* January 8, 2012; "TV Prices Fall, Squeezing Most Makers and Sellers," *New York Times,* December 26, 2011.

24 in today's dollars, would be $8,000
Richard Powelson, "First Color Television Sets Were Sold 50 Years Ago," Scripps Howard News Service, December 31, 2003, http://www.post-gazette.com/tv/20031231colortv1231p3.asp.

24 133 percent, even as jobs have decreased by 33 percent
Energy Information Administration, Annual Energy Review, October 19, 2011, http://www.eia.gov/totalenergy/data/annual/xls/stb0702.xls; Mine Safety and Health Administration, Table 3, "Average Number of Employees at Coal Mines in the United States, by Primary Activity, 1978–2008," http://www.msha.gov/STATS/PART50/WQ/1978/wq78cl03.asp.

24 increased significantly over much of that period
John E. Tilton and Hans H. Landsberg, September 1997, "Innovation, Produc-

tivity Growth, and the Survival of the U.S. Copper Industry," Resources for the Future, http://www.rff.org/RFF/Documents/RFF-DP-97-41.pdf.

25 **number of hours of labor required to produce a ton of copper fell by 50 percent**
Ibid.

25 **labor productivity in one of its largest mines by 400 percent**
Ibid.

25 **New sources of copper were developed in other countries**
Matthijs Randsdorp, "A Closer Look at Copper," November 3, 2011, TCW, https://www.tcw.com/News_and_Commentary/Market_Commentary/Insights/11 -03-11_A_Closer_Look_at_Copper.aspx.

25 **by 500 first-year associates**
John Markoff, "Armies of Expensive Lawyers, Replaced by Cheaper Software," *New York Times*, March 5, 2011.

25 **300,000 miles in all driving conditions without an accident**
Rebecca J. Rosen, "Google's Self-Driving Cars: 300,000 Miles Logged, Not a Single Accident Under Computer Control," *Atlantic*, August 9, 2012.

25 **employed in the United States alone as taxi drivers and chauffeurs**
U.S. Bureau of Labor Statistics, as cited in the *Statistical Abstract of the United States: 2010*, Table 640, http://www.census.gov/compendia/statab/.

27 **in part for cultural reasons—to go into savings instead of consumption**
Mauricio Cardenas, "Lower Savings in China Could Slow Down Growth in Latin America," Brookings Institution, February 11, 2011, http://www.brookings.edu/ research/opinions/2011/02/11-china-savings-cardenas-frank.

27 **developed through the much older technologies of metallurgy and ceramics**
Caltech Materials Science, "Welcome," 2012, http://www.matsci.caltech.edu/.

27 **"physical powers which will enable it to super-organize matter"**
Eric Steinhart, "Teilhard de Chardin and Transhumanism," *Journal of Evolution and Technology* 20, no. 1 (December 2008): 1–22.

28 **the molecular economy**
Christopher Meyer and Stan Davis, *It's Alive: The Coming Convergence of Information, Biology and Business* (New York: Crown Business, 2003), p. 4.

28 **experiments in the real world**
Ibid., pp. 3–6, 66–67.

28 **molecules when they are clustered in bulk**
John F. Sargent Jr., "Nanotechnology: A Policy Primer," Congressional Research Service, April 13, 2012, http://www.fas.org/sgp/crs/misc/RL34511.pdf.

28 **resistance to stains, wrinkles, and fire**
Ibid.

28 **hospitals guarding against infections**
"Nanotech-Enabled Consumer Products Continue to Rise," *ScienceDaily*, March 13, 2011, http://www.sciencedaily.com/releases/2011/03/110310101351 .htm.

28 **copper emerged in numerous locations in the same era**
Miljana Radivojevića et al., "On the Origins of Extractive Metallurgy: New Evidence from Europe," *Journal of Archaeological Science*, November 2010.

29 **combines high temperatures and some pressurization**
Richard Cowen, "Chapter 5: The Age of Iron," April 1999, http://mygeologypage
.ucdavis.edu/cowen/~GEL115/115CH5.html.

29 **more than 1,000 years later in Britain**
"Bronze Age," *Encyclopaedia Britannica*, http://www.britannica.com/EBchecked/
topic/81017/Bronze-Age.

29 **4,500 years ago in northern Turkey**
Cowen, "Chapter 5: The Age of Iron."

29 **from which it could be made into tools and weapons**
Ibid.

29 **harder and stronger than bronze**
Ibid.

29 **not made until the middle of the nineteenth century**
Ibid.

29 **create an entirely new category of products, including**
Jeremy Rifkin, *The Third Industrial Revolution: How Lateral Power Is Transforming
Energy, the Economy, and the World* (New York: Palgrave Macmillan, 2011).

29 **store energy and manifest previously unimaginable properties**
Pulickel M. Ajayan and Otto Z. Zhou, "Applications of Carbon Nanotubes," *Topics in Applied Physics* 80 (2001): 391–425; Eliza Strickland, "9 Ways Carbon Nanotubes Just Might Rock the World," *Discover Magazine*, August 6, 2009.

29 **already replacing steel in some niche applications**
Corie Lok, "Nanotechnology: Small Wonders," *Nature*, September 1, 2010, pp.
18–21.

29 **expected to have wide applications in industry**
Dmitri Kopeliovich, "Ceramic Matrix Composites (Introduction)," SubsTech,
http://www.substech.com/dokuwiki/doku.php?id=ceramic_matrix_composites
_introduction.

29 **already known processes, mostly in the health and fitness category**
"Nanotech-Enabled Consumer Products Continue to Rise," *ScienceDaily*,
March 13, 2011, http://www.sciencedaily.com/releases/2011/03/110310101351
.htm; Sargent, "Nanotechnology: A Policy Primer"; Project on Emerging Nanotechnologies, 2012, http://www.fas.org/sgp/crs/misc/RL34511.pdf.

30 **which opens a variety of useful applications**
A. K. Geim, "Graphene: Status and Prospects," *Science* 324, no. 5934 (June 19,
2009): 1530–34; Matthew Finnegan, "Graphene Nanoribbons Could Extend
Moore's Law by 10 Years," Techeye.com, September 28, 2011; "Adding Hydrogen
Triples Transistor Performance in Graphene," *ScienceDaily*, September 4, 2011.

30 **much debate in the first years of the twenty-first century**
Robert F. Service, "Nanotechnology Grows Up," *Science* 304, no. 5678 (June 18,
2004): 1732–34.

30 **consequent cell damage—are taken more seriously**
Ibid.

30 **"nothing about their synergistic impacts"**
Ibid.

30 certainly since the discovery of the double helix in 1953
National Research Council, Nanotechnology in Food Products: Workshop Summary (Leslie Pray and Ann Yaktine, rapporteurs, 2009).

30 application of nanotechnology to the development of new materials
Ibid.

30 fibers with 100 times the strength and one sixth the weight of steel
Lok, "Nanotechnology: Small Wonders," pp. 18–21.

30 until the object is formed in three-dimensional space
"The Printed World: Three-Dimensional Printing from Digital Designs Will Transform Manufacturing and Allow More People to Start Making Things," Economist, February 10, 2011.

31 different kind of material can be used
Ibid.

31 Model T, manufacturing has been dominated by mass production
"The Third Industrial Revolution," Economist, April 21, 2012; Peter Day, "Will 3D Printing Revolutionise Manufacturing?," BBC, July 27, 2011, http://www.bbc.co.uk/news/business-14282091.

31 manufacturing as profoundly as mass production did
"The Third Industrial Revolution," Economist; Day, "Will 3D Printing Revolutionise Manufacturing?"

31 later produce en masse in more traditional processes
Day, "Will 3D Printing Revolutionise Manufacturing?"; Neil Gershenfeld, "How to Make Almost Anything," Foreign Affairs, September 27, 2012.

31 prototyped as 3D models for wind tunnel testing
"The Printed World," Economist.

31 builds $2,000 models and completes them overnight
Ashlee Vance, "3-D Printing Spurs a Manufacturing Revolution," New York Times, September 14, 2010.

31 the expense of employing large numbers of people
Day, "Will 3D Printing Revolutionise Manufacturing?"; "The Third Industrial Revolution," Economist.

31 material that is used in the mass production process
"The Printed World," Economist; Jeremy Rifkin, "The Third Industrial Revolution: How the Internet, Green Electricity, and 3-D Printing Are Ushering in a Sustainable Era of Distributed Capitalism," Huffington Post, March 28, 2012, http://www.huffingtonpost.com/jeremy-rifkin/the-third-industrial-revo_1_b_1386430.html.

31 not to mention a small fraction of the energy costs
"The Printed World," Economist; Rifkin, "The Third Industrial Revolution."

31 even as their value has increased more than threefold
Diane Coyle, introduction to The Weightless World: Strategies for Managing the Digital Economy (Oxford: Capstone, 1997).

31 unsatisfactory for many kinds of specialized products
"The Printed World," Economist.

31 delivery of parts to the factory and finished products to distant markets
Day, "Will 3D Printing Revolutionise Manufacturing?"

32 **each product to widely dispersed 3D printers**
"The Third Industrial Revolution," *Economist;* Gershenfeld, "How to Make Almost Anything."

32 **"warehouses waiting to be printed locally when required"**
Day, "Will 3D Printing Revolutionise Manufacturing?"

32 **prints an entire house in only twenty hours**
Vance, "3-D Printing Spurs a Manufacturing Revolution"; Behrokh Khoshnevis, TEDx Conference presentation, February 2012.

32 **in some cases, 1,000 items**
"The Printed World," *Economist.*

32 **turning out hundreds of thousands of identical parts and products**
Ibid.

32 **do not have protection against replication under "useful" copyright laws**
Michael Weinberg, "The DIY Copyright Revolution," *Slate*, February 23, 2012, http://www.slate.com/articles/technology/future_tense/2012/02/_3_d_printing_copyright_and_intellectual_property_.html; "The Third Industrial Revolution," *Economist;* Peter Marsh, "Made to Measure," *Financial Times*, September 7, 2012.

32 **United States, China, and Europe are working hard to exploit its potential**
"The Printed World," *Economist.*

32 **printing prosthetics and other devices with medical applications**
Vance, "3-D Printing Spurs a Manufacturing Revolution"; "Transplant Jaw Made by 3D Printer Claimed as First," BBC News, February 6, 2012, http://www.bbc.co.uk/news/technology-16907104; "Engineers Pioneer Use of 3D Printer to Create New Bones," BBC News, November 30, 2011, http://www.bbc.com/news/technology-15963467; Joann Pan, "3D Printer Creates 'Magic Arms' for Two-Year-Old Girl," Mashable, August 3, 2012, http://mashable.com/2012/08/03/3d-printed-magic-arms/; "Artificial Blood Vessels Created on a 3D Printer," BBC News, September 16, 2011, http://www.bbc.co.uk/news/technology-14946808.

32 **Inexpensive 3D printers have already found their way**
Vance, "3-D Printing Spurs a Manufacturing Revolution."

32 **"Something seismic is going on."**
Bob Parks, "Creation Engine: Autodesk Wants to Help Anyone, Anywhere, Make Anything," *Wired*, September 21, 2012.

32 **advocates of more widespread gun ownership**
"3D Printers Could 'Print Ammunition for an Army,'" *Dezeen Magazine*, October 3, 2012, http://www.dezeen.com/2012/10/03/3d-printers-could-print-ammunition-for-an-army/.

33 **guns used in crimes could be easily melted down**
Nick Bilton, "Disruptions: With a 3-D Printer, Building a Gun with the Push of a Button," *New York Times*, October 7, 2012.

33 **some of the jobs they had originally outsourced to low-wage countries**
"The Third Industrial Revolution," *Economist.*

33 **less willing than their global counterparts to endorse either conclusion**
Boston Consulting Group, press release, "Nearly a Third of Companies Say Sustainability Is Contributing to Their Profits, Says MIT Sloan Management

Review–Boston Consulting Group Report," January 24, 2012, http://www.bcg
.com/media/PressReleaseDetails.aspx?id=tcm:12-96246.

34 **"higher-income individuals consume, as a fraction of their income"**
Joseph E. Stiglitz, "The 1 Percent's Problem," *Vanity Fair,* May 31, 2012.

34 **in "extreme poverty"—defined as having an income less than $1.25 per day**
World Bank, World Development Indicators, 2010 annual report, http://data
.worldbank.org/sites/default/files/wdi-final.pdf.

34 **every twenty-four hours into the planet's atmosphere**
Drew Shindell, phone interview with author, September 1, 2009.

35 **average holding period for stocks**
James Montier, *Behavioural Investing: A Practitioner's Guide to Applying Behavioural Finance* (Chichester, UK: Wiley, 2007), p. 277.

35 **over a business cycle and a half, roughly seven years**
Richard Dobbs, Keith Leslie, and Lenny T. Mendonca, "Building the Healthy Corporation," *McKinsey Quarterly,* August 2005; Roger A. Morin and Sherry L. Jarrell, *Driving Shareholder Value: Value-Building Techniques for Creating Shareholder Wealth* (New York: McGraw-Hill, 2001), p. 56; Roland J. Burgman, David J. Adams, David A. Light, and Joshua B. Bellin, "The Future Is Now," *MIT Sloan Management Review,* October 26, 2007.

35 **holding period for stocks is less than seven months**
Henry Blodget, "You're an Investor? How Quaint," *Business Insider,* August 8, 2009, http://www.businessinsider.com/henry-blodget-youre-an-investor-how-quaint -2009-8.

36 **"is expensive and painstaking and offers far less potential for speedy returns"**
Jon Gertner, "Does America Need Manufacturing?," *New York Times Magazine,* August 28, 2011.

36 **Eighty percent said no**
Tilde Herrera, "BSR 2011: Al Gore Says Short-Term Thinking Is 'Functionally Insane,'" GreenBiz, November 2, 2011, http://www.greenbiz.com/blog/2011/11/ 02/bsr-2011-al-gore-says-short-term-thinking-functionally-insane.

37 **almost 200 millennia**
Sileshi Semaw et al., "2.6-Million-Year-Old Stone Tools and Associated Bones from OGS-6 and OGS-7, Gona, Afar, Ethiopia," *Journal of Human Evolution* 45 (2003): 169–77.

37 **took less than eight millennia**
Graeme Barker, *The Agricultural Revolution in Prehistory: Why Did Foragers Become Farmers?* (New York: Oxford University Press, 2009), p. v ("Ten thousand years ago there were few if any societies which can properly be described as agricultural. Five thousand years ago large numbers of the world's population were farmers. . . .").

37 **from 90 to 2 percent of the workforce**
Ibid.; Claude Fischer, "Can You Compete with A.I. for the Next Job?," *Fiscal Times,* April 14, 2011; Carolyn Dimitri, Anne Effland, and Neilson Conklin, Economic Research Service, U.S. Department of Agriculture, "The 20th Century Transformation of U.S. Agriculture and Farm Policy," June 2005, http://www.ers .usda.gov/publications/eib3/eib3.htm; United Nations Social Policy and Development Division, *Report on the World Social Situation 2007: The Employment Imperative,*

2007, http://www.un.org/esa/socdev/rwss/docs/2007/chapter1.pdf ("Agriculture still accounts for about 45 per cent of the world's labour force, or about 1.3 billion people").

37 **less than half of all jobs worldwide are now on farms**
United Nations Social Policy and Development Division, *Report on the World Social Situation 2007: The Employment Imperative.*

38 **the Industrial Revolution took only 150 years**
Barker, *The Agricultural Revolution in Prehistory*, p. v.

38 **"indistinguishable from magic"**
"Clarke's Third Law," in *Brave New Words: The Oxford Dictionary of Science Fiction*, edited by Jeff Prucher (New York: Oxford University Press, 2007), p. 22.

38 **different from that of our ancestors 200,000 years ago**
"Human Brains Enjoy Ongoing Evolution," *New Scientist*, September 9, 2005.

38 **in the world thirty years ago, the Cray-2**
John Markoff, "The iPad in Your Hand: As Fast as a Supercomputer of Yore," *New York Times*, May 9, 2011.

39 **jobs of weavers obsolete**
Steven E. Jones, *Against Technology: From the Luddites to Neo-Luddism* (New York: Routledge, 2006), pp. 54–55.

39 **"Luddite fallacy"**
Ford, *Lights in the Tunnel*, pp. 95–100.

40 **technologies as "extensions" of basic human capacities**
Marshall McLuhan, *Understanding Media: The Extensions of Man* (Cambridge, MA: MIT Press, 1994).

CHAPTER 2: THE GLOBAL MIND

45 **to serve primarily as a distribution service for advertisements and junk mail**
Steven Greenhouse, "Postal Service Is Nearing Default as Losses Mount," *New York Times*, September 5, 2011.

45 **phenomena-driven by the connection of two billion people (thus far) to the Internet**
International Telecommunication Union, "The World in 2011: ICT Facts and Figures," 2011, http://www.itu.int/ITU-D/ict/facts/2011/material/ICTFactsFigures 2011.pdf.

45 **with no human being involved—already exceeds the population of the Earth**
Dave Evans, "The Internet of Things," Cisco Blog, July 15, 2011, http://blogs.cisco.com/news/the-internet-of-things-infographic/.

45 **connected to the Internet and exchanging information on a continuous basis**
Jessi Hempel, "The Hot Tech Gig of 2022: Data Scientist," *Fortune*, January 6, 2012; Evans, "The Internet of Things."

45 **the number of "connected things" is already much larger**
Maisie Ramsay, "Cisco: 1 Trillion Connected Devices by 2013," *Wireless Week*, March 25, 2010.

45 **RFID tags in an effort to combat truancy**
David Rosen, "Big Brother Invades Our Classrooms," *Salon*, October 8, 2012, http://www.salon.com/2012/10/08/big_brother_invades_our_classrooms/.

45 "The round globe is a vast brain, instinct with intelligence"
Nathaniel Hawthorne, *The House of the Seven Gables* (Boston: Ticknor, Reed, & Fields, 1851), p. 283.

45 "where knowledge and ideas are received, sorted"
H. G. Wells, *World Brain* (London: Ayer, 1938).

45 World Wide Web on Google for some of the estimated one trillion web pages
Jesse Alpert and Nissan Hajaj, "We Knew the Web Was Big . . . ," Google Official Blog, July 25, 2008, http://googleblog.blogspot.com/2008/07/we-knew-web-was-big.html.

46 network of human thoughts that he termed the "Global Mind"
Pierre Teilhard de Chardin, *The Future of Man* (1964), chap. 7, "The Planetisation of Man."

46 "We shape our tools, and thereafter, our tools shape us"
McLuhan, *Understanding Media*.

47 "a very complex organism that often follows its own urges"
Kevin Kelly, *What Technology Wants* (New York: Penguin, 2010).

47 we are spending more and more time "alone together"
Sherry Turkle, *Alone Together: Why We Expect More from Technology and Less from Each Other* (New York: Basic Books, 2011); Robert Kraut et al., "Internet Paradox: A Social Technology That Reduces Social Involvement and Psychological Well-Being?," *American Psychologist* 53, no. 9 (September 1998): 1017–31; Stephen Marche, "Is Facebook Making Us Lonely?," *Atlantic*, May 2012.

47 "Internet Use Disorder" in its appendix for the first time
Tony Dokupil, "Is the Web Driving Us Mad?," *Daily Beast*, July 8, 2012.

47 estimated 500 million people
Jane McGonigal, "Video Games: An Hour a Day Is Key to Success in Life," *Huffington Post*, February 15, 2012, http://www.huffingtonpost.com/jane-mcgonigal/video-games_b_823208.html.

47 as much time playing online games as they spend in classrooms
Ibid.

47 the average online social games player
Mathew Ingram, "Average Social Gamer Is a 43-Year-Old Woman," GigaOM, February 17, 2010, http://gigaom.com/2010/02/17/average-social-gamer-is-a-43-year-old-woman/.

47 55 percent of those playing social games
Ibid.

47 generate 60 percent of the comments and post 70 percent of the pictures on Facebook
Robert Lane Greene, "Facebook: Like?," Intelligent Life, May/June 2012, http://moreintelligentlife.com/content/ideas/robert-lane-greene/facebook?page=full.

47 to the amount of time we are spending online
Nicholas Carr, *The Shallows: What the Internet Is Doing to Our Brains* (New York: Norton, 2010).

48 control group not informed that the facts could be found online
John Bohannon, "Searching for the Google Effect on People's Memory," *Science*, July 15, 2011.

48 **began to lose some of their innate sense of direction**
Alex Hutchinson, "Global Impositioning Systems," *Walrus*, November 2009.

48 **studies indicate that it is a literal reallocation of mental energy**
Carr, *The Shallows*.

48 **"Never memorize what you can look up in books"**
Library of Congress, World Treasures of the Library of Congress, July 29, 2010, http://www.loc.gov/exhibits/world/world-record.html.

48 **the disuse of neuron "trees" leads to their shrinkage**
Walter J. Freeman, *How Brains Make Up Their Minds* (New York: Columbia University Press, 2000), pp. 37–43, 81–82; Society for Neuroscience, "Brain Plasticity and Alzheimer's Disease," 2010, http://web.archive.org/web/20101225174414/http://sfn.org/index.aspx?pagename=publications_rd_alzheimers.j.

48 **connecting our brains seamlessly to the enhanced capacity**
McLuhan, *Understanding Media*.

48 **"calling things to remembrance"**
Plato, *Plato's Phaedrus*, translated by Reginald Hackforth (Cambridge, UK: Cambridge University Press, 1972), p. 157.

48n **TCP/IP protocol**
Kleinrock Internet History Center at UCLA, "The IMP Log: October 1969 to April 1970," September 21, 2011, http://internethistory.ucla.edu/2011/09/imp-log-october-1969-to-april-1970.html; Jim Horne, "What Hath God Wrought," *New York Times*, Wordplay blog, September 8, 2009, http://wordplay.blogs.nytimes.com/2009/09/08/wrought/; George P. Oslin, *The Story of Telecommunications* (Macon, GA: Mercer University Press, 1999), pp. 2, 219.

49 **and the less one relies on memories stored in the brain itself**
Carr, *The Shallows*, pp. 191–97.

49 **life-forms on Earth is our capacity for complex and abstract thought**
Michael S. Gazzaniga, *Human: The Science Behind What Makes Us Unique* (New York: HarperCollins, 2008), p. 199.

49 **neocortex in roughly its modern form around 200,000 years ago**
R.I.M. Dunbar, "Coevolution of Neocortical Size, Group Size and Language in Humans," *Behavioral and Brain Sciences* 16, no. 4 (1993): 681–735.

49 **with a genetic mutation or whether it developed more gradually**
Constance Holden, "The Origin of Speech," *Science* 303, no. 5662 (February 27, 2004): 1316–19.

49 **to communicate more intricate thoughts from one person to others**
John Noble Wilford, "Who Began Writing? Many Theories, Few Answers," *New York Times*, April 6, 1999.

49 **hunter-gatherer period is associated with oral communication**
Nicholas Wade, "Phonetic Clues Hint Language Is Africa-Born," *New York Times*, April 14, 2011.

49 **language is associated with the early stages of the Agricultural Revolution**
Wilford, "Who Began Writing?"

49 **Mesopotamia, Egypt, China and India, the Mediterranean, and Central America**
William J. Duiker and Jackson J. Spielvogel, *World History*, 6th ed., vol. 1 (Boston: Wadsworth/Cengage Learning, 2010), p. 43.

49 the emergence of sophisticated concepts like democracy
Carr, *The Shallows*, pp. 50–57.

50 Their relative powerlessness was driven by their ignorance
Marshall McLuhan, *The Gutenberg Galaxy* (Toronto: University of Toronto Press, 1962).

50 written in a language that for the most part only the monks could understand
Burnett Hillman Streeter, *The Chained Library: A Survey of Four Centuries in the Evolution of the English Library* (New York: Cambridge University Press, 2011).

50 eleven print editions of the account of his journey captivated Europe
"The Diffusion of Columbus's Letter through Europe, 1493–1497," University of Southern Maine, Osher Map Library, http://usm.maine.edu/maps/web-document/1/5/sub-/5-the-diffusion-of-columbuss-letter-through-europe-1493-1497.

50 bringing artifacts and knowledge
Laurence Bergreen, *Over the Edge of the World: Magellan's Terrifying Circumnavigation of the Globe* (New York: William Morrow, 2004).

51 including the exciting new derivatives product: indulgences
Hans J. Hillerbrand, *The Protestant Reformation*, rev. ed. (New York: HarperCollins, 2009), pp. ix–xiii, 66–67.

51 but thousands of copies distributed to the public were printed in German
"How Luther Went Viral," *Economist*, December 17, 2011.

51 more than a quarter of them written by Luther himself
Ibid.

51 beginning a wave of literacy that began in Northern Europe and moved southward
Tom Head, *It's Your World, So Change It: Using the Power of the Internet to Create Social Change* (Indianapolis, IN: Que, 2010), p. 115.

51 the printing press was denounced as "the work of the Devil"
Charles Coffin, *The Story of Liberty* (New York: Harper & Brothers, 1879), p. 77.

51 with the publication of Nicolaus Copernicus's *Revolution of the Spheres*
William T. Vollmann, *Uncentering the Earth: Copernicus and the Revolutions of the Heavenly Spheres* (New York: Norton, 2006).

52 At the beginning of January 1776
"Jan 9, 1776: Thomas Paine Publishes Common Sense," History.com, http://www.history.com/this-day-in-history/thomas-paine-publishes-common-sense.

52 ignite the American War of Independence that July
David McCullough, *1776* (New York: Simon & Schuster, 2005), p. 112.

52 codified by Adam Smith in the same year
Adam Smith, *An Inquiry into the Nature and Causes of the Wealth of Nations* (London, 1776).

52 *Decline and Fall of the Roman Empire* was also published in the same year
Edward Gibbon, *The History of the Decline and Fall of the Roman Empire* (London, 1776).

52 a counterpoint to the prevailing exhilaration about the future
T. H. Breen, "Making History," *New York Times Book Review*, May 7, 2000.

53 quantum computing
Michio Kaku, *Physics of the Future: How Science Will Shape Human Destiny and Our Daily Lives by the Year 2100* (New York: Doubleday, 2011), Chapter 1.

53 **digital data by companies and individuals**
McKinsey Global Institute, *Big Data: The Next Frontier for Innovation, Competition, and Productivity*, May 2011.

53 **grown by a factor of nine in just five years**
"The 2011 Digital Universe Study: Extracting Value from Chaos," IDC, June 2011, http://idcdocserv.com/1142.

53 **telephone call grew shorter by almost half**
Tom Vanderbilt, "The Call of the Future," *Wilson Quarterly*, Spring 2012.

53 **double between 2005 and 2010**
International Telecommunications Union, "The World in 2010: ICT Facts and Figures," http://www.itu.int/ITU-D/ict/material/FactsFigures2010.pdf.

53 **in 2012 reached 2.4 billion users globally**
Mary Meeker and Liang Wu, "2012 Internet Trends (Update)," December 3, 2012, http://kpcb.com/insights/2012-internet-trends-update.

53 **as many mobile devices as there are people**
Cisco Systems, Inc., Cisco Visual Networking Index: Global Mobile Data Traffic Forecast Update, 2010–2015, February 1, 2011, http://newsroom.cisco.com/ekits/Cisco_VNI_Global_Mobile_Data_Traffic_Forecast_2010_2015.pdf.

53 **Internet users is expected to increase 56-fold over the next five years**
Ibid.

53 **smartphones is projected to increase 47-fold over the same period**
Ibid.

53 **half of the mobile phone market in the United States**
Aaron Smith, Pew Internet & American Life Project, "Nearly Half of American Adults Are Smartphone Owners," March 1, 2012, http://pewinternet.org/Reports/2012/Smartphone-Update-2012.aspx.

53 **More than 5 billion of the 7 billion**
International Telecommunications Union, "ICT Facts and Figures: The World in 2011."

54 **1.1 billion active smartphone users worldwide**
Meeker and Wu, "2012 Internet Trends (Update)."

54 **3.2 billion people have their own devices**
"SIM Earth," *Economist*, October 19, 2012, http://www.economist.com/blogs/babbage/2012/10/global-mobile-usage.

54 **low-end smartphones that will soon be nearly ubiquitous**
Christina Bonnington, Wired Gadget Lab, "Global Smartphone Adoption Approaches 30 Percent," November 28, 2011, http://www.wired.com/gadgetlab/2011/11/smartphones-feature-phones/; Juro Osawa and Paul Mozur, "The Battle for China's Low-End Smartphone Market," *Wall Street Journal*, June 22, 2012.

54 **Internet access as a new "human right" in a United Nations report**
David Kravets, "U.N. Report Declares Internet Access a Human Right," *Wired*, June 3, 2011.

54 **computer or tablet to every child in the world who does not have one**
"Nicholas Negroponte and One Laptop Per Child," Public Radio International, April 29, 2009, http://www.pri.org/stories/business/social-entrepreneurs/one-laptop-per-child.html.

54 subsidized the connection of every school and library to the Internet
Austan Goolsbee and Jonathan Guryan, "World Wide Wonder?," *Education Next* 6, no. 1 (Winter 2006).

54 immediately upon awakening—even before they get out of bed
Kevin J. O'Brien, "Top 1% of Mobile Users Consume Half of World's Bandwidth, and Gap Is Growing," *New York Times*, January 5, 2012.

54 simultaneously trying to operate their cars and trucks
Matt Richtel, "U.S. Safety Board Urges Cellphone Ban for Drivers," *New York Times*, December 13, 2011.

54 before the distracted pilots finally disengaged from their computers
Micheline Maynard and Matthew L. Wald, "Off-Course Pilots Cite Computer Distraction," *New York Times*, October 27, 2009.

55 "'FaceTime Facelift' effect"
Jason Gilbert, "FaceTime Facelift: The Plastic Surgery Procedure for iPhone Users Who Don't Like How They Look on FaceTime," *Huffington Post*, February 27, 2012.

55 "Internet of Everything."
Dave Evans, "How the Internet of Everything Will Change the World . . . for the Better," Cisco Blog, November 7, 2012, http://blogs.cisco.com/news/how-the-internet-of-everything-will-change-the-worldfor-the-better-infographic/.

55 voluminous new quantities of data
McKinsey Institute, "Big Data: The Next Frontier for Innovation, Competition, and Productivity," May 2011, http://www.mckinsey.com/Insights/MGI/Research/Technology_and_Innovation/Big_data_The_next_frontier_for_innovation.

55 without being processed by computers for patterns and meaning
Al Gore, "The Digital Earth: Understanding Our Planet in the 21st Century," speech at the California Science Center, January 31, 1998, http://portal.opengeospatial.org/files/?artifact_id=6210&version=1&format=doc.

55 actuators has been disposed of soon after it is collected
Michael Chui, Markus Löffler, and Roger Roberts, "The Internet of Things," *McKinsey Quarterly*, 2010.

55 promote efficiency in industry and business
McKinsey Institute, "Big data."

55 information collected during the seconds prior to and during
"In-Car Camera Records Accidents," BBC News, October 14, 2005, http://news.bbc.co.uk/2/hi/uk_news/england/southern_counties/4341342.stm.

55 airplanes and most security cameras in buildings
Kevin Bonsor, "How Black Boxes Work," HowStuffWorks, http://science.howstuffworks.com/transport/flight/modern/black-box3.htm.

56 twice the amount of information presently generated
Tony Hoffman, "IBM Preps Hyper-Fast Computing System for World's Largest Radiotelescope," *PC Magazine*, April 2, 2012.

56 billions of messages posted each day on social networks
Chui, Löffler, and Roberts, "The Internet of Things"; McKinsey Institute, "Big Data."

56 **Twitter Earthquake Detector**
Tim Lohman, "Twitter to Detect Earthquakes, Tsunamis," *Computer World*, June 1, 2011.

56 **the Global Pulse program Ban Ki-moon launched**
Steve Lohr, "The Internet Gets Physical," *New York Times*, December 17, 2011.

56 **predict social unrest in countries and regions of particular interest**
John Markoff, "Government Aims to Build a 'Data Eye in the Sky,'" *New York Times*, October 10, 2011.

56 **predict how well Hollywood—and Bollywood—movies will perform**
Ibid.

57 **dominant content on the Internet is printed words**
Roger E. Bohn and James E. Short, "How Much Information? 2009 Report on American Consumers," December 2009, http://hmi.ucsd.edu/pdf/HMI_2009 _ConsumerReport_Dec9_2009.pdf.

57 **massive crowds of election protesters in Moscow**
Alissa de Carbonnel, "Social Media Makes Anti-Putin Protests 'Snowball,'" Reuters, December 7, 2011.

57 **spotlighting the excesses of elites**
Thomas Friedman, "This Is Just the Start," *New York Times*, March 1, 2011.

57 **used by rebels in Misrata to guide their mortars**
Tom Coghlan, "Google and a Notebook: The Weapons Helping to Beat Gaddafi in Libya," *Times* (London), June 16, 2011.

58 **across the border to collaborators in the diaspora living in Thailand**
Mridul Chowdhury, Berkman Center for Internet & Society, "The Role of the Internet in Burma's Saffron Revolution," September 2008, http://cyber.law.harvard .edu/sites/cyber.law.harvard.edu/files/Chowdhury_Role_of_the_Internet_in _Burmas_Saffron_Revolution.pdf_0.pdf.

58 **completely blacking out the Internet inside the country's borders**
Ibid.

58 **Aung San Suu Kyi, from her long house arrest**
Tim Johnson, "Aung San Suu Kyi Freed," *Financial Times*, November 13, 2010.

58 **destined to take control of the government**
Dean Nelson, "Aung San Suu Kyi 'Wins Landslide Landmark Election' as Burma Rejoices," *Telegraph*, April 1, 2012.

58 **protest against the fraudulent presidential election**
Bruce Etling, Robert Faris, and John Palfrey, "Political Change in the Digital Age: The Fragility and Promise of Online Organizing," *SAIS Review* 30, no. 2 (2010).

58 **controlling Internet use by the protesters**
Ibid.

59 **the tragic death of Neda Agha-Soltan**
Ibid.

59 **protest movement were almost completely shut down**
Ibid.

59 **the government simply blacked it out**
Ibid.

59 stifle any effective resistance to the dictatorship's authority
Will Heaven, "Iran and Twitter: The Fatal Folly of the Online Revolutionaries,"
Telegraph, December 29, 2009; Christopher Williams, "Iran Cracks Down on Web
Dissident Technology," *Telegraph*, March 18, 2011.

59 Iran and the retro-Stalinist dictatorship of Belarus
Larry Diamond, "Liberation Technology," *Journal of Democracy* 21, no. 3
(July 2010).

59 turn the Internet within China into a national intranet
Ibid.

59 was censored and made unavailable to the people of China
Josh Chin, "Netizens React: Premier's Interview Censored," China Real Time
Report blog, *Wall Street Journal*, October 7, 2010.

59 open values of the world's largest search engine, Google
Clive Thompson, "Google's China Problem (and China's Google Problem)," *New
York Times Magazine*, April 23, 2006.

60 "in certain areas the genie *has* been put back in the bottle"
Tim Carmody, "Google Co-Founder: China, Apple, Facebook Threaten the
'Open Web,'" *Wired*, April 16, 2012.

60 "It's hopeless to try to control the Internet"
Ian Katz, "Web Freedom Faces Greatest Threat Ever, Warns Google's Sergey
Brin," *Guardian*, April 15, 2012.

60 more than 500 million people, 40 percent of its total population
Matt Silverman, "China: The World's Largest Online Population," Mashable,
April 10, 2012; Jon Russell, "Internet Usage in China Surges 11%," *USA Today*,
July 19, 2012.

60 to take to the Internet themselves in order to respond to public controversies
Lye Liang Fook and Yang Yi, EAI Background Brief No. 467, "The Chinese Leadership and the Internet," July 27, 2009, http://www.eai.nus.edu.sg/BB467.pdf.

60 Dmitri Medvedev also felt the pressure to engage personally on the Internet
"Medvedev Believes Internet Best Guarantee Against Totalitarianism," Itar-Tass
News Agency, July 30, 2012, http://www.itar-tass.com/en/c154/484098.html.

60 four out of every ten Tunisians were connected to the Internet
Zahera Harb, "Arab Revolutions and the Social Media Effect," *M/C Journal*
[Media/Culture Journal] 14, no. 2 (2011).

60 with almost 20 percent of them on Facebook
Ibid.

60 80 percent of the Facebook users were under the age of thirty
Ibid.

60 as censoring political dissent on the Internet
Reporters without Borders, "Enemies of the Internet," March 12, 2010, http://en
.rsf.org/IMG/pdf/Internet_enemies.pdf.

60 It was the downloaded video that ignited the Arab Spring
John D. Sutter, "How Smartphones Make Us Superhuman," CNN, September 10, 2012.

60 **In Saudi Arabia, Twitter has facilitated public criticism**
Robert F. Worth, "Twitter Gives Saudi Arabia a Revolution of Its Own," *New York Times*, October 20, 2012.

61 **feisty and relatively independent satellite television channel Al Jazeera**
Jon Alterman, "The Revolution Will Not Be Televised," Middle East Notes and Comment, Center for Strategic and International Studies, March 2011; Heidi Lane, "The Arab Spring's Three Foundations," *per Concordiam*, March 2012.

61 **even in countries where they are technically illegal**
Angelika Mendes, "Media in Arab Countries Lack Transparency, Diversity and Independence," Konrad-Adenauer-Stiftung, June 25, 2012, http://www.kas.de/wf/en/33.31742/; Lin Noueihed and Alex Warren, *The Battle for the Arab Spring: Revolution, Counter-Revolution and the Making of a New Era* (New Haven, CT: Yale University Press, 2012), p. 50; Lane, "The Arab Spring's Three Foundations."

61 **the Internet had spread throughout Egypt and the region**
Harb, "Arab Revolutions and the Social Media Effect"; Alterman, "The Revolution Will Not Be Televised."

61 **Al Jazeera and its many siblings were the more important factor**
Alterman, "The Revolution Will Not Be Televised."

61 **"All that trouble from this little matchbox?"**
"Special Report: Al Jazeera's News Revolution," Reuters, February 17, 2011.

61 **shut down access to the Internet in the way Myanmar and Iran had**
Harb, "Arab Revolutions and the Social Media Effect."

61 **the public's reaction was so strong that the fires of revolt grew even hotter**
Ibid.

61 **including Malcolm Gladwell**
Malcolm Gladwell, "Small Change: Why the Revolution Will Not Be Tweeted," *New Yorker*, October 4, 2010.

61 **actually represented a tiny fraction of Egypt's huge population**
Noah Shachtman, "How Many People Are in Tahrir Square? Here's How to Tell," Wired Danger Room blog, February 1, 2011, http://www.wired.com/dangerroom/2011/02/how-many-people-are-in-tahrir-square-heres-how-to-tell/.

61 **new political consensus around what kind of government**
David D. Kirkpatrick, "Named Egypt's Winner, Islamist Makes History," *New York Times*, June 25, 2012.

62 **from those advocated by most of the Internet-inspired reformers**
Ibid.

62 **when the Ottoman Empire banned the printing press**
Fatmagul Demirel, *Encyclopedia of the Ottoman Empire*, edited by Gabor Agoston and Bruce Masters (New York: Facts on File, 2009), p. 130.

62 **they had deprived themselves of the fruits of the Print Revolution**
Ishtiaq Hussain, "The Tanzimat: Secular Reforms in the Ottoman Empire," Faith Matters, February 5, 2011, http://faith-matters.org/images/stories/fm-publications/the-tanzimat-final-web.pdf.

62 **depending on how they are used and who uses them to greatest effect**
Evgeny Morozov, "The Dark Side of Internet for Egyptian and Tunisian Protesters," *Globe and Mail*, January 28, 2011; Louis Klaveras, "The Coming Twivolutions? Social

Media in the Recent Uprisings in Tunisia and Egypt," *Huffington Post*, January 31, 2011, http://www.huffingtonpost.com/louis-klarevas/post_1647_b_815749.html.

63 **have even experimented with Internet voting in elections and referenda**
Sutton Meagher, "Comment: When Personal Computers Are Transformed into Ballot Boxes: How Internet Elections in Estonia Comply with the United Nations International Covenant on Civil and Political Rights," *American University International Law Review* 23 (2008).

63 **proposals placed by citizens on a government website**
Freedom House—Latvia, 2012, http://www.freedomhouse.org/report/nations-transit/2012/latvia.

63 **achieve higher levels of quality in the services they deliver**
Tina Rosenberg, "Armed with Data, Fighting More Than Crime," *New York Times*, Opinionator blog, May 2, 2012, http://opinionator.blogs.nytimes.com/2012/05/02/armed-with-data-fighting-more-than-crime/.

63 **productive dialogues and arguments about issues and legislation**
Clay Shirky, "How the Internet Will (One Day) Transform Government," TEDGlobal 2012, June 2012.

63 **watch Internet videos on television screens**
Jenna Wortham, "More Are Watching Internet Video on Actual TVs, Research Shows," *New York Times*, September 26, 2012.

63 **"be appropriated into the realm of the digital"**
William Gibson, "Back from the Future," *New York Times Magazine*, August 19, 2007.

64 **any other activity besides sleeping and working**
Joe Light, "Leisure Trumps Learning in Time-Use Survey," *Wall Street Journal*, June 22, 2011.

64 **watches television more than five hours per day**
Nielsen, "State of the Media: Consumer Usage Report," 2011, p. 3. The American Video Viewer, 32 hours, 47 minutes of TV viewing weekly = 4.7 hours per day, http://www.nielsen.com/content/dam/corporate/us/en/reports-downloads/2011-Reports/StateofMediaConsumerUsageReport.pdf.

64 **spends 80 percent of his or her campaign money**
eMarketer, "Are Political Ad Dollars Going Online?," May 14, 2008, http://www.emarketer.com/Article.aspx?id=1006271&R=1006271.

64 **Philadelphia and easily find several low-cost print shops**
Christopher Munden, "A Brief History of Early Publishing in Philadelphia," Philly Fiction, http://phillyfiction.com/more/brief_history_of_early_days_of_philadelphia_publishing.html.

65 **destructive trend is likely to get much worse before it gets better**
Citizens United v. FEC, 130 S. Ct. 876 (2010); Adam Liptak, "Justices, 5-4, Reject Corporate Spending Limit," *New York Times*, January 22, 2010.

65 **television is much more tightly controlled than the Internet**
Charles Clover, "Internet Subverts Russian TV's Message," *Financial Times*, December 1, 2011.

65 **"There is one face: Putin"**
David M. Herszenhorn, "Putin Wins, but Opposition Keeps Pressing," *New York Times*, March 4, 2012.

65 people aged sixty-five and older watch, on average
Alana Semuels, "Television Viewing at All-Time High," *Los Angeles Times*, February 24, 2009.

66 most major cities that people used to read
Donald A. Ritchie, *Reporting from Washington: The History of the Washington Press Corps* (New York: Oxford University Press, 2005), p. 131.

66 the morning newspapers began to go bankrupt as well
Mark Fitzgerald, "How Did Newspapers Get in This Pickle?," *Editor & Publisher*, March 18, 2009.

66 digital news stories already reach more people
David Carr, "Tired Cries of Bias Don't Help Romney," *New York Times*, October 1, 2012.

67 staring at chalk on a blackboard
"Why Do 60% of Students Find Their Lectures Boring?," *Guardian*, May 11, 2009.

67 sharp declines in budgets for public education
"Education Takes a Beating Nationwide," *Los Angeles Times*, July 31, 2011.

68 college-level instruction on the Internet
Tamar Lewin, "Questions Follow Leader of For-Profit Colleges," *New York Times*, May 27, 2011; Tamar Lewin, "For-Profit College Group Sued as U.S. Lays Out Wide Fraud," *New York Times*, August 9, 2011.

68 The school was later prosecuted and shut down
"Degrees for Sale at Spam U.," CBS News, February 11, 2009, http://www.cbsnews.com/2100-205_162-659418.html; "Diploma Mill Operators Hit with Court Judgments," *Consumer Affairs*, March 18, 2005, http://www.consumeraffairs.com/news04/2005/diploma_mill.html.

68 emergence of chronic disease states that account for most medical problems
"Counting Every Moment," *Economist*, March 3, 2012.

68 beginning to improve the allocation and deployment of public health resources
Andrea Freyer Dugas et al., "Google Flu Trends: Correlation with Emergency Department Influenza Rates and Crowding Metrics," *Clinical Infectious Diseases* 54, no. 4 (January 8, 2012).

68 insurance companies have begun to use data mining techniques
"Very Personal Finance," *Economist*, June 2, 2012.

69 for customers whose data profiles classify them as low-risk
Ibid.

69 the legend of Doctor Faust first appeared
Christopher Marlowe, *The Tragical History of Doctor Faustus*, 1604, edited by Rev. Alexander Dyce, http://www.gutenberg.org/files/779/779-h/779-h.htm.

69 historians claim that Faust was based
Philip B. Meggs and Alston W. Purvis, *Meggs' History of Graphic Design*, 5th ed. (Hoboken, NJ: Wiley, 2012), p. 76–77.

69 "Faustian bargains"
Herman Kahn, "Technology and the Faustian Bargain," January 1, 1976, http://www.hudson.org/index.cfm?fuseaction=publication_details&id=2218; Lance Morrow, "The Faustian Bargain of Stem Cell Research," *Time*, July 12, 2001.

70 read by the government without a warrant
John Seabrook, "Petraeus and the Cloud," *New Yorker*, November 14, 2012.

70 reliance on the cloud creates new potential choke points
Nicole Perlroth, "Amazon Cloud Service Goes Down and Takes Popular Sites with It," *New York Times*, October 22, 2012.

71 "So you have these two fighting against each other"
Richard Siklos, "Information Wants to Be Free . . . and Expensive," CNN, July 20, 2009, http://tech.fortune.cnn.com/2009/07/20/information-wants-to-be-free-and-expensive/

72 on servers based in Sweden, Iceland, and possibly other locations
Andy Greenberg, "Wikileaks Servers Move to Underground Nuclear Bunker," *Forbes*, August 30, 2010.

73 broke into numerous other government and corporate
"WikiLeaks Backlash: The First Global Cyber War Has Begun, Claim Hackers," *Guardian*, December 11, 2010.

73 Independent groups of hacktivists
Hayley Tsukayama, "Anonymous Claims Credit for Crashing FBI, DOJ Sites," *Washington Post*, January 20, 1012; Ellen Nakashima, "CIA Web Site Hacked; Group LulzSec Takes Credit," *Washington Post*, June 15, 2011; Thom Shanker and Elisabeth Bumiller, "Hackers Gained Access to Sensitive Military Files," *New York Times*, July 14, 2011; David E. Sanger and John Markoff, "I.M.F. Reports Cyberattack Led to 'Very Major Breach,'" *New York Times*, June 11, 2011; David Batty, "Vatican Becomes Latest Anonymous Hacking Victim," *Guardian*, March 7, 2012; Melanie Hick, "Anonymous Hacks Interpol Site After 25 Arrests," *Huffington Post*, January 3, 2012, http://www.huffingtonpost.co.uk/2012/03/01/anonymous-hacks-interpol-_n_1312544.html; Martin Beckford, "Downing Street Website Also Taken Down by Anonymous," *Telegraph*, April 8, 2012; Tom Brewster, "Anonymous Strikes Downing Street and Ministry of Justice," *TechWeek Europe*, April 10, 2012, http://www.techweekeurope.co.uk/news/anonymous-government-downing-street-moj-71979; "NASA Says Was Hacked 13 Times Last Year," Reuters, March 2, 2012.

73 hackers recorded the call and put it on the web
Duncan Gardham, "'Anonymous' Hackers Intercept Conversation Between FBI and Scotland Yard on How to Deal with Hackers," *Telegraph*, February 3, 2012.

73 penetrated by a cyberattack believed to have originated in China
Michael Joseph Gross, "Enter the Cyber-Dragon," *Vanity Fair*, September 2011.

73 "fifth domain" for potential military conflict
Susan P. Crawford, "When We Wage Cyberwar, the Whole Web Suffers," Bloomberg, April 25, 2012.

73 "a global cyber arms race"
David Alexander, "Global Cyber Arms Race Engulfing Web—Defense Official," Reuters, April 11, 2012.

73 cybersecurity technology, offense has the advantage over defense
Ibid.; Ron Rosenbaum, "Richard Clarke on Who Was Behind the Stuxnet Attack," *Smithsonian*, April 2, 2012.

73 which prevented ancient Greece's conquest by Persia
Simon Singh, *The Code Book: The Science of Secrecy from Ancient Egypt to Quantum Cryptography* (New York: Doubleday, 1999).

74 on the messenger's scalp, and then "waited for the hair to regrow"
Ibid.

74 cryptography in its various forms
Ibid.; Andrew Lycett, "Breaking Germany's Enigma Code," BBC, February 17, 2011, http://www.bbc.co.uk/history/worldwars/wwtwo/enigma_01.shtml.

74 "The system kind of got loose"
Michael Joseph Gross, "World War 3.0," *Vanity Fair*, May 2012.

75 four trends have converged to make cybersecurity a problem
James Kaplan, Shantnu Sharma, and Allen Weinberg, "Meeting the Cybersecurity Challenge," *McKinsey Quarterly*, June 2011.

75 corporations, government agencies, and organizations
Gross, "Enter the Cyber-Dragon."

75 "We don't do that"
Rosenbaum, "Richard Clarke on Who Was Behind the Stuxnet Attack."

75 373,000 jobs each year—and $16 billion in lost earnings—from the theft of intellectual property
Richard Adler, Report of the 26th Annual Aspen Institute Conference on Communications Policy, *Updating Rules of the Digital Road: Privacy, Security, Intellectual Property*, 2012, p. 14.

75 worth $1 billion—in a single night
Richard A. Clarke, "How China Steals Our Secrets," *New York Times*, April 3, 2012.

76 examined one yet that has not been infected
Nicole Perlroth, "How Much Have Foreign Hackers Stolen?," *New York Times*, Bits blog, February 14, 2012, http://bits.blogs.nytimes.com/2012/02/14/how-much-have-foreign-hackers-stolen/?scp=7&sq=cyber%20security&st=cse.

76 "nearly four times the amount of data"
Ibid.

76 "cyberthreat will be the number one threat to the country"
J. Nicholas Hoover, "Cyber Attacks Becoming Top Terror Threat, FBI Says," *Information Week*, February 1, 2012.

76 thirteen U.S. defense contractors, and a large number of other corporations
Michael Joseph Gross, "Exclusive: Operation Shady Rat—Unprecedented Cyber-Espionage Campaign and Intellectual-Property Bonanza," *Vanity Fair*, August 2, 2011.

76 six weeks' worth of emails between the Chamber
Nicole Perlroth, "Traveling Light in a Time of Digital Thievery," *New York Times*, February 10, 2012.

76 still sending information over the Internet to China
Ibid.

76 individual packages containing the products they produce
Organisation for Economic Co-operation and Development, "Machine-to-Machine Communications: Connecting Billions of Devices," OECD Digital Economy Papers, No. 192, 2012, http://dx.doi.org/10.1787/5k9gsh2gp043-en.

76 dairy farmers in Switzerland are even connecting
John Tagliabue, "Swiss Cows Send Texts to Announce They're in Heat," *New York Times*, October 2, 2012.

77 "control systems that run these facilities, a nearly fivefold increase from 2010"
John O. Brennan, "Time to Protect Against Dangers of Cyberattack," *Washington Post*, April 15, 2012.

77 repeated cyberattacks from an unknown source
Thomas Erdbrink, "Iranian Officials Disconnect Some Oil Terminals from Internet," *New York Times*, April 24, 2012.

77 Aramco, was the victim of cyberattacks
Thom Shanker and David E. Sanger, "U.S. Suspects Iran Was Behind a Wave of Cyberattacks," *New York Times*, October 14, 2012.

77 The attack on Aramco
Nicole Perlroth, "In Cyberattack on Saudi Firm, U.S. Sees Iran Firing Back," *New York Times*, October 23, 2012.

77 Iranian gas centrifuges that were enriching uranium
William J. Broad, John Markoff, and David E. Sanger, "Israeli Test on Worm Called Crucial in Iran Nuclear Delay," *New York Times*, January 15, 2011.

77 began infecting computers in Iran and several other nations
"'Flame' Computer Virus Strikes Middle East; Israel Speculation Continues," Associated Press, May 29, 2012.

77 destructive attacks against Internet-connected machinery
William J. Broad, John Markoff, and David E. Sanger, "Israeli Test on Worm Called Crucial in Iran Nuclear Delay," *New York Times*, January 15, 2011.

78 inadvertently infected by Stuxnet
Rachel King, "Virus Aimed at Iran Infected Chevron Network," *Wall Street Journal*, November 9, 2012.

78 Leon Panetta publicly warned that a "cyber–Pearl harbor"
Elisabeth Bumiller and Thom Shanker, "Panetta Warns of Dire Threat of Cyberattack on U.S.," *New York Times*, October 11, 2012.

78 "only they're making it 30 percent cheaper"
Perlroth, "Traveling Light in a Time of Digital Thievery."

78 then steal some of its most valuable customers
Steve Fishman, "Floored by News Corp.: Who Hacked a Rival's Computer System?," *New York*, September 28, 2011.

78 emails of individuals to gather information for news stories
Sarah Lyall and Ravi Somaiya, "British Broadcaster with Murdoch Link Admits to Hacking," *New York Times*, April 5, 2012.

78 hacking into the telephone voicemails
Don Van Natta Jr., Jo Becker, and Graham Bowley, "Tabloid Hack Attack on Royals, and Beyond," *New York Times*, September 1, 2010.

78 hack into supposedly secure videoconferences
Nicole Perlroth, "Cameras May Open Up the Board Room to Hackers," *New York Times*, January 22, 2012.

78 theft of important information because they have a financial incentive
James Kaplan, Shantnu Sharma, and Allen Weinberg, "Meeting the Cybersecurity Challenge," *McKinsey Quarterly*, June 2011.

79 targets have failed to take action to protect themselves
Michaela L. Sozio, "Cyber Liability—a Real Threat to Your Business," *California*

Business Law Confidential, March 2012; Preet Bharara, "Asleep at the Laptop," *New York Times,* June 4, 2012.

79 **collecting information about their own customers and users**
Alexis Madrigal, "I'm Being Followed: How Google—and 104 Other Companies—Are Tracking Me on the Web," *Atlantic,* February 29, 2012.

79 **tailor advertising to match each person's individual collection of interests**
Ibid.

79 **online interests without offering them an opportunity to give their consent**
Riva Richmond, "As 'Like' Buttons Spread, So Do Facebook's Tentacles," *New York Times,* Bits blog, September 27, 2011, http://bits.blogs.nytimes.com/2011/09/27/as-like-buttons-spread-so-do-facebooks-tentacles/.

79 **"simply tools to improve the grip strength of the Invisible Hand"**
Madrigal, "I'm Being Followed."

79 **discover information that one would not necessarily want**
Jeffrey Rosen, "The Web Means the End of Forgetting," *New York Times Magazine,* July 21, 2010.

79 **their Facebook accounts so that private sites can also be accessed**
Michelle Singletary, "Would You Give Potential Employers Your Facebook Password?," *Washington Post,* March 29, 2012.

80 **policy is to never give out such passwords**
Joanna Stern, "Demanding Facebook Passwords May Break Law, Say Senators," ABC News, March 26, 2012, http://abcnews.go.com/Technology/facebook-passwords-employers-schools-demand-access-facebook-senators/story?id=16005565#.UCPKWY40jdk.

80 **many employees have been subjected to cybersurveillance**
Tam Harbert, "Employee Monitoring: When IT Is Asked to Spy," *Computer World,* June 16, 2010.

80 **especially the Internet, increases exponentially as more people connect to it**
James Hendler and Jennifer Golbeck, "Metcalfe's Law, Web 2.0, and the Semantic Web," *Web Semantics* 6, no. 1 (February, 2008): 14–20.

80 **actually increases as the *square* of the number of people who connect to it**
Ibid.

80 **options for changing settings that some sites offer**
Alexis Madrigal, "Reading the Privacy Policies You Encounter in a Year Would Take 76 Work Days," *Atlantic,* March 1, 2012; Elaine Rigoli, "Most People Worried About Online Privacy, Personal Data, Employer Bias, Privacy Policies," *Consumer Reports,* April 25, 2012.

81 **But users who try to opt out of the tracking itself**
Julia Angwin and Emily Steel, "Web's Hot New Commodity: Privacy," *Wall Street Journal,* February 28, 2011.

81 **due to persistent lobbying pressure from the advertising industry**
Tanzina Vega, "Opt-Out Provision Would Halt Some, but Not All, Web Tracking," *New York Times,* February 26, 2012; Madrigal, "I'm Being Followed."

81 **but there are so many clicks that billions of dollars are at stake**
Madrigal, "I'm Being Followed"; Vega, "Opt-Out Provision Would Halt Some, but Not All, Web Tracking."

81 **report information about a user's online activities**
"What They Know" Series, *Wall Street Journal*, http://online.wsj.com/public/page/what-they-know-digital-privacy.html.

81 **about the user's online activity to advertisers and others who purchase the data**
Julia Angwin, "The Web's New Gold Mine: Your Secrets," *Wall Street Journal*, July 30, 2010.

81 **matching individual computer numbers with the name, address, and telephone numbers**
Madrigal, "I'm Being Followed."

81 **has spoken out against the use of DPI**
Olivia Solon, "Tim Berners-Lee: Deep Packet Inspection a 'Really Serious' Privacy Breach," *Wired*, April 18, 2012.

81 **tragically, the gay student committed suicide soon after**
Ian Parker, "The Story of a Suicide: Two College Roommates, a Webcam, and a Tragedy," *New Yorker*, February 6, 2012.

82 **tag people when they appear in photos on the site**
"Facebook 'Face Recognition' Feature Draws Privacy Scrutiny," Bloomberg News, June 8, 2011.

82 **now used by many sites to identify people when they speak**
Natasha Singer, "The Human Voice, as Game Changer," *New York Times*, March 31, 2012.

82 **improve the accuracy with which the machine interprets**
Ibid.

82 **which information can be delivered with relevance to the user's location**
"Privacy Please! U.S. Smartphone App Users Concerned with Privacy When It Comes to Location," Nielsen, April 21, 2011, http://blog.nielsen.com/nielsenwire/online_mobile/privacy-please-u-s-smartphone-app-users-concerned-with-privacy-when-it-comes-to-location/.

82 **25,000 U.S. citizens are also victims of "GPS stalking" each year**
Justin Scheck, "Stalkers Exploit Cellphone GPS," *Wall Street Journal*, August 3, 2010.

82 **1,200 pages of information, most of which he thought he had deleted**
Kevin J. O'Brien, "Austrian Law Student Faces Down Facebook," *New York Times*, February 5, 2012.

82 **designed to steal information from the user's computer or mobile device**
Matt Richtel and Verne G. Kopytoff, "E-Mail Fraud Hides Behind Friendly Face," *New York Times*, June 2, 2011.

82 **all the private information about individuals**
Ann Carrns, "Careless Social Media Use May Raise Risk of Identity Fraud," *New York Times*, February 29, 2012.

82 **which have reported large losses as a result of cybercrime**
"IMF Is Victim of 'Sophisticated Cyberattack,' Says Report," *IDG Reporter*, June 13, 2011; "US Senate Orders Security Review After LulzSec Hacking," *Guardian*, June 14, 2011; Julianne Pepitone and Leigh Remizowski, "'Massive' Credit Card Data Breach Involves All Major Brands," CNN, April 2, 2012, http://money.cnn.com/2012/03/30/technology/credit-card-data-breach/index.htm; "Heartland Pay-

ment Systems Hacked," Associated Press, January 20, 2009; Bianca Dima, "Top 5: Corporate Losses Due to Hacking," HOT for Security, May 17, 2012.

83 **more than $7.2 million, with the cost increasing each year**
"The Real Cost of Cyber Attacks," *Atlantic,* February 16, 2012.

83 **"more than the annual global market for marijuana, cocaine, and heroin combined"**
Symantec, press release, "Norton Study Calculates Cost of Global Cybercrime: $114 Billion Annually," September 7, 2011, http://www.symantec.com/about/news/release/article.jsp?prid=20110907_02. However, some analysts note that some estimates of cybercrime are unreliable. Dinei Florêncio and Cormac Herley, "The Cybercrime Wave That Wasn't," *New York Times,* April 14, 2012.

83 **LinkedIn**
Ian Paul, "LinkedIn Confirms Account Passwords Hacked," *PC World,* June 6, 2012.

83 **eHarmony**
Salvador Rodriguez, "Like LinkedIn, eHarmony Is Hacked; 1.5 Million Passwords Stolen," *Los Angeles Times,* June 6, 2012.

83 **Google's Gmail**
Nicole Perlroth, "Yahoo Breach Extends Beyond Yahoo to Gmail, Hotmail, AOL Users," *New York Times,* July 12, 2012.

83 **Bank of America, JPMorgan Chase, Citigroup, U.S. Bank, Wells Fargo, and PNC**
David Goldman, "Major Banks Hit with Biggest Cyberattacks in History," CNN, September 28, 2012; "Week-Long Cyber Attacks Cripple US Banks," Associated Press, September 29, 2012.

83 **store Internet and telephone communications**
Brian Wheeler, "Communications Data Bill Creates 'a Virtual Giant Database,'" BBC, July 19, 2012, http://www.bbc.co.uk/news/uk-politics-18884460.

83 **already installed 60,000 security cameras**
Heather Brooke, "Investigation: A Sharp Focus on CCTV," *Wired UK,* April 1, 2010.

83 **"restrictions that fence in even the most disinterested"**
Justice Felix Frankfurter, Concurring Opinion, *Youngstown Sheet & Tube Co. v. Sawyer,* 343 U.S. 579 (1952).

83 **"Knowledge is power"**
Georg Henrik Wright, *The Tree of Knowledge and Other Essays* (Leiden: Brill, 1993), p. 127–28.

84 **ability to eavesdrop on telephone calls as they were taking place**
James Bradford, "The NSA Is Building the Country's Biggest Spy Center (Watch What You Say)," *Wired,* March 15, 2012.

84 **"began to rapidly turn the United States of America"**
Jason Reed, "NSA Whistleblowers: Government Spying on Every Single American," Reuters, July 25, 2012.

84 **has intercepted "between 15 and 20 trillion" communications**
Bradford, "The NSA Is Building the Country's Biggest Spy Center (Watch What You Say)."

84 **The formal state of emergency**
President of the United States, "Notice—Continuation of the National Emergency with Respect to Certain Terrorist Attacks," September 11, 2012.

85 **"I think there's really something at a deep level creepy"**
Matt Sledge, "Warrantless Electronic Surveillance Surges Under Obama Justice Department," *Huffington Post*, September 28, 2012.

85 **riding a bicycle with a defective "audible bell"**
Brief for the Petitioner in the United States Supreme Court, *Albert W. Florence v. Board of Chosen Freeholders of the County of Burlington et al.*, No. 10-945, http://www.americanbar.org/content/dam/aba/publishing/previewbriefs/Other_Brief_Updates/10-945_petitioner.authcheckdam.pdf. Justice Stephen Breyer dissented from the strip search decision in a powerful rebuke of the expansive powers granted to law enforcement by the Court's action. See *Florence v. Board of Chosen Freeholders*, April 2, 2012, http://www.supremecourt.gov/opinions/11pdf/10-945.pdf.

85 **other digital device when he or she reenters the country**
Glenn Greenwald, "U.S. Filmmaker Repeatedly Detained at Border," *Salon*, April 8, 2012, http://www.salon.com/2012/04/08/u_s_filmmaker_repeatedly_detained_at_border/.

86 **and no reasonable cause for allowing the search is required**
Ibid.

86 **whose digital information has been searched and seized**
Ibid.

86 **"marketing a catalog of 'surveillance fees'"**
Eric Lichtblau, "Police Are Using Phone Tracking as a Routine Tool," *New York Times*, April 1, 2012.

86 **plans to sell the data to private investigators, insurers, and others**
Julia Angwin and Jennifer Valentino-Devries, "New Tracking Frontier: Your License Plates," *Wall Street Journal*, October 2, 2012.

86 **The market for these technologies has grown**
Nicole Perlroth, "Software Meant to Fight Crime Is Used to Spy on Dissidents," *New York Times*, August 31, 2012.

86 **including Iran, Syria, and China**
Rebecca MacKinnon, "Internet Freedom Starts at Home," *Foreign Policy*, April 3, 2012; Cindy Cohn, Trevor Timm, and Jillian C. York, Electronic Frontier Foundation, "Human Rights and Technology Sales: How Corporations Can Avoid Assisting Repressive Regimes," April 2012, https://www.eff.org/document/human-rights-and-technology-sales; Jon Evans, "Selling Software That Kills," TechCrunch, May 26, 2012, http://techcrunch.com/2012/05/26/selling-software-that-kills/.

86 **video cameras will become commonplace tools**
Francis Fukuyama, "Why We All Need a Drone of Our Own," *Financial Times*, February 24, 2012.

87 **sixty-three active drone sites in the U.S.**
"Is There a Drone in Your Neighbourhood? Rise of Spy Planes Exposed After FAA Is Forced to Reveal 63 Launch Sites Across U.S.," *Daily Mail*, April 24, 2012.

87 **awareness—even if the device has been turned off**
David Kushner, "The Hacker Is Watching," *GQ*, January 2012.

87 **have also been used to monitor the conversations of some suspects**
Declan McCullagh, "Court to FBI: No Spying on In-Car Computers," CNET,

November 19, 2003, http://news.cnet.com/2100-1029_3-5109435.html. The Ninth Circuit Court of Appeals has ruled that this instance of surveillance is illegal.

87 **other confidential information as it is typed**
Nicole Perlroth, "Malicious Software Attacks Security Cards Used by Pentagon," *New York Times*, Bits blog, January 12, 2012, http://bits.blogs.nytimes.com/2012/01/12/malicious-software-attacks-security-cards-used-by-pentagon/.

87 **powerful data collection system that the world has ever known**
Bamford, "The NSA Is Building the Country's Biggest Spy Center (Watch What You Say)."

87 **caused public outrage and resulted in congressional action**
American Civil Liberties Union, "Congress Dismantles Total Information Awareness Spy Program; ACLU Applauds Victory, Calls for Continued Vigilance Against Snoop Programs," September 25, 2003, http://www.aclu.org/national-security/congress-dismantles-total-information-awareness-spy-program-aclu-applauds-victory-.

88 **against these proposals, resulted in the withdrawal of both**
Jonathan Weisman, "After an Online Firestorm, Congress Shelves Antipiracy Bills," *New York Times*, January 21, 2012.

88 **eavesdrop on any online communication**
Robert Pear, "House Votes to Approve Disputed Hacking Bill," *New York Times*, April 27, 2012.

88 **"habits of the heart and resisting the allure of the ideology of technology"**
Michael Sacasas, "Technology in America," *American*, April 13, 2012.

88 **norms and values that reflect the American tradition of free speech and robust free markets**
Gross, "World War 3.0."

89 **Brazil, India, and South Africa are following**
Georgina Prodhan, "BRIC Nations Push for Bigger Say in Policing of Internet," *Globe and Mail*, September 6, 2012.

89 **as a last resort for protecting confidential, high-value information**
Gross, "World War 3.0."

89 **"walled garden" approach**
Ryan Nakashima, "Ex-AOL Exec Calls Facebook New 'Walled Garden,'" Associated Press, May 1, 2012.

89 **attempted to slow down or make more expensive**
Claire Cain Miller and Miguel Helft, "Web Plan from Google and Verizon Is Criticized," *New York Times*, August 10, 2010.

89 **laws that protect free speech and free competition**
"Protecting the Internet," editorial, *New York Times*, December 18, 2010.

CHAPTER 3: POWER IN THE BALANCE

93 **"closing of a 500-year cycle in economic history"**
"China Became World's Top Manufacturing Nation, Ending 110 Year US Leadership," MercoPress, March 15, 2011, http://en.mercopress.com/2011/03/15/china-became-world-s-top-manufacturing-nation-ending-110-year-us-leadership.

93 **first time since 1890 that any economy**
Charles Kenny, "China vs. the U.S.: The Case for Second Place," *BloombergBusinessweek*, October 13, 2011.

95 **International Monetary Fund require support from 85 percent**
"Profile: IMF and World Bank," BBC News, April 17, 2012, http://news.bbc.co.uk/2/hi/americas/country_profiles/3670465.stm; Thomas J. Bollyky, "How to Fix the World Bank," op-ed, *New York Times*, April 9, 2012; David Bosco, "A Primer on World Bank Voting Procedures," *Foreign Policy*, March 28, 2012.

95 **it has effective veto power over their decisions**
BBC News, "Profile: IMF and World Bank," BBC News; World Bank, "World Bank Group Voice Reform: Enhancing Voice and Participation in Developing and Transition Countries in 2010 and Beyond," April 25, 2010, http://siteresources.worldbank.org/NEWS/Resources/IBRD2010VotingPowerRealignmentFINAL.pdf.

95 **members of the U.N. Security Council when Brazil**
CIA, *The World Factbook*, https://www.cia.gov/library/publications/the-world-factbook/rankorder/2001rank.html.

95 **some have already preemptively labeled it the "G2"**
C. Fred Bergsten, "Two's Company," *Foreign Affairs*, September/October 2009.

96 **there were episodic warnings that American power was waning**
Josef Joffe, "Declinism's Fifth Wave," *American Interest*, January/February 2012; Samuel P. Huntington, "The U.S.—Decline or Renewal?," *Foreign Affairs*, Winter 1988/1989; Victor Davis Hanson, "Beware the Boom in American 'Declinism,'" CBS News, November 14, 2011, http://www.cbsnews.com/8301-215_162-57324071/beware-the-boom-in-american-declinism/.

96 **the U.S. was in danger of quickly falling**
Joffe, "Declinism's Fifth Wave"; Huntington, "The U.S.—Decline or Renewal?"; Victor Hanson, "Beware the Boom in American 'Declinism.'"

96 **1940s to roughly 25 percent in the early 1970s**
Stephen M. Walt, "The End of the American Era," *National Interest*, October 25, 2011; Robert Kagan, "Not Fade Away," *New Republic*, January 11, 2012.

97 **remained at that same level for the last forty years**
Kagan, "Not Fade Away."

97 **largely at the expense of Europe, not of the United States**
Ibid.

97 **when it first became the world's largest economy**
Charles Kenny, "China vs. the U.S.: The Case for Second Place," *BloombergBusinessweek*, October 13, 2011.

97 **suffered casualties 100 times greater than those of the United States**
Irina Titova, "Medvedev Orders Precise Soviet WWII Death Toll," Associated Press, January 27, 2009; Anne Leland, Mari-Jana Oboroceanu, Congressional Research Service, "American War and Military Operations Casualties: Lists and Statistics," February 2010, http://www.fas.org/sgp/crs/natsec/RL32492.pdf.

97 **Stalin's 1939 pact with Hitler**
"The Day in History: August 23rd, 1939. The Hitler-Stalin Pact," History.com, 2012.

97 which formalized the U.S. dollar as the world's reserve currency
    "Beyond Bretton Woods 2," *Economist*, November 4, 2010.

97 which later evolved into the Common Market and the European Union
    European Commission, "Treaty Establishing the European Coal and Steel Community, ECSC Treaty," October 15, 2010, http://europa.eu/legislation_summaries/institutional_affairs/treaties/treaties_ecsc_en.htm.

97 "the father of the United Nations"
    Cordell Hull Foundation, "Cordell Hull Biography," http://www.cordellhull.org/english/About_Us/Biography.asp.

97 "when goods cross borders, armies do not"
    Jill Lerner, "Free Trade's Champion," *Atlanta Business Chronicle*, February 13, 2006.

98 "the end of history"
    Francis Fukuyama, *The End of History and the Last Man* (New York: Harper Perennial, 1993).

99 three waves of democracy that spread throughout the world
    U.S. Department of State, Bureau of International Information Programs, "Democracy's Third Wave," http://www.4uth.gov.ua/usa/english/politics/whatsdem/whatdm13.htm.

99 aftermath of the American Revolution, produced twenty-nine democracies
    Ibid.

99 carried a picture of George Washington in his breast pocket
    John F. Kennedy, "Remarks at an Independence Day Celebration with the American Community in Mexico City," June 30, 1962, American Presidency Project, http://www.presidency.ucsb.edu/ws/?pid=8748.

99 twelve by the beginning of World War II
    U.S. Department of State, "Democracy's Third Wave."

99 the number of democracies to thirty-six
    Ibid.

99 decline to thirty from 1962 until the mid-1970s
    Ibid.

99 with the collapse of communism in 1989
    Ibid.

100 decline in the number of democratic nations in the world
     Economist Intelligence Unit, Democracy Index 2010, 2010, http://www.eiu.com/democracy.

100 fourth wave of democratization
     Larry Diamond, "A Fourth Wave or False Start?," *Foreign Affairs*, May 22, 2011.

100 increased in absolute terms to the highest level since 1945
     Kagan, "Not Fade Away," January 11, 2012; Todd Purdum, "One Nation, Under Arms," *Vanity Fair*, January 2012.

101 almost equal to the military spending of the entire rest of the world
     Purdum, "One Nation, Under Arms."

101 more pilots for unmanned vehicles than it trains pilots of manned fighter jets
     Christian Caryl, "Predators and Robots at War," *New York Review of Books*, Au-

gust 30, 2011; Elisabeth Bumiller, "Air Force Drone Operators Report High Levels of Stress," *New York Times,* December 19, 2011.

101 **drone pilots suffer post-traumatic stress disorder**
Caryl, "Predators and Robots at War."

101 **U.S. stealth drone and commanded it to land**
Scott Peterson, "Downed US Drone: How Iran Caught the 'Beast,'" *Christian Science Monitor,* December 9, 2011; Rick Gladstone, "Iran Shows Video It Says Is of U.S. Drone," *New York Times,* December 9, 2011; "Insurgents Hack U.S. Drones," *Wall Street Journal,* December 17, 2009; "Iran 'Building Copy of Captured US Drone' RQ-170 Sentinel," BBC, April 22, 2012.

102 **More than fifty countries**
David Wood, "American Drones Ignite New Arms Race from Gaza to Iran to China," *Huffington Post,* November 27, 2012, http://www.huffingtonpost.com/2012/11/27/american-drones_n_2199193.html.

102 **right to unleash deadly fire when threatened**
Ibid.

102 **doubt that the social, political, and economic foundations in China**
Walt, "The End of the American Era"; Thair Shaikh, "When Will China Become a Global Superpower?," CNN, June 10, 2011, http://www.cnn.com/2011/WORLD/asiapcf/06/10/china.military.superpower/index.html.

102 **experts warn that the lack of free speech**
Walt, "The End of the American Era"; Kagan, "Not Fade Away," January 11, 2012; Martin Feldstein, "China's Biggest Problems Are Political, Not Economic," *Wall Street Journal,* August 2, 2012; Frank Rich, "Mayberry R.I.P.," *New York,* July 22, 2012.

102 **the concentrated autocratic power in Beijing**
Ibid.

102 **high levels of corruption throughout China**
Feldstein, "China's Biggest Problems Are Political, Not Economic."

102 **an estimated 64 million empty apartments in China**
"Crisis in China: 64 Million Empty Apartments," *Asia News,* September 15, 2010, http://www.asianews.it/news-en/Crisis-in-China:-64-million-empty-apartments-19459.html.

103 **windmills constructed by China are not connected to the electrical grid**
"Weaknesses in Chinese Wind Power," *Forbes,* July 20, 2009.

103 **largest internal migration in history**
"The Largest Migration in History," *Economist,* February 24, 2012.

103 **"180,000 protests, riots and other mass incidents"**
Tom Orlik, "Unrest Grows as Economy Booms," *Wall Street Journal,* September 26, 2011.

103 **fourfold increase from 2000**
Ibid.

103 **building in response to economic inequality**
Feldstein, "China's Biggest Problems Are Political, Not Economic"; Wendy Dobson, *Gravity Shift: How Asia's New Economic Powerhouses Will Shape the Twenty-First Century* (Toronto: University of Toronto Press, 2009).

103 intolerable environmental conditions
Dobson, *Gravity Shift*.

103 autocratic local and regional leaders
Orlik, "Unrest Grows as Economy Booms."

103 wages have been increasing significantly in the last two years
David Leonhardt, "In China, Cultivating the Urge to Splurge," *New York Times Magazine,* November 28, 2010.

103 sources besides the participatory nature of their system
Daniel Bell, "Real Meaning of the Rot at the Top of China," *Financial Times,* April 23, 2012.

104 "form the fundamental principle of Mao Zedong Thought?"
Deng Xiaoping, "Speech at the All-Army Conference on Political Work: June 2, 1978," in *Selected Works of Deng Xiaoping,* vol. 2 (Beijing: Foreign Languages Press, 1984), p. 132.

104 corporate lobbies to sit in the actual drafting sessions
Laura Sullivan, "Shaping State Laws with Little Scrutiny," NPR, October 29, 2010, http://www.npr.org/2010/10/29/130891396/shaping-state-laws-with-little-scrutiny; Mike McIntire, "Conservative Nonprofit Acts as a Stealth Business Lobbyist," *New York Times*, April 22, 2012.

105 routinely rubber-stamp laws
Sullivan, "Shaping State Laws with Little Scrutiny"; McIntire, "Conservative Nonprofit Acts as a Stealth Business Lobbyist"; John Cassidy, "America's Class War," *New Yorker* blog, June 8, 2012, http://www.newyorker.com/online/blogs/comment/2012/06/wisconsin-scott-walker-class-war.html.

105 longest running corporation was created in Sweden in 1347
"Sweden: The Oldest Corporation in the World," *Time*, March 15, 1963.

105 common until the seventeenth century, when the Netherlands
"The taste of adventure," *Economist*, December 17, 1998.

106 United Kingdom
Joel Bakan, *The Corporation* (New York: Free Press, 2004), p. 6.

106 South Sea Company scandal
Ibid., p. 7.

106 England banned corporations in 1720
Ibid., p. 6.

106 The prohibition was not lifted until 1825
Ibid., p. 9.

106 civic and charitable purposes, for limited periods of time
Justin Fox, "What the Founding Fathers Really Thought About Corporations," *Harvard Business Review*, April 1, 2010, http://blogs.hbr.org/fox/2010/04/what-the-founding-fathers-real.html.

106 "bid defiance to the laws of our country"
Thomas Jefferson, "To George Logan," November 12, 1816.

106 expanded by an order of magnitude, from 33 to 328
Bakan, *The Corporation*, p. 9.

106 New York State enacted the first of many statutes
Linda Smiddy and Lawrence Cunningham, "Corporations and Other Business Organizations: Cases, Materials, Problems," LexisNexis, 2010, p. 16.

106 increased considerably with the mobilization of Northern industry
David C. Korten, *When Corporations Rule the World* (Bloomfield, CT: Kumarian Press, 1995), http://www.thirdworldtraveler.com/Korten/RiseCorpPower_WCRW.html.

106 huge government procurement contracts
Ibid.

106 building of the railroads
Ibid.

106 corporate role in American life grew quickly
Ibid.

106 decisions in Congress and state legislatures grew as well
Ibid.

106 The tainted election of 1876
"Compromise of 1877," History.com, http://www.history.com/topics/compromise-of-1877.

106 wealth and power played the decisive role
Korten, *When Corporations Rule the World.*

106 "government of corporations, by corporations, and for corporations"
Ibid.

107 Between 1888 and 1908, 700,000 American workers
Ibid.

107 approximately 100 every day
Ibid.

107 lawyers and lobbyists flooded the U.S. Capitol and state legislatures
Ibid.

107 U.S. Supreme Court voided and made unenforceable
Jack Maskell, "Lobbying Congress: An Overview of Legal Provisions and Congressional Ethics Rules," CRS Report for Congress, September 14, 2011, http://digital.library.unt.edu/ark:/67531/metacrs1903/m1/1/high_res_d/RL31126_2001Sep14.pdf.

107 "all the injurious effects of a direct fraud on the public"
Ibid.

107 "measuring the decay of the public morals"
Lawrence Lessig, *Republic, Lost: How Money Corrupts Congress—and a Plan to Stop It* (New York: Twelve, 2011), p. 101.

107 Georgia's new constitution explicitly banned the lobbying
Ibid., p. 101.

107 "where the price of votes was haggled over, and laws"
Matthew Josephson, *The Robber Barons: The Great American Capitalist 1861–1901* (New Brunswick, NJ: Transaction, 2010), p. 168.

108 *Santa Clara County v. Southern Pacific Railroad Company*
Bakan, *The Corporation*, p. 16.

108 some historians believe was written by Justice Stephen Field
Joshua Holland, "The Supreme Court Sold Out Our Democracy—How to Fight the Corporate Takeover of Elections," *AlterNet*, October 25, 2010.

108 court reporter, who was the former president of a railway company
Ibid.

108 "the court does not wish to hear"
Pamela Karlan, "Me, Inc.," *Boston Review*, July 2011.

108 "We are all of the opinion that it does"
Santa Clara County v. Southern Pacific, Justia.com, 1886, http://supreme.justia.com/cases/federal/us/118/394/.

108 laid the first transoceanic telegraph cable in 1858
"Cyrus W. Field," *Encyclopaedia Britannica*, http://www.britannica.com/EBchecked/topic/206188/Cyrus-W-Field.

180 resulted in Stephen's appointment to the Supreme Court
Lincoln Institute, "David Dudley Field (1805–1894)," Mr. Lincoln and New York, http://www.mrlincolnandnewyork.org/inside.asp?ID=56&subjectID=3.

108 subsequent massacre back to the United States in real time
Mike Sacks, "Corporate Citizenship: How Public Dissent in Paris Sparked Creation of the Corporate Person," *Huffington Post*, October 12, 2011, http://www.huffingtonpost.com/2011/10/12/corporate-citizenship-corporate-personhood-paris-commune_n_1005244.html.

108 followed in the United States, as it unfolded, on a daily basis
Ibid.

108 Franco-Prussian War that month and the struggle
Alice Bullard, *Human Rights and Revolutions*, edited by Jeffrey N. Wasserstrom, Lynn Hunt, and Marilyn B. Young (Oxford: Rowan & Littlefield, 2000), pp. 81–83.

108 first symbolic clash between communism and capitalism
Marx, however, wrote in *The Communist Manifesto* that the 1848 French revolution was the first "class struggle."

108 "forever celebrated as the glorious harbinger of a new society"
Karl Marx, "The Fall of Paris," May 1871, http://www.marxists.org/archive/marx/works/1871/civil-war-france/ch06.htm.

109 white flag that had been flown by Parisians
Alistair Horne, *The Fall of Paris: The Siege and the Commune 1870–71* (New York: Penguin Books, 2007), p. 433.

109 obsessively following the daily reports
Sacks, "Corporate Citizenship."

109 than any other story that year besides government corruption
John Harland Hicks and Robert Tucker, *Revolution & Reaction: The Paris Commune, 1871* (Amherst: University of Massachusetts Press, 1973), p. 60; Jack Beatty, *Age of Betrayal: The Triumph of Money in America, 1865–1900* (New York: Vintage Books, 2008), p. 153.

109 bankruptcy of financier and railroad entrepreneur Jay Cooke
"The Panic of 1873," *The American Experience, Ulysses S. Grant*, PBS, http://www.pbs.org/wgbh/americanexperience/features/general-article/grant-panic/.

109 "opportunity or the incentive to spread abroad"
"The Communists," *New York Times*, January 20, 1874.

109 decided to make it his mission to strengthen corporations
Sacks, "Corporate Citizenship."

110 Theodore Roosevelt unexpectedly became president, and the following year
"Domestic Politics," *The American Experience, TR*, PBS, http://www.pbs.org/wgbh/
americanexperience/features/general-article/tr-domestic/.

110 inside his new Department of Commerce and Labor
Ibid.

110 break up J. P. Morgan's Northern Securities Corporation
Ibid.

110 112 corporations worth a combined $571 billion (in 2012 dollars)
Korten, *When Corporations Rule the World*, p. 67.

110 "twice the total assessed value of all property in thirteen states"
Ibid.

110 This was followed by forty more antitrust suits
"Domestic Politics," PBS.

110 protected more than 230 million acres of land
Ibid.

110 winning the Nobel Peace Prize
Historians now believe that, while Roosevelt was no doubt essential to the broker-
ing of an effective deal, he was not truly a neutral arbiter and tilted heavily toward
Japan in private. See James Bradley, "Diplomacy That Will Live in Infamy," *New
York Times*, December 6, 2009.

110 "wise custom" by serving only two terms
Edmund Morris, *Theodore Rex* (New York: Random House, 2002), p. 364.

110 William Howard Taft, abandoned many of TR's reforms
"American President: William Howard Taft," Miller Center, University of Virginia, http://millercenter.org/president/taft/essays/biography/1.

110 "corrupt the men and methods of government for their own profit"
Theodore Roosevelt, "The New Nationalism," August 31, 1910, http://www.pbs
.org/wgbh/americanexperience/features/primary-resources/tr-nationalism/.

111 given privileges to which they are not entitled
Lessig, *Republic, Lost*, p. 4.

111 "twist the methods of free government"
Ibid., p. 5. Theodore Roosevelt, "From the Archives: President Teddy Roosevelt's
New Nationalism Speech," August 31, 2010, http://www.whitehouse.gov/blog/
2011/12/06/archives-president-teddy-roosevelts-new-nationalism-speech.

111 "prohibit the use of corporate funds"
Roosevelt, "From the Archives: President Teddy Roosevelt's New Nationalism
Speech."

111 "does not give the right of suffrage to any corporation"
Ibid.

111 secretly bribed Harding administration officials
United States Senate, "1921–1940: Senate Investigates the 'Teapot Dome'

Scandal," http://www.senate.gov/artandhistory/history/minute/Senate_Investigates
_the_Teapot_Dome_Scandal.htm.

111 "menace to the welfare of the nation"
Jeffrey Rosen, "POTUS v. SCOTUS," *New York*, March 17, 2010.

111 Historians differ on whether Roosevelt's threat was the cause
"Presidential Politics," *American Experience, FDR*, PBS, http://www.pbs.org/wgbh/
americanexperience/features/general-article/fdr-presidential/.

111 began approving the constitutionality
Ibid.

112 return court rulings to the philosophy that existed prior to the New Deal
Jeffrey Rosen, "Second Opinions," *New Republic*, May 4, 2012.

112 Powell, a Richmond lawyer
Jim Hoggan, "40th Anniversary of the Lewis Powell Memo Launching Corporate Propaganda Infrastructure," DeSmogBlog, August 23, 2011, http://www
.desmogblog.com/40th-anniversary-lewis-powell-memo-launching-corporate
-propaganda-infrastructure; John Jeffries, *Justice Lewis F. Powell, Jr.: A Biography*
(New York: Fordham University Press, 2001), p. 4.

112 state legislatures, and the judiciary in order to tilt
Lewis F. Powell, "The Powell Memo," August 23, 1971, http://reclaimdemocracy
.org/powell_memo_lewis/.

113 "corporate speech," which he found to be protected by the First Amendment
Jeffrey Clements, "The Real History of 'Corporate Personhood': Meet the Man to
Blame for Corporations Having More Rights Than You," *AlterNet*, December 6, 2011.

113 that the law violated the free speech of "corporate persons"
Ibid.

113 have revenues larger than many of the world's nation-states
Vincent Trivett, "25 US Mega Corporations: Where They Rank If They Were
Countries," *Business Insider*, June 27, 2011, http://www.businessinsider.com/25
-corporations-bigger-tan-countries-2011-6?op=1.

114 "I'm not a U.S. company and I don't make decisions based"
Steve Coll, *Private Empire: ExxonMobil and American Power* (New York: Penguin
Press, 2012), p. 71.

114 "dependence upon the positions of particular governments"
Bakan, *The Corporation*, p. 25.

114 "Nobody tells those guys what to do"
Coll, *Private Empire*, p. 257.

114 political action committees exploded
Federal Election Commission, "The Growth of Political Action Committees,
1974–1998," http://www.voteview.com/Growth_of_PACs_by_Type.htm.

114 corporations with registered corporate lobbyists
Jacob S. Hacker and Paul Pierson, *Winner-Take-All Politics: How Washington Made
the Rich Richer—And Turned Its Back on the Middle Class* (New York: Simon & Schuster, 2011), p. 118.

115 from $100 million in 1975 to $3.5 billion per year in 2010
Robert G. Kaiser, "Citizen K Street: Introduction," *Washington Post*, March 2007,
http://blog.washingtonpost.com/citizen-k-street/chapters/introduction/;   Bennett

Roth and Alex Knott, "Lobby Dollars Dip for First Time in Years," *Roll Call*, February 1, 2011.

115 **top the list of lobbying expenditures**
Roth and Knott, "Lobby Dollars Dip for First Time in Years."

115 **more than all lobbyist expenditures combined when the Powell Plan**
Kaiser, "Citizen K Street: Introduction."

115 **only 3 percent of retiring members of Congress**
Lessig, *Republic, Lost*, p. 123.

115 **more than 50 percent of retiring senators and more than 40 percent**
Ibid.

115 **"ideological warfare"**
Powell, "The Powell Memo."

115 **change the character of American government**
Timothy Noah, "Think Cranks," *New Republic*, March 30, 2012.

115 **principal as quickly as possible in order to have maximum impact**
Ibid.

115 **Lynde and Harry Bradley Foundation**
Ibid.

115 **Adolph Coors Foundation**
David Brock, *The Republican Noise Machine: Right-Wing Media and How It Corrupts Democracy* (New York: Random House, 2005), p. 43.

115 **"if, in turn, business is willing to provide the funds"**
Powell, "The Powell Memo."

116 **seminars organized by wealthy corporate interests**
Eric Lichtblau, "Advocacy Group Says Justices May Have Conflict in Campaign Finance Case," *New York Times*, January 19, 2011.

117 **basic functions that had been performed by democratically**
Emily Thornton, "Roads to Riches," *BusinessWeek*, May 6, 2007; Jonathan Hoenig, "Opportunities in Infrastructure: Should We Privatize Bridges and Roads?," Fox News, August 5, 2007, http://www.foxnews.com/story/0,2933,253438,00.html.

118 **highest inequality of incomes and the highest poverty rate**
James Gustave Speth, "America the Possible: A Manifesto, Part I," *Orion*, March/April 2012.

118 **lowest "material well-being"**
Ibid.

118 **highest child poverty rate**
Ibid.

118 **highest infant mortality rate**
Ibid.

118 **biggest prison population and the highest homicide rate**
Ibid.

118 **largest percentage of its citizens unable to afford health care**
Ibid.

119 "should be monitored in the same way that textbooks"
Powell, "The Powell Memo."

119 "opportunity for supporters of the American system"
Ibid.

120 "vex and oppress each other than to co-operate for their common good"
James Madison, *Federalist* No. 10, "The Same Subject Continued: The Union as a Safeguard Against Domestic Faction and Insurrection," November 23, 1787, http://thomas.loc.gov/home/histdox/fed_10.html.

120 "unfriendly passions and excite their most violent conflicts"
Ibid.

120 "the various and unequal distribution of property"
Ibid.

120 property, and income in the United States is now larger
Timothy Noah, "Introducing the Great Divergence," *Slate*, September 3, 2010.

121 who are relatively more tolerant and others who are relatively less tolerant
Jonathan Haidt, "Born This Way? Nature, Nurture, Narratives, and the Making of Our Political Personalities," *Reason*, May 2012.

121 value liberty and fairness but think about them differently
Ibid.

121 differences are reinforced by social feedback loops
Ibid.; Sasha Issenberg, "Born This Way: The New Weird Science of Hardwired Political Identity," *New York*, April 16, 2012.

122 what language would most trigger feelings of outrage
Frank Luntz, *Words That Work: It's Not What You Say, It's What People Hear* (New York: Hyperion, 2006), p. 165.

122 strategy known as "starve the beast"
Bruce Bartlett, "'Starve the Beast': Origins and Development of a Budgetary Metaphor," *Independent Review*, Summer 2007.

123 U.S. debt-to-GDP ratio will be 70 percent in 2013
Congressional Budget Office, "The 2012 Long-Term Budget Outlook," http://www.cbo.gov/sites/default/files/cbofiles/attachments/06-05-Long-Term_Budget_Outlook.pdf.

123 exceeds GDP if money the government owes to itself is added to the debt
Agence France-Presse, "US Borrowing Tops 100% of GDP: Treasury," August 3, 2011; Matt Phillips, "The U.S. Debt Load: Big and Cheap," *Wall Street Journal*, July 25, 2012.

123 no perceptible effect on the demand for U.S. bonds
Tim Mullaney, "A Year After Downgrade, S&P's View on Washington Unchanged," *USA Today*, August 7, 2012.

123 add approximately $1 trillion to interest payments over the next decade
Jeanne Sahadi, "Washington's $5 Trillion Interest Bill," CNN Money, March 12, 2012, http://money.cnn.com/2012/03/05/news/economy/national-debt-interest/index.htm.

124 "the diffusion of power away from government"
Joseph S. Nye, "Cyber War and Peace," Project Syndicate, April 10, 2012, http://www.project-syndicate.org/commentary/cyber-war-and-peace.

125 not even close to the fiscal conditions that would have reduced the risk
David Marsh, "The Euro's Lost Promise," *New York Times*, March 17, 2010; Sven Boll, "New Documents Shine Light on Euro Birth Defects," *Der Spiegel*, May 8, 2012.

126 for the two decades since reunification—at an estimated cost of $2.17 trillion
Katrin Bennhold, "What History Can Explain About the Greek Crisis," *New York Times*, May 21, 2012.

127 plant varieties that were especially suitable for cultivation
Jared Diamond, "What Makes Countries Rich or Poor?," *New York Review of Books*, June 7, 2012.

127 Fertile Crescent (and in nearby Crete)
Ibid.

127 several other areas of the world, including Mexico, the Andes, and Hawaii
Ibid.

127 shared form of national languages stimulated
Benedict Anderson, *Imagined Communities*, new ed. (New York: Verso, 2006), pp. 39–48.

127 difficulty communicating with speakers of other forms
Ibid.

128 further strengthened national identities
Ibid.

128 while often neglecting to include narratives
Ibid.

128 China and Korea have regularly become sources of tension in Northeast Asia
"Japan Textbook Angers Chinese, Korean Press," BBC News, April 6, 2005, http://news.bbc.co.uk/2/hi/asia-pacific/4416593.stm.

129 Google Translate
Franz Och, Google Official Blog, "Breaking Down the Language Barrier—Six Years In," April 26, 2012, http://googleblog.blogspot.com/2012/04/breaking-down -language-barriersix-years.html.

129 more documents, articles, and books in one day
Ibid.

129 Seventy-five percent of the web pages translated
Personal correspondence with Franz Och, Google.

129 Chinese language users of the Internet
Matt Silverman, "China: The World's Largest Online Population," Mashable, April 10, 2012, http://mashable.com/2012/04/10/china-largest-online-population/; David Teegham, "Chinese to Be Most Popular Language on the Internet," *Discovery News*, January 2, 2011, http://news.discovery.com/tech/chinese-to-be-most -popular-language-on-internet.html.

129 once vested in its national government
U.S. Department of State, "Background Note: Belgium," March 22, 2012, http:// www.state.gov/r/pa/ei/bgn/2874.htm.

130 an "assembly of men"
Thomas Hobbes, "Of the Natural Condition of Mankind as Concerning Their Felicity and Misery," *The Leviathan*, 1651; ibid., "Of the Rights of Sovereign by Institution."

131 **1,500 years ago between the Western and Eastern Roman empires**
"Bosnia and Hercegovina," Lonely Planet, 2008, http://www.lonelyplanet.com/shop
_pickandmix/previews/mediterranean-europe-8-bosnia-hercegovina-preview.pdf.

131 **went to the disputed territory of Kosovo**
Barney Petrovic, "Serbia Recalls an Epic Defeat," *Guardian*, June 29, 1989.

131 **launched genocidal violence against both Bosnians and Croats**
Ibid.

131 **20 percent of all land in the world, putting 150 million people**
"The United States Becomes a World Power," Digital History, http://www
.digitalhistory.uh.edu/disp_textbook_print.cfm?smtid=2&psid=3158; Saul David,
"Slavery and the 'Scramble for Africa,'" BBC, February 17, 2011, http://www.bbc
.co.uk/history/british/abolition/scramble_for_africa_article_01.shtml.

131 **nineteen of the forty-three known terrorist groups**
Lieutenant Colonel David A. Haupt, U.S. Air Force, "Narco-Terrorism: An Increasing Threat to U.S. National Security," Joint Forces Staff College, Joint Advanced Warfighting School, 2009.

132 **larger than the national economies of 163 of the world's 184 nations**
Ibid.

132 **Hundreds of thousands died unnecessarily, $3 trillion was wasted**
Joseph E. Stiglitz and Linda J. Bilmes, *The Three Trillion Dollar War: The True Cost
of the Iraq Conflict* (New York: Norton, 2008).

133 **amount of television watched and the decline of support for fundamentalism**
Mansoor Moaddel and Stuart A. Karabenick, "Religious Fundamentalism among
Young Muslims in Egypt and Saudi Arabia," *Social Forces* 86, no. 4 (June 2008).

133 **popularity of its movies and television programs**
Thomas Seibert, "Turkey Has a Star Role in More Than Just TV Drama," *National*, February 8, 2012.

133 **decline in the number of people killed in wars**
Joshua S. Goldstein, *Winning the War on War: The Decline of Armed Conflict Worldwide*
(New York: Dutton/Penguin, 2011), pp. 5–6.

133 **decline in the number of wars in every category**
Ibid.

134 **"by arbitration under International Law"**
Bulletin of the Pan-American Union 38, nos. 244–49 (1914), p. 79.

134 **in spite of U.S. superiority in conventional and nuclear weaponry**
Richard Clarke, "China's Cyberassault on America," *Wall Street Journal*, June 15,
2011.

135 **more than on any other program besides Social Security**
John Mueller, "Think Again: Nuclear Weapons," *Foreign Policy* no. 177 (January/
February 2010); Peter Passell, "The Flimsy Accounting in Nuclear Weapons De-
cisions," *New York Times*, July 9, 1998.

136 **President Barack Obama reversed U.S. policy four years ago**
"A Treaty on Conventional Arms," editorial, *New York Times*, July 9, 2012.

136 **some of which end up being trafficked in black markets**
C. J. Chivers, "Small Arms, Big Problems," *Foreign Affairs* 90, no. 1 (January/
February 2011): 110–21; Richard F. Grimmett, Congressional Research Service,

"Conventional Arms Transfers to Developing Nations, 2003–2010," September 22, 2011, http://fpc.state.gov/documents/organization/174196.pdf.

136 **warned the United States about the "military industrial complex"**
Sam Roberts, "In Archive, New Light on Evolution of Eisenhower Speech," *New York Times*, December 11, 2010.

136 **all of the military weapons sold to countries around the world originate in the United States**
Grimmett, "Conventional Arms Transfers to Developing Nations, 2003–2010."

137 **countries with the potential to build nuclear bombs**
Polly M. Holdorf, "Limited Nuclear War in the 21st Century," Center for Strategic and International Studies, 2010, http://csis.org/files/publication/110916_Holdorf.pdf.

137 **could over time result in the ability to project intercontinental power**
Graham Allison, "Nuclear Disorder," *Foreign Affairs* 89, no. 1 (January/February 2010): 74–85.

137 **many believe it is capable of selling nuclear weapons components**
William J. Broad, James Glanz, and David E. Sanger, "Iran Fortifies Its Arsenal with the Aid of North Korea," *New York Times*, November 29, 2010.

138 **"correlation between high levels of development and stable democracy"**
Francis Fukuyama, "The Future of History: Can Liberal Democracy Survive the Decline of the Middle Class?," *Foreign Affairs* 91, no. 1 (January/February 2012).

138 **will reach almost five billion people by 2030**
European Strategy and Policy Analysis System, *Global Trends 2030—Citizens in an Interconnected and Polycentric World*, http://www.iss.europa.eu/uploads/media/ESPAS_report_01.pdf.

138 **"economic and social rights and, increasingly, environmental issues"**
Ibid.

CHAPTER 4: OUTGROWTH

143 **"the personal distribution of income" or "a variety of costs that must be recognized"**
Simon Kuznets, *National Income 1929–1932*, Report to the U.S. Senate, 73rd Congress, 2nd Session (Washington, DC: U.S. Government Printing Office, 1934), pp. 5–6.

143 **three billion people in just the next seventeen years**
Homi Kharas, OECD Development Center, "The Emerging Middle Class in Developing Countries," January 2010, www.oecd.org/dataoecd/54/62/44798225.pdf.

143 **standards that are more common in the wealthiest nations**
Ibid.

143 **exceeding the rate of growth in the number of people in the world**
Jeremy Grantham, "Time to Wake Up: Days of Abundant Resources and Falling Prices Are Over Forever," *GMO Quarterly Letter*, April 2011.

144 **"furnished by the arts and manufactures to be debarred the use of them"**
"Letter from Thomas Jefferson to George Washington, 15 March 1784," Library of Virginia, http://www.lva.virginia.gov/lib-edu/education/psd/nation/gwtj.htm.

144 **general public's happiness or sense of well-being**
Institute for Studies in Happiness, Economy and Society, interview with Lester Brown, November 7, 2011, http://ishesorg/en/interview/itv02_01html.

144 increases in consumption do not enhance a sense of well-being
Daniel Kahneman and Angus Deaton, "High Income Improves Evaluation of Life but Not Emotional Well-Being," *Proceedings of the National Academy of Sciences*, September 7, 2010, http://www.pnas.org/content/early/2010/08/27/1011492107 .abstract.

144 a dangerous "planetary scale 'tipping point'"
Anthony Barnosky et al., "Approaching a State Shift in Earth's Biosphere," *Nature*, June 7, 2012.

144 "It's either got to be deflated gently, or it's going to burst"
Justin Gillis, "Are We Nearing a Planetary Boundary?," *New York Times*, Green blog, June 6, 2012, http://green.blogs.nytimes.com/2012/06/06/are-we-nearing-a -planetary-boundary/.

145 high levels in 2008 and again in 2011
United Nations Food and Agriculture Organization, "FAO Initiative on Soaring Food Prices," http://www.fao.org/isfp/en/; Annie Lowrey, "Experts Issue a Warning as Food Prices Shoot Up," *New York Times*, September 4, 2012.

145 political upheavals struck several countries
Jack Farchy and Gregory Meyer, "World Braced for New Food Crisis," *Financial Times*, July 19, 2012; Evan Fraser and Andrew Rimas, "The Psychology of Food Riots," *Foreign Affairs*, January 30, 2011.

145 northern China, India, and the Western United States
Li Jiao, "Water Shortages Loom as Northern China's Aquifers Are Sucked Dry," *Science*, June 18, 2010; "Groundwater Depletion Rate Accelerating Worldwide," *ScienceDaily*, September 23, 2010, http://www.sciencedaily.com/releases/2010/09/ 100923142503.htm.

145 countries where 50 percent of the world's people live
Lester Brown, Earth Policy Institute, *Plan B 3.0: Mobilizing to Save Civilization* (New York: Norton, 2008), http://www.earth-policy.org/images/uploads/book _files/pb3book.pdf.

145 depressing crop yields in several important food-growing regions
John Vidal, "Soil Erosion Threatens to Leave Earth Hungry," *Guardian*, December 14, 2010.

145 surged simultaneously in the last eleven years
Grantham, "Time to Wake Up."

145 increases larger than those that accompanied either World War I or II
Ibid.

145 the danger that we may soon reach "peak everything"
Ibid.

146 have been at least three times faster than those in the industrial world
Ibid.

146 world's iron ore, coal, pigs, steel, and lead—and roughly 40 percent
Ibid.; Scott Neuman, "World Starts to Worry as Chinese Economy Hiccups," NPR News, December 2, 2011, http://www.npr.org/2011/12/02/143048898/world -starts-to-worry-as-chinese-economy-hiccups; presentation by Robert Zoellick, World Bank Spring Meetings 2012, http://siteresources.worldbank.org/NEWS/ Resources/RBZ-SM12-for-Print-FINAL.pdf.

146 **being produced each year are now made in China**
Charles Riley, "Obama Hits China with Trade Complaint," CNN Money, September 17, 2012, http://money.cnn.com/2012/09/17/news/economy/obama-china-trade-autos/index.html.

146 **automobiles in China than in its home country**
Alisa Priddle, "GM's Big Plans for China Includes More Cadillac Models," *USA Today*, April 25, 2012.

146 **250 million to slightly over one billion in 2013**
"One Billion Vehicles Now Cruise the Planet," *Discovery News*, August 18, 2011, http://news.discovery.com/autos/one-billion-cars-cruise-planet-110818.html.

146 **double again in the next thirty years**
ExxonMobil, "The Outlook for Energy: A View to 2040," 2012, http://www.exxonmobil.com/Corporate/files/news_pub_eo.pdf.

146 **"All of the net growth"**
International Energy Agency, "World Energy Outlook," 2011.

146 **countries may be slowing, and in some cases may have peaked**
See, for example, U.S. Energy Information Administration, press release, "EIA examines alternate scenarios for the future of U.S. energy," June 25, 2012, http://www.eia.gov/pressroom/releases/press361.cfm, and "U.S. Energy-Related Carbon Dioxide Emissions, 2011," August 14, 2012, http://www.eia.gov/environment/emissions/carbon/.

146 **world's population of cars and trucks would be 5.5 billion**
Justin Lahart, "What If the Rest of World Had as Many Cars as U.S.?," *Wall Street Journal* blog, November 12, 2011, http://blogs.wsj.com/economics/2011/11/12/number-of-the-week-what-if-rest-of-world-had-as-many-cars-as-u-s/.

147 **U.S. oil production may soon edge back slightly above the 1970 peak**
Ronald D. White and Tiffany Hsu, "U.S. to Become World's Largest Oil Producer by 2020, Report Says," *Los Angeles Times*, November 13, 2012.

147 **the Arab members of OPEC implemented the first oil embargo**
U.S. Department of State, Office of the Historian, "OPEC Oil Embargo, 1973–1974," 2012, http://history.state.gov/milestones/1969-1976/OPEC.

147 **China and other emerging markets portend further significant increases**
International Energy Agency, Key World Energy Statistics, 2011, http://www.iea.org/publications/freepublications/publication/key_world_energy_stats-1.pdf.

147 **China's coal imports have already increased**
Kevin Jianjun Tu, Carnegie Endowment for International Peace Policy Outlook, "Understanding China's Rising Coal Imports," February 2012.

147 **and will double again by 2015**
Rebekah Kebede and Michael Taylor, "China Coal Imports to Double in 2015, India Close Behind," Reuters, May 30, 2011.

147 **increase in both coal and oil consumption through the next two decades**
International Energy Agency, "World Energy Outlook," 2011.

147 **to exploit the newly discovered abundance of deep shale gas**
Chrystia Freeland, "The Coming Oil Boom," *New York Times*, August 9, 2012.

147 **At current levels of growth**
International Energy Agency, "World Energy Outlook," 2011.

148 seems to have peaked more than thirty years ago
Grantham, "Time to Wake Up."

148 more expensive unconventional onshore sources
Ibid.

148 unforgiving and environmentally fragile Arctic Ocean
Guy Chazan, "Total Warns Against Oil Drilling in Arctic," *Financial Times*, September 25, 2012.

148 more expensive oil than the world has enjoyed in the past
Jeff Rubin, "How High Oil Prices Will Permanently Cap Economic Growth," Bloomberg View, September 23, 2012, http://www.bloomberg.com/news/2012 -09-23/how-high-oil-prices-will-permanently-cap-economic-growth.html; Bryan Walsh, "There Will Be Oil—and That's the Problem," *Time*, March 29, 2012.

148 methane for 90 percent of their fertilizer costs
Maria Blanco, Agronomos Etsia Upm, "Supply of and Access to Key Nutrients NPK for Fertilizers for Feeding the World in 2050," November 28, 2011, http:// eusoils.jrc.ec.europa.eu/projects/NPK/Documents/Madrid_NPK_supply_report _FINAL_Blanco.pdf, p. 26.

148 "more than a calorie of fossil fuel energy to produce a calorie of food"
Michael Pollan, *The Omnivore's Dilemma: A Natural History of Four Meals* (New York: Penguin, 2006), p. 46.

148 spend 50 to 70 percent of their income on food
Lester Brown, *Full Planet, Empty Plates: The New Geopolitics of Food Scarcity* (New York: Norton, 2012), ch. 1, http://www.energybulletin.net/stories/2012-09-17/full -planet-empty-plates-new-geopolitics-food-scarcity-new-book-chapter.

149 diminishes grain yields by 6 percent
Jims Vincent Capuno, "Soil Erosion: The Country's Unseen Enemy," Edge Davao, July 11, 2011, http://www.edgedavao.net/index.php?option=com_content &view=article&id=4801:soil-erosion-the-countrys-unseen-enemy&catid=51:on -the-cover&Itemid=83; Lester Brown, *Eco-Economy: Building an Economy for the Earth* (New York: Norton, 2001), ch. 3, http://www.earth-policy.org/books/eco/ eech3_ss5.

149 50 percent reduces many crop yields by 25 percent
Vidal, "Soil Erosion Threatens to Leave Earth Hungry."

149 Increasing desertification of grasslands
Judith Schwartz, "Saving US Grasslands: A Bid to Turn Back the Clock on Desertification," *Christian Science Monitor*, October 24, 2011.

149 45 percent more water
"No Easy Fix: Simply Using More of Everything to Produce More Food Will Not Work," *Economist*, February 24, 2011.

149 from 3.5 percent annually three decades ago to a little over one percent
Grantham, "Time to Wake Up."

149 three quarters of all plant genetic diversity may have already been lost
United Nations Food and Agriculture Organization, "Building on Gender, Agrobiodiversity and Local Knowledge," 2004, ftp://ftp.fao.org/docrep/fao/007/y5609e/ y5609e00.pdf.

149 "some form of export ban in an effort to increase domestic food security"
Toni Johnson, Council on Foreign Relations, "Food Price Volatility and Inse-curity," August 9, 2011, http://www.cfr.org/food-security/food-price-volatility -insecurity/p16662.

149 less frequent but larger downpours
Kevin Trenberth, "Changes in Precipitation with Climate Change," *Climate Re-search* 47 (2010): 123–38.

149 produce a 10 percent decline in crop yields
Wolfram Schlenker and Michael Roberts, "Nonlinear Temperature Effects Indi-cate Severe Damages to U.S. Crop Yields under Climate Change," *Proceedings of the National Academy of Sciences* 106, no. 37 (October 2008): 15594–98.

149 increasing global preference for resource-intensive meat consumption
Johnson, "Food Price Volatility and Insecurity."

149 food crops to crops suitable for biofuel
Ibid.

149 Conversion of cropland to urban and suburban sprawl
Lester Brown, *Plan B 4.0: Mobilizing to Save Civilization* (New York: Norton, 2009), http://www.earth-policy.org/images/uploads/book_files/pb4book.pdf.

150 with 67 percent of its people under the age of twenty-four
John Ishiyama et al., "Environmental Degradation and Genocide, 1958–2007," *Ethnopolitics* 11 (2012): 141–58.

150 "don't succeed in solving them by our own actions"
Jared Diamond, "Malthus in Africa, Rwanda's Genocide," ch. 10 in *Collapse: How Societies Choose to Fail or Succeed* (New York: Viking, 2005).

150 run into a wall
"Groundwater Depletion Rate Accelerating Worldwide," *ScienceDaily*, Septem-ber 23, 2010, http://www.sciencedaily.com/releases/2010/09/100923142503.htm.

150 "when the balloon bursts, untold anarchy"
Fred Pearce, "Asian Farmers Sucking the Continent Dry," *New Scientist*, Au-gust 2004.

150 capital city of Sana'a only one day in four
Lester Brown, "This Will Be the Arab World's Next Battle," *Guardian*, April 22, 2011.

150 declined more than 30 percent in the last four decades
Ibid.

150 "a hydrological basket case"
Ibid.

151 dramatically overminimize the future effects of choices
David Laibson, "Golden Eggs and Hyperbolic Discounting," *Quarterly Journal of Economics* 112 (May 1997): 443–78.

151 more than 95 percent of the new additions will be in developing countries
United Nations Department of Economic and Social Affairs, "World Population Prospects: The 2010 Revision," 2011, http://esa.un.org/wpp/Documentation/pdf/WPP2010_Highlights.pdf.

151 global population will take place in cities
United Nations Department of Economic and Social Affairs, "World Urbaniza-

tion Prospects: The 2011 Revision," March 2012, http://esa.un.org/unpd/wup/pdf/WUP2011_Highlights.pdf.

151 **population of the world at the beginning of the 1990s**
Ibid.; U.S. Census Bureau, "Total Midyear Population for the World: 1950–2050," http://www.census.gov/population/international/data/worldpop/table_population.php.

151 **increased tenfold over the last forty years**
United Nations Department of Economic and Social Affairs, "World Urbanization Prospects: The 2011 Revision."

151 **see a reduction in their share of the world's urban population**
Ibid.

152 **no more than 15 percent of people ever lived in urban areas**
Susan Thomas, "Urbanization as a Driver of Change," *Arup Journal*, 2008.

152 **still only 13 percent at the beginning of the twentieth century**
Sukkoo Kim, "Urbanization," *The New Palgrave Dictionary of Economics* (New York: Palgrave Macmillan, 2008); Thomas, "Urbanization as a Driver of Change."

152 **for the first time, more than half of us lived in cities**
United Nations Department of Economic and Social Affairs, "World Urbanization Prospects: The 2011 Revision."

152 **64 percent of the population in less developed countries living in cities**
Ibid.

152 **in 2013, twenty-three cities will have more than that number**
Ibid.

152 **thirty-seven such megacities will sprawl across the Earth**
Ibid.

152 **a projected increase of 175 percent between 2000 and 2030**
United Nations Population Fund (UNFPA), *State of World Population 2007: Unleashing the Potential of Urban Growth*, http://www.unfpa.org/swp/2007/english/introduction.html.

152 **from 11 million today to just under 19 million in 2025**
United Nations Department of Economic and Social Affairs, "World Urbanization Prospects: The 2011 Revision."

152 **projected to have a population of almost 33 million people by 2025**
Ibid.

152 **By 2050, almost 70 percent of the world's population will be city dwellers**
Ibid.

152 **roughly one out of every three inhabitants of cities**
Most of us have an image of what a "slum" is, but in fact slums come in different shapes and sizes. What they share in common, according to the United Nations definition: "lacking at least one of the basic conditions of decent housing: adequate sanitation, improved water supply, durable housing or adequate living space." UNFPA, *State of World Population 2007: Unleashing the Potential of Urban Growth*.

152 **two billion people within the next seventeen years**
Ben Sutherland, "Slum Dwellers 'to Top 2 Billion,'" BBC, June 20, 2006, http://news.bbc.co.uk/2/hi/in_depth/5099038.stm.

152 **growing at a rate even faster than the overall urban growth rate**
United Nations Department of Economic and Social Affairs, World Population Monitoring: Focusing on Population Distribution, Urbanization, Internal Migration, and Development, 2009.

152 **especially in developing countries—do so to earn higher incomes**
David Satterthwaite et al., "Urbanization and Its Implications for Food and Farming," *Philosophical Transactions of the Royal Society B* 365, no. 1554 (2010): 2809–20.

153 **and into the middle class—particularly in Asia**
European Strategy and Policy Analysis System, *Global Trends 2030—Citizens in an Interconnected and Polycentric World*, http://www.iss.europa.eu/uploads/media/ESPAS_report_01.pdf.

153 **growing global middle class will live in cities**
Ibid., p. 19.

153 **more than 80 percent of global production takes place in cities**
Richard Dobbs, Jaana Remes, and Charles Roxburgh, "Boomtown 2025: A Special Report," *Foreign Policy*, March 24, 2011.

153 **significantly higher than in rural areas**
David Owen, *Green Metropolis: Why Living Smaller, Living Closer, and Driving Less Are the Keys to Sustainability* (New York: Riverhead Trade, 2010); Qi Jingmei, "Urbanization Helps Consumption," *China Daily*, December 15, 2009.

153 **per capita consumption of meat**
United Nations Food and Agriculture Organization, "Livestock in the Balance," *The State of Food and Agriculture 2009*.

153 **nine kilograms of plant protein are consumed**
"Mankind Benefits from Eating Less Meat," PhysOrg, April 6, 2006, http://phys.org/news63547941.html.

153 **even as more than 900 million people**
United Nations, Millennium Development Goals Report 2011.

153 **by approximately twenty pounds in the last forty years**
Claudia Dreifus, "A Mathematical Challenge to Obesity," *New York Times*, May 14, 2012.

153 **half the adult population of the United States will be obese by 2030, with one quarter of them "severely obese"**
Eric Finkelstein et al., "Obesity and Severe Obesity Forecasts Through 2030," *American Journal of Preventive Medicine*, June 2012; "Most Americans May Be Obese by 2030, Report Warns," ABC News, September 18, 2012; "Fat and Getting Fatter: U.S. Obesity Rates to Soar by 2030," Reuters, September 18, 2012.

153 **steadily decreasing the number of people suffering from chronic hunger**
United Nations, Millennium Development Goals Report 2011.

153 **obesity has more than doubled in the last thirty years**
World Health Organization Media Centre, "Obesity and Overweight," May 2012, http://www.who.int/mediacentre/factsheets/fs311/en/index.html.

153 **more than a third of them are classified as obese**
Ibid.

154 **obese and overweight than from conditions related to being underweight**
Ibid.

154 suffering from diabetes die from either stroke or heart disease
Centers for Disease Control and Prevention, National Diabetes Statistics, 2011, http://diabetes.niddk.nih.gov/dm/pubs/statistics/.

154 almost 17 percent of U.S. children are obese today
Tara Parker-Pope, "Obesity Rates Stall, but No Decline," *New York Times*, Well blog, January 17, 2012, http://well.blogs.nytimes.com/2012/01/17/obesity-rates -stall-but-no-decline/.

154 almost 7 percent of all children in the world
ProCor, "Global: Childhood Obesity Rate Higher Than 20 Years Ago," September 28, 2010, http://www.procor.org/prevention/prevention_show.htm?doc_id =1367793.

154 will continue to grow in the future, both in the U.S. and globally
Parker-Pope, "Obesity Rates Stall, but No Decline."

154 trigger brain systems that increase the desire to eat more
Tara Parker-Pope, "How the Food Makers Captured Our Brains," *New York Times*, June 23, 2009.

155 "salt and sugars but low in vitamins, minerals and other micronutrients"
World Health Organization Media Centre, "Obesity and Overweight."

155 separated more people from reliable sources of fresh fruit and vegetables
"If You Build It, They May Not Come," *Economist*, July 7, 2011.

155 calories per gram in sweets and foods abundant in starch
David Bornstein, "Time to Revisit Food Deserts," *New York Times*, Opinionator blog, April 25, 2012, http://opinionator.blogs.nytimes.com/2012/04/25/time-to -revisit-food-deserts/.

155 while prices of fats declined by 15 percent and sugared soft drinks by 25 percent
Ibid.

155 knowledge necessary for food preparation both also play a role
Ibid.

155 threw the healthier food away
Vivian Yee, "No Appetite for Good-for-You School Lunches," *New York Times*, October 5, 2012.

155 introduction of American fast food outlets and climbing obesity rates
Jeannine Stein, "Wealthy Nations with a Lot of Fast Food: Destined to Be Obese?," *Los Angeles Times*, December 22, 2011.

155 Consequently, food prices went down significantly
Charles Kenny, "The Global Obesity Bomb," *BloombergBusinessweek*, June 4, 2012.

155 precisely with the large average weight gains and increased obesity
Dreifus, "A Mathematical Challenge to Obesity."

156 skimpily clad sex symbol washing a car
Eric Noe, "How Well Does Paris Sell Burgers?," ABC News, June 29, 2005, http:// abcnews.go.com/Business/story?id=893867&page=1#.UGMPQI40jdk.

156 equivalent of adding an extra one billion people
Matt McGrath, "Global Weight Gain More Damaging Than Rising Numbers," BBC, June 20, 2012.

156 first national circulation magazines and the first silent films
Johannes Malkmes, *American Consumer Culture and Its Society: From F. Scott Fitzgerald's 1920s Modernism to Bret Easton Ellis' 1980s Blank Fiction* (Hamburg: Diplomica, 2011), p. 44.

156 new products like automobiles and radios
Jeremy Rifkin, *The End of Work* (New York: Putnam, 1995), p. 22.

156 70 percent of U.S. homes by the end of the 1920s
Stephen Moore and Julian L. Simon, "The Greatest Century That Ever Was: 25 Miraculous Trends of the Past 100 Years," Cato Policy Analysis No. 364, Cato Institute, December 15, 1999, http://www.cato.org/pubs/pas/pa364.pdf, p. 20.

156 manufacturers and merchants in the emerging science of mass marketing
Rifkin, *The End of Work*, pp. 20–22.

156 industry entered a new and distinctly different role
Daniel Pope, "Making Sense of Advertisements," History Matters: The U.S. Survey on the Web, http://historymatters.gmu.edu/mse/ads/ads.pdf.

156 many of the other most prominent intellectuals in America
Russell Jacoby, "Freud's Visit to Clark U," *Chronicle of Higher Education*, September 2009.

157 American Psychoanalytic Society was founded two years after Freud's visit
Leon Hoffman, "Freud's Adirondack Vacation," *New York Times*, August 29, 2009.

157 Committee on Public Information
Woodrow Wilson: Executive Order 2594—Creating Committee on Public Information, April 13, 1917, American Presidency Project, http://www.presidency.ucsb.edu/ws/?pid=75409.

157 set out to introduce the techniques into mass marketing
Institute for Studies in Happiness, Economy, and Society, Alternatives and Complements to GDP-Measured Growth as a Framing Concept for Social Progress, 2012.

157 in order to avoid using the word "propaganda"
I explain the history of the word "propaganda"—and its meaning in the United States—in a previous book. See *The Assault on Reason* (New York: Penguin Press, 2007), pp. 93–96.

157 by Germany to describe its mass communications strategy
Sam Pocker, *Retail Anarchy: A Radical Shopper's Adventures in Consumption* (Philadelphia: Running Press, 2009), p. 122.

157 in their subconscious minds that might be relevant
Larry Tye, "The Father of Spin: Edward L. Bernays and the Birth of P.R.," PR Watch, 1999, http://www.prwatch.org/prwissues/1999Q2/bernays.html.

157 "Man's desires must overshadow his needs"
Paul Mazur, as quoted in *Century of the Self*, BBC Four, April–May 2002.

158 "It is they who pull the wires which control the public mind"
Edward Bernays, *Propaganda* (New York: Horace Liveright, 1928), p. 38.

158 sinister association of smoking with women's rights
William E. Geist, "Selling Soap to Children and Hairnets to Women," *New York Times*, March 27, 1985.

158 "new economic gospel of consumption"
Robert LaJeunesse, *Work Time Regulation as Sustainable Full Employment Strategy: The Social Effort Bargain* (New York: Routledge, 2009), pp. 37–38.

158 "part of the greater work of regeneration and redemption"
James B. Twitchell, *Adcult USA: The Triumph of Advertising in American Culture* (New York: Columbia University Press, 1996).

159 "it would seem that we can go on with increasing activity"
Benjamin Hunnicutt, *Work Without End: Abandoning Shorter Hours for the Right to Work* (Philadelphia: Temple University Press, 1988), p. 44.

159 "needs and wants of people have to be continuously stirred up"
"Retail Therapy," *Economist*, December 17, 2011.

159 use of Bernays's book *Propaganda* in organizing Hitler's genocide
Dennis W. Johnson, *Routledge Handbook of Political Management* (New York: Routledge, 2009), p. 314 n. 3; see Edward Bernays, *Biography of an Idea: Memoirs of Public Relations Counsel Edward L. Bernays* (New York: Simon & Schuster, 1965).

160 "infinitely more significant than any shifting of economic power"
Walter Lippmann, *Public Opinion* (New York: Harcourt, Brace, 1922), p. 248.

160 reinvigorated the use of subconscious analysis in the field of neuromarketing
Natasha Singer, "Making Ads That Whisper to the Brain," *New York Times*, November 14, 2010.

160 an average of 2,000 commercial messages per day thirty-five years ago
Louise Story, "Anywhere the Eye Can See, It's Likely to See an Ad," *New York Times*, January 15, 2007.

160 the average city dweller now sees 5,000 commercial messages per day
Ibid.

161 the total volume is projected to increase by 70 percent in a dozen years
Daniel Hoornweg and Perinaz Bhada-Tata, World Bank, "What a Waste: A Global Review of Solid Waste Management," March 2012.

161 $375 billion per year—with most of the increase in developing countries
Ibid.

161 a .69 percent increase in municipal solid waste in developed countries
Antonis Mavropoulos, "Waste Management 2030+," http://www.waste-management-world.com.

161 produced each day is more than the body weight of all seven billion people
Alexandra Sifferlin, "Weight of the World: Globally, Adults Are 16.5 Million Tons Overweight," *Time*, June 18, 2012; Paul Hawken, "Resource Waste," *Mother Jones*, March/April 1997; U.S. Environmental Protection Agency (EPA), "Municipal Solid Waste," http://www.epa.gov/epawaste/nonhaz/municipal/index.htm.

161 increased by more than 250 percent in the last decade
Mavropoulos, "Waste Management 2030+."

161 much larger volumes are on land in millions of waste dumps
EPA, "Municipal Solid Waste Generation, Recycling, and Disposal in the United States: Facts and Figures for 2010," November 2011, http://www.epa.gov/epawaste/nonhaz/municipal/pubs/msw_2010_rev_factsheet.pdf; NOAA Marine Debris Program, "De-mystifying the 'Great Pacific Garbage Patch,'" http://marinedebris.noaa.gov/info/patch.html.

161 **decomposes to produce 4 percent of all the global warming pollution**
Ian Williams, University of Southampton, "Future of Waste: Initial Perspectives," in Tim Jones and Caroline Dewing, eds., *Future Agenda: Initial Perspectives* (Newbury, UK: Vodafone Group, 2009), pp. 84–89.

162 **including pesticides, arsenic, cadmium, and flame retardants**
Centers for Disease Control and Prevention, Fourth National Report on Human Exposure to Environmental Chemicals, 2009, http://www.cdc.gov/exposurereport/pdf/FourthReport.pdf.

162 **smokers who fall asleep and drop their lit cigarettes**
Michael Hawthorne, "Testing Shows Treated Foam Offers No Safety Benefit," *Chicago Tribune*, May 6, 2012.

162 **enough influence to require the addition of dangerous chemicals**
Nicholas D. Kristof, "Are You Safe on That Sofa?," *New York Times*, May 19, 2012.

162 **exposure with evidence of cancer, reproductive disorders, and damage to fetuses**
Hawthorne, "Testing Shows Treated Foam Offers No Safety Benefit."

163 **added to the foam in the furniture didn't work**
Ibid.

163 **Toxic Substances Control Act, has never been truly implemented**
Bryan Walsh, "The Perils of Plastic," *Time*, April 1, 2010.

163 **relevant information about these chemicals from regulators**
Ibid.

163 **was also the inventor of synthetic nitrogen fertilizer**
Diarmuid Jeffreys, *Hell's Cartel: IG Farben and the Making of Hitler's War Machine* (New York: Metropolitan Books, 2008).

163 **South Asia, Africa, and portions of the Middle East**
John Cameron, Paul Hunter, Paul Jagals, and Katherine Pond, eds., *Valuing Water, Valuing Livelihoods*, World Health Organization, http://whqlibdoc.who.int/publications/2011/9781843393108_eng.pdf.

163 **"one-half of the world's major rivers"**
World Water Council, "Water and Nature," http://www.worldwatercouncil.org/index.php?id=21.

163 **"This is asymmetric accounting"**
Jorgen Randers, *2052: A Global Forecast for the Next Forty Years* (White River Junction, VT: Chelsea Green, 2012), p. 75.

164 **"Of course from the government's point of view"**
Keith Bradsher, "A Chinese City Moves to Limit Cars," *New York Times*, September 4, 2012.

164 **exposed to chemical waste or other health threats in their drinking water**
Charles Duhigg, "Clean Water Laws Are Neglected, at a Cost in Suffering," *New York Times*, September 13, 2009.

164 **"and 2.5 billion lack improved sanitation"**
World Health Organization, "Progress on Drinking Water and Sanitation: 2012 Update," http://www.wssinfo.org/fileadmin/user_upload/resources/JMP-report-2012-en.pdf.

164 **"2.4 billion people will lack access to improved sanitation facilities"**
Ibid.

164 ill each year due to their drinking water, and tens of thousands die
Jane Qiu, "China to Spend Billions Cleaning Up Groundwater," *Science*, November 2011, p. 745.

165 growing craze for deep shale gas
Chesapeake Energy, "Water Use in Deep Shale Gas Exploration," 2012, http://www.chk.com/Media/Educational-Library/Fact-Sheets/Corporate/Water_Use_Fact_Sheet.pdf; Jack Healy, "Struggle for Water in Colorado with Rise in Fracking," *New York Times*, September 5, 2012.

165 supplies in regions that were already experiencing shortages
Chesapeake Energy, "Water Use in Deep Shale Gas Exploration"; Healy, "Struggle for Water in Colorado with Rise in Fracking."

165 water for energy production is projected to grow
International Energy Agency, *World Energy Outlook 2012* (Paris: International Energy Agency, 2012).

165 opening new fissures and modifying underground flow patterns
Abrahm Lustgarten, "Are Fracking Wastewater Wells Poisoning the Ground beneath Our Feet?," *Scientific American*, June 21, 2012.

165 leaked waste upward into regions containing drinking water aquifers
Ibid.

165 one percent represented by all of the surface freshwater
"Groundwater Depletion Rate Accelerating Worldwide," *ScienceDaily*, September 23, 2010, http://www.sciencedaily.com/releases/2010/09/100923142503.htm.

165 rate of shrinkage in groundwater aquifers has doubled
Ibid.

165 increases have proceeded at a much faster pace
Ibid.

165 have been drilled by the 100 million Indian farmers
Lester Brown, *Plan B 4.0: Mobilizing to Save Civilization* (New York: Norton, 2009), http://www.earth-policy.org/images/uploads/book_files/pb4book.pdf.

165 farmers must rely on increasingly unpredictable rainfall
Ibid.

166 Australia, the Yangtze and Yellow rivers in China, and the Elbe
Geoffrey Lean, "Rivers: A Drying Shame," *Independent*, March 12, 2006.

166 estimated at between 10 and 15 billion
United Nations Department of Economic and Social Affairs, "World Population to Reach 10 Billion by 2100 If Fertility in All Countries Converges to Replacement Level," May 3, 2011, http://esa.un.org/wpp/Other-Information/Press_Release_WPP2010.pdf.

166 that the most likely range is slightly above 10 billion
Ibid.; United Nations Department of Economic and Social Affairs, "World Population Prospects: The 2010 Revision," 2011, http://esa.un.org/unpd/wpp/Analytical-Figures/htm/fig_1.htm.

166 at least the balance of the century as the most populous
United Nations Department of Economic and Social Affairs, "World Population Prospects: The 2010 Revision," 2011, http://esa.un.org/unpd/wpp/unpp/panel_population.htm.

166 end of the century is projected to have more people than both combined
Ibid.

166 to an astonishing 3.6 billion by the end of this century
David E. Bloom, "Africa's Daunting Challenges," *New York Times*, May 5, 2011.

167 fertility rate in scores of less developed countries—the majority of them in Africa
United Nations Department of Economic and Social Affairs, "World Population to Reach 10 Billion by 2100 If Fertility in All Countries Converges to Replacement Level."

167 management knowledge and technology available to women
Malcolm Potts and Martha Campbell, "The Myth of 9 Billion," *Foreign Policy*, May 9, 2011; Justin Gillis and Celia W. Dugger, "U.N. Forecasts 10.1 Billion People by Century's End," *New York Times*, May 4, 2011.

167 lowest level since the Great Depression
Bonnie Kavousi, "Birth Rate Plunges, Projected to Reach Lowest Level in Decades," *Huffington Post*, July 26, 2012.

167 creation of social conditions that can and do have an impact on population
T. Paul Shultz, Yale Economic Growth Center, "Fertility and Income," October 2005, www.econ.yale.edu/~pschultz/cdp925.pdf.

167 thirteen of the fourteen are in sub-Saharan Africa
Bloom, "Africa's Daunting Challenge."

168 ability of girls to become literate and to obtain a good education
United Nations, Report of the International Conference on Population and Development, Cairo, September 5–13, 1994, http://www.un.org/popin/icpd/conference/offeng/poa.html.

168 about family size and other issues
Ibid.

168 how many children they wish to have and the spacing
Ibid.

168 "the most powerful contraceptive"
Ibid.

168 including 98 percent of sexually active Catholic women
Nicholas D. Kristof, "Beyond Pelvic Politics," *New York Times*, February 11, 2012.

168 where thirty-nine out of the fifty-five African countries have high levels of fertility
United Nations Department of Economic and Social Affairs, "World Population to Reach 10 Billion by 2100 If Fertility in All Countries Converges to Replacement Level."

168 population will triple in the balance of this century
Ibid.

169 2.5 children during their childbearing years
Bloom, "Africa's Daunting Challenge."

169 average is almost 4.5 children per woman
Ibid.

169 leading to disruptive and unsustainable population growth
Ibid.

169 by the end of the century to an estimated 129 million
Gillis and Dugger, "U.N. Forecasts 10.1 Billion People by Century's End."

169 to more than 730 million people by 2100
Ibid.

169 would put Nigeria's population at the level of China in the mid-1960s
"Total Population, CBR, CDR, NIR and TFR of China (1949–2000)," *China Daily*, August 20, 2010.

170 significant declines in child and infant mortality
Potts and Campbell, "The Myth of 9 Billion"; Robert Kunzig, "Population 7 Billion," *National Geographic*, January 2011.

170 beginning of the nineteenth century—from thirty-five to seventy-seven years
Kunzig, "Population 7 Billion."

170 compared to 8 percent in 1970—of college students in Saudi Arabia were women
UNESCO Institute for Statistics, *Global Education Digest 2009: Comparing Education Statistics Across the World*, 2009, http://www.uis.unesco.org/template/pdf/ged/2009/GED_2009_EN.pdf, p. 227.

170 Arab states is now 48 percent; in Iran 51 percent
UNESCO Institute for Statistics, *Global Education Digest 2011*, 2011, http://www.uis.unesco.org/Education/Pages/ged-2011.aspx.

170 67 of the 120 nations for which statistics are available
Gary S. Becker, William H. J. Hubbard, and Kevin M. Murphy, "The Market for College Graduates and the Worldwide Boom in Higher Education of Women," *American Economic Review* 100, no. 2 (2010): 229–33.

170 The world average is 51 percent
Ibid.; World Bank, *The Road Not Traveled: Education Reform in the Middle East and North Africa*, MENA Development Report, 2008, http://siteresources.worldbank.org/INTMENA/Resources/EDU_Flagship_Full_ENG.pdf, p. 171.

170 61 percent of master degrees, and 51 percent of doctoral degrees
U.S. Department of Education, National Center for Education Statistics, "Fast Facts," 2010, http://nces.ed.gov/fastfacts/display.asp?id=72.

170 has announced plans to allow women to vote beginning in 2015
"Saudi Women to Receive Right to Vote—in 2015," NPR, September 26, 2011, http://www.npr.org/2011/09/26/140818249/saudi-women-get-the-vote.

170 only 18 percent of the gap in political participation
Ricardo Hausmann, Laura D. Tyson, and Saadia Zahidi, "The Global Gender Gap Index 2010," *Global Gender Gap Report 2010*, 2010, http://www3.weforum.org/docs/WEF_GenderGap_Report_2010.pdf.

171 two women have entered the workplace for every man
"A Guide to Womenomics," *Economist*, April 12, 2006.

171 with 83 women in the workforce for every 100 men
Ibid.

171 filling between 60 and 80 percent of the jobs
Ibid.

171 "has contributed much more to global growth than China has"
Ibid.

171 responsible for producing slightly less than 40 percent of GDP
Ibid.

171 contribution of women to GDP would be well over 50 percent
Ibid.

171 who work outside the home skyrocketed from 12 percent to 55 percent
Robert R. Reich, *Aftershock: The Next Economy and America's Future* (New York: Knopf, 2010), p. 61.

171 rose during the same three decades from 20 to 60 percent
Ibid.

171 all adds up to what Kessler calls "conditioned hyper-eating"
Tara Parker-Pope, "How the Food Makers Captured Our Brains," *New York Times*, June 23, 2009.

172 playing outside in neighborhoods that, relatively speaking, are prone to more violence
Rebecca Cecil-Carb and Andrew Grogan-Kaylor, "Childhood Body Mass Index in Community Context: Neighborhood Safety, Television Viewing and Growth Trajectories of BMI," *Health and Social Work* 34 (March 2009): 169–77.

172 partly because of the increased participation of women in the workforce
United Nations Division for Social Policy and Development Division, Family Unit, 2003–2004, Major Trends Affecting Families, "Introduction," http://social.un.org/index/LinkClick.aspx?fileticket=LJsVbHQC7Ss%3d&tabid=282.

172 between 20 and 30 percent of all divorces
Carl Bialik, "Irreconcilable Claim: Facebook Causes 1 in 5 Divorces," *Wall Street Journal*, March 12, 2011; Carl Bialik, "Divorcing Hype from Reality in Facebook Stats," *Wall Street Journal* blog, March 11, 2011, http://blogs.wsj.com/numbersguy/divorcing-hype-from-reality-in-facebook-stats-1046/.

172 Now, only one quarter are
Pew Research Center, "The Decline of Marriage and Rise of New Families," November 18, 2010, http://pewresearch.org/pubs/1802/decline-marriage-rise-new-families.

172 and having children—without getting married
Ibid.

172 are now born to unmarried women
Ibid.

172 were born to unmarried mothers
Ibid.

172 among mothers under thirty is 50 percent
Jason DeParle and Sabrina Tavernise, "For Women Under 30, Most Births Occur Outside Marriage," *New York Times*, February 17, 2012.

172 Among African American mothers of all ages
Ibid.

172 the percentage is now 73 percent
Ibid.

172 Iceland, Norway, Finland, and Sweden; the lowest rank goes to Yemen
Hausmann, Tyson, and Zahidi, "The Global Gender Gap Index 2010."

172 the lowest percentage (11.4 percent) in the Arab states
Inter-Parliamentary Union, "Women in National Parliaments," April 30, 2011,
http://www.ipu.org/wmn-e/world.htm.

173 a constitutional requirement that a minimum of 30 percent
Catherine Rampell, "A Female Parliamentary Majority in Just One Country:
Rwanda," *New York Times*, Economix blog, March 9, 2010, http://economix.blogs
.nytimes.com/2010/03/09/women-underrepresented-in-parliaments-around-the
-world/; Inter-Parliamentary Union, "Women in National Parliaments."

173 only 7 percent of corporate boards in the world
"A Guide to Womenomics," *Economist*.

173 have also fallen below the replacement rate
Steven Philip Kramer, "Baby Gap: How to Boost Birthrates and Avoid Demographic Decline," *Foreign Affairs*, May/June 2012.

173 The U.S. birthrate fell to an all-time low in 2011
Terence P. Jeffrey, "CDC: U.S. Birth Rate Hits All-Time Low; 40.7% of Babies
Born to Unmarried Women," CNS News, October 31, 2012, http://cnsnews.com/
news/article/cdc-us-birth-rate-hits-all-time-low-407-babies-born-unmarried
-women.

173 64 million by 2100
Bryan Walsh, "Japan: Still Shrinking," *Time*, August 28, 2006.

173 career paths after having children, and other benefits
Kramer, "Baby Gap."

173 now once again nearly at their replacement rate of fertility
Ibid.

173 not yet been able to slow their fertility declines
Ibid.

174 greater per capita expense of U.S. health care
Simon Rogers, "Healthcare Spending Around the World, Country by Country,"
Guardian, June 30, 2012; Harvey Morris, "U.S. Healthcare Costs More Than 'So-
cialized' European Medicine," *International Herald Tribune*, June 28, 2012.

174 year 2000 are projected to live past the age of 100
"Most Babies Born Today May Live Past 100," ABC News, October 1, 2009,
http://abcnews.go.com/Health/WellnessNews/half-todays-babies-expected-live
-past-100/story?id=8724273.

174 will live to be more than 104
Ibid.

174 less than thirty years; some believe much less
Nicholas Wade, "Genetic Data and Fossil Evidence Tell Differing Tales of
Human Origins," *New York Times*, July 27, 2012; Sonia Arrison, "Average Life Ex-
pectancy Through History," *Wall Street Journal*, August 27, 2011.

174 but not until the middle of the nineteenth century
Arrison, "Average Life Expectancy Through History."

174 and in most industrial countries are now in the high seventies
United Nations Department of Economic and Social Affairs, *World Population
Prospects: The 2010 Revision;* Arrison, "Average Life Expectancy Through History."

174 **aged sixty-five and older within the next quarter century**
Ted C. Fishman, "As Populations Age, a Chance for Younger Nations," *New York Times Magazine*, October 17, 2010.

175 **and by 2050 fully one third of Chinese will be sixty or older**
Ibid.; Joseph Chamie, former director of the United Nations Population Division, "The Battle of the Billionaires: China vs. India," *Globalist*, October 4, 2010.

175 **percentage of the elderly will still be half that in China**
Chamie, "The Battle of the Billionaires: China vs. India."

175 **the Japanese bought more adult diapers than baby diapers**
Sam Jones and Ben McLannahan, "Hedge Funds Say Shorting Japan Will Work," *Financial Times*, November 29, 2012.

175 **increase from twenty-eight today to forty by midcentury**
Ibid.

175 **contributed to the pressures that resulted in the French Revolution**
NPR, "In Arab Conflicts, the Young Are the Restless," NPR, February 8, 2012, http://www.npr.org/2011/02/09/133567583/in-arab-conflicts-the-young-are-the-restless.

175 **majority of the revolutions in developing countries**
Jack Goldstone, "Population and Security: How Demographic Change Can Lead to Violent Conflict," *Journal of International Affairs* 56 (2002).

175 **coincided with the young adulthood of the post–World War II**
Kenneth Weiss, "Runaway Population Growth Often Fuels Youth-Driven Uprisings," *Los Angeles Times*, July 22, 2012.

176 **twice the rate of countries generally**
"In Arab Conflicts, the Young Are the Restless," NPR.

176 **have been in nations with youth bulges**
"The Hazards of Youth," *WorldWatch*, October 2004.

176 **during a period of food price hikes around the world**
Joseph Chamie, "A 'Youth Bulge' Feeds Arab Discontent," *Daily Star*, April 15, 2011; Ashley Fantz, "Tunisian on Life One Year Later: No Fear," CNN, December 16, 2011, http://www.cnn.com/2011/12/16/world/meast/tunisia-immolation-anniversary/index.html.

176 **number of jobs available to them is exceptionally low**
Madawi Al-Rasheed, "Yes, It Could Happen Here: Why Saudi Arabia Is Ripe for Revolution," *Foreign Policy*, February 28, 2011.

176 **will reach only 40 by midcentury**
Fishman, "As Populations Age, a Chance for Younger Nations."

176 **fertility of immigrant populations**
Ibid.

176 **population of developed countries to 10 percent**
United Nations Department of Economic and Social Affairs, "Trends in International Migrant Stock: Migrants by Age and Sex," http://esa.un.org/MigAge/index.asp?panel=8; United Nations Department of Social and Economic Affairs, "Trends in International Migrant Stock: The 2008 Revision," July 2009, http://www.un.org/esa/population/migration/UN_MigStock_2008.pdf.

176 **increase from 7.2 percent twenty years earlier**
Ibid.

176 moved from one region to another inside countries
Fiona Harvey, "Climate Change Could Trap Hundreds of Millions in Disaster Areas, Report Claims," *Guardian*, October 20, 2011.

176 developing country to developed regions of the world
Report of the Secretary-General, United Nations General Assembly, "International Migration and Development," May 18, 2006.

176 "are about as numerous as those moving 'South-to-North'"
Ibid.

177 including Afghanistan, Pakistan, and Algeria
Anne-Sophie Labadie, "Greek Far-Right Rise Cows Battered Immigrants," *Daily Star*, May 25, 2012.

177 the trans-Caucasus region, where there are significant Muslim populations
Atryom Liss, "Neo-Nazi Skinheads Jailed in Russia for Racist Killings," BBC, February 25, 2010, http://news.bbc.co.uk/2/hi/europe/8537861.stm; Mansur Mirovalev, "Russia: Far-Right Nationalists and Neo-Nazis March in Moscow," Associated Press, November 4, 2011.

177 three quarters of them have less than one million people
Report of the Secretary-General, "International Migration and Development."

177 make up 10 percent of the population or more
United Nations Department of Social and Economic Affairs, "Trends in International Migrant Stock: The 2008 Revision."

177 2,100-mile-long, 2.5-meter-high iron fence
Kurt M. Campbell et al., "The Age of Consequences: The Foreign Policy and National Security Implications of Global Climate Change," Center for Strategic & International Studies, November 2007, http://www.climateactionproject.com/docs/071105_ageofconsequences.pdf.

177 a surge of internal migration from low-lying coastal areas and offshore islands
Ibid.

177 in the Bay of Bengal, where four million people currently live
Ibid.

177 population of Bangladesh
Ibid.

178 even though it has only 5 percent of the world's population
Global Migration Group, "International Migration and Human Rights," October 2008, http://www.unhcr.org/cgi-bin/texis/vtx/home/opendocPDFViewer.html?docid=49e479cf0&query=migration.

178 outnumbered Caucasian babies for the first time
Conor Dougherty and Miriam Jordan, "Minority Births Are New Majority," *Wall Street Journal*, May 17, 2012.

178 experts as a major factor causing the surge of hate groups
Colleen Curry, "Hate Groups Grow as Racial Tipping Point Changes Demographics," ABC News, May 18, 2012, http://abcnews.go.com/US/militias-hate-groups-grow-response-minority-population-boom/story?id=16370136#.T7Zy1O2I3dl.

178 "nation's population growth in the decade that ended in 2010"
Sabrina Tavernise, "Whites Account for Under Half of Births in U.S.," *New York Times*, May 17, 2012.

178 **as the number of Hispanic and Asian children increased by 5.5 million**
William H. Frey, "America's Diverse Future: Initial Glimpses at the U.S. Child Population from the 2010 Census," Brookings Institution, April 6, 2011, http://www.brookings.edu/papers/2011/0406_census_diversity_frey.aspx.

178 **Hispanics at 26 percent and African Americans at 22 percent**
William H. Frey, "Melting Pot Cities and Suburbs: Racial and Ethnic Change in Metro America in the 2000s," Brookings Institution, May 4, 2011, http://www.brookings.edu/papers/2011/0504_census_ethnicity_frey.aspx.

178 **represent the largest minority group in the United States**
Dennis Cauchon and Paul Overberg, "Census Data Shows Minorities Now a Majority of U.S. Births," *USA Today*, May 17, 2012.

178 **bombing of the federal office building in Oklahoma City**
Brian Levin, "U.S. Hate and Extremist Groups Hit Record Levels, New Report Says," *Huffington Post*, March 8, 2012, http://www.huffingtonpost.com/brian-levin-jd/hate-groups-splc_b_1331318.html.

178 **renewed upsurge in 2009–12**
Ibid.

178 **"precisely coinciding with Obama's first three years as president"**
Curry, "Hate Groups Grow as Racial Tipping Point Changes Demographics."

178 **immigration from several other countries has continued**
Jeffrey Passel, D'Vera Cohn, and Ana Gonzalez-Barrera, Pew Research Center, "Net Migration from Mexico Falls to Zero—and Perhaps Less," May 3, 2012, http://www.pewhispanic.org/2012/04/23/net-migration-from-mexico-falls-to-zero-and-perhaps-less/.

178 **Flows of Asian immigrants to the U.S. overtook Hispanics**
"Asians Overtake Hispanics as Largest US Immigration Group," *Telegraph*, June 20, 2009.

178 **"by around 2023 if current immigration trends continue"**
William H. Frey, "A Demographic Tipping Point Among America's Three-Year-Olds," Brookings Institution, February 7, 2011, http://www.brookings.edu/research/opinions/2011/02/07-population-frey.

179 **democratic principle of majority rule**
"Arab Majority in 'Historic Palestine' After 2014: Survey," Agence France-Presse, December 30, 2010.

179 **higher standards of living in developed countries**
Report of the Secretary-General, "International Migration and Development."

179 **results in less support for public school budgets**
Tavernise, "Whites Account for Under Half of Births in U.S."

179 **United States, Australia, and the United Kingdom**
Report of the Secretary-General, "International Migration and Development."

179 **totaled $351 billion in 2011 and is projected to reach $441 billion**
Dipil Ratha, World Bank, "Outlook for Migration and Remittances 2012–14," February 9, 2012.

179 **back home from the cities where they work**
Overseas Development Institute, "Internal Migration, Poverty and Development in Asia, October 2006," http://www.odi.org.uk/resources/download/29.pdf.

179 as much as 60 percent of their income
Ibid.

180 majority of money flowing into those three states
Ibid.

180 persecution to new communities within their own country
United Nations Refugee Agency, "UNHCR: Global Trends," 2010.

180 "it's becoming more and more difficult to find solutions for them"
"UN Report Predicts Increase in World's Displaced," Associated Press, June 1, 2012.

180 meaning they have no place to go home to
United Nations, Millennium Development Goals Report 2011.

180 more refugees moved to cities than to refugee camps
United Nations High Commissioner for Refugees (UNHCR), "2009 Global Trends: Refugees, Asylum-Seekers, Returnees, Internally Displaced and Stateless Persons," June 15, 2010, http://www.unhcr.org/refworld/docid/4caee6552 .html.

180 80 percent of refugees live in poor regions of the world
Antoine Pécoud and Paul de Guchteneire, UNESCO, "International Migration, Border Controls and Human Rights: Assessing the Relevance of a Right to Mobility," *Journal of Borderlands Studies* 21, no. 1 (Spring 2006).

180 Myanmar, Colombia, and Sudan
UNHCR, "2009 Global Trends."

180 countries for refugees are Afghanistan and Iraq
Ibid.

180 mostly to Pakistan (1.9 million) and Iran (one million)
UNHCR, "Global Trends 2010," http://www.unhcr.org/4dfa11499.pdf.

180 Iraq have also gone mostly to neighboring countries
Ibid.

180 are hosted in nations neighboring their country of origin
"The Impacts of Refugees on Neighboring Countries: A Development Challenge," World Development Report 2011 Background Note, July 29, 2010, http://wdronline.worldbank.org/worldbank/a/nonwdrdetail/199.

181 the Middle East and North Africa (another 1.9 million)
UNHCR, "Global Trends 2010."

181 Muslims already make up 5 percent of Europe's population
Ibid.; Kurt M. Campbell et al., "The Age of Consequences."

181 nativist groups exploit the public's uneasiness
Peter Walker and Matthew Taylor, "Far Right on Rise in Europe, Says Report," *Guardian,* November 6, 2011.

181 "to the relentless advance of climate change"
"UN Report Predicts Increase in World's Displaced," Associated Press.

181 to protect against a predicted wave of climate refugees
Sharon Udasin, "Defending Israel's Borders from 'Climate Refugees,'" *Jerusalem Post,* May 15, 2012.

181 Israeli environmental protection minister Gilad Erdan
Ibid.

181 "where it is possible to escape this"
Ibid.

182 "they shoot, in Japan they shoot"
Ibid.

182 many climate refugees eastward into the Darfur region
Ibid.

182 Palestinian territories, Syria, and the Nile Delta in Egypt
Ibid.

182 "exactly as Europe is doing now"
Ibid.

182 "spur additional migration from Africa and South Asia"
Campbell et al., "The Age of Consequences."

182 the dangerous journey across to the Canaries
"Canaries Migrant Surge Tops 1,400," BBC, September 4, 2006, http://news.bbc
.co.uk/2/hi/europe/5310412.stm.

182 one meter or less higher than the current sea level
"Sea Levels May Rise by as Much as One Meter Before the End of the Century,"
ScienceDaily, June 10, 2012.

182 abandon the places they call home
Hugo Ahlenius, "Population, Area and Economy Affected by 1m Sea Level Rise,"
UNEP/GRID-Arendal, 2007, http://www.grida.no/graphicslib/detail/population
-area-and-economy-affected-by-a-1-m-sea-level-rise-global-and-regional
-estimates-based-on-todays-situation_d4fe.

183 rate of less than one half of one percent per year
WorldWatch Institute, "World Population, Agriculture, and Malnutrition," 2011.

183 approximately 2.5 centimeters every 500 years
Pete Miller and Laura Westra, *Just Ecological Integrity: The Ethics of Maintaining a
Planetary Life* (Lanham, MD: Rowman & Littlefield, 2002), p. 124.

183 productivity on almost one third of the arable land on Earth
Jims Vincent Capuno, "Soil Erosion: The Country's Unseen Enemy," Edge Davao,
July 11, 2011, http://www.edgedavao.net/index.php?option=com_content&view
=article&id=4801:soil-erosion-the-countrys-unseen-enemy&catid=51:on-the
-cover&Itemid=83.

183 ten times faster than it can be replenished
Tom Paulson, "The Lowdown on Topsoil: It's Disappearing," *Seattle Post-
Intelligencer*, January 21, 2008.

183 erodible soils down the steep slopes of its terrain
Lester Brown, "Civilization's Founding Eroding," September 28, 2010, http://
www.earth-policy.org/book_bytes/2010/pb4ch02_ss2.

183 have many experts beginning to get very worried
"Groundwater Depletion Rate Accelerating Worldwide," *ScienceDaily*, September 23, 2010, http://www.sciencedaily.com/releases/2010/09/100923142503.htm.

183 many aquifers are now falling several meters per year
"No Easy Fix: Simply Using More of Everything to Produce More Food Will Not Work," *Economist,* February 24, 2011.

184 "costs and the benefits in separate accounts for comparison"
Jorgen Randers, *2052: A Global Forecast for the Next Forty Years* (White River Junction, VT: Chelsea Green, 2012), p. 75.

184 "practical confusion between income and capital"
R. H. Parker and G. C. Harcourt, *Readings in the Concept and Measurement of Income* (Cambridge, UK: Cambridge University Press, 1969), p. 81.

184 holds true for nations and for the world as a whole
Kevin Holmes, *The Concept of Income: A Multi-disciplinary Analysis* (Amsterdam: IBFD, 2001), p. 109.

184 "system of environmental-economic accounts"
Janez Potočnik, "Our Natural Capital Is Endangered," European Union press release, June 20, 2012.

184 argument is often contingent on oversimplification
Simon Kuznets, *National Income 1929–1932,* Report to the U.S. Senate, 73rd Congress, 2nd Session (Washington, DC: U.S. Government Printing Office, 1934), www.nber.org/chapters/c2258.pdf.

185 "the center of conflict of opposing social groups"
Ibid.

185 "farmers did not get enough scientific guidance"
Li Jiao, "Water Shortages Loom as Northern China's Aquifers Are Sucked Dry," *Science,* June 2010.

185 U.S. depends far less on irrigation
Brown, *Plan B 4.0.*

185 all twenty-one of the world's longest rivers
"Dams Control Most of the World's Large Rivers," Environmental News Service, April 2005, http://www.ens-newswire.com/ens/apr2005/2005-04-15-04.asp.

186 when it was built seventy years ago
U.S. Bureau of Reclamation, "What Is the Biggest Dam in the World?," June 2012, http://www.usbr.gov/lc/hooverdam/History/essays/biggest.html.

186 global freshwater was used for agriculture
"No Easy Fix," *Economist.*

186 780 million people in the world still lack access to safe drinking water
Ibid.; UNICEF, "Water, Sanitation and Hygiene: Introduction," March 2012; World Health Organization, "Progress on Drinking Water and Sanitation: 2012 Update," 2012, http://whqlibdoc.who.int/publications/2012/9789280646320_eng_full_text.pdf.

186 has water found to be one million years old
Jack Eggleston, U.S. Geological Survey, "Million Year Old Groundwater in Maryland Water Supply," June 2012, http://www.usgs.gov/newsroom/article.asp?ID=3246#.UGS3kRh9lbo.

186 all have water more than one million years old
Ibid.

186 classic case of "out of sight, out of mind"
Jiao, "Water Shortages Loom as Northern China's Aquifers Are Sucked Dry."

186 will dramatically increase sea level rise later in this century
"Groundwater Depletion Rate Accelerating Worldwide," *ScienceDaily*.

186 Central Valley of California, and northeastern China
Ibid.

186 unsustainably to irrigate crops in dryland areas
Jiao, "Water Shortages Loom as Northern China's Aquifers Are Sucked Dry."

187 intended to remedy water shortages in northern China
Edward Wong, "Plan for China's Water Crisis Spurs Concern," *New York Times*, June 1, 2011.

187 "where the world's major irrigated lands are located"
UNEP, "Water Withdrawal and Consumption: The Big Gap," 2008, http://www.unep.org/dewa/vitalwater/article42.html.

187 water withdrawal in the coming decades
Ibid.; Matthew Power, "Peak Water: Aquifers and Rivers Are Running Dry. How Three Regions Are Coping," *Wired*, April 21, 2008.

187 Europe is consuming only a slightly larger percentage
Ibid.

187 are already experiencing severe shortages
Paul Quinlan, "US-Mexico Pact Hailed as Key Step Towards Solving Southwest Water Supply Woes," *New York Times*, December 22, 2010.

187 herded north from Texas to wetter, cooler pastures
Drover's Cattle, "More Than 150,000 Breeding Cattle Leave Texas in 2011 Drought," February 2012, http://www.cattlenetwork.com/e-newsletters/drovers-daily/More-than-150000-breeding-cattle-leave-Texas-in-2011-drought-138513934.html.

187 will run completely dry before the end of this decade
"Dry Lake Mead? 50-50 Chance by 2021 Seen," MSNBC, February 2008, http://www.msnbc.msn.com/id/23130256/ns/us_news-environment/t/dry-lake-mead--chance-seen/#.UGSvsBh9lbo.

187 has dropped more than 100 feet
Brown, *Plan B 4.0*.

187 two minutes on average, twenty-four hours a day
Charles Duhigg, "Saving US Water Systems Could Be Costly," *New York Times*, March 14, 2010.

187 like groundwater resources—"out of sight, out of mind"
Ibid.

187 vast new quantities of needed freshwater
Power, "Peak Water."

187 agricultural irrigation practices are still extremely wasteful
T. Marc Schober, "Irrigation: Yield Enhancer or Farmland Destroyer?," Seeking Alpha, July 11, 2011, http://seekingalpha.com/instablog/362794-t-marc-schober/194359-irrigation-yield-enhancer-or-farmland-destroyer; "No Easy Fix," *Economist;* World Health Organization, "Progress on Drinking Water and Sanitation: 2012 Update."

188 **farmers have been slow to make the change**
Sandra Postel, "Drip Irrigation Expanding Worldwide," *National Geographic*, June 25, 2012.

188 **amounts of salt that build up with continued use**
World Wildlife Fund, "Farming: Wasteful Water Use," 2005, http://wwf.panda .org/what_we_do/footprint/agriculture/impacts/water_use/.

188 **safe for watering plants**
Nancy Farghalli, "Recycling 'Grey Water' Cheaply," NPR News, June 2009, http://www.npr.org/templates/story/story.php?storyId=105089381.

188 **purify it, and put it into drinking water systems**
Kate Galbraith, "Taking the Ick Factor out of Recycled Water," *New York Times*, July 25, 2012.

188 **communities have successfully implemented the approach**
Ibid.

188 **more of the rainfall and store it for drinking water**
Peter Gleick and Matthew Herberger, "Devastating Drought Seems Inevitable in American West," *Scientific American*, January 2012.

188 **roughly 10 percent of the Earth's surface**
Susan Lang, "'Slow Insidious' Soil Erosion Threatens Human Health and Welfare as Well as the Environment, Cornell Study Asserts," *Cornell Chronicle*, March 2006.

189 **accelerate the emission of carbon dioxide into the atmosphere**
Personal conversation with Rattan Lal.

189 **increasing the fertility of the topsoil**
David R. Huggins and John P. Reganold, "No-Till: The Quiet Revolution," *Scientific American*, July 2008, pp. 70–77.

189 **replenish soil carbon and nitrogen**
Michael Pollan, *The Omnivore's Dilemma: A Natural History of Four Meals* (New York: Penguin, 2006), p. 42.

189 **expensive liability instead of a valued asset**
Ibid., p. 78.

189 **nontoxic manure as fertilizer and a three-year crop rotation**
Mark Bittman, "A Simple Fix for Farming," *New York Times*, October 19, 2012.

189 **virtually all of the nitrogen is derived**
U.S. Government Accountability Office, "Domestic Nitrogen Fertilizer Depends On Natural Gas Availability and Prices," 2003, p. 1, http://www.gao.gov/products/ GAO-03-1148.

189 **use per acre has been increasing dramatically**
Jeremy Grantham, "Time to Wake Up: Days of Abundant Resources and Falling Prices Are Over Forever," *GMO Quarterly Letter*, April 2011.

189 **devoid of life, which are growing in several ocean regions**
Robert Diaz and Rutger Rosenberg, "Spreading Dead Zones and Consequences for Marine Ecosystems," *Science*, April 15, 2008.

190 **recent spectacular algae blooms in Chinese**
"No Easy Fix," *Economist*.

190 **U.S., China, Southeast Asia, and parts of Latin America**
"Nitrogen Pollution an Increasing Problem Globally," PRI's The World, January 27, 2009, http://www.pri.org/stories/science/environment/nitrogen-pollution-an-increasing-problem-globally-8166.html.

190 **tripled the depletion of phosphorus from cropland**
David Vaccari, "Phosphorus: A Looming Crisis," *Scientific American,* June 2009.

190 **where 65 percent of U.S. production now takes place**
Ibid.; James Elser and Stuart White, "Peak Phosphorus," *Foreign Policy*, April 20, 2010.

190 **search for new reserves is beginning again**
Ibid.

190 **the "Saudi Arabia of phosphorus"**
Ibid.

190 **exports during the 2008 food price crisis**
Elser and White, "Peak Phosphorus."

191 **order to extend the supplies of phosphorus for fertilizers**
Mara Grunbaum, "Gee Whiz: Human Urine Is Shown to Be an Effective Agricultural Fertilizer," *Scientific American*, July 23, 2010.

191 **soil fertility and enhance the sequestration of soil carbon**
Rifat Hayat et al., "Soil Beneficial Bacteria and Their Role in Plant Growth Promotion: A Review," *Annals of Microbiology* 60, no. 4 (December 2010): 579–98; Tim J. LaSalle, *Regenerative Organic Farming: A Solution to Global Warming*, Rodale Institute, July 30, 2008, pp. 2–3, http://www.rodaleinstitute.org/files/Rodale_Research_Paper-07_30_08.pdf.

191 **soil and further protect against erosion**
J. Paul Mueller, Denise Finney, and Paul Hepperly, "The Field System," in *The Sciences and Art of Adaptive Management: Innovating for Sustainable Agriculture and Natural Resource Management*, edited by Keith M. Moore (Ankeny, IA: Soil and Water Conservation Society, 2009).

191 **fertility of the soil while diminishing erosion**
Huggins and Reganold, "No-Till: The Quiet Revolution."

191 **carefully managed way can also improve yields and soil quality**
David Laird and Jeffrey Novak, "Biochar and Soil Quality," *Encyclopedia of Soil Science*, 2nd ed. (New York: Taylor & Francis, 2011), pp. 1–4.

191 **when Victory Gardens were planted during World War II**
National WWII Museum, "Victory Gardens at a Glance," 2009, http://www.nationalww2museum.org/learn/education/for-students/ww2-history/at-a-glance/victory-gardens.html.

191 **keep up with the extra food production needed**
David Pimentel et al., "Impact of a Growing Population on Natural Resources: The Challenge for Environmental Management," *Frontiers* 3 (1997).

191 **(approximately 25 million acres) every year**
Lang, "'Slow Insidious' Soil Erosion."

192 **mostly in Kazakhstan (1954)—and created their own Dust Bowl**
Lester Brown, *World on the Edge* (New York: Norton, 2011), http://www.earthpolicy.org/books/wote/wotech3.

192 Aral Sea, almost completely disappeared
NASA, "A Shrinking Sea, Aral Sea," July 23, 2012, http://www.nasa.gov/mission
_pages/landsat/news/40th-top10-aralsea.html.

192 eroded land to grassland and a nationwide effort to fight soil erosion
Andrew Glass, "FDR Signs Soil Conservation Act, April 27, 1935," *Politico*, http://
www.politico.com/news/stories/0410/36362.html.

192 "Drylands are on the front line"
Alister Doyle, "World Urged to Stop Net Desertification by 2030," Reuters,
June 14, 2011.

192 way of life for an estimated one billion people in 100 countries
Ibid.

192 "dust storm moved through a large swatch of Arizona"
"Historic Dust Storm Sweeps Across Arizona, Turns Day into Night," July 6, 2011,
Reuters.

193 unusually large number of them in recent years
"7 Haboobs Have Hit Arizona Since July," KVOA, September 28, 2011, http://
www.kvoa.com/news/7-haboobs-have-hit-arizona-since-july/.

193 describing what is in store for many regions of desertifying drylands
Joe Romm, "Desertification: The Next Dust Bowl," *Nature*, October 2011.

193 "China, western Mongolia, and central Asia; the other in central Africa"
Lester Brown, "The Great Food Crisis of 2011," *Foreign Policy*, January 10, 2011.

193 have increased tenfold during the last fifty years
Gaia Vince, "Dust Storms on the Rise Globally," *New Scientist*, August 2004.

193 "activity and 40 percent of the continent's population"
"Desertification Affects 70 Percent of Economic Activity in Africa," *Pana Press*,
October 24, 2011.

193 United States' Midwest just prior to the Dust Bowl
Rattan Lal, interview with author, July 2, 2009; Rattan Lal, "Global Potential of
Soil Carbon Sequestration to Mitigate the Greenhouse Effect," *Critical Reviews in
Plant Sciences* 22, no. 2 (2003): 151–84.

193 the number of livestock exploded
Brown, *Plan B 4.0*.

193 Muslims moving from the north into non-Muslim areas
Ibid.

193 surrounding China's Gobi Desert
Damien Currington, "Desertification Is the Greatest Threat to the Planet, Experts Warn," *Guardian*, December 15, 2010.

193 goats compared to less than 10 million in the United States
Ibid.

194 China is now losing almost 1,400 square miles of arable land
Ibid.

194 Inner Mongolia and Gansu Province are merging and expanding
Ibid.

194 Taklamakan and Kumtag deserts are also merging and expanding
Ibid.

194 abandoned in northern and western regions of China
Ibid.

194 have already abandoned many villages to the encroaching desert
Ibid.

194 "a savannah-like region bordering the basin on its south side"
Ibid.

194 Amazon rainforest, adding even more risk to the integrity
David Lapola et al., "Indirect Land-Use Changes Can Overcome Carbon Savings from Biofuels in Brazil," *Proceedings of the National Academy of Sciences*, January 2010.

194 suffered from two "hundred-year" droughts in the last seven years
Simon Lewis et al., "The 2010 Amazon Drought," *Science*, February 2011.

194 greatest tropical rainforest on Earth into a massive dryland region
Brown, *Plan B 4.0*.

194 90 percent of the people affected live in developing countries
Currington, "Desertification Is the Greatest Threat to the Planet, Experts Warn."

194 "The top 20 centimeters of soil"
Ibid.

195 to accommodate additional shelter for Egypt's fast-growing population
Metwali Salem, "UN Report: Egypt Sustains Severe Land Loss to Desertification and Development," *Egypt Independent*, June 17, 2011.

195 resulting in the loss of cropland to salinization
"Seawater Intrusion Is the First Cause of Contamination of Coastal Aquifers," *ScienceDaily*, July 31, 2007, http://www.sciencedaily.com/releases/2007/07/070727091903.htm.

195 Ganges Delta, the Mekong Delta, and in other so-called mega-deltas
K. Wium Olesen et al., "Mega Deltas and the Climate Change Challenges," Eleventh International Symposium on River Sedimentation, September 6–9, 2010, http://www.irtces.org/zt/11isrs/paper/Kim_Wium_Olesen.pdf.

195 from whence 40 percent of Egypt's food production comes
United Nations Development Programme, "Adaption to Climate Change in the Nile Delta Through Integrated Coastal Zone Management," 2009, p. 9, http://nile-delta-adapt.org/index.php?view=DownLoadAct&id=6.

195 increase its population 85 percent during the same period
Brown, *Plan B 4.0*.

195 complaints by Iraq and Syria that they are being treated unjustly
Brown, "This Will Be the Arab World's Next Battle."

195 will only get worse as populations in all the affected countries increase
"Thirsty South Asia's River Rifts Threaten 'Water Wars,'" *Alertnet*, July 23, 2012, http://www.trust.org/alertnet/news/thirsty-south-asias-river-rifts-threaten-water-wars/; "Southeast Asia Drought Triggers Debate Over Region's Water Resources," VOA News, March 24, 2010, http://www.voanews.com/content/southeast-asia-drought-triggers-debate-over-regions-water-resources—89114447/114686.html.

195 Colorado River system are being waged in court
Felicia Fonseca, "Arizona High Court Settles Water Rights Query," Associated Press, September 12, 2012; "Colorado Court Ruling Limits Water Transfer

Rights," American Water Intelligence, July 2011, http://www.americanwaterintel
.com/archive/2/7/opinion/colorado-court-ruling-limits-water-tranfers-rights.html;
"Pivotal Water Rights Case on Wastewater Rights," American Water Intelligence,
June 2011, http://www.americanwaterintel.com/archive/2/6/analysis/pivotal-water
-rights-case-wastewater-rights.html; "Navajo Lawmakers Approve Water Rights
Settlement," Associated Press, November 5, 2010; Jim Carlton, "Wet Winter
Can't Slake West's Thirst," *Wall Street Journal*, March 31, 2011.

196 **an official with a Kenyan NGO, Friends of Lake Turkana**
Kremena Krumova, "Land Grabs in Africa Threaten Greater Poverty," *Epoch
Times*, September 21, 2011.

196 **"There is no doubt"**
Anil Ananthaswamy, "African Land Grabs Could Lead to Future Water Conflicts," *New Scientist*, May 26, 2011.

196 **"Rich countries are eyeing Africa"**
John Vidal, "How Food and Water Are Driving a 21st-Century African Land
Grab," *Guardian*, March 6, 2010.

196 **an agricultural real estate boom in Africa**
Lorenzo Cotula, "Analysis: Land Grab or Development Opportunity?," BBC
News, February 21, 2012.

196 **Liberia's land, for example, has been sold to private investors**
Anjala Nayar, "African Land Grabs Hinder Sustainable Development," *Nature*,
February 1, 2012.

196 **signed deals with foreign growers for 21.1 percent of its land**
Cotula, "Analysis: Land Grab or Development Opportunity?"

196 **was sold to investors after the country won its independence**
Nayar, "African Land Grabs Hinder Sustainable Development."

196 **grow palm oil for biofuel on 2.8 million hectares of land**
Vidal, "How Food and Water Are Driving a 21st-Century African Land Grab."

196 **calculated that 44 percent was dedicated to biofuels**
Krumova, "Land Grabs in Africa Threaten Greater Poverty."

197 **United Arab Emirates purchased slightly more**
Vidal, "How Food and Water Are Driving a 21st-Century African Land Grab."

197 **"Thousands of people will be affected and people will go hungry"**
Ibid.

197 **area of the nation of Pakistan—and that two thirds**
W. Anseeuw et al., "Transnational Land Deals for Agriculture in the Global South.
Analytical Report Based on the Land Matrix Database," *CDE/CIRAD/GIGA*, 2012.

197 **people have claimed they were unjustly evicted**
Ibid.

197 **concerning difficulties in financing the projects**
International Land Coalition, "Land Rights and the Rush for Land Report," 2011.

198 **will rely entirely on wheat imports by 2016**
"Saudi Arabia Launches Tender to Buy 550,000 Tons of Wheat," *Saudi Gazette*,
August 30, 2012.

198 water from a deep nonrenewable aquifer
     Brown, "This Will Be the Arab World's Next Battle."

198 80 to 85 percent of that water comes from underground aquifers
     Reem Shamseddine and Barbara Lewis, "Saudi Arabia's Water Needs Eating into
     Oil Wealth," Reuters, September 9, 2011; Brown, *Plan B 4.0*.

198 will eventually involve desalination of seawater
     Shamseddine and Lewis, "Saudi Arabia's Water Needs Eating into Oil Wealth."

198 locked up in the ice and snow of Antarctica and Greenland
     Howard Perlman, U.S. Geological Survey, "Where Is Earth's Water Located?,"
     September 7, 2012, http://ga.water.usgs.gov/edu/earthwherewater.html.

198 even energy-rich Saudi Arabia cannot afford it
     Caline Malek, "Solar Desalination 'the Only Way' for Gulf to Sustainably Produce
     Water," *National*, April 24, 2012, http://www.thenational.ae/news/uae-news/solar
     -desalination-the-only-way-for-gulf-to-sustainably-produce-water.

198 purchase the use of water-rich land in Africa
     John Vidal, "What Does the Arab World Do When Its Water Runs Out?," *Guard-
     ian*, February 19, 2011.

199 many desalination plants in the world—including in Saudi Arabia
     "Saudi Arabia and Desalinisation," *Harvard International Review*, December 23,
     2010, http://hir.harvard.edu/pressing-change/saudi-arabia-and-desalination-0.

199 towing them to areas experiencing severe droughts
     Bob Yirka, "Simulation Shows It's Possible to Tow an Iceberg to Drought Areas,"
     PhysOrg, August 9, 2011, http://phys.org/news/2011-08-simulation-iceberg
     -drought-areas.html.

199 could supply 500,000 people with freshwater for a year
     Ibid.

199 supplied with ample amounts of water, nutrients, and sunlight
     "Does It Really Stack Up?," *Economist*, December 9, 2010, http://www.economist
     .com/node/17647627.

199 rely on fish for approximately 15 percent
     United Nations Food and Agriculture Organization, "The State of Fisheries and
     Aquaculture," 2012, p. 5, http://www.fao.org/docrep/016/i2727e/i2727e00.htm.

199 from twenty-two pounds per person per year to almost thirty-eight pounds
     Bryan Walsh, "The End of the Line," *Time*, July 7, 2011.

199 one third of fish stocks in the oceans
     Ibid.

199 reduced by 90 percent since the 1960s
     Ransom Myers and Boris Worm, "Rapid Worldwide Depletion of Predatory Fish
     Communities," *Nature*, May 15, 2003.

200 The world reached "peak fish" twenty-five years ago
     Brad Plumer, "The End of Fish, in One Chart," *Washington Post*, May 20, 2012.

200 "with 14% of assessed stocks collapsed in 2007"
     Convention on Biological Diversity, "Global Biodiversity Outlook 3: Biodiversity
     in 2010," 2010, http://www.cbd.int/gbo3/?pub=6667&section=6709.

200 **large marine area in the Pacific Ocean**
Suzanne Goldberg, "Bush Designates Ocean Conservation Areas in Final Week as President," *Guardian*, January 5, 2009.

200 **61 percent of which will occur in China**
OECD-FAO, "Agricultural Outlook 2011–2020."

200 **can be tainted by pollution, antibiotics, and antifungals**
Laurel Adams, Center for Public Integrity, "FDA Screening of Fish Imports Not Catching Antibiotics and Drug Residue," May 18, 2011, http://www.publicintegrity.org/environment/natural-resources?page=3; George Mateljan, "Is There Any Nutritional Difference Between Wild-Caught and Farm-Raised Fish? Is One Type Better for Me Than the Other?," World's Healthiest Foods, http://www.whfoods.com/genpage.php?tname=george&dbid=96.

200 **five pounds of wild fish for each pound of farmed salmon produced**
U.S. Department of Agriculture, "Trout-Grain Project," 2012.

200 **Over half of the fish food in agriculture**
NOAA Fisheries Service–National Marine Fisheries Service, "Feeds for Aquaculture," 2012.

201 **Although more than 10 percent of all cropland**
Elizabeth Weise, "More of World's Crops Are Genetically Engineered," *USA Today*, February 22, 2011.

CHAPTER 5: THE REINVENTION OF LIFE AND DEATH

204 **introduce human genes into other animals**
Richard Gray, "Genetically Modified Cows Produce 'Human' Milk," *Telegraph*, April 2, 2011.

204 **mixing the genes of spiders and goats**
Adam Rutherford, "Synthetic Biology and the Rise of the 'Spider-Goats,'" *Guardian*, January 14, 2012.

204 **computer chips into the gray matter of human brains**
Daniel H. Wilson, "Bionic Brains and Beyond," *Wall Street Journal*, June 1, 2012.

204 **parents who wish to design their own children**
Keith Kleiner, "Designer Babies—Like It or Not, Here They Come," Singularity Hub, February 25, 2009, http://singularityhub.com/2009/02/25/designer-babies-like-it-or-not-here-they-come/.

205 **sometimes noted, "There Be Monsters"**
H. P. Newquist, *Here There Be Monsters: The Legendary Kraken and the Giant Squid* (New York: Houghton Mifflin, 2010).

205 **they seized knowledge that had been forbidden them**
Genesis 3:16–19.

205 **so he could endure the same fate the next morning**
Thomas Chen and Peter Chen, "The Myth of Prometheus and the Liver," *Journal of the Royal Society of Medicine* 87 (December 1994): 754.

205 **replacement livers in their laboratory bioreactors**
Wake Forest Baptist Medical Center, 10-30-10, "Researchers Engineer Miniature Human Livers in the Lab," October 30, 2010, http://www.wakehealth.edu/News-Releases/2010/Researchers_Engineer_Miniature_Human_Livers_in_the_Lab.htm.

205 almost certainly become the model for medical care
"Personalized Medicine," *USA Today*, January 20, 2011.

206 volume of fine-grained information about every individual
"Do Not Ask or Do Not Answer?," *Economist*, August 23, 2007.

206 reduce medical errors and enhance the skills of physicians
Farhad Manjoo, "Why the Highest-Paid Doctors Are the Most Vulnerable to Automation," *Slate*, September 27, 2011, http://www.slate.com/articles/technology/robot_invasion/2011/09/will_robots_steal_your_job_3.html.

206 "all are subject to radical transformation"
Topol, *The Creative Destruction of Medicine*, p. 243.

206 unhealthy behaviors in order to manage chronic diseases
David H. Freeman, "The Perfected Self," *Atlantic*, June 2012; Mark Bowden, "The Measured Man," *Atlantic*, July/August 2012.

206 digital monitors that are on—and inside—the patient's body
Topol, *The Creative Destruction of Medicine*, pp. 59–76.

206 leads to an improvement in the amount of progress made
Janelle Nanos, "Are Smartphones Changing What It Means to Be Human?," *Boston*, February 28, 2012.

207 will see their progress or lack thereof
Freeman, "The Perfected Self."

207 global access to large-scale digital programs
John Havens, "How Big Data Can Make Us Happier and Healthier," Mashable, October 8, 2012, http://mashable.com/2012/10/08/the-power-of-quantified-self/.

207 living systems are also being applied to the human brain
Matthew Hougan and Bruce Altevogt, *From Molecules to Minds: Challenges for the 21st Century*, Board on Health Sciences Policy, Institute of Medicine, 2008.

207 prosthetic arms and legs with their brains
Associated Press, "Man with Bionic Leg Climbs Chicago Skyscraper," November 5, 2012.

207 curing some brain diseases
Meghan Rosen, "Beginnings of Bionic," *Science News* 182, no. 10 (November 17, 2012): 18.

207 mapping of what brain scientists call the "connectome"
Olaf Sporns, a professor of computational cognitive neuroscience at Indiana University, was the first to coin the word "connectome." The National Institutes of Health now have a "Human Connectome Project." Ian Sample, "Quest for the Connectome: Scientists Investigate Ways of Mapping the Brain," *Guardian*, May 7, 2012.

207 greater than that required for mapping the genome
Hougan and Altevogt, *From Molecules to Minds*.

207 technologies necessary to complete the map are still in development
"Brain Researchers Start Mapping the Human 'Connectome,'" *ScienceDaily*, July 2, 2012, http://www.sciencedaily.com/releases/2012/07/120702152652.htm.

207 complete the first "larger-scale maps of neural wiring"
Sample, "Quest for the Connectome."

207 "artificially perfect the thinking instrument itself"
Eric Steinhart, "Teilhard de Chardin and Transhumanism," *Journal of Evolution and Technology* 20, no. 1 (December 2008): 1–22.

207 for the brains of people who have Parkinson's disease
Wilson, "Bionic Brains and Beyond."

207 provide deep brain stimulation to alleviate their symptoms
Ibid.

208 activated in stages to give the brain a chance to adjust to them
Johns Hopkins Medicine, Cochlear Implant Information, http://www
.hopkinsmedicine.org/otolaryngology/specialty_areas/listencenter/cochlear_info
.html#activation.

208 "designer babies" may be highly appealing to some parents
Kleiner, "Designer Babies"; Mark Henderson, "Demand for 'Designer Babies' to Grow Dramatically," *Times* (London), January 7, 2010.

208 competitive parenting has already done for the test preparation industry
Jose Ferreira, "A Short History of the Standardized Test Prep Industry," Knewton Blog, February 17, 2010, http://www.knewton.com/blog/edtech/2010/02/17/a-short-history-of-the-standardized-test-prep-industry/; Julian Brookes, "Chris Hayes on the Twilight of the Elites and the End of Meritocracy," *Rolling Stone,* July 11, 2012.

208 other parents may feel that they have to do the same
Armand Marie Leroi, "The Future of Neo-Eugenics," *EMBO Reports* 7 (2006): 1184–87.

208 may trigger collateral genetic changes that are not yet fully understood
Mike Steere, "Designer Babies: Creating the Perfect Child," CNN, October 30, 2008.

209 "We will have converted old-style evolution into neo-evolution"
Harvey Fineberg, "Are We Ready for Neo-Evolution?," TED Talks, 2011.

209 the United States is decreasing its investment in biomedical research
Robert D. Atkinson et al., *Leadership in Decline: Assessing U.S. International Competitiveness in Biomedical Research* (Washington, DC: Information Technology and Innovation Foundation, 2012).

209 "It's time for everyone to pull their heads out of the sand"
"Designer Baby Row Over US Clinic," BBC, March 2, 2009.

209 "who deserves to be born?"
Andrew Pollack, "DNA Blueprint for Fetus Built Using Tests of Parents," *New York Times,* June 6, 2012.

209 "They feel disempowered"
Steere, "Designer Babies."

209 "would undermine humanity and create a techno-eugenic rat race"
Ibid.

209 will soon exceed the sequencing capacity of the entire United States
Japan External Trade Organization, "BGI, China's Leading Genome Research Institute, Has Established a Japanese Arm in Kobe," February 7, 2012; Fiona Tam, "Scientists Seek to Unravel the Mystery of IQ," translated by Steve Hsu, *South China Morning Post,* December 4, 2010.

210 occupations that make the best use of their capabilities
"The Dragon's DNA," *Economist,* June 17, 2010; Emily Chang, "In China, DNA Tests on Kids ID Genetic Gifts, Careers," CNN, August 5, 2009, http://edition .cnn.com/2009/WORLD/asiapcf/08/03/china.dna.children.ability/.

210 over $100 billion on life sciences research over just the last three years
Lone Frank, "High-Quality DNA," *Newsweek,* April 24, 2011.

210 "science discovery and innovation within the next decade"
Ibid.

210 industry will be one of the pillars
"China Establishes National Gene Bank in Shenzhe," Xinhua News Agency, June 18, 2011.

210 eventually sequence the genomes of almost every child in China
David Cyranoski, "Chinese Bioscience: The Sequence Factory," *Nature,* March 3, 2010.

210 first patent on a gene
Harriet A. Washington, *Deadly Monopolies: The Shocking Corporate Takeover of Life Itself—and the Consequences for Your Healthy and Our Medical Future* (New York: Doubleday, 2011), p. 181.

210 more than 40,000 gene patents have been issued, covering 2,000 human genes
Sharon Begley, "In Surprise Ruling, Court Declares Two Gene Patents Invalid," *Daily Beast,* March 29, 2010.

210 used for commercial purposes without their permission
Washington, *Deadly Monopolies,* chs. 1 and 7.

210 gene therapy drug, known as Glybera
Ben Hirschler, "Europe Approves High-Price Gene Therapy," Reuters, November 2, 2012.

210 treatment of a rare genetic disorder
Andrew Pollack, "European Agency Backs Approval of a Gene Therapy," *New York Times,* July 20, 2012.

210 approved a drug known as Crizotinib
Alice T. Shaw, "The Crizotinib Story: From Target to FDA Approval and Beyond," InforMEDical, 2012, http://www.informedicalcme.com/lucatoday/crizotinib-story -from-target-to-fda-approval.

211 "We now believe that Monsanto"
"Monsanto Strong-Arms Seed Industry," Associated Press, January 4, 2011.

211 "Could you patent the sun?"
"'Deadly Monopolies'? Patenting the Human Body," *Fresh Air,* NPR, October 24, 2011, http://www.npr.org/2011/10/24/141429392/deadly-monopolies-patenting -the-human-body. The groundbreaking work of Albert Sabin—whose vaccine became the most widely used—cannot be overlooked as well.

211 research into the genome was still in its early stages
Norman Borlaug, biography, http://www.nobelprize.org/nobel_prizes/peace/laureates/ 1970/borlaug-bio.html.

211 "God help us if that were to happen"
Vandana Shiva, "The Indian Seed Act and Patent Act: Sowing the Seeds of Dicta-

torship," ZNet, February 14, 2005, http://www.grain.org/article/entries/2166-india
-seed-act-patent-act-sowing-the-seeds-of-dictatorship.

211 "always stood for free exchange of germplasm"
Ibid.

211 courts continue to uphold the patentability of genes
Reuters, "Court Reaffirms Right of Myriad Genetics to Patent Genes," *New York Times*, August 16, 2012.

212 we gain the ability to understand and manipulate the reality
Michael S. Gazzaniga, *Human: The Science Behind What Makes Us Unique* (New York: HarperCollins, 2008), p. 199.

212 four letters: A, T, C, and G
"The four bases—ATCG," Scitable, Nature Education, 2012, http://www.nature.com/scitable/content/the-four-bases-atcg-6491969.

212 "device the size of your thumb could store"
Robert Lee Hotz, "Harvard Researchers Turn Book into DNA Code," *Wall Street Journal*, August 16, 2012.

212 discovered by James Watson, Francis Crick, and Rosalind Franklin
Lynne Osman Elkin, "Rosalind Franklin and the Double Helix," *Physics Today* 56, no. 3 (March 2003): 42–48.

212 exactly fifty years later, the human genome was sequenced
US Department of Energy, Office of Science, "History of the Human Genome Project," June 4, 2012, http://www.ornl.gov/sci/techresources/Human_Genome/project/hgp.shtml.

212 they are beginning to sequence RNA
Genetics Home Reference, "RNA," http://ghr.nlm.nih.gov/glossary=rna.

212 system to convey the information that is translated into proteins
"RNAi," Nova scienceNOW, PBS, July 26, 2005, http://www.pbs.org/wgbh/nova/body/rnai.html.

212 cells that make up all forms of life
Genetics Home Reference, "Protein," http://ghr.nlm.nih.gov/glossary=protein.

212 being analyzed in the Human Proteome Project
Human Proteome Organisation, "Human Proteome Project (HPP)," 2010, http://www.hupo.org/research/hpp/.

212 patterns that affect their function and role
ThermoScientific, "Overview of Post-Translational Modifications (PTMs)," http://www.piercenet.com.

212 ways that extend their range of functions and control their behavior
Ibid.

212 The Human Epigenome Project has made major advances
G. G. Sanghani et al., "Human Epigenome Project: The Future of Cancer Therapy," *Inventi Impact: Pharm Biotech & Microbio* 2012, http://www.inventi.in/Article/pbm/94/12.aspx.

213 epigenetic breakthroughs are already helping cancer patients
"Epigenetics Emerges Powerfully as a Clinical Tool," Medical Xpress, September 12, 2012, http://medicalxpress.com/news/2012-09-epigenetics-emerges-powerfully-clinical-tool.html.

213 **transform and commandeer them**
Denise Caruso, "Synthetic Biology: An Overview and Recommendations for Anticipating and Addressing Emerging Risks," *Science Progress*, November 12, 2008, http://scienceprogress.org/2008/11/synthetic-biology/.

213 **custom chemicals that have value in the marketplace**
Caruso, "Synthetic Biology."

213 **less effective insulin produced from pigs and other animals**
Lawrence K. Altman, "A New Insulin Given Approval for Use in U.S.," *New York Times*, October 30, 1982.

213 **significant improvements in artificial skin**
Charles Q. Choi, "Spider Silk May Provide the Key to Artificial Skin," MSNBC, August 9, 2011; Katharine Sanderson, "Artificial Skins Detect the Gentlest Touch," *Nature*, September 12, 2010.

213 **synthetic human blood**
Fiona Macrae, "Synthetic Blood Created by British Scientists Could Be Used in Transfusions in Just Two Years," *Daily Mail*, October 28, 2011.

213 **diverse as fuel for vehicles**
Michael Totty, "A Faster Path to Biofuels," *Wall Street Journal*, October 16, 2011.

213 **protein for human consumption**
Jeffrey Bartholet, "When Will Scientists Grow Meat in a Petri Dish?," *Scientific American*, May 17, 2011; H. L. Tuomisto, "Food Security and Protein Supply—Cultured Meat a Solution?," 2010, http://oxford.academia.edu/HannaTuomisto/Papers/740015/Food_Security_and_Protein_Supply_-Cultured_meat_a_solution.

213 **"a juggernaut already beyond the reach of governance"**
Caruso, "Synthetic Biology."

214 **"not only for one, but also for all humanity"**
Jun Wang, Science, "Personal Genomes: For One and for All," *Science*, February 11, 2011.

214 **junk DNA actually contains millions of "on-off switches"**
Gina Kolata, "Bits of Mystery DNA, Far from 'Junk,' Play Crucial Role," *New York Times*, September 6, 2012.

214 **"very complicated three-dimensional structure"**
Brandon Keim, "New DNA Encyclopedia Attempts to Map Function of Entire Human Genome," *Wired*, September 5, 2012.

215 **first human genome ten years ago was approximately $3 billion**
John Markoff, "Cost of Gene Sequencing Falls, Raising Hopes for Medical Advances," *New York Times*, March 8, 2012.

215 **to be available at a cost of only $1,000 per person**
Ibid.

215 **At that price, according to experts**
Ibid.

215 **"all important topics for future discussions"**
Ibid.

215 **gene-sequencing machine for less than $900**
Oxford Nanopore Technologies, "Oxford Nanopore Introduces DNA 'Strand Sequencing' on the High-Throughput GridION platform and Presents MinION,

a Sequencer the Size of a USB Memory Stick," February 17, 2012, http://www.nanoporetech.com/news/press-releases/view/39/.

215 **has long been measured by Moore's Law**
K. A. Wetterstrand, "DNA Sequencing Costs: Data from the NHGRI Large-Scale Genome Sequencing Program," www.genome.gov/sequencingcosts.

215 **cost for sequencing began to drop at a significantly faster pace**
Ibid.

215 **increases in the length of DNA strands that can be quickly analyzed**
Jeffrey Fisher and Mostafa Ronaghi, "The Current Status and Future Outlook for Genomic Technologies," National Academy of Engineering, Winter 2010; Neil Bowdler, "1000 Genomes project maps 95% of all gene variations," BBC, October 27, 2011.

215 **breakneck speed for the foreseeable future**
Ibid.

215 **producing synthetic genomes**
John Carroll, "Life Technologies Budgets $100M for Synthetic Biology Deals," *Fierce Biotech*, June 3, 2010, http://www.fiercebiotech.com/story/life-technologies-budgets-100m-synthetic-biology-deals/2010-06-03.

215 **introduction by Hammurabi of the first written set of laws**
Paul Halsall, "Code of Hammurabi, c. 1780 BCE," Internet Ancient History Sourcebook, March 1998, http://www.fordham.edu/halsall/ancient/hamcode.asp.

216 **"a new wave of organisms, an artificially provoked neo-life"**
Pierre Teilhard de Chardin, *The Phenomenon of Man* (New York: HarperCollins, 2008), p. 250.

216 **who had already made history by sequencing his own genome**
Emily Singer, "Craig Venter's Genome," *Technology Review*, September 4, 2007, http://www.technologyreview.com/news/408606/craig-venters-genome/.

216 **first live bacteria made completely from synthetic DNA**
Joe Palca, "Scientists Reach Milestone on Way to Artificial Life," NPR, May 20, 2010.

216 **Venter had merely copied the blueprint of a known bacterium**
Clive Cookson, "Synthetic Life," *Financial Times*, July 27, 2012.

216 **used the empty shell of another as the container for his new life-form**
Clive Cookson, "Scientists Create a Living Organism," *Financial Times*, May 20, 2010.

216 **others marked it as an important turning point**
Stuart Fox, "J. Craig Venter Institute Creates First Synthetic Life Form," *Christian Science Monitor*, May 21, 2010.

216 **free-living microbe known as *Mycoplasma genitalium***
John Markoff, "In First, Software Emulates Lifespan of Entire Organism," *New York Times*, July 21, 2012.

216 **minimum amount of DNA information necessary for self-replication**
Cookson, "Synthetic Life."

217 **"if there had been one," Venter said**
Ibid.

217 "The fault, dear Brutus, is not in our stars, but in ourselves"
William Shakespeare, *Julius Caesar,* 1.2.140–41.

217 E. O. Wilson, has been bitterly attacked
Jennifer Schuessler, "Lessons from Ants to Grasp Humanity," *New York Times,* April 8, 2012; Richard Dawkins, "The Descent of Edward Wilson," *Prospect,* May 24, 2012.

217 Wilson, who was but is no longer a Christian
Donna Winchester, "E.O. Wilson on Ants and God and Us," *Tampa Bay Times,* November 14, 2008.

218 "decade by decade, century by century"
"The 'Evidence for Belief': An Interview with Francis Collins," Pew Forum on Religion and Public Life, April 17, 2008, http://pewresearch.org/pubs/805/the -evidence-for-belief-an-interview-with-francis-collins.

218 "then we can decide what metabolism we want it to have"
Cookson, "Synthetic Life."

218 breakthroughs in health care
Warren C. Ruder, Ting Lu, and James J. Collins, "Synthetic Biology Moving into the Clinic," *Science,* September 2, 2011.

218 energy production
Cookson, "Synthetic Life."

218 environmental remediation
Caruso, "Synthetic Biology."

218 and many other fields
Stephen C. Aldrich, James Newcomb, and Robert Carlson, *Genome Synthesis and Design Futures: Implications for the U.S. Economy* (Cambridge, MA: Bio Economic Research Associates, 2007).

218 destroy or weaken antibiotic-resistant bacteria
Ruder, Lu, and Collins, "Synthetic Biology Moving into the Clinic."

218 killing other targeted bacteria until the infection subsides
Ibid.

218 vaccine development is also generating great hope
Cookson, "Synthetic Life."

218 bird flu (H5N1) of 2007 and the so-called swine flu (H1N1) of 2009
Ibid.

218 ability to pass from one human to another through airborne transmission
"Bird Flu Pandemic in Humans Could Happen Any Time," Reuters, June 21, 2012.

218 a new mutant of the virus begins spreading
Huib de Vriend, "Vaccines: The First Commercial Application of Synthetic Biology?," Rathenau Instituut, July 2011.

218 using the tools of synthetic biology
Ibid.

219 decrease the cost and time of manufacturing of vaccines
Vicki Glaser, "Quest for Fully Disposable Process Stream," *Genetic Engineering & Biotechnology News* 29, no. 5, March 1, 2009.

219 Some experts have also predicted
Aldrich, Newcomb, and Carlson, *Genome Synthesis and Design Futures*.

219 utilizing a "widely dispersed" strategy
Cookson, "Synthetic Life."

219 "would not appear to him as indecent and unnatural"
J. B. S. Haldane, "Daedalus of Science and the Future," February 4, 1923, http://www.psy.vanderbilt.edu/courses/hon182/Daedalus_or_SCIENCE_AND_THE_FUTURE_JBS_Haldane.pdf.

220 "We intuit and we feel"
Leon Kass, *Life, Liberty and the Defense of Dignity* (San Francisco: Encounter Books, 2004), p. 150.

220 describes a feeling that itself lacks precision
Alexis Madrigal, "I'm Being Followed: How Google—and 104 Other Companies—Are Tracking Me on the Web," *Atlantic*, February 29, 2012.

220 a method for producing spider silk
Rutherford, "Synthetic Biology and the Rise of the 'Spider-Goats.'"

220 five times stronger than steel by weight
Other scientists have mimicked the molecular design of spider silk by synthesizing their own from a commercially available substance (polyurethane elastomer) treated with clay platelets only one nanometer (a billionth of a meter) thick and only 25 nanometers across, then carefully processing the mixture to create synthetic spider silk. This work has been funded by the Institute for Soldier Nanotechnologies at the Massachusetts Institute of Technology because the military applications are considered of such high importance. Rutherford, "Synthetic biology and the rise of the 'spider-goats'"; "Nexia and US Army Spin the World's First Man-Made Spider Silk Performance Fibers," Eureka Alert, January 17, 2002, http://www.eurekalert.org/pub_releases/2002-01/nbi-nau011102.php.

220 because of their antisocial, cannibalistic nature
Rutherford, "Synthetic Biology and the Rise of the 'Spider-Goats.'"

221 became a threat to native trees and plants
Richard J. Blaustein, "Kudzu's Invasion into Southern United States Life and Culture," 2001, www.srs.fs.usda.gov/pubs/ja/ja_blaustein001.pdf.

221 chain reaction in the ocean and create an unimaginable ecological Armageddon
Al Gore, "Planning a New Biotechnology Policy," *Harvard Journal of Law and Technology* 5 (1991): 19–30.

221 who were confident that such an event was absurdly implausible
Ibid.

221 diversion of trillions of dollars into weaponry
Wil S. Hylton, "How Ready Are We for Bioterrorism?," *New York Times Magazine*, October 26, 2011.

221 threatened the survival of human civilization?
George P. Shultz, William J. Perry, Henry A. Kissinger, and Sam Nunn, "A World Free of Nuclear Weapons," *Wall Street Journal*, January 4, 2007.

221 are now often described as probably overblown
Wil S. Hylton, "Craig Venter's Bugs Might Save the World," *New York Times Magazine*, June 3, 2012.

221 "I don't think anyone knows"
Ibid.

221 is the possibility of a new generation of biological weapons
Alexander Kelle, "Synthetic Biology and Biosecurity," *EMBO Reports* 10 (2009): S23–S27.

222 Soviet Union in a secret biological weapons program
Ibid.

222 "to attack genetically specific sub-populations"
Ibid.

222 publishing the full genetic sequence that accompanied their papers
Ibid.

222 involved in monitoring genetic research that could lead to new bioweapons
National Institutes of Health, Office of Science Policy, "About NSABB," 2012, http://oba.od.nih.gov/biosecurity/about_nsabb.html.

222 research teams working on projects considered militarily sensitive
Sample, "*Nature* Publishes Details of Bird Flu Strain That Could Spread Among People."

222 federally funded research into the cloning of human beings
Center for Genetics and Society, "Failure to Pass Federal Cloning Legislation, 1997–2003," http://www.geneticsandsociety.org/article.php?id=305.

222 legal implications of human cloning
Mary Meehan, "Looking More Like America?," *Our Sunday Visitor*, November 3, 1996, http://www.ewtn.com/library/ISSUES/LOOKLIKE.TXT.

223 government-financed research program into ethics
Edward J. Larson, "Half a Tithe for Ethics," *National Forum* 73, no. 2 (Spring 1993): 15–18.

223 "possible copying mechanism for the genetic material"
J. D. Watson and F.H.C. Crick, "Molecular Structure of Nucleic Acids," *Nature*, April 25, 1953.

223 science of cloning, genetic engineering, and genetic screening
See, for example: Subcommittee on Investigations and Oversight and the Subcommittee on Science, Research, and Technology, Committee on Science and Technology, U.S. House of Representatives, "Commercialization of Academic Biomedical Research," June 8–9, 1981; Subcommittee on Investigations and Oversight, Committee on Science and Technology, U.S. House of Representatives, "Genetic Screening and the Handling of High-Risk Groups in the Workplace," October 14–15, 1981.

223 and fifteen years later they succeeded with Dolly
U.S. Department of Energy, Office of Science, Human Genome Project, "Cloning Fact Sheet," May 11, 2009, http://www.ornl.gov/sci/techresources/Human_Genome/elsi/cloning.shtml#animalsQ.

223 they have cloned many other livestock and other animals
Ibid.

223 ethical concerns that had prevented them from attempting such procedures
Dan W. Brock, "Cloning Human Beings: An Assessment of the Ethical Issues Pro and Con," in *Cloning Human Beings*, vol. 2, *Commissioned Papers* (Rockville, MD:

National Bioethics Advisory Commission, 1997), http://bioethics.georgetown.edu/nbac/pubs/cloning2/cc5.pdf.

223 **human cloning has been made illegal in almost every country in Europe**
Ibid.; "19 European Nations OK Ban on Human Cloning," *National Catholic Register,* April 18, 1999.

223 **"protection of the security of human genetic material"**
Brock, "Cloning Human Beings."

223 **clear form of harm to the individual who is cloned or to society**
Brian Alexander, "(You)²," *Wired,* February 2001; "Dolly's Legacy," *Nature,* February 22, 2007; Steve Connor, "Human Cloning Is Now 'Inevitable,'" *Independent,* August 30, 2000; John Tierney, "Are Scientists Playing God? It Depends on Your Religion," *New York Times,* November 20, 2007.

224 **a line of identical embryonic stem cells that reproduced themselves**
David Cyranoski, "Cloned Human Embryo Makes Working Stem Cells," *Nature,* October 5, 2011.

224 **Several countries**
Tierney, "Are Scientists Playing God?"

224 **has broken this modern taboo against human cloning**
Steve Connor, "'I Can Clone a Human Being'—Fertility Doctor," *New Zealand Herald,* April 22, 2009; Tierney, "Are Scientists Playing God?"

224 **There has yet been no confirmed birth of a human clone**
National Human Genome Research Institute, Cloning Fact Sheet.

224 **other forms of technological progress**
Brock, "Cloning Human Beings."

224 **that it is inevitable in any case**
Roman Altshuler, "Human Cloning Revisited: Ethical Debate in the Technological Worldview," *Biomedical Law & Ethics* 3, no. 2 (2009): 177–95.

224 **most experiments because of the medical benefits that can be gained**
Brock, "Cloning Human Beings."

224 **individuals and run the risk of "commoditizing" human beings**
Ibid.; Altshuler, "Human Cloning Revisited."

224 **views of the rights and protections due to every person**
Leon Kass and James Q. Wilson, *Ethics of Human Cloning* (Washington, DC: American Enterprise Institute, 1998).

224 **more generalized humanist assertion of individual dignity**
Brock, "Cloning Human Beings"; Altshuler, "Human Cloning Revisited."

225 **In yet another illustration**
"Meat on Drugs," *Consumer Reports,* June 2012.

225 **a truly shocking 80 percent of all U.S. antibiotics**
Gardiner Harris, "U.S. Tightens Rules on Antibiotics Use for Livestock," *New York Times,* April 11, 2012.

225 **new rule that will require a prescription from veterinarians**
"Meat on Drugs," *Consumer Reports.*

225 Since the discovery of penicillin in 1929 by Alexander Fleming
"A Brief History of Antibiotics," BBC News, October 8, 1999, http://news.bbc.co
.uk/2/hi/health/background_briefings/antibiotics/163997.stm.

225 Although Fleming said his discovery was "accidental"
Douglas Allchin, SHiPS Resource Center, "Penicillin and Chance," http://www1
.umn.edu/ships/updates/fleming.htm.

225 who first discovered that CO_2 traps heat
Spencer Weart, "The Discovery of Global Warming: The Carbon Dioxide Greenhouse Effect," February 2011, http://www.aip.org/history/climate/co2.htm.

226 was not used in a significant way until the early 1940s
"A Brief History of Antibiotics," BBC News.

226 many other potent antibiotics were discovered in the 1950s and 1960s
Ibid.

226 discoveries have slowed to a trickle
Ibid.

226 life-saving antibiotics is rapidly eroding their effectiveness
"The Spread of Superbugs," Economist, March 31, 2011.

226 ways that circumvent the effectiveness of the antibiotic
Brandon Keim, "Antibiotics Breed Superbugs Faster Than Expected," Wired,
February 11, 2010.

226 only when they are clearly needed
Alexander Fleming, "Penicillin," Nobel Lecture, December 11, 1945, http://
www.nobelprize.org/nobel_prizes/medicine/laureates/1945/fleming-lecture
.pdf; E. J. Mundell, "Antibiotic Combinations Could Fight Resistant Germs,"
ABC News, March 23, 2007, http://abcnews.go.com/Health/Healthday/story?id
=4506442&page=1#.UDVmwo40jdk.

226 they stumble upon new traits that make the antibiotics impotent
Keim, "Antibiotics Breed Superbugs Faster Than Expected."

226 Some antibiotics have already become ineffective against certain diseases
Katie Moisse, "Antibiotic Resistance: The 5 Riskiest Superbugs," ABC News,
March 27, 2012, http://abcnews.go.com/Health/Wellness/antibiotic-resistance
-riskiest-superbugs/story?id=15980356#.UC7l0UR9nMo.

226 rate that is frightening to many health experts
Moisse, "Antibiotic Resistance: The 5 Riskiest Superbugs."

226 multidrug-resistant tuberculosis
Ibid.

226 the FDA formed a new task force
Stephanie Yao, "New FDA Task Force Will Support Innovation in Antibacterial Drug Development," Food and Drug Administration press release, September 24, 2012.

226 in spite of these basic medical facts, many governments
Worldwatch Institute, "Global Meat Production and Consumption Continue to Rise," 2011, http://www.worldwatch.org/global-meat-production-and
-consumption-continue-rise-1; Philip K. Thornton, "Livestock Production: Recent Trends, Future Prospects," Philosophical Transactions of the Royal Society B,
September 27, 2010.

226 including, shockingly, the United States government
"Meat on Drugs," *Consumer Reports*.

226 but the impact on profits is very clear and sizable
Matthew Perrone, "Does Giving Antibiotics to Animals Hurt Humans?," Associated Press, April 20, 2012.

226 superbugs that are immune to the impact of antibiotics
Ibid.

226 the antibiotics are given in subtherapeutic doses
"Our Big Pig Problem," *Scientific American*, February 8, 2012.

226 not principally used for the health of the livestock anyway
Harris, "U.S. Tightens Rules on Antibiotics Use for Livestock."

227 dispute the science while handing out campaign contributions
Ibid.; 2012 PAC Summary Data, Open Secrets, http://www.opensecrets.org/pacs/lookup2.php?strID=C00028787&cycle=2012, accessed August 22, 2012; National Cattlemen's Beef Association lobbying expenses, Open Secrets, http://www.sourcewatch.org/index.php?title=National_Cattlemen's_Beef_Association#cite_note-1, August 22, 2012.

227 Last year, scientists confirmed that
Richard Knox, "How Using Antibiotics in Animal Feed Creates Superbugs," NPR, February 21, 2012, http://www.npr.org/blogs/thesalt/2012/02/21/147190101/how-using-antibiotics-in-animal-feed-creates-superbugs.

227 have thus far been successful in preventing a ban
Harris, "U.S. Tightens Rules on Antibiotics Use for Livestock."

227 until recently, a regulation limiting this insane practice
Ibid.

227 European Union has already banned antibiotics in livestock feed
Knox, "How Using Antibiotics in Animal Feed Creates Superbugs."

227 but in a number of other countries
Ibid.; "Meat on Drugs," *Consumer Reports;* Worldwatch Institute, "Global Meat Production and Consumption Continue to Rise"; Thornton, "Livestock Production."

227 only one of many bacteria that are now becoming resistant
Knox, "How Using Antibiotics in Animal Feed Creates Superbugs."

227 mad cow disease
"Bill Seeks Permanent Ban on Downer Slaughter at Meat Plants," *Food Safety News,* January 13, 2012.

228 infected by the pathogen (a misfolded protein, or prion) that causes the disease
World Health Organization, "Bovine Spongiform Encephalopathy," November 2002, http://www.who.int/mediacentre/factsheets/fs113/en/.

228 Animals with later stages of the disease
I. Ramasamy, M. Law, S. Collins, and F. Brook, "Organ Distribution of Prion Proteins in Variant Creutzfeldt-Jakob Disease," *Lancet Infectious Diseases* 3, no. 4 (April 2003): 214–22.

228 fifty times more likely to have the disease
"Bill Seeks Permanent Ban on Downer Slaughter at Meat Plants," *Food Safety News*.

228 should be diverted from the food supply
"Bill Seeks Permanent Ban on Downer Slaughter at Meat Plants," *Food Safety News*.

228 manifested those symptoms just before they were slaughtered
Ibid.

228 in order to protect a tiny portion of the industry's profits
Emad Mekay, "Beef Lobby Blocks Action on Mad Cow, Activists Say," Inter Press Service, January 8, 2004, http://www.monitor.net/monitor/0401a/copyright/madcow4.html; Charles Abbott, "Analysis: U.S. Mad Cow Find: Lucky Break or Triumph of Science?," Reuters, April 25, 2012.

228 a regulation that embodies the intent of the law rejected
"Obama Bans 'Downer' Cows from Food Supply," Associated Press, March 14, 2009.

228 could be reversed by Obama's successor
"Bill Seeks Permanent Ban on Downer Slaughter at Meat Plants," *Food Safety News*.

229 In 1922, a "model eugenical sterilization law"
Paul A. Lombardo, *Three Generations, No Imbeciles: Eugenics, the Supreme Court, and Buck v. Bell* (Baltimore: Johns Hopkins University Press, 2008), p. 91.

229 were sterilized under laws similar to Laughlin's design
Alex Wellerstein, "Harry Laughlin's 'Model Eugenical Sterilization Law,'" http://alexwellerstein.com/laughlin/.

229 were burdensome to the state because of the expense
Paul Lombardo, "Eugenic Sterilization Laws," Image Archive on the American Eugenics Movement, http://www.eugenicsarchive.org/html/eugenics/essay8text.html.

229 people who were reproducing at rates not possible in the past
Jonathan D. Moreno, *The Body Politic: The Battle Over Science in America* (New York: Bellevue Literary Press, 2011), p. 67.

229 he obviously believed they were heritable
Ibid., p. 67.

229 Laughlin was himself an epileptic
Wellerstein, "Harry Laughlin's 'Model Eugenical Sterilization Law.'"

230 Europe was influential in forming the highly restrictive quota system
Ibid.

230 influenced by deep confusion over what evolution really means
Moreno, *The Body Politic*, pp. 64–67.

230 Sir Francis Galton, and was then popularized by Herbert Spencer
Ibid., p. 65.

230 based on the crackpot ideas of Jean-Baptiste Lamarck
Ibid.

230 after their birth were genetically passed on to their offspring
Ibid.

230 was also promoted in the Soviet Union by Trofim Lysenko
"Trofim Denisovich Lysenko," *Encyclopaedia Britannica*, http://www.britannica.com/EBchecked/topic/353099/Trofim-Denisovich-Lysenko.

230 mainstream genetics during the three decades
Ibid.

230 Geneticists who disagreed with Lysenko were secretly arrested
Ibid.; Moreno, *The Body Politic*, p. 69.

230 some were found dead in unexplained circumstances
"Trofim Denisovich Lysenko," *Encyclopaedia Britannica;* Moreno, *The Body Politic*,
p. 69.

230 that biological theory conform with Soviet agricultural needs
"Trofim Denisovich Lysenko," *Encyclopaedia Britannica*.

230 rather those that were best adapted to their environments
Michael Shermer, "Darwin Misunderstood," February 2009, http://www
.michaelshermer.com/2009/02/darwin-misunderstood/.

230 "undesirables," and had enabled them to proliferate
Moreno, *The Body Politic*, pp. 67–68.

230 led to the proliferation of "undesirables" in the first place
Ibid., pp. 69–70.

230 quite a few reactionary advocates of eugenics
Ibid.

230 described as a hate group by the Southern Poverty Law Center
Ibid., p. 70; Southern Poverty Law Center, Intelligence Files, "Pioneer Fund,"
http://www.splcenter.org/get-informed/intelligence-files/groups/pioneer-fund.

231 founding president was none other than Harry Laughlin
Wellerstein, "Harry Laughlin's 'Model Eugenical Sterilization Law.'"

231 Eugenics also found support, historians say
Moreno, *The Body Politic*, pp. 67–70.

231 what was appropriate by way of state intervention in heredity
Ibid.

231 theories that were even vaguely similar to that of Nazism
Ibid., pp. 67–69.

231 debate over current proposals that some have labeled "neo-eugenics"
Leroi, "The Future of Neo-Eugenics."

231 half of all Americans still say they do not believe in evolution
Gallup, "In U.S., 46% Hold Creationist View of Human Origins," June 1, 2012,
http://www.gallup.com/poll/155003/Hold-Creationist-View-Human-Origins.aspx.

232 one of the more than two dozen state eugenics laws
Buck v. Bell, 274 U.S. 200, May 2, 1927.

232 the young woman, Carrie Buck, had already had a child
University of Virginia—Claude Moore Health Sciences Library, "Carrie Buck,
Virginia's Test Case," 2004, http://www.hsl.virginia.edu/historical/eugenics/3
-buckvbell.cfm.

232 "Three generations of imbeciles are enough"
Buck v. Bell, 274 U.S. 200, May 2, 1927.

232 which has never been overturned
Dan Vergano, "Re-Examining Supreme Court Support for Sterilization," *USA
Today*, November 19, 2008.

232 who had been forcibly sterilized, who was then in her eighties
Stephen Jay Gould, "Carrie Buck's Daughter," *Natural History*, July 1985.

232 Buck was lucid and of normal intelligence
Ibid.

232 who had been raped by a nephew of one of her foster parents
"Carrie Buck, Virginia's Test Case."

232 to avoid what they feared would otherwise be a scandal
Vergano, "Re-Examining Supreme Court Support for Sterilization."

232 had syphilis and was unmarried when she gave birth to Carrie
"Carrie Buck, Virginia's Test Case."

232 declared Buck "congenitally and incurably defective"
Vergano, "Re-Examining Supreme Court Support for Sterilization."

233 "The fix was in"
Ibid.

233 "shiftless, ignorant, and worthless"
"Carrie Buck, Virginia's Test Case."

233 "There is a look about it that is not quite normal"
Ibid.

233 was taken from her family and given to the family of Carrie's rapist
Vergano, "Re-Examining Supreme Court Support for Sterilization."

233 Carrie's sister, Doris, was also sterilized at the same institution
Gould, "Carrie Buck's Daughter."

233 which was the basis for the Virginia statute upheld
Alex Wellerstein, "Harry Laughlin's 'Model Eugenical Sterilization Law.'"

233 President Woodrow Wilson
Vergano, "Re-Examining Supreme Court Support for Sterilization."

233 Alexander Graham Bell
Glenn Kessler, "Herman Cain's Rewriting of Birth-Control History," *Washington Post*, Fact Checker blog, November 1, 2011, http://www.washingtonpost.com/blogs/fact-checker/post/herman-cains-rewriting-of-birth-control-history/2011/10/31/gIQAr53uaM_blog.html.

233 Margaret Sanger
Ibid.

234 remain celibates or have no children or only one or two
Harry Bruinius, *Better for All the World: The Secret History of Forced Sterilization and America's Quest for Racial Purity* (New York: Knopf, 2006), pp. 190–91.

234 "To assist the race toward the elimination of the unfit"
Lori Robertson, "Cain's False Attack on Planned Parenthood," FactCheck.org, November 1, 2011, http://factcheck.org/2011/11/cains-false-attack-on-planned-parenthood/.

234 "More children from the fit, less from the unfit"
Daniel J. Kevles, *In the Name of Eugenics: Genetics and the Uses of Human Heredity* (New York: Knopf, 1985), p. 90.

234 "mixed-race individuals, single mothers with many children"
Nicole Pasulka, "Forced Sterilization for Transgender People in Sweden," *Mother Jones*, January 25, 2012.

234 can officially change his or her gender identification
Nicole Pasulka, "Sweden Moves to End Forced Sterilization of Transgender People," *Mother Jones*, February 24, 2012.

234 in Sweden are debating whether to change the law
Ibid.

234 which dates back to 1972
Pasulka, "Forced Sterilization for Transgender People in Sweden."

234 political parties has made the repeal of the law impossible thus far
Pasulka, "Sweden Moves to End Forced Sterilization of Transgender People."

234 In Uzbekistan, forced sterilizations apparently began in 2004
Natalia Antelava, "Uzbekistan Carrying Out Forced Sterilisations, Say Women," *Guardian*, April 20, 2012.

234 became official state policy in 2009
Ibid.

234 allegations by escaped activist Chen Guangcheng
Ashley Hayes, "Activists Allege Forced Abortions, Sterilizations in China," CNN, April 30, 2012, http://articles.cnn.com/2012-04-30/asia/world_asia_china-forced-abortions_1_reggie-littlejohn-china-s-national-population-abortions?_s=PM:ASIA.

235 paid a bonus for each person who is sterilized
Gethin Chamberlain, "UK Aid Helps to Fund Forced Sterilisation of India's Poor," *Guardian*, April 14, 2012.

235 already completed the full genomes of fifty animal and plant species
"The Dragon's DNA," *Economist*.

235 But China's principal focus seems to be
Tam, "Scientists Seek to Unravel the Mystery of IQ."

235 China's National Gene Bank in Shenzhen
"China Establishes National Gene Bank in Shenzhen," Xinhua News Agency.

235 which genes are involved in determining intelligence
Tam, "Scientists Seek to Unravel the Mystery of IQ."

235 link genetic information about a child to intelligence
"Bob Abernathy's Interview with Francis Collins," *PBS Religion and Ethics Weekly*, November 7, 2008.

236 eventually genes associated with intelligence may well be identified
Moheb Costandia, "Genetic Variants Build a Smarter Brain," *Science*, June 19, 2012.

236 measured by Moore's Law
Ian H. Stevenson and Konrad P. Kording, "How Advances in Neural Recording Affect Data Analysis," *Nature Neuroscience* 14, no. 2 (February 2011): 139–42.

236 which has only 302 neurons, has been completed
Jonah Lehrer, "Neuroscience: Making Connections," *Nature*, January 28, 2009.

236 Nevertheless, with an estimated 100 billion neurons
Ibid.

236 **and at least 100 trillion synaptic connections**
"Scientists Have New Help Finding Their Way Around Brain's Nooks and Crannies," *ScienceDaily*, August 9, 2011, http://www.sciencedaily.com/releases/2011/08/110809184153.htm.

236 **mapping all of the proteins that are expressed by the genes**
Human Genome Project, "The Science Behind the Human Genome Project: From Genome to Proteome," March 26, 2008, http://www.ornl.gov/sci/techresources/Human_Genome/project/info.shtml.

236 **which themselves adopt multiple geometric forms**
Jie Lang et al., "Geometric Structures of Proteins for Understanding Folding, Discriminating Natives and Predicting Biochemical Functions," 2009, http://gila-fw.bioengr.uic.edu/lab/papers/2009/protein-liang.pdf.

236 **biochemical modifications after they are translated by the genes**
Christopher Walsh et al., "Protein Posttranslational Modifications: The Chemistry of Proteome Diversifications," *Angewandte Chemie*, International Edition 44 (2005): 7342–72.

236 **"in extraordinarily complex biochemical cascades"**
Evan R. Goldstein, "The Strange Neuroscience of Immortality," *Chronicle of Higher Education*, July 16, 2012.

236 **Exploiting advances in the new field of optogenetics**
Karl Deisseroth, "Optogenetics: Controlling the Brain with Light," *Scientific American*, October 20, 2010.

236 **corresponding genes, which then become optical switches**
Matthew Hougan and Bruce Altevogt, *From Molecules to Minds: Challenges for the 21st Century* (Washington, DC: National Academies Press, 2008).

236 **observe its effects on other neurons with a green light**
Ibid.; Carl E. Schoonover and Abby Rabinowitz, "Control Desk for the Neural Switchboard," *New York Times*, May 16, 2011.

237 **control of symptoms associated with Parkinson's disease**
Amy Barth, "Controlling Brains with a Flick of a Light Switch," *Discover Magazine*, September 2012.

237 **by having each category light up in a different color**
Hougan and Altevogt, *From Molecules to Minds;* Schoonover and Rabinowitz, "Control Desk for the Neural Switchboard."

237 **much more detailed visual map of neuronal connections**
Hougan and Altevogt, *From Molecules to Minds.*

237 **decoding other parts of the connectome is thereby accelerated**
Joshua T. Vogelstein, "Q&A: What Is the Open Connectome Project?," *Neural Systems & Circuits*, November 18, 2011.

237 **scans of body parts, tracks blood flow in the brain to neurons**
"Shiny New Neuroscience Technique (Optogenetics) Verifies a Familiar Method (fMRI)," *Discover Magazine*, May 17, 2010.

237 **take in blood containing the oxygen and glucose needed for energy**
Ibid.; Leonie Welberg, "Brain Metabolism: Astrocytes Bridge the Gap," *Nature Reviews Neuroscience* 10, no. 86 (February 2009): 86.

237 difference between oxygenated blood and oxygen-depleted blood
"Major Advance in MRI Allows Much Faster Brain Scans," *ScienceDaily*, January 5, 2011.

237 identify which areas of the brain are active at any given moment
"Shiny New Neuroscience Technique (Optogenetics) Verifies a Familiar Method (fMRI)," *Discover Magazine*.

237 discoveries about where specific functions are located in the brain
Pagan Kennedy, "The Cyborg in Us All," *New York Times Magazine*, September 18, 2011.

237 at the University of Cambridge in England
David Cyranoski, "Neuroscience: The Mind Reader," *Nature*, June 13, 2012.

238 select pictures that are then displayed on the iPhone's screen
Kennedy, "The Cyborg in Us All."

238 empower users to control objects on a computer screen
Katia Moskovitch, "Real-Life Jedi: Pushing the Limits of Mind Control," BBC, October 9, 2011.

238 "muscle rhythms rather than real neural activity"
Clive Cookson, "Healthcare: Into the Cortex," *Financial Times*, July 31, 2012.

238 headset to allow thought control of other electronic devices
Moskovitch, "Real-Life Jedi."

239 approach to build wheelchairs and robots controlled by thoughts
Cookson, "Healthcare: Into the Cortex."

239 Four other companies
Moskovitch, "Real-Life Jedi."

239 that enable soldiers to communicate telepathically
Kennedy, "The Cyborg in Us All."

239 devoting more than $6 million to the project
Ibid.

239 target date for completion of the prototype device is 2017
"Pentagon Plans for Telepathic Troops Who Can Read Each Others' Minds . . . and They Could Be in the Field within Five Years," *Daily Mail*, April 8, 2012.

239 According to Nick Bostrom
Nick Bostrom, "A History of Transhumanist Thought," 2005, http://www.nickbostrom.com/papers/history.pdf.

239 a ferment that continued into the twentieth century
Ibid.

239 First used by Teilhard de Chardin
Ibid.

240 "Shortly after, the human era will be ended"
Vernor Vinge, "The Coming Technological Singularity: How to Survive in the Post-Human Era," 1993, http://www-rohan.sdsu.edu/faculty/vinge/misc/singularity.html.

240 form that can be comprehended by and *contained* in advanced computers
Lara Farrar, "Scientists: Humans and Machines Will Merge in Future," CNN,

July 15, 2008, http://articles.cnn.com/2008-07-15/tech/bio.tech_1_emergent
-technologies-bostrom-human-life/2?_s=PM:TECH.

240 **"post-Singularity, between human and machine or between physical and virtual reality"**
Ibid.

240 **has challenged Kurzweil to a $20,000 bet**
"By 2029 No Computer—or 'Machine Intelligence'—Will Have Passed the Turing Test," A Long Bet, http://longbets.org/1/.

240 **before the computer-based "Technological Singularity" is ever achieved**
John Chelen, "Could the Organic Singularity Occur Prior to Kurzweil's Technological Singularity?," *Science Progress*, June 20, 2012.

241 **replace not only hips**
Ben Coxworth, "New Discovery Could Lead to Better Artificial Hips," *Gizmag*,
November 27, 2011, http://www.gizmag.com/artificial-hip-joint-lubrication-layer/
20949/.

241 **knees**
James Dao, "High-Tech Knee Holds Promise for Veterans," *New York Times*, August 18, 2010.

241 **legs**
Alexis Okeowo, "A Once-Unthinkable Choice for Amputees," *New York Times*,
May 14, 2012.

241 **arms**
Thomas H. Maugh II, "Two Paralyzed People Successfully Use Robot Arm," *Los
Angeles Times*, May 16, 2012.

241 **but also eyes**
Carl Zimmer, "'I See,' Said the Blind Man with an Artificial Retina," *Discovery
News*, September 15, 2011.

241 **replaceable with artificial substitutes**
Richard Yonck, "The Path to Future Intelligence," *Psychology Today*, May 13,
2011; Rob Beschizza, "Mechanical Fingers Give Strength, Speed to Amputees,"
Wired, July 2, 2007.

241 **Cochlear implants, as noted, are used to restore hearing**
"Cochlear Implants Restore Hearing in Rare Disorder," *Science Daily*, April 20, 2012.

241 **exoskeletons to enable paraplegics to walk**
Melissa Healy, "Body Suit May Soon Enable the Paralyzed to Walk," *Los Angeles
Times*, October 6, 2011.

241 **confer additional strength on soldiers**
Susan Karlin, "Raytheon Sarcos's Exoskeleton Nears Production," *IEEE Spectrum*, August 2011.

241 **bespoke in-ear hearing aids are already made with 3D printers**
Quest Means Business, CNN transcript, November 8, 2012, http://transcripts.cnn
.com/TRANSCRIPTS/1211/08/qmb.01.html; Nick Glass, "Pitch Perfect: The
Quest to Create the World's Smallest Hearing Aid," CNN, November 9, 2012,
http://www.cnn.com/2012/11/09/tech/hearing-aid-widex-3d-printing/index.html.

241 **woman who was not a candidate for traditional reconstructive surgery**
"Transplant Jaw Made by 3D Printer Claimed as First," BBC News, February 6,
2012, http://www.bbc.co.uk/news/technology-16907104.

241 field of transplantation because of the current shortage of organs
Ibid.

242 regenerative medicine scientists at Wake Forest University
Wake Forest Baptist Medical Center, press release, "Lab-Engineered Kidney Project Reaches Early Milestone," June 21, 2012; Wake Forest Baptist Medical Center, press release, "Researchers Engineer Miniature Human Livers in the Lab," October 30, 2010.

242 precisely copied the size and shape of the windpipe
Henry Fountain, "A First: Organs Tailor-Made with Body's Own Cells," *New York Times*, September 16, 2012.

242 sensed the matrix of the scaffolding being broken down
Henry Fountain, "Human Muscle, Regrown on Animal Scaffolding," *New York Times*, September 17, 2012.

242 developing silicon nanowires a thousand times smaller
Elizabeth Landieu, "When Organs Become Cyborgs," CNN, August 29, 2012.

242 which the U.S. shares with all other countries besides Iran
Stephen J. Dubner, "Human Organs for Sale, Legally, in . . . Which Country?," Freakonomics blog, April 29, 2008, http://www.freakonomics.com/2008/04/29/human-organs-for-sale-legally-in-which-country/.

242 transplantation into people living in wealthy countries
"Organ Black Market Booming," UPI, May 28, 2012.

242 "the longest chain of kidney transplants ever constructed"
Kevin Sack, "60 Lives, 30 Kidneys, All Linked," *New York Times*, February 19, 2012.

242 "organ donor" as one of the items to be updated
Matt Richtel and Kevin Sack, "Facebook Is Urging Members to Add Organ Donor Status," *New York Times*, May 1, 2012.

242 the process to print more advanced artificial limbs
Ashlee Vance, "3-D Printing Spurs a Manufacturing Revolution," *New York Times*, September 14, 2010.

242 using it to make numerous medical implants
"The Printed World," *Economist*, February 10, 2011.

243 print vaccines and pharmaceuticals from basic chemicals
Tim Adams, "The 'Chemputer' That Could Print Out Any Drug," *Guardian*, July 21, 2012.

243 essentially the same product
Eric Topol, *The Creative Destruction of Medicine: How the Digital Revolution Will Create Better Health Care* (New York: Basic Books, 2012), ch. 10.

243 activated by shining a laser light on them from outside the body
Avi Schroeder et al., "Remotely Activated Protein-Producing Nanoparticles," *Nano Letters* 2, no. 6 (2012): 2685–89; George Dvorsky, "Microscopic Machines Could Produce Medicine Directly Inside Your Body," io9, July 29, 2012, http://io9.com/5922447/microscopic-machines-could-produce-medicine-directly-inside-your-body.

243 Specialized prosthetics for the brain
Cookson, "Healthcare: Into the Cortex."

243 **digital devices on the surface of the brain and, in some cases, deeper within the brain**
Wilson, "Bionic Brains and Beyond"; Allison Abbott, "Brain Implants Have Long-Lasting Effect on Depression," *Nature*, February 7, 2011.

243 **activate and direct the movement of robots with their thoughts**
Cookson, "Healthcare: Into the Cortex."

244 **dispense with the wires connecting the chip to a computer**
Ibid.

244 **Scientists and engineers at the University of Illinois**
Ibid.

244 **"nanotechnology, micro-power generation—to provide therapeutic benefit"**
Ibid.

244 **rat's brain stem to interpret information**
Linda Geddes, "Rat Cyborg Gets Digital Cerebellum," *New Scientist*, September 27, 2011.

244 **"synthetic correlates before the end of the century"**
Ibid.

244 **in humans, including prosthetics for bladder control**
Monica Friedlander, Lawrence Livermore National Laboratory, "Neural Implants Come of Age," *Science and Technology Review*, June 2012.

244 **relief of spinal pain**
Ibid.

244 **remediation of some forms of blindness**
Wilson, "Bionic Brains and Beyond."

244 **and deafness**
Ibid.

244 **to enhance focus and concentration**
Ibid.

245 **people to enhance concentration at times of their choosing**
Ibid.

245 **Adderall, Ritalin, and Provigil to improve their test scores**
Margaret Talbot, "Brain Gain: The Underground World of 'Neuroenhancing' Drugs," *New Yorker*, April 27, 2009.

245 **"ranges from 15 percent to 40 percent"**
Alan Schwarz, "Risky Rise of the Good-Grade Pill," *New York Times*, June 10, 2012.

245 **doctors who work with low-income families have started prescribing Adderall**
Alan Schwarz, "Attention Disorder or Not, Pills to Help in School," *New York Times*, October 9, 2012.

245 **felt they improved their memory and ability to focus**
Drew Halley, "Brain-Doping at the Lab Bench," Project Syndicate, April 20, 2009.

246 **the promise of actually boosting intelligence**
Jamais Cascio, "Get Smarter," *Atlantic*, July/August 2009; Ross Anderson, "Why Cognitive Enhancement Is in Your Future (and Your Past)," *Atlantic*, February 6, 2012.

246 **as little stigma as cosmetic surgery does today**
V. Cakic, "Smart Drugs for Cognitive Enhancement: Ethical and Pragmatic Considerations in the Era of Cosmetic Neurology," *Journal of Medical Ethics* 35 (2009):

611–15; Anderson, "Why Cognitive Enhancement Is in Your Future (and Your Past)."

246 used for object recognition in order to improve the training of snipers
Sally Adee, "Zap Your Brain into the Zone: Fast Track to Pure Focus," *New Scientist*, February 6, 2012.

246 first double amputee track athlete ever to compete
Jere Longman, "After Long Road, Nothing Left to Do but Win," *New York Times*, August 5, 2012.

246 relay, in which the South African team reached the finals
David Trifunov, "Oscar Pistorius Eliminated in 400m Semifinal at London 2012 Olympics," *Global Post*, August 5, 2012.

246 "it is unfair to the able-bodied competitors"
Longman, "After Long Road, Nothing Left to Do but Win."

246 according to Pistorius
"Oscar Pistorius Apologizes for Timing of Paralympics Criticism," BBC Sport, September 3, 2012.

246 (EPO)—which regulates the production of red blood cells
"Genetically Modified Olympians?," *Economist*, July 31, 2008.

246 delivering more oxygen to the muscles for a longer period of time
Lana Bandoim, "Erythropoietin Abuse Among Athletes Can Lead to Vascular Problems," Yahoo, December 25, 2011, http://sports.yahoo.com/top/news?slug =ycn-10747311.

246 He has admitted use of EPO, along with other illegal enhancements
"Landis Admits EPO Use," ESPN, May 20, 2010, http://www.espn.co.uk/more/ sport/story/23635.html.

247 Armstrong was stripped of his championships and banned from cycling
Juliet Macur, "Lance Armstrong Is Stripped of His 7 Tour de France Titles," *New York Times*, October 22, 2012

247 use of EPO, steroids, and blood transfusions
"Lance Armstrong Won't Fight Charges," ESPN, August 24, 2012.

247 detecting new enhancements that violate the rules
Matthew Knight, "Hi-Tech Tests to Catch Olympics Drug Cheats at London 2012," CNN, July 31, 2012, http://edition.cnn.com/2012/04/12/sport/drugs-london -2012-olympics-laboratory/index.html; Andy Bull, "Ye Shiwen's World Record Olympic Swim 'Disturbing,' Says Top US Coach," *Guardian*, July 30, 2012.

247 produce more red blood cells
"Fairly Safe," *Economist,* July 31, 2008.

247 ruling without genetic testing of the athlete's relatives
"Genetically Modified Olympians?," *Economist.*

247 called myostatin that regulates the building of muscles
Aaron Saenz, "Super Strength Substance (Myostatin) Closer to Human Trials," Singularity Hub, December 8, 2009.

247 "sperm-making biological machine"
"Artificial Testicle, World's First to Make Sperm, Under Development by California Scientists," *Huffington Post*, January 19, 2012, http://www.huffingtonpost.com/ 2012/01/19/artificial-testicle_n_1215964.html.

248 the first so-called test tube baby in 1978, Louise Brown
Donna Bowater, "Lesley Brown, Mother of First Test Tube Baby Louise Brown,
Dies Aged 64," *Telegraph*, June 21, 2012.

248 debate about the ethics and propriety of the procedure
Robert Bailey, "The Case for Enhancing People," *New Atlantis*, June 20, 2012.

248 diminish parental love and weaken generational ties
Ibid.

248 "People want children"
Fiona Macrae, "Death of the Father: British Scientists Discover How to Turn
Women's Bone Marrow into Sperm," *Daily Mail*, January 31, 2008.

248 born to infertile people wanting children
Jeanna Bryner, "5 Million Babies Born from IVF, Other Reproductive Technologies," *Live Science*, July 3, 2012.

248 I saw this pattern repeated many times
See, for example: Subcommittee on Investigations and Oversight and the Sub-
committee on Science, Research, and Technology, Committee on Science and
Technology, U.S. House of Representatives, "Commercialization of Academic
Biomedical Research," June 8–9, 1981; Subcommittee on Investigations and
Oversight, Committee on Science and Technology, U.S. House of Representa-
tives, "Genetic Screening and the Handling of High-Risk Groups in the Work-
place," October 14–15, 1981.

248 Dr. Christiaan Barnard
Lawrence K. Altman, "Christiaan Barnard, 78, Surgeon for First Heart Transplant,
Dies," *New York Times*, September 3, 2001.

248 "My God, it's working!"
Personal conversation with author.

248 preimplantation genetic diagnosis
"Saviour Siblings—the Controversy and the Technique," *Telegraph*, May 6, 2011.

248 "savior sibling"
Ibid.

248 who can serve as an organ
Stephen Wilkinson, "'Saviour Siblings' as Organ Donors," Sveriges Yngre Läk-
ares Förening (Swedish junior doctors' association), November 2, 2012, http://
www.slf.se/SYLF/Moderna-lakare/Artiklar/Nummer-2-2012/Saviour-Siblings-as
-Organ-Donors/.

248 tissue
Robert Sparrow and David Cram, "Saviour Embryos? Preimplantation Genetic
Diagnosis as a Therapeutic Technology," Reproductive BioMedicine Online,
May 15, 2010, http://www.ivf.net/ivf/saviour-embryos-preimplantation-genetic
-diagnosis-as-a-therapeutic-technology-o5043.html.

248 bone marrow
Josephine Marcotty, "'Savior Sibling' Raises a Decade of Life-and-Death Questions," *Star Tribune*, September 22, 2010.

248 or umbilical cord stem cell donor
"Saviour Siblings—the Controversy and the Technique," *Telegraph*.

248 the instrumental purpose of such conceptions devalues the child
Stephen Wilkinson, *Choosing Tomorrow's Children: The Ethics of Selective Reproduction* (New York: Oxford University Press, 2010).

249 important cure for the first with the assistance of the second
K. Devolder, "Preimplantation HLA Typing: Having Children to Save Our Loved Ones," *Journal of Medical Ethics* 31 (January 2005): 582–86.

249 "three-parent babies"
David Derbyshire, "Babies with THREE Parents and Free of Genetic Disease Could Soon Be Born Using Controversial IVF Technique," *Daily Mail*, March 12, 2011.

249 genetic modification is one that will affect
James Gallagher, "Three-Person IVF 'Is Ethical' to Treat Mitochondrial Disease," BBC, June 11, 2012, http://www.bbc.co.uk/news/health-18393682.

249 be left to the pregnant woman herself, at least in the earlier stages of the pregnancy
World Public Opinion, "World Publics Reject Criminal Penalties for Abortion," June 18, 2008, http://www.worldpublicopinion.org/pipa/articles/btjusticehuman_rightsra/492.php.

249 that are designed to identify the gender of the embryo
Rachel Rickard Straus, "To Ensure Prized Baby Boy, Indians Flock to Bangkok," *Times of India*, December 27, 2010.

249 abortion of 500,000 female fetuses each year
Madeleine Bunting, "India's Missing Women," *Guardian*, July 22, 2011.

250 better enforce the prohibition against the sex selection of children
"Delhi Govt to Crack Down on Sex-Selection Tests," *Times of India*, January 5, 2012.

250 identification procedures in India utilize ultrasound machines
Bunting, "India's Missing Women."

250 Some couples from India
Straus, "To Ensure Prized Baby Boy, Indians Flock to Bangkok."

250 blood samples taken from pregnant mothers
"Baby Sex ID Test Won't Be Sold in China or India Due to Fears of 'Gender Selection,'" Associated Press, August 10, 2011.

250 sample from the pregnant woman and a saliva sample from the father
Andrew Pollack, "DNA Blueprint for Fetus Built Using Tests of Parents," *New York Times*, June 6, 2012.

250 an estimated $20,000 to $50,000 for one fetal genome
Ibid.

250 last year, the cost was $200,000 per test
Mara Hvistendahl, "Will Gattaca Come True?," *Slate*, April 27, 2012.

250 likely to continue falling very quickly
Ibid.

250 widely available within two years for an estimated $3,000
Stephanie M. Lee, "New Stanford Fetal DNA Test Adds to Ethical Issues," *San Francisco Chronicle*, July 26, 2012.

250 **serious disorders that might be treated through early detection**
Drew Halley, "Revolution In Newborn Screening Saves Newborn Lives," Singularity Hub, March 10, 2009.

251 **are terminating their pregnancies**
Ross Douthat, "Eugenics, Past and Future," *New York Times*, June 9, 2012.

251 **"sophisticated methods of eugenic selection"**
Leroi, "The Future of Neo-Eugenics."

251 **embryos for such traits as hair**
Hvistendahl, "Will Gattaca Come True?"; Kleiner, "Designer Babies."

251 **and eye color**
Kleiner, "Designer Babies."

251 **skin complexion**
Ibid.

251 **"98.1 percent of death-row inmates do"**
David Eagleman, "The Brain on Trial," *Atlantic*, July/August 2011.

251 **yet another round of difficult ethical choices**
Leroi, "The Future of Neo-Eugenics."

251 **markers associated with hundreds of diseases before implantation**
Drew Halley, "Prenatal Screening Could Eradicate Genetic Disease, Replace Natural Conception," Singularity Hub, July 21, 2009.

252 **preserved for potential later implantation**
Denise Grady, "Parents Torn over the Fate of Frozen Embryos," *New York Times*, December 4, 2008.

252 **women who undergo the in vitro fertilization procedure**
Ibid.; Laura Bell, "What Happens to Extra Embryos After IVF?," CNN, September 1, 2009, http://articles.cnn.com/2009-09-01/health/extra.ivf.embryos_1_embryos-fertility-patients-fertility-clinics?_s=PM:HEALTH.

252 **to improve the odds that one will survive**
Tiffany Sharples, "IVF Study: Two Embryos No Better Than One," *Time*, March 30, 2009.

252 **in vitro fertilization than in the general population**
U.S. Centers for Disease Control and Prevention, "Contribution of Assisted Reproductive Technology and Ovulation-Inducing Drugs to Triplet and Higher-Order Multiple Births—United States, 1980–1997," *MMWR*, June 23, 2000.

252 **The United Kingdom has set a legal limit on the number of embryos**
Sarah Boseley, "IVF Clinics Told to Limit Embryo Implants to Curb Multiple Births," *Guardian*, January 6, 2004.

252 **Auxogyn, is using digital imaging**
Reproductive Science Center, "Auxogyn," http://rscbayarea.com/for-physicians/auxogyn.

252 **select the embryo that is most likely to develop**
Yahoo Finance News, "Auxogyn and Hewitt Fertility Center Announce First Availability of New Non-Invasive Early Embryo Viability Assessment (Eeva) Test in the European Union," September 17, 2012, http://finance.yahoo.com/news/auxogyn-hewitt-fertility-center-announce-060000428.html.

252 to allow the pregnant woman to control the choice on abortion
Leroi, "The Future of Neo-Eugenics."

252 right to require a pregnant woman to have an abortion
Ibid.

253 so significant that they justify such experiments
Pew Forum on Religion and Public Life, "Stem Cell Research Around the World," July 17, 2012, http://www.pewforum.org/Science-and-Bioethics/Stem-Cell-Research-Around-the-World.aspx.

253 Shinya Yamanaka at Kyoto University
Alok Jha, "Look, No Embryos! The Future of Ethical Stem Cells," *Guardian*, March 12, 2011.

253 was awarded the 2012 Nobel Prize in Medicine
Nicholas Wade, "Cloning and Stem Cell Work Earns Nobel," *New York Times*, October 8, 2012.

253 unique qualities and potential that justify their continued use
Andrew Pollack, "Setback for New Stem Cell Treatment," *New York Times*, May 13, 2011.

253 restore some vision to mice with an inherited retinal disease
Sarah Boseley, "Medical Marvels: Drugs Treat Symptoms. Stem Cells Can Cure You. One Day Soon, They May Even Stop Us Ageing," *Guardian*, January 29, 2009.

253 forms of blindness in humans may soon be treatable
Fergus Walsh, "'Blind' Mice Eyesight Treated with Transplanted Cells," BBC, April 18, 2012, http://www.bbc.co.uk/news/health-17748165.

253 rebuild nerves in the ears of gerbils and restore their hearing
James Gallagher, "Deaf Gerbils 'Hear Again' After Stem Cell Cure," BBC, September 12, 2010.

253 sperm when transplanted into the testicles of mice that were infertile
Nick Collins, "Stem Cells Used to Make Artificial Sperm," *Telegraph*, August 4, 2011.

253 though the offspring had genetic defects
Roxanne Khamsi, "Bone Stem Cells Turned into Primitive Sperm Cells," *New Scientist*, April 13, 2007.

253 for infertile men to have biological children
Collins, "Stem Cells Used to Make Artificial Sperm."

253 lesbian couples to have children that are genetically and biologically their own
Macrae, "Death of the Father."

254 increase human lifespans by multiple centuries
Aubrey de Grey, "'We Will Be Able to Live to 1,000,'" BBC, December 3, 2004, http://news.bbc.co.uk/2/hi/uk_news/4003063.stm.

254 25 percent is more likely
Gary Taubes, "The Timeless and Trendy Effort to Find—or Create—the Fountain of Youth," *Discover Magazine*, February 7, 2011.

254 According to most experts, evolutionary theory
Nir Barzilai et al., "The Place of Genetics in Ageing Research," *Nature Reviews Genetics* 13 (August 2012): 589–94.

254 numerous studies in human
James W. Curtsinger, "Genes, Aging, and Prospects for Extended Life Span," *Minnesota Medicine*, October 2007.

254 and animal genetics
Ibid.

254 roughly three quarters to the aging process
Barzilai et al., "The Place of Genetics in Ageing Research."

254 somewhere between 20 and 30 percent
Ibid.

254 extreme caloric restriction extends the lives of rodents dramatically
Taubes, "The Timeless and Trendy Effort to Find—or Create—the Fountain of Youth."

254 adjustment has the same effect on longevity in humans
Gina Kolata, "Severe Diet Doesn't Prolong Life, at Least in Monkeys," *New York Times*, August 30, 2012.

254 rhesus monkeys do *not* live longer with severe caloric restrictions
Ibid.

254 experts on all sides point out, between longevity and aging
Roger B. McDonald and Rodney C. Ruhe, "Aging and Longevity: Why Knowing the Difference Is Important to Nutrition Research," *Nutrients* 3 (2011): 274–82.

254 to slow or reverse unwanted manifestations of the aging process
Gretchen Voss, "The Risks of Anti-Aging Medicine," CNN, March 30, 2012, http://www.cnn.com/2011/12/28/health/age-youth-treatment-medication/index .html; Dan Childs, "Growth Hormone Ineffective for Anti-Aging, Studies Say," ABC News, January 16, 2007, http://abcnews.go.com/Health/ActiveAging/story?id =2797099&page=1#.UGDZ3Y40jdk.

254 most prominently, testosterone
"Anti-Aging Hormones: Little or No Benefit and the Risks Are High, According to Experts," *ScienceDaily*, April 13, 2010, http://www.sciencedaily.com/releases/ 2010/04/100413121326.htm.

255 genetic factors that can be used to extend longevity in others
Barzilai et al., "The Place of Genetics in Ageing Research."

255 over the last century have come from improvements in sanitation
Robert Kunzig, "7 Billion: How Your World Will Change," *National Geographic*, November 1, 2011.

255 about one extra year per decade
Curtsinger, "Genes, Aging, and Prospects for Extended Life Span."

255 Much of this work is now focused on malaria, tuberculosis
United Nations, Millennium Development Goals Report 2011.

255 HIV/AIDS, influenza
George Verikios et al., "The Global Economic Effects of Pandemic Influenza," paper prepared for the 14th Annual Conference on Global Economic Analysis, Venice, June 16–18, 2011, https://www.gtap.agecon.purdue.edu/resources/ download/5291.pdf.

255 viral pneumonia
Olli Ruuskanen, Elina Lahti, Lance C Jennings, and David R Murdoch, "Viral Pneumonia," *Lancet* 377 (2011): 1264–75.

255 and multiple so-called "neglected tropical diseases"
Dr. Lorenzo Savioli, World Health Organization, "Neglected Tropical Diseases: Letter from the Director," 2011, http://www.who.int/neglected_diseases/director/en/index.html.

255 In 2012, the number fell to 1.7 million
Deena Beasley and Tom Miles, "AIDS Deaths Worldwide Dropping as Access to Drugs Improves," July 18, 2012, Reuters.

255 reduce the infection rate continue to be focused on preventive education
Avert, "Introduction to HIV Prevention," http://www.avert.org/prevent-hiv.htm.

255 the distribution of condoms in high-risk areas
United Nations Population Fund, Preventing HIV/AIDS, "Comprehensive Condom Programming: A Strategic Response to HIV and AIDS," http://www.unfpa.org/hiv/programming.htm.

255 and accelerated efforts to develop a vaccine
Beasley and Miles, "AIDS Deaths Worldwide Dropping as Access to Drugs Improves."

255 Malaria has also been reduced significantly
United Nations, Millennium Development Goals Report 2011.

256 Although an ambitious effort in the 1950s to eradicate malaria
"Malaria Eradication No Vague Aspiration, Says Gates," Reuters, October 18, 2011.

256 did succeed in eliminating the terrible scourge of smallpox in 1980
Katie Hafner, "Philanthropy Google's Way: Not the Usual," *New York Times*, September 14, 2006.

256 succeeded in eliminating a second disease, rinderpest
Donald G. McNeil Jr., "Rinderpest, Scourge of Cattle, Is Vanquished," *New York Times*, June 27, 2011.

256 chronic diseases that are not communicable
Ala Alwan, World Health Organization, "Monitoring and Surveillance of Chronic Non-Communicable Diseases: Progress and Capacity in High-Burden Countries," *Lancet* 376 (November 2010): 1861–68.

256 massive effort to create a "Cancer Genome Atlas"
Gina Kolata, "Genetic Aberrations Seen as Path to Stop Colon Cancer," *New York Times*, July 18, 2012.

256 possibilities for shutting off the blood supply to cancerous cells
Erika Check Hayden, "Cutting Off Cancer's Supply Lines," *Nature*, April 20, 2009.

256 dismantling their defense mechanisms
Nicholas Wade, "New Cancer Treatment Shows Promise in Testing," *New York Times*, June 28, 2009.

256 to identify and attack the cancer cells
Denise Grady, "An Immune System Trained to Kill Cancer," *New York Times*, September 9, 2011.

256 **that involve proteomics—the decoding**
Henry Rodriguez, "Fast-Tracking Personalized Medicine: The New Proteomics Pipeline," *R&D Directions*, 2012, http://www.pharmalive.com/magazines/randd/view.cfm?articleID=9178#.

256 **all of the proteins translated by cancer genes**
Danny Hillis, "Understanding Cancer Through Proteomics," TEDMED 2010, October 2010, http://www.ted.com/talks/danny_hillis_two_frontiers_of_cancer_treatment.html.

256 **it is actually more akin to a list of parts or ingredients**
Ibid.

257 **reprogramming cells to restore the health of heart muscles**
Leila Haghighat, "Regenerative Medicine Repairs Mice from Top to Toe," *Nature*, April 18, 2012.

257 **for combating chronic diseases is to make changes in lifestyles**
World Health Organization, World Health Statistics, 2011, p. 19.

257 **spreading from North America and Western Europe to the rest of the world**
Pedro C. Hallal et al., "Global Physical Activity Levels: Surveillance Progress, Pitfalls, and Prospects," *Lancet* 380, no. 9838 (2012): 247–57; Gretchen Reynolds, "The Couch Potato Goes Global," *New York Times*, Well blog, July 18, 2012, http://well.blogs.nytimes.com/2012/07/18/the-couch-potato-goes-global/.

257 **from conditions linked with physical inactivity than die from smoking**
Pamela Das and Richard Horton, "Rethinking Our Approach to Physical Activity," *Lancet* 380, no. 9838 (2012): 189–90; Reynolds, "The Couch Potato Goes Global."

257 **ten deaths worldwide is now due to diseases caused by persistent inactivity**
For various articles on physical activity and inactivity see issue of *Lancet* 380, no. 9838 (2012): i, 187–306; Matt Sloane, "Physical Inactivity Causes 1 in 10 Deaths Worldwide, Study Says," CNN, July 18, 2012, http://www.cnn.com/2012/07/18/health/physical-inactivity-deaths/index.html.

257 **apps that assist those who wish to keep track of how many calories**
David H. Freeman, "The Perfected Self," *Atlantic*, June 2012.

258 **some new headbands**
Mark Bowden, "The Measured Man," *Atlantic*, July/August 2012.

258 **Mood disorders**
"Counting Every Moment," *Economist*, March 3, 2012.

258 **genetic analyses designed to improve their individual nutritional needs**
April Dembosky, "Olympians Trade Data for Tracking Devices," *Financial Times*, July 22, 2012.

258 **Personal digital monitors of patients' heart rates**
Gary Wolf, "The Data-Driven Life," *New York Times Magazine*, April 28, 2010; "Counting Every Moment," *Economist*, March 3, 2012; Freeman, "The Perfected Self."

258 **reporting information on a constant basis**
Sharon Gaudin, "Nanotech Could Make Humans Immortal by 2040, Futurist Says," *Computerworld*, October 1, 2009; Bowden, "The Measured Man."

258 **"Constant monitoring is a recipe"**
Bowden, "The Measured Man."

258 resulting interventions were apparently doing more harm than good
Gardiner Harris, "U.S. Panel Says No to Prostate Screening for Healthy Men,"
New York Times, October 7, 2011.

258 seen as highly valuable to insurance companies
"Do Not Ask or Do Not Answer?," *Economist*, August 23, 2007.

258 and employers
Adam Cohen, "Can You Be Fired for Your Genes?," *Time*, February 2, 2012.

259 fear that they will lose their jobs and/or their health insurance
Amy Harmon, "Insurance Fears Lead Many to Shun DNA Tests," *New York Times*,
February 24, 2008.

259 prohibits the disclosure or improper use of genetic information
Cohen, "Can You Be Fired for Your Genes?"

259 But enforcement is difficult
Amy Harmon, "Congress Passes Bill to Bar Bias Based on Genes," *New York Times*,
May 2, 2008; Cohen, "Can You Be Fired for Your Genes?"

259 trust in the law's protection is low
Eric A. Feldman, "The Genetic Information Nondiscrimination Act (GINA):
Public Policy and Medical Practice in the Age of Personalized Medicine," *Journal
of General Internal Medicine* 27, no. 6 (June 2012): 743–46.

259 employers usually pay for the majority of health care expenditures
Harmon, "Insurance Fears Lead Many to Shun DNA Tests."

259 fails to guarantee patient access to records
Amy Dockser Marcus and Christopher Weaver, "Heart Gadgets Test Privacy-Law
Limits," *Wall Street Journal*, November 28, 2012.

259 outside of an institutional setting
Freeman, "The Perfected Self."

259 personalized medicine continues to move forward
Chad Terhune, "Spending on Genetic Tests Is Forecast to Rise Sharply by 2021,"
Los Angeles Times, March 12, 2012.

259 For example, many health care
Since people sometimes switch from one insurance company to another, companies paying for prevention may end up benefiting a competitor.

259 required coverage of preventive care
"Preventive Services Covered under the Affordable Care Act," Healthcare.gov,
2012.

259 As everyone knows, the U.S. spends
Simon Rogers, "Healthcare Spending Around the World, Country by Country,"
Guardian, June 30, 2012; Harvey Morris, "U.S. Healthcare Costs More Than 'So-
cialized' European Medicine," *International Herald Tribune*, June 28, 2012.

259 many other countries that pay far less
Morris, "U.S. Healthcare Costs More Than 'Socialized' European Medicine."

259 still, tens of millions do not have reasonable access to health care
Emily Smith and Caitlin Stark, "By the Numbers: Health Insurance," CNN,
June 28, 2012, http://edition.cnn.com/2012/06/27/politics/btn-health-care/index
.html.

260 **where the cost of intervention is highest**
Sarah Kliff, "Romney Was Against Emergency Room Care Before He Was for It," *Washington Post*, Ezra Klein's Wonkblog, September 24, 2012, http://www.washingtonpost.com/blogs/ezra-klein/wp/2012/09/24/romney-was-against-emergency-room-care-before-he-was-for-it/.

260 **chance of success is lowest**
Sarah Kliff, "The Emergency Department Is Not Health Insurance," *Washington Post*, Ezra Klein's Wonkblog, September 24, 2012, http://www.washingtonpost.com/blogs/ezra-klein/wp/2012/09/24/the-emergency-department-is-not-health-insurance/.

260 **reforms will significantly improve some of these defects**
Emily Oshima Lee, Center for American Progress, "How ObamaCare Is Benefitting Americans," July 12, 2012, http://www.americanprogress.org/issues/healthcare/news/2012/07/12/11843/update-how-obamacare-is-benefiting-americans/.

260 **but the underlying problems are likely to grow worse**
U.S. Government Accountability Office, "Federal Government Long-Term Fiscal Outlook: Spring 2012," April 2, 2012, http://www.gao.gov/products/GAO-12-521SP.

260 **The business of insurance began as far back as ancient Rome**
LifeHealthPro, "Timeline: The History of Life Insurance," 2012, http://www.lifehealthpro.com/interactive/timeline/history/.

260 **and Greece**
American Bank, "A Brief History of Insurance," June 2011, http://www.american-bank.com/insurance/a-brief-history-of-insurance-part-3-roman-life-insurance/.

260 **were similar to what we now know as burial insurance**
Ibid.

260 **not offered until the seventeenth century in England**
Habersham Capital, "The History of Life Insurance and Life Settlements," 2012, http://www.habershamcapital.com/brief-no2-history.

260 **development of extensive railroad networks**
"Health Insurance," *Encarta*, 2009, http://www.webcitation.org/5kwqZV6V7.

260 **drive costs above what many patients could pay on their own**
Timothy Noah, "A Short History of Health Care," *Slate*, March 13, 2007, http://www.slate.com/articles/news_and_politics/chatterbox/2007/03/a_short_history_of_health_care.single.htm.

260 **Blue Cross for hospital charges**
"Health Insurance," *Encarta*.

260 **preexisting conditions**
Noah, "A Short History of Health Care."

260 **twice took preliminary steps—in 1935**
Kyle Noonan, New America Foundation, "Health Reform through History: Part I: The New Deal," May 26, 2009, http://www.newamerica.net/blog/new-health-dialogue/2009/health-reform-through-history-part-i-new-deal-11961.

260 **feared the political opposition of the American Medical Association**
Ibid.

260 regarded as more pressing priorities
Paul Starr, "In Sickness and in Health," On the Media, August 21, 2009, http://www.onthemedia.org/2009/aug/21/in-sickness-and-in-health/transcript/.

260 offered a quixotic third opportunity to proceed but Roosevelt
Noonan, "Health Reform through History: Part I: The New Deal."

261 During World War II, with wages (and prices) controlled
Noah, "A Short History of Health Care."

261 extensive health insurance as part of their negotiated contracts
"Health Insurance," Encarta.

261 revive the idea for national health insurance, but the opposition in Congress
Starr, "In Sickness and in Health"; Noonan, "Health Reform through History: Part I: The New Deal."

261 employer-based health insurance became the primary model
Noah, "A Short History of Health Care."

261 new government programs were implemented to help both groups
"Health Insurance," Encarta.

261 needed health insurance the most had a difficult time obtaining it
Noah, "A Short History of Health Care"; "Health Insurance," Encarta.

261 believe that genetically engineered food should be labeled
Gary Langer, "Poll: Skepticism of Genetically Modified Foods," ABC News, June 19, 2011, http://abcnews.go.com/Technology/story?id=97567&page=1#.UGIUS7S1Ndx.

261 adopted the point of view advocated by large agribusiness
Tom Philpott, "Congress' Big Gift to Monsanto," Mother Jones, July 2, 2012.

262 However, most European countries
Amy Harmon and Andrew Pollack, "Battle Brewing Over Labeling of Genetically Modified Food," New York Times, May 25, 2012.

262 genetically engineered alfalfa
Ibid.

262 which plants twice as many acres in GM crops as any other country
International Service for the Acquisition of Agri-Biotech Applications, ISAAA Brief 43-2011, Global Status of Commercialized Biotech/GM Crops: 2011, http://www.isaaa.org/resources/publications/briefs/43/executivesummary/default.asp.

262 California defeated a referendum in 2012 to require such labeling
Andrew Pollack, "After Loss, the Fight to Label Modified Food Continues," New York Times, November 7, 2012.

262 approximately 70 percent of the processed foods
Harmon and Pollack, "Battle Brewing Over Labeling of Genetically Modified Food"; Richard Shiffman, "How California's GM Food Referendum May Change What America Eats," Guardian, June 13, 2012; Center for Food Safety, "Genetically Engineered Crops," http://www.centerforfoodsafety.org/campaign/genetically-engineered-food/crops/.

262 as enthusiastic advocates often emphasize, hardly new
Michael Antoniou, Claire Robinson, and John Fagan, "GMO Myths and Truths, Version 1.3," June 2012, http://earthopensource.org/files/pdfs/GMO_Myths_and_Truths/GMO_Myths_and_Truths_1.3a.pdf, p. 21; Council for Biotechnology In-

formation, "Myths & Facts: Plant Biotechnology," http://www.whybiotech.com/resources/myths_plantbiotech.asp#16.

262 **genetically modified during the Stone Age by careful selective breeding**
Council for Biotechnology Information, "Myths & Facts: Plant Biotechnology."

262 **"plants in the process of domesticating our food crop species"**
Anne Cook, "Borlaug: Will Farmers Be Permitted to Use Biotechnology?," Knight Ridder/Tribune, June 14, 2001.

262 **merely accelerating and making more efficient a long-established practice**
Council for Biotechnology Information, "Myths & Facts: Plant Biotechnology."

262 **has ever produced any increase in the intrinsic yields**
Doug Gurian-Sherman, *Failure to Yield* (Cambridge, MA: Union of Concerned Scientists, 2009).

262 **ecosystem concerns that are not so easily dismissed**
Michael Faure and Andri Wibisana, "Liability for Damage Caused by GMOs: An Economic Perspective," *Georgetown International Environmental Law Review* 23, no. 1 (2010): 1–69.

262 **normal pattern of the organism's genetic code and can cause**
Antoniou, Robinson, and Fagan, "GMO Myths and Truths, Version 1.3."

262 **FLAVR SAVR**
G. Bruening and J. M. Lyons, "The Case of the FLAVR SAVR Tomato," *California Agriculture*, July–August 2000.

262 **remain firm for a longer period of time after it ripened**
Ibid.

263 **consumer resistance**
Ibid.

263 **less rounded bottom to accommodate**
"Square Tomato," Davis Wiki, 2012, http://daviswiki.org/square_tomato.

263 **catastrophic loss of flavor in modern tomatoes**
Dan Charles, "How the Taste of Tomatoes Went Bad (and Kept On Going)," NPR, June 28, 2012, http://www.npr.org/blogs/thesalt/2012/06/28/155917345/how-the-taste-of-tomatoes-went-bad-and-kept-on-going; Kai Kupferschmidt, "How Tomatoes Lost Their Taste," *ScienceNOW*, June 28, 2012, http://news.sciencemag.org/sciencenow/2012/06/how-tomatoes-lost-their-taste.html.

263 **Almost 11 percent of all the world's farmland**
Matthew Weaver, "Report: World Embraces Biotech Crops," *Capital Press*, March 1, 2012.

263 **number of acres planted in GM crops**
International Service for the Acquisition of Agri-Biotech Applications, "Pocket K No. 16: Global Status of Commercialized Biotech/GM Crops in 2011," http://www.isaaa.org/resources/publications/pocketk/16/default.asp.

263 **Although the United States is by far the largest grower**
International Service for the Acquisition of Agri-Biotech Applications, ISAAA Brief 43-2011, Global Status of Commercialized Biotech/GM Crops: 2011, http://www.isaaa.org/resources/publications/briefs/43/executivesummary/default.asp.

264 **Monsanto's Roundup herbicide, are the largest GM crop globally**
Ibid.; "Monsanto Strong-Arms Seed Industry," Associated Press, January 4, 2011.

264 **Corn is the second most widely planted GM crop**
International Service for the Acquisition of Agri-Biotech Applications, ISAAA Brief 43-2011, Global Status of Commercialized Biotech/GM Crops: 2011.

264 **In the U.S., 95 percent of soybeans planted**
"Monsanto Strong-Arms Seed Industry," Associated Press, January 4, 2011.

264 **must purchase from Monsanto or one of their licensees**
International Service for the Acquisition of Agri-Biotech Applications, ISAAA Brief 43-2011, Global Status of Commercialized Biotech/GM Crops: 2011.

264 **known as "rapeseed" outside the United States**
E. S. Oplinger et al., "Canola (Rapeseed)," *Alternative Field Crops Manual*, 1989, http://www.hort.purdue.edu/newcrop/afcm/canola.html.

264 **three generations, or waves, of the technology**
J. Fernandez-Cornejo and M. Caswell, "The First Decade of Genetically Engineered Crops in the United States," U.S. Department of Agriculture, Economic Research Service, 2006; Gurian-Sherman, *Failure to Yield*.

264 **The introduction of genes that give corn**
Gurian-Sherman, *Failure to Yield*.

264 **Genes introduced into corn**
Ibid. "Monsanto Strong-Arms Seed Industry," Associated Press; Beverly Bell, "Haitian Farmers Commit to Burning Monsanto Hybrid Seeds," *Huffington Post*, May 17, 2010, http://www.huffingtonpost.com/beverly-bell/haitian-farmers -commit-to_b_578807.html.

264 **designed to enhance the survivability of crops during droughts**
Fernandez-Cornejo and Caswell, "The First Decade of Genetically Engineered Crops in the United States." It is worth noting that plant geneticists have also engineered a new variety of rice designed to survive complete submergence in water for more than two weeks; it is now being tested in rice fields in the Philippines that have been hit hard by flooding.

264 **report initial reductions in their cost of production**
National Research Council, "Impact of Genetically Engineered Crops on Farm Sustainability in the United States," 2010.

264 **strain that is engineered to produce its own insecticide**
Ibid.; Calestous Juma, "Agricultural Biotechnology: Benefits, Opportunities and Leadership," Testimony to the U.S. House of Representatives, Committee on Agriculture, Subcommittee on Rural Development, Research, Biotechnology and Foreign Agriculture, June 23, 2011, http://belfercenter.ksg.harvard.edu/files/juma -house-testimony-june-23-2011-rev.pdf.

264 **In India the new Bt cotton made the nation a net exporter**
Juma, "Agricultural Biotechnology."

264 **have begun to protest the high cost of the GM seeds**
Gargi Parsai, "Protests Mark 10th Anniversary of Bt Cotton," *Hindu*, March 27, 2012; Zia Haq, "Ministry Blames Bt Cotton for Farmer Suicides," *Hindustan Times*, March 26, 2012.

265 **that field trials of GM crops "under any garb"**
Pallava Bagla, "India Should Be More Wary of GM Crops, Parliamentary Panel Says," *ScienceInsider*, August 2012.

265 **the *intrinsic* yields of the crops themselves are not increased at all**
National Research Council, "Impact of Genetically Engineered Crops on Farm Sustainability in the United States," 2010.

265 **unexpected collateral changes in the plants' genetic code**
Gurian-Sherman, *Failure to Yield*.

265 **both greater yields and greater resistance to the effects of drought**
Personal conversation with author.

265 **offer the promise of increased yields during dry periods**
Union of Concerned Scientists, "High and Dry," May 2012, http://www.ucsusa .org/assets/documents/food_and_agriculture/high-and-dry-summary.pdf.

265 **tremendous interest in drought-resistant strains, especially for maize**
Gurian-Sherman, *Failure to Yield;* Andrew Pollack, "Drought Resistance Is the Goal, but Methods Differ," *New York Times,* October 23, 2008.

265 **genes working together in complicated ways**
"Why King Corn Wasn't Ready for the Drought," *Wired,* August 9, 2012.

265 **"limited at best"**
Union of Concerned Scientists, "High and Dry."

265 **introduction of genes that enhance the nutrient value**
Fernandez-Cornejo and Caswell, "The First Decade of Genetically Engineered Crops in the United States."

266 **higher protein content in corn (maize) that is used primarily for livestock**
Calestous Juma, *The New Harvest: Agricultural Innovation in Africa* (New York: Oxford University Press, 2011).

266 **new strain of rice that produces extra vitamin A**
Juma, "Agricultural Biotechnology: Benefits, Opportunities and Leadership."

266 **enhance the resistance of plants to particular fungi and viruses**
Pamela C. Ronald and James E. McWilliams, "Genetically Engineered Distortions," *New York Times,* May 14, 2010.

266 **The third wave of GM crops**
Fernandez-Cornejo and Caswell, "The First Decade of Genetically Engineered Crops in the United States."

266 **with high cellulose and lignin**
National Research Council, "Impact of Genetically Engineered Crops on Farm Sustainability in the United States"; Fernandez-Cornejo and Caswell, "The First Decade of Genetically Engineered Crops in the United States"; Fuad Hajji, "Engineering Renewable Cellulosic Thermoplastics," *Reviews in Environmental Science and Biotechnology* 10, no. 1 (2011): 25–30.

266 **in a world with growing population and food consumption**
Hajji, "Engineering Renewable Cellulosic Thermoplastics."

266 **because of the unprecedented complexity of the challenge**
Matt Ridley, "Getting Crops Ready for a Warmer Tomorrow," *Wall Street Journal,* July 6, 2012.

266 **temporary reduction in losses to pests**
Gurian-Sherman, *Failure to Yield;* Antoniou, Robinson, and Fagan, "GMO Myths and Truths, Version 1.3."

266 **National Bioeconomy Blueprint**
Andrew Pollack, "White House Promotes a Bioeconomy," *New York Times*, April 26, 2012.

267 **make themselves impervious to the herbicides and insecticides**
National Research Council, "Impact of Genetically Engineered Crops on Farm Sustainability in the United States"; Faure and Wibisana, "Liability for Damage Caused by GMOs"; Antoniou, Robinson, and Fagan, "GMO Myths and Truths, Version 1.3."

267 **forcing the mutation of new strains of pests that are highly resistant**
Faure and Wibisana, "Liability for Damage Caused by GMOs"; Antoniou, Robinson, and Fagan, "GMO Myths and Truths, Version 1.3."

267 **genetically engineered to survive application of the herbicide**
National Research Council, "Impact of Genetically Engineered Crops on Farm Sustainability in the United States."

267 **increases among weeds and insects, the overall use of both herbicides**
Antoniou, Robinson, and Fagan, "GMO Myths and Truths, Version 1.3."

267 **though advocates of GM crops dispute their analysis**
Council for Biotechnology Information, "Myths & Facts: Plant Biotechnology," http://www.whybiotech.com/resources/myths_plantbiotech.asp.

267 **and more dangerous—herbicides**
Antoniou, Robinson, and Fagan, "GMO Myths and Truths, Version 1.3."

267 **$17.5 billion and both insecticides and fungicides representing**
Clive Cookson, "Agrochemicals: Innovation Has Slowed Since Golden Age of the 1990s," *Financial Times*, October 13, 2011.

267 **U.S. Air Force to clear jungles and forest cover during the Vietnam War**
"'Agent Orange Corn' Debate Rages as Dow Seeks Approval of New Genetically Modified Seed," *Huffington Post*, April 26, 2012, http://www.huffingtonpost.com/2012/04/26/enlist-dow-agent-orange-corn_n_1456129.html.

268 **"endocrine disruption, reproductive problems, neurotoxicity, and immunosuppression"**
Ibid.

268 **U.S. farm belt by almost 60 percent over the last decade**
Tom Philpott, "Researchers: GM Crops Are Killing Monarch Butterflies, After All," *Mother Jones*, March 21, 2012.

268 **cropland dedicated to crop varieties engineered to be tolerant of Roundup**
Ned Potter, "Are Monarch Butterflies Threatened by Genetically Modified Crops?," ABC News, July 13, 2011, http://abcnews.go.com/Technology/monarch-butterflies-genetically-modified-gm-crops/story?id=14057436#.UA2kPUQ-KF4; Philpott, "Researchers: GM Crops Are Killing Monarch Butterflies, After All."

268 **harmful impact on at least one subspecies of monarchs**
Faure and Wibisana, "Liability for Damage Caused by GMOs."

268 **Although proponents of GM crops have minimized**
Potter, "Are Monarch Butterflies Threatened by Genetically Modified Crops?"; Monsanto, "Frequently Asked Questions," http://www.monsanto.com/hawaii/Pages/faqs-hawaii.aspx.

268 **a new group of pesticides known as neonicotinoids**
Elizabeth Kolbert, "Silent Hives," *New Yorker*, April 20, 2012.

268 since the affliction first appeared in 2006
Ibid.

268 "About one mouthful in three"
U.S. Department of Agriculture, Agricultural Research Service, "Questions and Answers: Colony Collapse Disorder," December 17, 2010, http://www.ars.usda .gov/News/docs.htm?docid=15572.

268 because the engineered seeds must be purchased annually by farmers
Science Museum (UK), "Who Benefits from GM?," http://www.sciencemuseum .org.uk/antenna/futurefoods/debate/debateGM_CIPbusiness.asp.

268 can introduce genes that do not fit into the seed company's design
Miriam Jordan, "The Big War Over a Small Fruit," *Wall Street Journal*, July 13, 2012.

269 with pollen from citrus varieties that have seeds
Ibid.

269 all of the major commodity crops
Union of Concerned Scientists, "Industrial Agriculture: Features and Policy," May 17, 2007, http://www.ucsusa.org/food_and_agriculture/science_and_impacts/ impacts_industrial_agriculture/industrial-agriculture-features.html.

269 reliance on monocultures makes agriculture highly vulnerable to pests
Ibid.

269 stem rust began attacking wheat fields in Uganda
"Scientists in Kenya Try to Fend Off Disease Threatening World's Wheat Crop," PBS *NewsHour*, December 28, 2011, http://www.pbs.org/newshour/bb/ globalhealth/july-dec11/wheat_12-28.html.

269 Similarly, cassava
Donald G. McNeil Jr., *New York Times*, "Virus Ravages Cassava Plants in Africa," June 1, 2010.

269 "The speed is just unprecedented"
Ibid.

270 Ireland's heavy reliance on a monocultured potato strain from the Andes
Nicholas Wade, "Testing Links Potato Famine to an Origin in the Andes," *New York Times*, June 7, 2011.

270 destroyed in 1970 by a new variety of Southern corn leaf blight
Union of Concerned Scientists, "Industrial Agriculture: Features and Policy."

270 robust global work under way to genetically modify trees
Clive Cookson, "Barking Up the Right GM Tree?," *Financial Times*, July 20, 2012.

270 genes from one species into the genome of another
National Research Council, "Emerging Technologies to Benefit Farmers in Sub-Saharan Africa and South Asia," 2009.

271 synthetic growth hormone in dairy cattle
Carina Storrs, "Hormones in Food: Should You Worry?," Health.com/*Huffington Post*, January 19, 2011.

271 elevated levels of IGF and a significantly higher risk of prostate cancer
Ibid.

271 for the labeling of milk with bovine growth hormone
Andrew Martin, "Consumers Won't Know What They're Missing," *New York Times*, November 11, 2007.

271 has significantly decreased its use
Dan Shapley, "Eli Lilly Buys Monsanto's Dairy Hormone Business," *Daily Green*, August 20, 2008, http://www.thedailygreen.com/healthy-eating/eat-safe/rbst -hormones-milk-470820; "Safeway Milk Free of Bovine Hormone," Associated Press, January 21, 2007.

271 into the embryos of dairy cows
Haze Fan and Maxim Duncan, "Cows Churn Out 'Human Breast Milk,'" Reuters, June 16, 2011.

271 at the National Institute of Agribusiness Technology
Robin Yapp, "Scientists Create Cow That Produces 'Human' Milk," *Telegraph*, June 11, 2011.

271 intended for direct consumption by human beings
Harmon and Pollack, "Battle Brewing Over Labeling of Genetically Modified Food."

271 a salmon modified with an extra growth hormone gene
Andrew Pollack, "Panel Leans in Favor of Engineered Salmon," *New York Times*, September 20, 2010.

271 as fast as a normal salmon
Randy Rieland, "Food, Modified Food," *Smithsonian*, June 29, 2012.

271 about the possibility of increased levels of insulin-like growth factor
Storrs, "Hormones in Food: Should You Worry?"

272 changing the species in an unintended way
Pollack, "Panel Leans in Favor of Engineered Salmon"; Bill Chameides, "Genetically Modified Salmon: The Meta-Question," *New Scientist*, November 23, 2010.

272 to reduce the amount of phosphorus in their feces
Andrew Pollack, "Move to Market Gene-Altered Pigs in Canada Is Halted," *New York Times*, April 4, 2012.

272 They called their creation Enviropigs
University of Guelph, "Enviropig™," http://www.uoguelph.ca/enviropig/index .shtml.

272 because phosphorus is a source of algae blooms
University of Guelph, "Environmental Benefits," http://www.uoguelph.ca/ enviropig/environmental_benefits.shtml.

272 abandoned their project and euthanized the pigs
Pollack, "Move to Market Gene-Altered Pigs in Canada Is Halted."

272 because scientists elsewhere engineered an enzyme, phytase
Clive Cookson, "Agrochemicals: Innovation Has Slowed Since Golden Age of the 1990s," *Financial Times*, October 13, 2011.

272 which, when added to pig feed
Pollack, "Move to Market Gene-Altered Pigs in Canada Is Halted."

272 genetically engineer insects, including bollworms
Henry Nicholls, "Swarm Troopers: Mutant Armies Waging War in the Wild," *New Scientist*, September 12, 2011.

272 **and mosquitoes**
Michael Specter, "The Mosquito Solution," *New Yorker*, July 9 and 16, 2012, pp. 38–46.

272 **species of mosquito that carries dengue fever**
Nicholls, "Swarm Troopers."

272 **The larvae, having no access to tetracycline**
Andy Coghlan, "Genetically Altered Mosquitoes Thwart Dengue Spreaders," *New Scientist*, November 11, 2010; Nicholls, "Swarm Troopers."

272 **proposed the release of large numbers of their mosquitoes**
Specter, "The Mosquito Solution."

272 **potentially disruptive effects on the ecosystem**
Nicholls, "Swarm Troopers"; Specter, "The Mosquito Solution."

272 **small number of the offspring do in fact survive**
Nicholls, "Swarm Troopers"; Specter, "The Mosquito Solution"; Andrew Pollack, "Concerns Are Raised About Genetically Engineered Mosquitoes," *New York Times*, October 31, 2011.

273 **spread their adaptation to the rest of the mosquito population**
Nicholls, "Swarm Troopers"; Specter, "The Mosquito Solution."

273 **"may have profound impacts on the ecology of certain infectious diseases"**
Tim Sandle, "Link between Dengue Fever and Climate Change in the US," *Digital Journal*, July 7, 2012, http://digitaljournal.com/print/article/328094.

273 **Dengue, which now afflicts up to 100 million people each year**
World Health Organization, Dengue and Severe Dengue Fact Sheet, January 2012, http://www.who.int/mediacentre/factsheets/fs117/en/.

273 **causes thousands of fatalities**
Yenni Kwok, "Across Asia, Dengue Fever Cases Reach Record Highs," *Time*, September 24, 2010.

273 **"breakbone fever"**
Gardiner Harris, "As Dengue Fever Sweeps India, a Slow Response Stirs Experts' Fears," *New York Times*, November 6, 2012.

273 **the extreme joint pain that is one of its worst symptoms**
Margie Mason, "Dengue Fever Outbreak Hits Parts of Asia," Associated Press, October 26, 2007.

273 **Simultaneous outbreaks emerged in Asia, the Americas, and Africa**
Suzanne Moore Shepherd, "Dengue," Medscape Reference, http://emedicine.medscape.com/article/215840-overview.

273 **the disease was largely contained until World War II**
Ibid.

273 **inadvertently spread by people during and after the war**
Ibid.; Thomas Fuller, "The War on Dengue Fever," *New York Times*, November 3, 2008.

273 **In 2012, there were an estimated 37 million cases in India alone**
Harris, "As Dengue Fever Sweeps India, a Slow Response Stirs Experts' Fears."

273 **dengue's range was still limited to tropical and subtropical regions**
Jennifer Kyle and Eva Harris, "Global Spread and Persistence of Dengue," *Annual Review of Microbiology* 62 (2008): 71–92.

273 **dengue is likely to spread throughout the Southern United States**
Sandle, "Link between Dengue Fever and Climate Change in the US."

274 **including HIV/AIDS**
Jim Robbins, "The Ecology of Disease," *New York Times*, July 15, 2012. The expansion of livestock farming into areas where wild animals are in close proximity has been implicated in the spreading of diseases from wildlife to domesticated animals and from there to people. The bird flu, for example, evolves in domesticated animals when it spreads from wild animals. HIV/AIDS spread to humans ninety years ago when African hunters killed chimpanzees and sold the meat for human consumption. The extremely deadly Ebola virus, first identified in the border regions of western South Sudan and northeastern Democratic Republic of the Congo in 1976, originated in chimpanzees, gorillas, monkeys, forest antelope, and fruit bats.

274 **brought into close proximity with livestock**
Ibid.

274 **60 percent of the new infectious diseases**
Sonia Shah, "The Spread of New Diseases: The Climate Connection," *Yale Environment 360*, October 15, 2009.

274 **that outnumber the cells of our bodies**
Robert Stein, "Finally, a Map of All the Microbes on Your Body," NPR, June 13, 2012, http://www.npr.org/blogs/health/2012/06/13/154913334/finally-a-map-of-all-the-microbes-on-your-body.

274 **with approximately 100 trillion microbes**
Carl Zimmer, "Tending the Body's Microbial Garden," *New York Times*, June 19, 2012.

274 **with 3 million nonhuman genes**
"Microbes Maketh Man," *Economist*, April 21, 2012, http://www.economist.com/node/21560559.

274 **published the genetic sequencing of this community of bacteria**
Human Microbiome Project Consortium, "A Framework for Human Microbiome Research," *Nature*, June 14, 2012.

274 **much like blood types—that exist in all races and ethnicities**
Robert T. Gonzalez, "10 Ways the Human Microbiome Project Could Change the Future of Science and Medicine," io9, June 25, 2012, http://io9.com/5920874/10-ways-the-human-microbiome-project-could-change-the-future-of-science-and-medicine.

274 **All told, the team identified eight million**
Rosie Mestel, "Microbe Census Maps Out Human Body's Bacteria, Viruses, Other Bugs," *Los Angeles Times*, August 13, 2012.

274 **acquired immune system, particularly during infancy and childhood**
James Randerson, "Antibiotics Linked to Huge Rise in Allergies," *New Scientist*, May 27, 2004, http://www.newscientist.com/article/dn5047-antibiotics-linked-to-huge-rise-in-allergies.html.

274 "The microbial gut flora is an arm of the immune system"
Ibid.

275 which needs to learn to distinguish invaders from cells of the body itself
National Institute of Arthritis and Musculoskeletal and Skin Diseases, Understanding Autoimmune Diseases, September 2010, http://www.niams.nih.gov/health_info/autoimmune/default.asp.

275 contributing to the apparent rapid rise of numerous diseases
Martin Blaser, "Antibiotic Overuse: Stop the Killing of Beneficial Bacteria," *Nature*, August 25, 2011; Mette Nørgaard et al., Aarhus University Hospital, "Use of Penicillin and Other Antibiotics and Risk of Multiple Sclerosis: A Population-Based Case-Control Study," *American Journal of Epidemiology* 174, no. 8 (2011): 945–48.

275 type 1 diabetes
Blaser, "Antibiotic Overuse."

275 multiple sclerosis
Nørgaard et al., "Use of Penicillin and Other Antibiotics and Risk of Multiple Sclerosis."

275 Crohn's disease, and ulcerative colitis
"Antibiotic Use Tied to Crohn's, Ulcerative Colitis," Reuters, September 27, 2011.

275 human immune system is not fully developed at birth
Zimmer, "Tending the Body's Microbial Garden."

275 develops and matures after passage through the birth canal
Ibid.

275 humans have the longest period of infancy and helplessness of any animal
Alison Gopnik, *The Philosophical Baby: What Children's Minds Tell Us About Truth, Love, and the Meaning of Life* (New York: Farrar, Straus & Giroux, 2009).

275 development of the brain following birth
David F. Bjorklund, *Why Youth Is Not Wasted on the Young: Immaturity in Human Development* (Malden, MA: Blackwell, 2007).

275 development and learning taking place in interaction with the environment
Gopnik, *The Philosophical Baby*.

275 to destroy invading viruses or bacteria
National Institute of Arthritis and Musculoskeletal and Skin Diseases, September 2010, Understanding Autoimmune Diseases.

275 do not discriminate between harmful bacteria and beneficial bacteria
Zimmer, "Tending the Body's Microbial Garden."

275 Julie Segre, a senior investigator
Ibid.

275 human stomach that are involved in energy balance and appetite
Blaser, "Antibiotic Overuse."

275 *H. pylori* has lived inside us in large numbers for 58,000 years
Kate Murphy, "In Some Cases, Even Bad Bacteria May Be Good," *New York Times*, October 31, 2011.

275 single most common microbe in the stomachs of most human beings
Martin Blaser, "Antibiotic Overuse."

276 "may also eradicate *H. pylori* in 20–50% of cases"
Ibid.

276 has been found to play a role in both gastritis
Murphy, "In Some Cases, Even Bad Bacteria May Be Good."

276 "more likely to develop asthma, hay fever or skin allergies in childhood"
Blaser, "Antibiotic Overuse."

276 Its absence is also associated with increased acid reflux
Ibid.

276 *H. pylori* into the guts of mice serves to protect them against asthma
Murphy, "In Some Cases, Even Bad Bacteria May Be Good."

276 approximately 160 percent throughout the world in the last two decades
Randerson, "Antibiotics Linked to Huge Rise in Allergies."

276 ghrelin, is one of the keys to appetite
Murphy, "In Some Cases, Even Bad Bacteria May Be Good."

276 caused by harmful microbes normally kept in check by beneficial microbes
Zimmer, "Tending the Body's Microbial Garden."

277 when the balance of their internal microbiome was restored
Ibid.

CHAPTER 6: THE EDGE

281 into the extraordinarily thin shell of atmosphere
Glen Peters et al., "Rapid Growth in CO_2 Emissions After the 2008–2009 Global Financial Crisis," *Nature Climate Change* 2 (2012): 2–4.

281 Industrial Revolution at a rate
Original calculations were derived from: Scott Mandia, "Global Warming: Man or Myth: And You Think the Oil Spill Is Bad?," June 17, 2010, http://profmandia .wordpress.com/2010/06/17/and-you-think-the-oil-spill-is-bad/. Mandia's original calculations were revised to reflect later scientific estimates of the number of barrels per day. Source: Marcia McKnutt et al., "Review of Flow Rate Estimates of the Deepwater Horizon Oil Spill," *Proceedings of the National Academy of Sciences*, December 20, 2011.

281 prospective extinction of 20 to 50 percent of all the living species
Nicholas Stern, *The Economics of Climate Change: The Stern Review* (New York: Cambridge University Press, 2007).

281 400,000 Hiroshima atomic bombs
James Hansen, "Why I Must Speak Out About Climate Change," TED Talks, February 2012.

282 already competitive with the average grid price for electricity
"Commercial Solar Now Cost-Competitive in US," CleanTechnica, June 20, 2012, http://cleantechnica.com/2012/06/20/commercial-solar-now-cost-competitive-us/; "Wind Innovations Drive Down Costs, Stock Prices," Bloomberg, March 14, 2012, http://go.bloomberg.com/multimedia/wind-innovations-drive-down-costs-stock -prices/; "Grid Parity and Beyond: Brazilian Wind Energy Supported by Turbines Manufactured at 'Chinese Prices,'" CleanTechInvestor, August 29, 2011, http://

www.cleantechinvestor.com/events/es/bwec-blog/301-grid-parity-and-beyond
-brazilian-wind-energy-supported-by-turbines-manufactured-at-chinese-prices
-.html.

282 renewables will be the second-largest source of power generation by 2015
International Energy Agency, *World Energy Outlook 2012*.

282 each and every hour than would be needed for all of the world's energy consumption
Nathan Lewis and Daniel Nocera, "Powering the Planet: Chemical Challenges in
Solar Energy Utilization," *Proceedings of the National Academy of Sciences* 103 (October 2006): 15729–35.

282 wind energy also exceeds
Xi Lu et al., "Global Potential for Wind-Generated Electricity," *Proceedings of the
National Academy of Sciences* 106 (June 2009): 10933–38.

282 there were periods when Germany
Reuters, "Solar Power Generation World Record Set in Germany," *Guardian*,
May 28, 2012.

282 entire world's additional electricity generation
Fiona Harvey, "Renewable Energy Can Power the World, Says Landmark IPCC
Study," *Guardian*, May 9, 2011.

282 exceeded those in fossil fuels ($187 billion, compared to $157 billion)
Alex Morales, "Renewable Power Trumps Fossils for First Time as UN Talks
Stall," Bloomberg News, November 25, 2011.

282 102 percent over those installed just one year earlier
Climate Guest Blogger, "Solar Is the 'Fastest Growing Industry in America' and
Made Record Cost Reductions in 2010," Think Progress ClimateProgress, September 16, 2011, http://thinkprogress.org/climate/2011/09/16/321131/solar-fastest
-growing-industry-in-america-and-made-record-cost-reductions/.

283 approximately 30 percent of all CO_2 emissions come from buildings
Harvard Center for Health and the Global Environment, "The Built Environment," http://chge.med.harvard.edu/topic/built-environment.

283 of all buildings needed by 2050, two thirds have yet to be built
Alexis Biller and Chris Phillips, "The Role of Engineering in the Built Environment," Institution of Engineering and Technology lecture, London, November 26, 2009.

283 "30 percent of the energy consumed in commercial buildings is wasted"
A Better Building. A Better Bottom Line. A Better World, Environmental Protection
Agency brochure (2010), http://www.energystar.gov/ia/partners/publications/
pubdocs/C+I_brochure.pdf.

284 more concern about global warming than most elected officials
"Climate Change May Challenge National Security, Classified Report Warns,"
ScienceDaily, June 26, 2008, http://www.sciencedaily.com/releases/2008/06/080625
090302.htm.

285 "our government, our institutions and our borders"
Don Belt, "The Coming Storm: Bangladesh," *National Geographic*, May 2005.

285 "direct cause of large-scale human crises"
David Zhang and Harry Lee, "The Causality Analysis of Climate Change and

Large-Scale Human Crisis," *Proceedings of the National Academy of Sciences* 108 (March 2011): 17296–301.

285 **Central America and the temporary colonization of southern Greenland**
Scott Mandia, Suffolk University, "Vikings During the Medieval Warm Period," http://www2.sunysuffolk.edu/mandias/lia/vikings_during_mwp.html; Brian Fagan, *The Long Summer: How Climate Changed Civilization* (New York: Basic Books, 2004), p. 236.

285 **paddled their kayaks to Scotland; farther south, millions died**
Scott Mandia, Suffolk University, "The Little Ice Age in Europe," http://www2.sunysuffolk.edu/mandias/lia/little_ice_age.html.

285 **a chain of events leading to the Black Death**
Lei Xu et al., "Nonlinear Effect of Climate on Plague During the Third Pandemic in China," *Proceedings of the National Academy of Sciences*, May 4, 2011.

285 **unusually large eruption of the Tambora volcano**
"Volcanic Eruption, Tambora," *Encyclopedia of Global Environmental Change* (Chichester, UK: Wiley, 2002), pp. 737–38.

286 **An estimated 25 percent of the CO_2**
David Archer and Victor Brovkin, "The Millennial Atmospheric Lifetime of Anthropogenic CO_2," *Climatic Change* 90 (2008): 283–97; personal correspondence with Daniel Schrag, January 19, 2011.

286 **have occurred in the last ten years**
NASA, "NASA Finds 2011 Ninth-Warmest Year on Record," January 19, 2012, http://www.nasa.gov/topics/earth/features/2011-temps.html.

286 **flooding in Pakistan that displaced 20 million people**
"Pakistan Floods Leave 20 Million Homeless," CBC News, August 14, 2010, http://www.cbc.ca/news/world/story/2010/08/14/pakistan-floods-homeless.html.

286 **unprecedented heat waves in Europe in 2003**
J. Robine et al., "Death Toll Exceeded 70,000 in Europe During Summer of 2003," *Comptes Rendus Biologies*, February 2008.

286 **Russia in 2010 that led to 55,000 deaths**
"World Disasters Report: 2010 Death Toll Highest in Decade," Red Cross, September 22, 2011, http://www.redcross.org.au/world-disasters-report-2010-death-toll-highest-in-decade.aspx.

286 **massive fires, and crop damage that pushed global food prices**
"World Food Prices at Fresh High, Says UN," BBC, January 5, 2011, http://www.bbc.co.uk/news/business-12119539.

286 **the flooding of northeastern Australia in 2011**
J. David Goodman, "Australia Flooding Displaces Thousands," *New York Times*, December 31, 2010.

286 **the huge droughts in southern China**
Edward Wong, "Drought Leaves 14 Million Chinese and Farmland Parched," *New York Times*, September 9, 2010.

286 **southwestern North America in 2011**
Kim Severson and Kirk Johnson, "14 States Suffering Under Drought," *New York Times*, July 12, 2011.

286 **Superstorm Sandy**
James Barron, "After the Devastation, a Daunting Recovery," *New York Times*, October 30, 2012.

286 **warmer air *holds* more water vapor**
Kevin Trenberth, "Changes in Precipitation with Climate Change," *Climate Research* 47 (2010): 123–38.

287 **has a large effect on the hydrological cycle**
Ibid.

287 **funnel it inward into the regions where storm conditions trigger a downpour**
Kevin Trenberth, "Conceptual Framework for Changes of Extremes of the Hydrological Cycle with Climate Change," *Climatic Change* 42 (1999): 327–39.

287 **down through the soil to recharge the underground aquifers**
Intergovernmental Panel on Climate Change, Working Group 2, "3.4.2 Groundwater," 2007, http://www.ipcc.ch/publications_and_data/ar4/wg2/en/ch3s3-4-2.html.

287 **local temperatures rise higher still**
Ben Brabson et al., "Soil Moisture and Predicted Spells of Extreme Temperatures in Britain," *Journal of Geophysical Research* 110 (2004).

287 **topsoil becomes more vulnerable to wind erosion**
New South Wales Government, "Wind Erosion," March 2, 2011, http://www.environment.nsw.gov.au/soildegradation/winder.htm.

287 **"not understanding the highly dangerous situation we are in"**
Justin Gillis, "A Warming Planet Struggles to Feed Itself," *New York Times*, June 6, 2011.

288 **record one-month price increase for food**
Yaneer Bar-Yam and Greg Lindsay, "The Real Reason for Spikes in Food Prices," Reuters, October 25, 2012.

288 **record price hikes predicted for 2013**
Emma Rowley and Garry White, "World on Track for Record Food Prices 'Within a Year' Due to US Drought," *Telegraph*, September 23, 2012.

288 **More than 65 percent of the U.S. suffered from drought conditions**
Michael Pearson and Melissa Abbey, "U.S. Drought Biggest Since 1956, Climate Agency Says," CNN, July 17, 2012, http://www.cnn.com/2012/07/16/us/us-drought/index.html.

288 **"but it will rain in the non-rainy season"**
Gillis, "A Warming Planet Struggles to Feed Itself."

288 **"an under-recognition of just how sensitive crops are to heat"**
Justin Gillis, "Food Supply Under Strain on a Warming Planet," *New York Times*, June 4, 2011.

288 **that the CO_2 fertilization effect is much smaller than predicted**
Ibid.

288 **weeds appear to benefit from extra CO_2 much more**
Tim Christopher, "Can Weeds Help Solve the Climate Crisis?," *New York Times*, June 9, 2008.

288 **above a threshold of 84 degrees**
Schlenker and Roberts, "Nonlinear Temperature Effects Indicate Severe Damages to U.S. Crop Yields under Climate Change."

289 yield declines plummet further with every degree added
Ibid.

289 disruption of precipitation patterns taking a large toll still
Ibid.

289 same accelerated drops in yields begin when temperatures reach and exceed
Ibid.

289 spring is arriving about a week earlier (and fall about a week later)
Alexander Stine et al., "Changes in the Phase of the Annual Cycle of Surface
Temperature," *Nature*, January 22, 2009.

289 depriving these regions of water
Thomas Karl et al., *Global Climate Change Impacts in the United States* (Washington,
DC: U.S. Climate Change Science Program, 2009), p. 41.

289 nighttime temperatures are at least as important
Christopher Mims, "Why 107-Degree Overnight Temperatures Should Freak You
Out," Grist, July 21, 2011, http://grist.org/list/2011-07-21-nyc-mayor-bloomberg
-gives-50-million-to-fight-coal-michael-bloom/.

289 increases nighttime temperatures more than daytime temperatures
Intergovernmental Panel on Climate Change, "WG1: FAQ 3.3," 2007, http://www
.ipcc.ch/publications_and_data/ar4/wg1/en/faq-3-3.html.

289 corresponds with a linear decrease in wheat yields
PV Prasad et al., "Impact of Nighttime Temperature on Physiology and Growth of
Spring Wheat," *Crop Science* 48 (2008): 2372–80.

289 fell due to climate-related factors by 5.5 percent
David Lobell et al., "Climate Trends and Global Crop Production Since 1980,"
Science, July 2011.

289 declined by 10 percent with each one degree Celsius
Shaobing Peng et al., "Rice Yields Decline with Higher Night Temperature from
Global Warming," *Proceedings of the National Academy of Sciences*, July 2004.

289 "downstream effects of changes in crop yield variability"
Noah Diffenbaugh et al., "Global Warming Presents New Challenges for Maize
Pest Management," *Environmental Research Letters*, 2008.

290 aphids and Japanese beetles
Orla Demody et al., "Effects of Elevated CO_2 and O_3 on Leaf Damage and Insect
Abundance in a Soybean Agroecosystem," *Anthropod-Plant Interactions*, July 2008.

290 "That means crop losses may go up in the future"
Union of Concerned Scientists, "Crops, Beetles and Carbon Dioxide," May 11,
2010, http://www.ucsusa.org/global_warming/science_and_impacts/impacts/Global
-warming-insects.html.

290 deactivating other genes used by soybeans
Jorge Zavala et al., "Anthropogenic Increase in Carbon Dioxide Compromises
Plant Defense Against Invasive Insects," *Proceedings of the National Academy of Sci-
ences*, January 2008.

290 "appear to be helpless against herbivores"
Union of Concerned Scientists, "Crops, Beetles and Carbon Dioxide."

290 "formidable enemies affecting food crops"
CGIAR, "Climate Change Puts Southeast Asia's Billion Dollar Cassava Indus-

try on High Alert for Pest and Disease Outbreaks," April 13, 2012, http://ccafs
.cgiar.org/news/press-releases/climate-change-puts-southeast-asia%E2%80%99s
-billion-dollar-cassava-industry-high-alert.

290 "five additional lifecycles per season"
Shyam S. Yadav et al., *Crop Adaptation to Climate Change* (Ames, IA: Wiley-
Blackwell, 2011), p. 419.

290 pests and plant diseases that expand with warmer temperatures
CGIAR, "Climate Change Puts Southeast Asia's Billion Dollar Cassava Industry
on High Alert for Pest and Disease Outbreaks."

290 "Southeast Asia, Southern China and the cassava-growing areas of Southern India"
Ibid.

291 "increases in disease spread or incidence with climate warming"
"Global Warming May Spread Diseases," CBS News, February 11, 2009, http://
www.cbsnews.com/2100-205_162-512920.html.

291 spread of diseases like dengue fever, West Nile virus, and others
Sonia Shah, "The Spread of New Diseases: The Climate Connection," *Yale Environment 360*, October 15, 2009; Nicole Heller, "The Climate Connection to Dengue Fever," Climate Central, May 12, 2010, http://www.climatecentral.org/blogs/
the-climate-connection-to-dengue-fever/.

291 "hosts and pathogens, the length of the transmission season, and the timing and intensity"
Union of Concerned Scientists, "Early Warning Signs of Global Warming: Spreading
Disease," http://www.ucsusa.org/global_warming/science_and_impacts/impacts/
early-warning-signs-of-global-9.html.

291 "epidemics could be characterized by high levels of sickness and death"
Ibid.

291 worst outbreak of West Nile virus
Thomas Maugh, "West Nile Outbreak Worst Ever, CDC Says," *Los Angeles Times*,
September 5, 2012.

291 aerial spraying of the city for the first time since 1966
"Dallas West Nile Virus Outbreak Leads Texas City's Mayor to Approve Aerial
Spraying," *Huffington Post*, August 15, 2012.

292 public safety officials issued an appeal
"Health Officials: No Need to Call 911 for Mosquito Bites," CBS DFW, August 24, 2012, http://dfw.cbslocal.com/2012/08/24/health-officials-no-need-to-call
-911-for-mosquito-bites/.

292 The disease eventually spread
Centers for Disease Control and Prevention, "West Nile Virus," November 20,
2012, http://www.cdc.gov/ncidod/dvbid/westnile/index.htm.

292 "long-term extreme weather phenomena associated with climate change"
Paul Epstein, "West Nile Virus and the Climate," *Journal of Urban Health* 78
(2001): 367–71.

292 have been predicting the effects of climate change on West Nile
Christie Wilcox, "Is Climate to Blame for This Year's West Nile Outbreak?," *Scientific American*, August 22, 2012.

292 hottest decade ever measured
NASA GISS, "2009: Second Warmest Year on Record; End of Warmest Decade," January 21, 2010, http://www.giss.nasa.gov/research/news/20100121/.

292 October 2012 was the 332nd month in a row
National Climatic Data Center, National Oceanic and Atmospheric Administration, "State of the Climate : Global Analysis, October 2012," http://www.ncdc .noaa.gov/sotc/global/2012/10.

292 toxins in corn and other crops unable to process nitrogen fertilizer
"After Drought Blights Crops, US Farmers Face Toxin Threats," Reuters, August 16, 2012.

293 a groundbreaking statistical analysis
James Hansen et al., "Perception of Climate Change," *Proceedings of the National Academy of Sciences*, August 2012.

293 on only 0.1 to 0.2 percent of the Earth's surface
Ibid.

293 seasons that are way outside the boundary of the statistical range that used to prevail
Ibid.

294 frequency of extremely hot temperatures has gone up dramatically
Ibid.

294 temperatures will likely rise by 4 degrees C (7.2 degrees F)
Potsdam Institute for Climate Impact Research and Climate Analytics, *Turn Down the Heat: Why a 4 Degree C Warmer World Must Be Avoided*, Report for the World Bank, November 2012, http://climatechange.worldbank.org/sites/default/ files/Turn_Down_the_heat_Why_a_4_degree_centrigrade_warmer_world_must_ be_avoided.pdf.

294 "no certainty that adaptation to a 4 degree C world is possible"
Brad Plumer, "We're on Pace for 4°C of Global Warming. Here's Why That Terrifies the World Bank," *Washington Post*, November 19, 2012.

295 "worst case" future projections are the ones most likely to occur
Brian Vastag, "Warmer Still: Extreme Climate Predictions Appear Most Accurate, Report Says," *Washington Post*, November 8, 2012.

295 49 percent loss
Joanna Zelman and James Gerken, "Arctic Sea Ice Levels Hit Record Low, Scientists Say We're 'Running Out of Time,'" *Huffington Post*, September 19, 2012.

295 100 percent loss in as little as a decade
Muyin Wang and James Overland, "A Sea Ice Free Summer Arctic within 30 Years?," *Geophysical Research Letters* 36 (2009).

295 A Chinese ship, the *Snow Dragon*, traversed the North Pole to Iceland and back
Jon Viglundson and Alister Doyle, "First Chinese Ship Crosses Arctic Ocean Amid Record Melt," Reuters, August 17, 2012.

295 so that computer-driven trades can be executed more quickly
Christopher Mims, "How Climate Change Is Making the Internet Faster," Grist, March 29, 2012, http://grist.org/list/how-climate-change-is-making-the-internet -faster/.

295 Arctic Ocean, which until now have been protected by the ice
Ivan Semeniuk, "Scientists Call for No-Fishing Zone in Arctic Waters," Nature

News Blog, April 23, 2012, http://blogs.nature.com/news/2012/04/scientists-call-for-no-fishin-zone-in-arctic-waters.html.

295 movement of military assets into the region
"Arctic Climate Change Opening Region to New Military Activity," Associated Press, April 16, 2012.

295 new drilling opportunities
"Shell Starts Preparatory Drilling for Offshore Oil Well off Alaska," CNN, September 9, 2012, http://articles.cnn.com/2012-09-09/us/us_arctic-oil_1_sea-ice-beaufort-sea-ice-data-center.

295 consequences of an accidental wellhead blowout
Jim Kollewe and Terry Macalister, "Arctic Oil Rush Will Ruin Ecosystem, Warns Lloyd's of London," *Guardian*, April 12, 2012.

296 oil in the Arctic Ocean posed unacceptable ecological risks
Guy Chazan, "Total Warns Against Oil Drilling in Arctic," *Financial Times*, September 25, 2012.

296 enough sunlight was penetrating to the ocean below
Kevin Arrigo et al., "Massive Phytoplankton Blooms Under Arctic Sea Ice," *Science*, June 15, 2012.

296 consequences for the location and pattern
Jennifer Francis and Stephen Vavrus, "Evidence Linking Arctic Amplification to Extreme Weather in Mid-Latitudes," *Geophysical Research Letters* 39 (2012).

296 perhaps beyond
Petr Chylek et al., "Arctic Air Temperature Change Amplification and the Atlantic Multidecadal Oscillation," *Geophysical Research Letters* 36 (2009).

296 Microbes turn the carbon into CO_2 or methane
Natalia Shakhova et al., "Extensive Methane Venting to the Atmosphere from Sediments of the East Siberian Arctic Shelf," *Science*, March 5, 2010.

296 shallow frozen lakes and ponds surrounding the Arctic
David Archer, "Methane Hydrate Stability and Anthropogenic Climate Change," *Biogeosciences* 4 (2007).

296 the heat absorption by the water
Katey Walter Anthony et al., "Geologic Methane Seeps Along Boundaries of Arctic Permafrost Thaw and Melting Glaciers," *Nature Geoscience* 5 (June 2012).

297 outgassing under way that exceeded what they expected
Shakhova et al., "Extensive Methane Venting to the Atmosphere from Sediments of the East Siberian Arctic Shelf."

297 methane underneath the Antarctic ice sheet
J. L. Wadham et al., "Potential Methane Reservoirs Beneath Antarctica," *Nature*, August 2012.

297 losing mass at an increasing rate
Eric Rignot et al., "Acceleration of the Contribution of the Greenland and Antarctic Ice Sheets to Sea Level Rise," *Geophysical Research Letters* 38 (2011).

297 west Antarctica and Greenland, however, already confirm
Ibid.

297 "It shocked the hell out of us"
Personal conversation with Bob Corell.

297 doubling time of the observed loss

James Hansen and Miki Sato, "Paleoclimate Implications for Human-Made Climate Change," in *Climate Change: Inferences from Paleoclimate and Regional Aspects*, edited by A. Berger, F. Mesinger, and D. Šijački (New York: Springer, 2012).

298 "multi-meter" sea level rise in this century

Ibid.

298 twenty to thirty feet higher than the present—although it took millennia

Aradhna Tripati et al., "Coupling of CO_2 and Ice Sheet Stability over Major Climate Transitions of the Last 20 Million Years," *Science*, December 2009.

298 50 percent of the world's population

"CO_2 Emissions to Cause Catastrophic Rise in Sea Levels, Warns Top NASA Climatologist," *Natural News*, January 15, 2007.

298 "In much of the developing world, coastal populations are exploding"

National Academies, "Coastal Hazards: Highlights of the National Academies Reports," 2009, http://www.oceanleadership.org/wp-content/uploads/2009/08/OHH.pdf.

298 recent study by Deborah Balk

Gordon McGranahan et al., "The Rising Tide: Assessing the Risks of Climate Change and Human Settlements in Low Coastal Elevation Zones," *Environment and Urbanization* 19 (2007).

298 are already beginning to relocate

Brian Reed, "Preparing for Sea Level Rise, Islanders Leave Home," NPR, February 17, 2011.

298 populations are also at risk in the Philippines and Indonesia

McGranahan et al., "The Rising Tide."

298 The number of climate refugees

Stern, *The Economics of Climate Change*.

298 more than 200 million people

Neil MacFarquhar, "Refugees Join List of Climate-Change Issues," *New York Times*, May 28, 2009.

298 mega-deltas of South Asia, Southeast Asia, China, and Egypt

Robert Nicholls, *IPCC 2007*, "Chapter 6: Coastal and Low-Lying Ecosystems," 2007, http://www.ipcc.ch/publications_and_data/ar4/wg2/en/ch6.html.

298 many have moved farther north across the border

Erik German and Solana Pyne, "Disasters Drive Mass Migration to Dhaka," *Global Post*, September 8, 2010.

299 text messaging and email into cities throughout India

Vikas Bajaj, "Internet Analysts Question India's Efforts to Stem Panic," *New York Times*, August 21, 2012.

299 stronger cyclones (known as hurricanes in the U.S.) gain energy from warmer seas

Kerry Emanuel et al., "Hurricanes and Global Warming: Results from Downscaling IPCC AR4 Simulations," *American Meteorological Society* 89 (March 2008): 347–67.

299 magnified by storm surges that carry the ocean inland

Claudia Tebaldi et al., "Modeling Sea Level Rise Impacts on Storms Surges Along US Coasts," *Environmental Research Letters* 7 (2012).

299 **New York City was put on emergency alert**
James Barron, "With Hurricane Irene Near, 370,000 in New York City Get Evacuation Order," *New York Times*, August 26, 2011.

299 **that can be closed to protect the city against such surges**
Steve Connor, "Sea Levels Rising Too Fast for Thames Barrier," *Independent*, March 22, 2008.

299 **cities with the highest population at risk from rising seas**
Susan Hanson et al., "A Global Ranking of Port Cities with High Exposure to Climate Extremes," *Climatic Change* 104 (December 2010).

299 **cities with the most exposed assets vulnerable to sea level rise**
Ibid.

299 **into areas from which they may once again become climate refugees**
Dizery Salim, United Nations Office for Disaster Risk Reduction, "Climate Migrants Risk More Harm in New Surroundings," 2012, http://www.unisdr.org/archive/28113.

299 **are slowly sinking, in a kind of seesaw effect**
Michael Lemonick, "The Secret of Sea Level Rise: It Will Vary Greatly by Region," March 22, 2010, *Yale Environment 360*.

299 **for a mixture of complicated reasons**
OurAmazingPlanet Staff, "City of Venice Still Sinking, Study Says," March 21, 2010, http://www.cbsnews.com/8301-205_162-57401506/city-of-venice-still-sinking-study-says/; Forrest Wilder, "That Sinking Feeling," *Texas Observer*, November 1, 2007.

300 **between South Carolina and Rhode Island, for example**
Asbury Sallenger, "Hotspot of Accelerated Sea-Level Rise on the Atlantic Coast of North America," *Nature Climate Change* 2 (May 2012).

300 **saltwater intrusion into drinking water wells and aquifers**
Cameron McWhirter and Mike Esterl, "Saltwater in Mississippi Taints Drinking Supply," *Wall Street Journal*, August 17, 2012.

300 **Approximately 30 percent of human-caused CO_2 emissions**
C. L. Sabine et al., "The Oceanic Sink for Anthropogenic CO_2," *Science*, July 16, 2004.

300 **oceans more acidic than at any time in the last 55 million years**
Andy Ridgwell and Daniela Schmidt, "Past Constraints on the Vulnerability of Marine Calcifiers to Massive Carbon Dioxide Release," *Nature Geoscience* 3 (February 2010).

300 **faster than at any time in the last 300 million years**
Bärbel Hönisch et al., "The Geologic Record of Ocean Acidification," *Science*, March 2012.

300 **ocean acidification global warming's "evil twin"**
"Ocean Acidification Is Climate Change's 'Equally Evil Twin,' NOAA Chief Says," Associated Press, July 12, 2012.

301 **several events in the space of a few years can and do kill the reefs**
K. Frieler et al., "Limiting Global Warming to 2°C Is Unlikely to Save Most Coral Reefs," *Nature Climate Change*, September 2012.

301 one quarter of all ocean species spend
Elizabeth Kolbert, "The Acid Sea," *National Geographic*, April 2011.

301 in danger of killing almost all of the coral reefs
David Jolly, "Oceans at Dire Risk, Team of Scientists Warns," *New York Times*, Green blog, June 21, 2011, http://green.blogs.nytimes.com/2011/06/21/oceans-are -at-dire-risk-team-of-scientists-warns/.

301 80 percent of the coral reefs in the Caribbean were lost
T. Gardner et al., "Long-Term Region-Wide Declines in Caribbean Corals," *Science*, July 2003.

301 same fate threatens reefs in every ocean
Frieler et al., "Limiting Global Warming to 2°C Is Unlikely to Save Most Coral Reefs."

301 Great Barrier Reef corals had died
Glenn De'ath et al., "The 27-Year Decline of Coral Cover on the Great Barrier Reef and Its Causes," *Proceedings of the National Academy of Sciences*, October 1, 2012.

301 a cold container of soda stays more carbonated
Brian Palmer, "Does Soda Taste Different in a Bottle Than a Can?," *Slate*, July 23, 2009, http://www.slate.com/articles/news_and_politics/explainer/2009/07/does_ soda_taste_different_in_a_bottle_than_a_can.html.

301 many of the cold-water reefs may be in even greater danger
"Oceans and Shallow Seas," *IPCC 2007*, http://www.ipcc.ch/publications_and _data/ar4/wg2/en/ch4s4-4-9.html.

301 very thin shells that play an important role
Anthony Richardson, "In Hot Water: Zooplankton and Climate Change," *ICES Journal of Marine Science* 65 (March 2008).

301 Southern California that have been sampled, are actually *corrosive*
Richard Feely et al., "Evidence for Upwelling of Corrosive 'Acidified' Water onto the Continental Shelf," *Science*, June 13, 2008.

301 killing commercially valuable shellfish
Alan Barton et al., "The Pacific Oyster, Crassostrea gigas, Shows Negative Correlation to Naturally Elevated Carbon Dioxide Levels: Implications for Near-Term Acidification Effects," *Limnology and Oceanography* 57, no. 3 (2012): 698–710.

301 oceans returned to a state comparable
Kolbert, "The Acid Sea."

302 almost a third of all fish species are presently overexploited
United Nations Food and Agriculture Organization, "The State of World Fisheries and Aquaculture 2010," 2010, http://www.fao.org/docrep/013/i1820e/i1820e .pdf.

302 depletion of up to 90 percent of large fish like tuna, marlin, and cod
Ransom Myers and Boris Worm, "Rapid Worldwide Depletion of Predatory Fish Communities," *Nature*, May 2003.

302 critical ocean habitats like mangrove forests
Beth Polidoro et al., "The Loss of Species: Mangrove Extinction Risk and Geographic Areas of Global Concern," *PLoS ONE* 5 (2010).

302 **sea grass meadows are also at risk**
Frederick Short et al., "Extinction Risk Assessment of the World's Seagrass Species," *Biological Conservation* 144 (July 2011).

302 **near the mouths of major river systems is doubling every decade**
National Science Foundation, "SOS: Is Climate Change Suffocating Our Seas?," 2009, http://www.nsf.gov/news/special_reports/deadzones/climatechange.jsp.

302 **large dead zone spreading from the mouth of the Mississippi**
"Good News from the Bad Drought: Gulf 'Dead Zone' Smallest in Years," *ScienceDaily*, August 23, 2012, http://www.sciencedaily.com/releases/2012/08/120824093519.htm.

302 **"when coupled with current rates of population increase"**
A. Rogers et al., "International Earth System Expert Workshop on Ocean Stresses and Impacts. Summary Report," IPSO Oxford, 2011, http://www.stateoftheocean.org/pdfs/1906_IPSO-LONG.pdf.

303 **"We have spent our entire existence adapting"**
Council on Foreign Relations, "The New North American Energy Paradigm: Reshaping the Future," June 27, 2012.

303 **damaged by extreme downpours and resulting floods and mud slides**
Patrick Rucker and Mica Rosenberg, "Analysis: Storms Damage Budgets in Central America, Mexico," Reuters, November 12, 2010.

303 **skyrocketing expenditures for food imports**
United Nations Food and Agriculture Organization, "One Trillion Food Import Bill as Prices Rise," November 17, 2010, http://www.fao.org/news/story/en/item/47733/icode/; United Nations Food and Agriculture Organization, "Agricultural Impacts Surge in Developing Countries," 2011, http://www.fao.org/docrep/014/i1952e/i1952e00.htm.

304 **nations are struggling to integrate arriving refugee groups**
Joanna Kakissis, "Environmental Refugees Unable to Return Home," *New York Times*, January 3, 2010.

304 **another degree Fahrenheit of warming is already "in the pipeline"**
James Hansen et al., "Earth's Energy Imbalance: Confirmation and Implications," *Science*, June 2005.

304 **large-scale changes in atmospheric circulation patterns**
Jian Lu et al., "Expansion of the Hadley Cell Under Global Warming," *Geophysical Research Letters* 34 (2007).

305 **feeds the shallower Humboldt current**
Erich Hoyt, *Marine Protected Areas for Whales, Dolphins and Porpoises: A World Handbook for Cetacean Habitat Conservation* (Oxford: Earthscan Publications Ltd., 2004), p. 397.

305 **giant pipelines through which the trade winds**
Henry Diaz and Raymond Bradley, *The Hadley Circulation: Present, Past and Future* (London: Kluwer Academic Publishers, 2005), p. 9.

305 **moisture they carried upward has fallen back**
Ibid.

306 **laden once more with heat and water vapor**
Ibid.

306 are located under these dry downdrafts
Ibid.

306 the "rain shadows" of mountain ranges
Brian Brinch, "How Mountains Influence Rainfall Patterns," *USA Today*, November 1, 2007.

306 what geographers call continentality
"Continental Climate and Continentality," *Encyclopedia of World Climatology*, p. 303.

306 downdraft of the Hadley cell
Personal correspondence with Dargan Frierson, September 24, 2012.

306 south of the equator has also moved poleward
Celeste Johanson and Qiang Fu, "Hadley Cell Widening: Model Simulations Versus Observations," *American Meteorological Society* 22 (May 2009): 2713–25.

306 theories for why global warming is causing a shift in the Hadley cells
Lu et al., "Expansion of the Hadley Cell Under Global Warming."

307 the difference in average temperatures
Jennifer Francis and Stephen Vavrus, "Evidence Linking Arctic Amplification to Extreme Weather in Mid-Latitudes," *Geophysical Research Letters* 39 (2012).

307 The widening of the Hadley cells
Lu et al., "Expansion of the Hadley Cell Under Global Warming."

307 on the edge of persistent water shortages anyway
Personal communication with Dargan Frierson, May 25, 2012.

307 human alteration of the same natural climate feature
Rudolph Kuper and Stefan Kröpelin, "Climate-Controlled Holocene Occupation in the Sahara: Motor of Africa's Evolution," *Science*, August 11, 2006.

308 barrel loop atmospheric currents known as the Ferrel cells
National Oceanic and Atmospheric Administration, "JetStream-Online School for Weather," October 2011, http://www.srh.noaa.gov/jetstream/global/circ.htm.

308 pulling cold Arctic air southward in winter, and disrupting
Frances and Vavrus, "Evidence Linking Arctic Amplification to Extreme Weather in Mid-Latitudes."

309 begun to produce a slight thinning of ozone
U.S. Environmental Protection Agency, "Environmental Indicators: Ozone Depletion," August 2010, http://www.epa.gov/ozone/science/indicat/index.html.

309 dangerous ozone hole above the Arctic might form on a more regular basis
Tim Flannery, *Here on Earth: A National History of the Planet* (New York: Atlantic Monthly Press, 2010), ch. 14, "The Eleventh Hour?"

309 interacting with the unique atmospheric conditions
Mario Molina and Sherwood Rowland, "Stratospheric Sink for Chlorofluoromethanes: Chlorine Atomic Catalyzed Destruction of Ozone," *Nature*, June 28, 1974.

310 radiates the reflected sunlight back into space more powerfully
Australian Government, Antarctic Division, "Environment—Land, Sea and Air," http://www.antarctica.gov.au/about-antarctica/fact-files.

310 **like Australia and Patagonia to high levels of ultraviolet radiation**
J. Ajtić et al., "Dilution of the Antarctic Ozone Hole into Southern Midlatitudes, 1998–2000," *Journal of Geophysical Research* 109 (2004).

310 **when air with low concentrations of ozone is no longer able**
Ibid.

310 **injecting water vapor in the stratosphere**
James Anderson et al., "UV Dosage Levels in Summer: Increased Risk of Ozone Loss from Convectively Injected Water Vapor," *Science*, August 2012.

311 **the Earth's atmosphere attempting to maintain its energy "balance"**
V. Ramaswamy et al., "Anthropogenic and Natural Influences in the Evolution of Lower Stratospheric Cooling," *Science* 311, no. 5764 (February 24, 2006): 1138–41.

311 **"Some say the world will end in fire; some say in ice"**
Robert Frost, "Fire and Ice," *Harper's Magazine*, December 1920.

311 **"future decades if industrial fuel combustion continues to rise exponentially"**
Roger Revelle and Hans Suess, "Carbon Dioxide Exchange Between Atmosphere and Ocean and the Question of an Increase of Atmospheric CO_2 During the Past Decades," *Tellus* 9 (February 1957).

311 **began more than 150 years ago**
NASA, "John Tyndall (1820–1893)," http://earthobservatory.nasa.gov/Features/ Tyndall/.

311 **first oil well by Colonel Edwin Drake in Pennsylvania**
Judah Ginsberg, "The Development of the Pennsylvania Oil Industry," American Chemistry Society, http://portal.acs.org.

312 **doubling of CO_2 concentrations**
Svante Arrhenius, "On the Influence of Carbonic Acid in the Air upon the Temperature of the Ground," *Philosophical Magazine and Journal of Science* 41 (April 1896).

312 **concentration was increasing steadily by a significant amount**
Spencer Weart, "The Discovery of Global Warming: Money for Keeling: Monitoring CO_2," 2003, http://www.aip.org/history/climate/Kfunds.htm.

312 **CO_2 throughout this yearly seasonal cycle was being shifted steadily upward**
Ibid.

312 **there are sixty other "distributed cooperative" sets of measurements**
"Tracking Long-Term Measurements of Gases and Aerosols That Contribute to Climate Change," *NOAA Magazine*, July 15, 2004, http://www.magazine.noaa.gov/ stories/mag140.htm.

312 **which has long predicted this result, and an effective cross-check**
"Atmospheric Oxygen Research: Research Overview," Scripps Institute of Oceanography, http://scrippso2.ucsd.edu/research-overview.

313 **the national academies of the G8 nations**
Coral Davenport, "Heads in the Sand," *National Journal*, December 2, 2011, http://www.nationaljournal.com/magazine/heads-in-the-sand-20111201.

313 **"need for urgent action to address climate change is now indisputable"**
National Academies of Science, "G8+5 Academies' Joint Statement: Climate Change and the Transformation of Energy Technologies for a Low Carbon Future," May 2009, www.nasonline.org/about-nas/leadership/president/statement -climate-change.pdf.

313 "97–98 percent of the climate researchers"
William Anderegg et al., "Expert Credibility in Climate Change," *Proceedings of the National Academy of Sciences*, 2010.

315 "Denial can be conscious or unconscious refusal"
Elisabeth Kübler-Ross Foundation, "Five Stages of Grief," 2012.

315 "refusal to acknowledge painful realities, thoughts, or feelings"
"Denial," *Stedman's Medical Dictionary*, http://dictionary.reference.com/browse/denial.

315 and stronger attacks on those who insist that we must take action
Chris Mooney, "The Science of Why We Don't Believe Science," *Mother Jones*, June 2011.

315 Two social scientists
Jane Risen and Clayton Critcher, "Visceral Fit: While in a Visceral State, Associated States of the World Seem More Likely," *Journal of Personality and Social Psychology* 100, no. 5 (2012).

316 automatically rejecting any potential alternative
"System Justification Theory," *Encyclopedia of Peace Psychology*, 2011.

317 "In our obsession with antagonisms of the moment"
Ronald Reagan, Speech to the United Nations General Assembly, September 21, 1987.

317 As E. O. Wilson recently wrote
E. O. Wilson, "Why Humans, Like Ants, Need a Tribe," *Daily Beast*, April 1, 2012.

319 "reposition global warming as theory not fact"
Matthew Wald, "Pro-Coal Ad Campaign Disputes Warming Idea," *New York Times*, July 8, 1991.

320 The large public multinational fossil fuel companies
John Fullerton, Capital Institute, "The Big Choice," July 19, 2011, http://capitalinstitute.org/blog/big-choice-0.

320 global scientific consensus is accepted
Ibid.

321 adds up to a total of $27 trillion
Ibid.

321 Saudi Arabia to convert its domestic energy use to 100 percent renewables
Fiona Harvey, "Saudi Arabia Reveals Plans to be Powered Entirely by Renewable Energy, *Guardian*, October 19, 2012.

322 in the aggregate, 7.5 million of them
Ben Bernanke, Federal Reserve Annual Conference on Bank Structure and Competition, "The Subprime Mortgage Market," May 17, 2007, http://www.federalreserve.gov/newsevents/speech/bernanke20070517a.htm.

323 "attack false enemies and deny real problems than find solutions"
Jay Rockefeller, Statement on Inhofe Resolution Vote, June 20, 2012.

323 four anti-climate lobbyists for every single member of the U.S. Senate and House
Marianne Lavelle, Center for Public Integrity, "The Climate Change Lobby Explosion," February 24, 2009, http://www.publicintegrity.org/node/4593.

323 **become one of the largest sources of campaign contributions**
Center for Responsive Politics, "Oil and Gas," http://www.opensecrets.org/industries/indus.php?ind=E01.

324 **One right-wing state attorney general**
John Rudolf, "A Climate Skeptic with a Bully Pulpit in Virginia Finds an Ear in Congress," *New York Times*, February 22, 2011.

324 **Right-wing legal foundations and think tanks**
Tom Clynes, "The Battle over Climate Science," *Popular Science*, June 21, 2012.

324 **Right-wing members of Congress have repeatedly**
Kate Sheppard, "Taking Climate Denial to New Extremes," *Mother Jones*, February 11, 2011.

325 **essential monitoring satellites being delayed or canceled**
Ledyard King, "Report Warns of Weather Satellites 'Rapid Decline,'" Gannett News, May 2, 2012.

325 **four separate independent investigations**
John Cook, Skeptical Science, "What Do the ClimateGate Emails Tell Us?," http://www.skepticalscience.com/Climategate-CRU-emails-hacked.htm.

326 *The Frozen Planet* **was edited before the Discovery Network**
Brian Stelter, "No Place for Heated Opinions," *New York Times*, April 20, 2012.

326 **"showing a documentary on lung cancer and leaving out the part about the cigarettes"**
Ibid.

326 **"the last best hope of earth"**
Abraham Lincoln, "Annual Remarks to Congress," December 1, 1862.

326 **"for good men to do nothing"**
Quote Investigator, December 4, 2010.

327 **support for actions to reduce greenhouse gas**
Connie Roser-Renouf et al., Yale Project on Climate Communication, "The Political Benefits of Taking a Pro-Climate Stand in 2012," 2012, http://environment.yale.edu/climate/files/Political-Benefits-Pro-Climate-Stand.pdf.

327 **stimulus bill put a major emphasis on green provisions**
Michael Grunwald, "The 'Silent Green Revolution' Underway at the Department of Energy," *Atlantic*, September 9, 2012.

327 **"The biggest single step that any nation has taken"**
Christopher Mims, "Efficiency Standards Are the Single Biggest Climate Deal Ever," Grist, December 5, 2011, http://grist.org/list/2011-12-05-efficiency-standards-are-the-single-biggest-climate-deal-ever/.

328 **well below the cost of production in the United States**
"Solar Prices Expected to Keep Falling in 2012," Associated Press, June 26, 2010.

329 **with the average cost of shale gas going up significantly in the process**
U.S. Energy Information Agency, "Annual Energy Outlook 2012: Market Trends—Natural Gas," June 25, 2012, http://www.eia.gov/forecasts/aeo/MT_naturalgas.cfm.

329 **which is more than** *seventy-two times* **as potent as** CO_2 **in trapping heat**
Intergovernmental Panel on Climate Change, "IPCC Fourth Assessment Report: Climate Change 2007," http://www.ipcc.ch/publications_and_data/ar4/wg1/en/tssts-2-5.html.

329 buy time for the implementation
Drew Shindell et al., "Simultaneously Mitigating Near-Term Climate Change and Improving Human Health and Food Security," *Science*, January 2012.

329 settle on the surface of ice and snow
V. Ramanathan and G. Carmichael, "Global and Regional Climate Changes Due to Black Carbon," *Nature Geoscience* 1 (April 2008).

329 the majority of smaller "wildcat" drillers do not
Robert Howarth et al., "Venting and Leaking of Methane from Shale Gas Development: Response to Cathles et al.," *Climatic Change* 113 (July 2012).

330 virtually all of the benefit natural gas might have
Nathan Myhrvold and Ken Caldeira, "Greenhouse Gases, Climate Change and the Transition from Coal to Low-Carbon Electricity," *Environmental Research Letters*, March 2012.

330 an average of five million gallons of water for each well
Chesapeake Energy, "Water Use in Deep Shale Gas Exploration," 2012, http://www.chk.com/Media/Educational-Library/Fact-Sheets/Corporate/Water_Use_Fact_Sheet.pdf; Jack Healy, "Struggle for Water in Colorado with Rise in Fracking," *New York Times*, September 5, 2012.

330 acute even before the spread of the thirsty fracking process
Healy, "Struggle for Water in Colorado with Rise in Fracking."

330 fracking wells are being drilled in communities where
Russell Gold and Ana Campoy, "Oil's Growing Thirst for Water," *Wall Street Journal*, December 6, 2011.

330 as chimneys for the upward migration of both methane and drilling fluids
Ian Urbina, "Tainted Water Well, and Concern There May Be More," *New York Times*, August 3, 2011.

330 cause of pollution in the aquifer above the area that was fracked
Tenille Tracy, "EPA Says Wyoming Fracking Results Are Consistent," *Wall Street Journal*, September 26, 2012.

330 at the behest of then vice president Dick Cheney
Abraham Lustgarten, "Hydrofracked: One Man's Quest for Answers About Natural Gas Drilling," *ProPublica*, June 27, 2011.

331 "The consequences of a misstep in a well"
Council on Foreign Relations, "The New North American Energy Paradigm."

331 political resistance from landowners
Inae Oh, "New York Fracking Protest Urges Cuomo to Ban Controversial Drilling," *Huffington Post*, August 22, 2012, http://www.huffingtonpost.com/2012/08/22/new-york-fracking-protest-cuomo-photos_n_1822575.html.

331 has caused multiple small (usually harmless) earthquakes
Charles Choi, "Fracking Earthquakes: Injection Practice Linked to Scores of Tremors," *Livescience*, August 7, 2012.

331 alleged to have infiltrated water aquifers
Abraham Lustgarten and ProPublica, "Are Fracking Wastewater Wells Poisoning the Ground Beneath Our Feet?," *Scientific American*, June 21, 2012.

331 more common source of complaints than the initial injections
Rachel Ehrenberg, "The Facts Behind the Frack," *ScienceNews*, September 8, 2012,

http://www.sciencenews.org/view/feature/id/343202/title/The_Facts_Behind_the _Frack.

331 spread on roads, ostensibly for dust control
National Resources Defense Council, "Report: Five Primary Disposal Methods for Fracking Wastewater All Fail to Protect Public Health and Environment," May 9, 2012, http://www.nrdc.org/media/2012/120509.asp.

331 "something wrong and dangerous, they should punish them"
Christopher Helman, "Billionaire Father of Fracking Says Government Must Step Up Regulation," *Forbes*, July 19, 2012.

332 a potential tipping point
Joe Romm, ThinkProgress, "Gas Emissions Reduction Target for 2020," January 13, 2009, http://www.americanprogress.org/issues/green/report/2009/01/13/5472/the-united-states-needs-a-tougher-greenhouse-gas-emissions-reduction-target-for-2020/.

332 frenzy of exploration for shale gas in China, Europe
Bryan Walsh, "In Hunt for Energy, China and Europe Explore Fracking," *Time*, May 21, 2012.

332 Africa, and elsewhere
Ruona Agbroko, "S Africa Lifts Fracking Ban," *Financial Times*, September 7, 2012.

333 U.S. horizontal drilling and hydraulic fracturing technologies to China
Jerry Mandel, "Will U.S Shale Technology Make the Leap Across the Pacific?," *E&E News*, July 17, 2012, http://www.eenews.net/public/energywire/2012/07/17/1.

333 particularly in northern and northwestern China
"Water, Water Everywhere," *China Economic Review*, July 26, 2012.

333 the switch from coal to natural gas by electric utilities
Kevin Begos, "CO_2 Emissions in US Drop to 20-Year Low," Associated Press, August 17, 2012.

333 Coal has the highest carbon content
U.S. Energy Information Agency, "How Much Carbon Dioxide Is Produced When Different Fuels Are Burned?," 2012, http://www.eia.gov/tools/faqs/faq.cfm?id =73&t=11; U.S. Environmental Protection Agency, "Air Emissions," 2007, http://www.epa.gov/cleanenergy/energy-and-you/affect/air-emissions.html.

334 Harriman, Tennessee, in my home state four years ago
Bobby Allyn, "TVA Held Responsible for Massive Coal Ash Spill," *Tennessean*, August 23, 2012.

334 principal source of human-caused mercury in the environment
U.S. Environmental Protection Agency, "Mercury: Basic Information," February 7, 2012, http://www.epa.gov/hg/about.htm.

334 include at least some amount of methyl-mercury
U.S. Environmental Protection Agency, "What You Need To Know About Mercury in Fish and Shellfish," June 20, 2012, http://water.epa.gov/scitech/swguidance/fishshellfish/outreach/advice_index.cfm#isthere.

334 cancellation of 166 new coal plants
Mark Hertsgaard, "How a Grassroots Rebellion Won the Nation's Biggest Climate Victory," *Mother Jones*, April 2, 2012.

334 **1,200 new coal plants are now planned in 59 countries**
Ailun Yang and Yiyun Cui, *Global Coal Risk Assessment: Data Analysis and Market Research* (Washington, DC: World Resources Institute, 2012).

334 **65 percent in the next two decades**
International Energy Agency, "World Energy Outlook: Executive Summary," 2011, http://www.worldenergyoutlook.org/publications/weo-2011/.

334 **the CO_2 emissions would continue destroying the Earth's ecosystem**
Kurt Kleiner, "Coal to Gas: Part of a Low-Emissions Future?," *Nature*, February 28, 2008.

334 **70 to 75 percent of the carbon in coal for each unit of energy produced**
U.S. Environmental Protection Agency, "Air Emissions," December 2007.

334 **more expensive to produce and carry even harsher impacts for the environment**
Bryan Walsh, "There Will Be Oil and That's the Problem," *Time*, March 29, 2012.

335 **three quarters of the amount in the atmosphere**
Rattan Lal, "Carbon Sequestration," *Philosophical Transactions of the Royal Society B*, February 2008.

335 **his successor has made policy changes that are reversing**
Jeffrey T. Lewis, "Pace of Deforestation in Brazil's Amazon Falls," *Wall Street Journal*, November 28, 2012.

335 **dramatic "dieback" of the Amazon at mid-century**
Justin Gillis, "The Amazon Dieback Scenario," *New York Times*, Green blog, October 7, 2011.

335 **Indonesia and Malaysia—in order to establish palm oil plantations**
Brad Plumer, "EPA Faces Crucial Climate Decision on Diesel Made from Palm Oil," *Washington Post*, April 27, 2012.

335 **peatlands contain more than one third of all the global soil carbon**
Reynaldo Victoria et al., United Nations Environment Programme, "UNEP Yearbook: The Benefits of Soil Carbon," 2012.

335 **80 percent of global forest cover is in publicly owned forests**
United Nations Food and Agriculture Organization, "Global Forest Resources Assessment 2010," 2010, p. xxiv, http://www.fao.org/forestry/fra/fra2010/en/.

335 **Papua New Guinea, Indonesia, Borneo, and the Philippines**
United Nations Food and Agriculture Organization, "State of the World's Forests 2011," 2011, http://www.fao.org/docrep/013/i2000e/i2000e00.htm.

336 **especially cattle ranching**
Doug Boucher et al., Union of Concerned Scientists, "Solutions for Deforestation-Free Meat," 2012, http://www.ucsusa.org/global_warming/solutions/forest _solutions/solutions-for-deforestation-free-meat.html.

336 **as much as 22 percent of all carbon stored**
Sharon Oosthoek, "Boreal Forests Ignored in Climate Change Fight," CBC News, November 12, 2009, http://www.cbc.ca/news/technology/story/2009/11/11/ boreal-carbon-climate-change.html.

336 **symbiosis between the larch and the tundra is thereby disrupted**
Douglas Fischer and Daily Climate, "Shift in Northern Forests Could Increase Global Warming," *Scientific American*, March 28, 2011.

336 reproducing three generations per summer rather than one
Noah S. Diffenbaugh et al., "Global Warming Presents New Challenges for Maize Pest Management," *Environmental Research Letters* 3 (2008).

336 "unprecedented outbreak of the mountain pine beetle"
David A. Gabel, "Expanding Forests in the Northern Latitudes," Environmental News Network, March 23, 2011, http://www.enn.com/ecosystems/article/42501.

337 "the real culprit is water stress caused by climate change"
Justin Gillis, "With Deaths of Forests, a Loss of Key Climate Protectors," *New York Times*, October 1, 2011.

337 direct proportion to the rising temperatures
"More Large Forest Fires Linked to Climate Change," *ScienceDaily*, July 10, 2006, http://www.sciencedaily.com/releases/2006/07/060710084004.htm; Gillis, "With Deaths of Forests, a Loss of Key Climate Protectors."

337 in the trees and plants themselves
Gillis, "With Deaths of Forests, a Loss of Key Climate Protectors."

337 rather than a net "sink," withdrawing CO_2 as the trees grow
Ben Bond-Lamberty et al., "Fire as the Dominant Driver of Central Canadian Boreal Forest Carbon Balance," *Nature*, November 1, 2007; "Wildfires Turning Northern Forests into Carbon-Dioxide Sources," CBC News, October 31, 2007, http://www.cbc.ca/news/technology/story/2007/10/31/boreal-forests.html.

337 eastern North America, Europe, the Caucasus, and Central Asia
Gabel, "Expanding Forests in the Northern Latitudes."

337 $3.7 trillion in environmental costs
Pavan Sukhdev et al., *The Economics of Ecosystems and Biodiversity: Mainstreaming the Economics of Nature: A Synthesis of the Approach.* Bonn: TEEB, 2010.

337 40 percent as many trees as the rest of the world put together
United Nations Food and Agriculture Organization, "State of the World's Forests 2009," 2009, http://www.fao.org/docrep/011/i0350e/i0350e00.htm.

337 required to plant at least three trees per year
"China's Hu Takes Part in Tree Planting," UPI, April 5, 2009.

337 approximately 100 million acres of new trees
Gillis, "With Deaths of Forests, a Loss of Key Climate Protectors."

338 in trees include the U.S., India, Vietnam, and Spain
United Nations Food and Agriculture Organization, "State of the World's Forests 2011."

338 compared to the rich variety supported by a healthy, multispecies primary forest
Jianchu Xu, "China's New Forests Aren't as Green as They Seem," *Nature*, September 21, 2011.

338 carbon sequestered in the first few feet of soil
Damian Carrington, "Desertification Is Greatest Threat to Planet, Expert Warns," *Guardian*, December 15, 2010.

338 10.57 percent of the Earth's land surface covered by arable land
CIA World Factbook, https://www.cia.gov/library/publications/the-world-factbook/geos/xx.html

338 health of the soil and interfere with the normal sequestration
Tom Philpott, "New Research: Synthetic Nitrogen Destroys Soil Carbon, Un-

dermines Soil Health," Grist, February 24, 2010, http://grist.org/article/2010-02-23
-new-research-synthetic-nitrogen-destroys-soil-carbon-undermines/.

338 by pushing subsistence farmers to clear more forests
David Lapola et al., "Indirect Land-Use Changes Can Overcome Carbon Savings from Biofuels in Brazil," *Proceedings of the National Academy of Sciences*, January 2010.

338 have proven that assumption to be wrong
Claude Mandil and Adnan Shihab-Eldin, International Energy Forum, "Assessment of Biofuels: Potential and Limitations," February 2010, http://www.ief.org/news/news-details.aspx?nid=311.

339 20 to 50 percent of all living species on Earth within this century
Nicholas Stern, *The Economics of Climate Change*.

339 Approximately one third of all amphibian species
Camila Ruz, "Amphibians Facing 'Terrifying' Rate of Extinction," *Guardian*, November 2011.

339 hit by a spreading fungal disease
Michelle Nijhuis, "A Rise in Fungal Diseases Is Taking Growing Toll on Wildlife," *Yale Environment 360*, October 24, 2011.

339 the encroachment of human activities
Owen Clyke, "The Militarization of Africa's Animal Poachers," *Atlantic*, July 31, 2012; David Braun, "Human Encroachment Threatens Thousands of Gorillas in African Swamp," *National Geographic*, November 24, 2009; Yaa Ntiamoa-Baidu, United Nations Food and Agriculture Organization, "West African Wildlife: A Resource in Jeopardy," 1998, http://www.fao.org/docrep/s2850e/s2850e05.htm.

339 caused by a large asteroid crashing into the Earth
Anthony Barnosky et al., "Has the Earth's Sixth Mass Extinction Already Arrived?," *Nature*, March 2011.

339 "precipitated entirely by man"
Richard Leakey and Roger Lewin, *The Sixth Extinction: Patterns of Life and the Future of Humankind* (New York: Anchor Books, 1995), p. 235.

339 average 3.8 miles per decade toward the poles
Camille Parmesan and Gary Yohe, "A Globally Coherent Fingerprint of Climate Change Impacts Across Natural Systems," *Nature*, January 2003.

340 half of the mountain species had moved, on average
Craig Moritz et al., "Impact of a Century of Climate Change on Small-Mammal Communities in Yosemite National Park, USA," *Science*, October 2008.

340 reach the poles and the mountaintops and can go no farther
Elisabeth Rosenthal, "Climate Threatens Birds from Tropics to Mountaintops," *New York Times*, January 21, 2011.

340 at risk because they cannot adapt to climate change quickly enough
Kai Zhu et al., "Failure to Migrate: Lack of Tree Range Expansion in Response to Climate Change," *Global Change Biology* 18 (November 2011).

340 according to scientists, are facing a rising risk of extinction
Lucas Joppa et al., "How Many Species of Flowering Plants Are There?," *Proceedings of the Royal Society B*, July 2010.

340 **United Nations Convention on Biological Diversity notes**
United Nations Convention on Biological Diversity, "Global Biodiversity Outlook 3," January 2010, http://www.unep-wcmc.org/gbo-3_90.html, p. 51.

340 **as a precautionary measure for the future of mankind**
John Roach, "'Doomsday' Vault Will End Crop Extinction, Expert Says," *National Geographic*, December 27, 2007.

342 **carbon tax to the taxpayers but would have applied one third**
Matt Kasper, "Rep. Jim McDermott Introduces Carbon Tax Law," August 6, 2012, ClimateProgress, http://thinkprogress.org/climate/2012/08/06/641831/rep-jim-mcdermott-introduces-carbon-tax-law/.

342 **go to carbon fuel companies**
John Broder, "Obama's Bid to End Oil Subsidies Revives Debate," *New York Times*, January 31, 2011.

342 **the dirtiest liquid fuel, kerosene, is heavily subsidized**
Narasimha Rao, "Kerosene Subsidies in India: When Energy Policy Fails as Social Policy," *Energy for Sustainable Development*, March 2012.

342 **only a few years away from reaching that threshold**
Alex Morales and Jacqueline Simmons, "Renewables from Vestas to Suntech Plan Profit without Subsidy," Bloomberg, January 26, 2012.

343 **utility sector oppose such measures**
Joe Romm, "Who Killed the Senate RPS?," ClimateProgress, June 27, 2007, http://thinkprogress.org/climate/2007/06/27/201573/who-killed-the-senate-rps/.

343 **including, most prominently, California**
State of California, "California Renewables Portfolio Standard," 2012, http://www.cpuc.ca.gov/PUC/energy/Renewables/index.htm.

343 **install 500 megawatts of solar energy by 2020**
Dave Roberts, "Why Do 'Experts' Always Lowball Clean-Energy Projections?," Grist, July 19, 2012, http://grist.org/renewable-energy/experts-in-2000-lowballed-the-crap-out-of-renewable-energy-growth/.

343 **China installed double that amount by 2010**
Ibid.

343 **goal was exceeded 22-fold and is expected**
Ibid.

344 **"weren't just off, they were *way* off"**
Ibid.

345 **alternative to government regulations mandating reductions**
John Fialka, "How a Republican Anti-Pollution Measure, Expanded by Democrats, Got Roots in Europe and China," *E&E News*, November 17, 2011.

345 **originally supported the idea**
Ibid.

346 **energy during their own past periods**
International Energy Agency, "Energy Poverty," 2012, http://www.iea.org/topics/energypoverty/.

346 **even if it means that they too must shoulder**
Arthur Max, "Developing Nations Pledge Actions to Curb Climate Change," Associated Press, March 22, 2011.

346 costs from climate disruption will be borne by developing countries
World Development Report 2010: Development and Climate Change (Washington, DC: World Bank, 2010)

346 world now exceed those in rich countries
Alex Morales, "Renewable Power Trumps Fossils for First Time as UN Talks Stall," Bloomberg News, November 25, 2011.

346 more than half of the installed global renewable energy capacity
Charles Kenny, "Greening It Alone," *Foreign Policy*, August 1, 2011.

346 hampered the ability of many communities
Rick Jervis and Gregory Korte, "FEMA Could Run Out of Money over Stalemate," *USA Today*, September 25, 2011.

346 caused more than $15 billion in damage
"Hurricane Irene 2011: One Year Anniversary of East Coast Storm," *Huffington Post*, August 24, 2012, http://www.huffingtonpost.com/2012/08/24/hurricane-irene -2011-2012_n_1826060.html.

346 and wildfires in 240 of its 242 counties
Patrik Jonsson, "Texas Wildfire Chief: Wildfires Still Raging, but 'We Are Making Successes,'" *Christian Science Monitor*, April 21, 2011.

346 all-time-high temperature records were broken or tied
Andrew Freedman, "Hot Summer of 2011 Rewrites Record Books," Climate Central, September 8, 2011, http://www.climatecentral.org/blogs/a-record-hot -summer-interactive-map/.

346 seven of them caused more than $1 billion in damage
National Oceanic and Atmospheric Administration, "Extreme Weather 2012," January 19, 2012.

346 more than half of the counties in the U.S.
David Ariosto and Melissa Abbey, "Historical Drought Puts Over Half of US Counties in Disaster Zones, USDA Says," CNN, August 1, 2012.

346 Hurricane Sandy cost at least $71 billion
Matthew Craft, "Hurricane Sandy's Economic Damage Could Reach $50 Billion, Eqecat Estimates," Associated Press, November 1, 2012.

347 report supporting such border adjustments
United Nations Environment Programme and World Trade Organization, *Trade and Climate Change*, 2009, http://www.wto.org/english/res_e/booksp_e/trade_climate _change_e.pdf.

347 has been a success in most of the nations, provinces, and regions
Janet Raloff, "Kyoto Climate Treaty's Greenhouse 'Success,'" *ScienceNews*, November 3, 2009, http://www.sciencenews.org/view/generic/id/49058/title/Science _%2B_the_Public__Kyoto_climate_treatys_greenhouse_success.

347 "The carbon market may be complex, but we live in a complex world"
Marton Kruppa and Andrew Allan, "Carbon Trading May Be Ready for Its Next Act," Reuters, November 13, 2011.

348 will imminently begin cap and trade systems
Ibid.

348 Most significantly, California began implementing its system in 2012
Jason Dearen, "California's Cap-and-Trade System to Launch with First Pollution Permits Auction," Associated Press, November 12, 2012.

348 "it may become the biggest in the world, and allowances in that system would then give a global price signal"
Kruppa and Allan, "Carbon Trading May Be Ready for Its Next Act."

348 experience that will be used to implement a nationwide cap and trade system by 2015
Lan Lan, "Beijing Preparing for Carbon Trading System," *China Daily*, April 20, 2012.

348 almost 20 percent of the Chinese population
Alexandre Kossoy and Pierre Gioan, World Bank, "State and Trends of the Carbon Market 2012," May 2012, p. 99, http://siteresources.worldbank.org/INTCARBONFINANCE/Resources/State_and_Trends_2012_Web_Optimized_19035_Cvr&Txt_LR.pdf.

349 China for allegedly providing unfair subsidies to its wind and solar manufacturers
Keith Bradsher, "200 Chinese Subsidies Violate Rules, US Says," *New York Times*, October 6, 2011.

349 European Union began its consideration of a similar complaint
"US Imposes Import Tariffs on Chinese Solar Panels," BBC News, May 17, 2012, http://www.bbc.co.uk/news/business-18112983.

349 overtake the United States as the largest global warming polluter
Chris Buckley, "China Says Is World's Top Greenhouse Gas Emitter," Reuters, November 23, 2010.

349 China against dirty energy projects are growing
Keith Bradsher, "Budding Environmental Movement Finds Resonance Across China," *New York Times*, July 4, 2012.

349 gone up more than 150 percent, surpassing
Goldman Sachs, "Sustainable Growth in China: Spotlight on Energy," August 13, 2012, http://www.goldmansachs.com/our-thinking/topics/environment-and-energy/sustainable-growth-china.html.

349 70 percent of its energy from coal
"Coal Industry in China—Coal Accounts for About 70% of China's Total Energy," BusinessWire, December 14, 2011.

349 to a level three times that of U.S. coal consumption
Goldman Sachs, "Sustainable Growth in China: Spotlight on Energy," August 13, 2012, http://www.goldmansachs.com/our-thinking/topics/environment-and-energy/sustainable-growth-china.html.

349 two and a half times more than the U.S.
Osamu Tsukimori, "China Overtakes Japan as World's Top Coal Importer," Reuters, January 26, 2012.

349 is equivalent to all of the U.S. annual consumption
Mikkal Herberg, New America Foundation, "China's Energy Rise and the Future of U.S.-China Energy Relations," June 21, 2011, http://newamerica.net/publications/policy/china_s_energy_rise_and_the_future_of_us_china_energy_relations.

349 though many experts are skeptical about their ability to stay within the cap
Susan Kraemer, "China to Simply Cap Coal Use Within 3 Years," Clean Technica,

March 8, 2012, http://cleantechnica.com/2012/03/08/china-to-simply-cap-coal-use -within-3-years/.

350 first decade of this century, and is now second only
Herberg, "China's Energy Rise and the Future of U.S.-China Energy Relations."

350 Saudi Arabia's oil exports to China
Ibid.

350 will import three quarters of its oil within the next two decades
U.S. Energy Information Agency, "Country Analysis: China," September 4, 2012, http://www.eia.gov/countries/cab.cfm?fips=CH.

350 engagement with oil-rich countries in the Middle East and Africa
Ibid.

350 Chinese became the largest investor in Iraq's oilfields
Ibid.

350 energy consumption in China is only a fraction
Heather Billings and Sisi Wei, "China's Energy Grab," *Washington Post*, October 30, 2011.

350 per capita CO_2 emissions are approaching those of Europe
Duncan Clark, "Average Chinese Person's Carbon Footprint Now Equals European's," *Guardian*, July 18, 2012.

350 letting all energy prices float further upward to global market levels
Keith Bradsher, "China Sharply Raises Energy Prices," *New York Times*, June 20, 2008.

350 lagging behind other leading global economies
Danielle Kurtzlaben, "China, European Countries Best U.S. on Energy Efficiency," *U.S. News & World Report*, July 12, 2012.

350 it will invest almost $500 billion in clean energy
Esther Tanquintic-Misa, "China Leads Global Investments in Renewable Energy," *IB Times*, December 5, 2011, http://au.ibtimes.com/articles/261083/20111205/china -leads-global-investments-renewable-energy.htm#.UFJWkhg-KP0.

350 Chinese make use of "feed-in tariffs"
Coco Liu, "China Uses Feed-in Tariff to Build Domestic Solar Market," ClimateWire, September 14, 2011.

350 subsidy plan that worked extremely well in Germany
Cristoph Stefes, "Room for Debate—The German Solution: Feed in Tariffs," *New York Times*, September 21, 2011.

350 renewable energy percentage targets on utilities
Pew Charitable Trusts, "Global Clean Power: A $2.3 Trillion Opportunity— Appendix: China," December 8, 2010, p. 48.

350 targets for the reduction of CO_2 emissions per unit of economic growth
Bill McKibben, "Can China Go Green?," *National Geographic*, June 2011.

351 "miracle will end soon, because the environment can no longer keep pace"
"The Chinese Miracle Will End Soon," *Der Spiegel*, March 7, 2005.

351 factories and even blackouts in order to ensure that the goals were met
Jonathan Watts, "China Resorts to Blackouts in Pursuit of Energy Efficiency," *Guardian*, September 19, 2010.

351 central government has linked promotions
Alexandre Kossoy and Pierre Gioan, World Bank, "State and Trends of the Carbon Market 2012," May 2012, pp. 96–99, http://siteresources.worldbank.org/INTCARBONFINANCE/Resources/State_and_Trends_2012_Web_Optimized_19035_Cvr&Txt_LR.pdf.

351 exports 95 percent of the solar panels it produces
David Pierson, "China Offers Measured Response to U.S. Tariffs on Solar Panels," Los Angeles Times, March 21, 2012.

351 50 percent of all the windmills installed globally were in China
Global Wind Energy Council, "China Wind Energy Development Update 2010," 2010, http://www.gwec.net/china-wind-energy-development-update-2012/.

351 connected to lines that cannot handle the electricity flow
Mat McDermott, "One Quarter of China's Wind Power Still Not Connected to Electricity Grid," TreeHugger, March 7, 2011, http://www.treehugger.com/corporate-responsibility/one-quarter-of-chinas-wind-power-still-not-connected-to-electricity-grid.html.

351 "almost the equivalent of rebuilding"
Jeff St. John, "HVDC Grows on the Grid from China to Oklaunion," August 28, 2012, Greentech Media, http://www.greentechmedia.com/articles/read/hvdc-grows-in-smart-grid-from-china-to-oklaunion/.

351 the Middle East to large electricity consumers in Europe
Beth Gardiner, "An Energy Supergrid for Europe Faces Big Obstacles," New York Times, January 16, 2012.

351 Mexico can easily provide all of the electricity needed
Thomas L. Friedman, "This Is a Big Deal," New York Times, December 4, 2011.

352 India
Brad Gammons, "India Set to Leap-Frog Ahead with 'Smart Grid' Energy Strategy," International Business Times, September 8, 2011.

352 Australia
Fran Foo, "'EnergyAustralia' Bags $93m Smart Grid Contract," Australian, October 8, 2010.

352 more than $200 billion per year
National Energy Technology Laboratory, "Modern Grid Benefits," 2007, p. 14, www.netl.doe.gov/smartgrid/referenceshelf/whitepapers/Modern%20Grid%20Benefits_Final_v1_0.pdf.

352 problems in managing electricity flows through the antiquated grid
Simon Denyer and Rama Lakshmi, "India Blackout, on Second Day, Leaves 600 Million without Power," Washington Post, August 1, 2012.

352 majority of their time in garages or parking spaces
Matthew L. Wald, "Better Batteries: Not Just for Cars Anymore," New York Times, Green blog, October 31, 2011, http://green.blogs.nytimes.com/2011/10/31/better-batteries-not-just-for-cars-any-more/?scp=14&sq=energy%20storage&st=cse.

352 impressive movement by many companies to take advantage
Amory B. Lovins and the Rocky Mountain Institute, Reinventing Fire: Bold Business Solutions for the New Energy Era (White River Junction, VT: Chelsea Green Publishing, 2011).

352 **wave and tidal energy are both being explored**
U.S. Department of Energy, "DOE Reports Show Major Potential for Wave and Tidal Energy Production Near U.S. Coasts," January 18, 2012, http://apps1.eere .energy.gov/news/progress_alerts.cfm/pa_id=664.

352 **that they may have great potential in the future**
Elisabeth Rosenthal, "Tidal Power: The Next Wave?," *New York Times*, October 20, 2010.

352 **"unlikely to significantly contribute to global energy supply before 2020"**
O. Edenhofer et al., Intergovernmental Panel on Climate Change, "Special Report on Renewable Energy Sources and Climate Change Mitigation—Press Release," 2011, http://srren.ipcc-wg3.de/press/content/potential-of-renewable -energy-outlined-report-by-the-intergovernmental-panel-on-climate-change.

353 **significant contribution in nations like Iceland**
Christopher Mims, "One Hot Island: Iceland's Renewable Geothermal Power," *Scientific American*, October 20, 2008.

353 **New Zealand**
New Zealand Geothermal Association, "Geothermal Energy & Electricity Generation," http://www.nzgeothermal.org.nz/elec_geo.html.

353 **Philippines, where there is an abundance**
Dan Jennejohn et al., Geothermal Energy Association, "Geothermal: International Market Overview Report," May 2012, http://www.geo-energy.org/pdf/ reports/2012-GEA_International_Overview.pdf.

353 **serious ecological risks in particular locations**
Arun Kumar, Intergovernmental Panel on Climate Change, "Special Report on Renewable Energy Sources and Climate Change Mitigation-Hydropower," 2011, pp. 437–96.

353 **use of biomass is expanding**
Toby Price, "Power Generation from Biomass Booms Worldwide," *Renewable Energy*, September 13, 2012.

353 **though enforcement of this mandate has been lagging**
"An Overview of China's Renewable Energy Market," China Briefing, June 16, 2011, http://www.china-briefing.com/news/2011/06/16/an-overview-of-chinas-renewable -energy-market.html.

354 **storage of the CO_2 that would otherwise be released into the atmosphere**
Barbara Freese, Steve Clemmer, and Alan Nogee, Union of Concerned Scientists, "Coal Power in a Warming World: A Sensible Transition to Cleaner Energy Options," October 2008, p. 18, http://www.ucsusa.org/assets/documents/clean _energy/Coal-power-in-a-warming-world.pdf.

354 **because it begins to be absorbed into the geological formation itself**
James Katzer, ed., Massachusetts Institute of Technology, "The Future of Coal: Options for a Carbon-Constrained World," 2007, p. 44, http://web.mit.edu/coal/ The_Future_of_Coal.pdf.

354 **political paralysis that characterizes the present state of democracy in the United States**
Jeff Tollefson and Richard Van Noorden, "Slow Progress to Cleaner Coal," *Nature*, April 2012.

354 **Norway, the United Kingdom, Canada, and Australia**
Damien Carrington, "Q&A: Carbon Capture and Storage," *Guardian*, May 10, 2012.

354 **is to put a price on carbon**
David Talbot, "Needed: A Price on Carbon," *Technology Review*, August 14, 2006.

355 **has had difficulties with its new generation of reactors**
Liam Moriarty, "French Sour on Nuclear Power," *PRI The World*, April 24, 2012, http://www.theworld.org/2012/04/france-nuclear-power/.

355 **with a design that many experts believe is promising**
Korea Herald, "S. Korea to Proceed with Two New Reactors," *Jakarta Post*, May 6, 2012.

355 **smaller and hopefully safer reactors**
Clay Dillow, "Can Next-Generation Reactors Power a Safe Nuclear Future?," *Popular Science*, March 17, 2011.

355 **ingrained preference for single solutions**
Jon Gertner, "Why Isn't the Brain Green?," *New York Times*, April 16, 2009.

355 **tiny strips of tinfoil in orbit around the Earth**
U.S. National Academy of Science, "Policy Implications of Greenhouse Warming: Mitigation, Adaptation and the Science Base," 1992, http://books.nap.edu/openbook.php?isbn=0309043867.

355 **giant space parasol, also intended to block incoming sunlight**
Robert Kunzig, "A Sunshade for Planet Earth," *Scientific American*, November 2008.

355 **massive quantities of sulfur dioxide into the upper atmosphere**
Ibid.

356 **existence on Earth would no longer be blue—or at least no longer be *as* blue**
Ben Kravitz et al., "Geoengineering: Whiter Skies?," *Geophysical Research Letters* 39 (2012).

356 **color of the night sky from black to reddish black**
C. C. M. Kyba et al., "Red Is the New Black: How the Colour of Urban Skyglow Varies with Cloud Cover," *Monthly Notices of the Royal Astronomical Society* 425 (August 2012).

357 **melting of glaciers and snowpacks**
Tim Wall, "Peru's Peaks Go White to Guard Glaciers," *Discovery News*, December 5, 2011.

357 **"We must disenthrall ourselves, and then we shall save our country"**
Abraham Lincoln, "Annual Remarks to Congress," December 1, 1862.

357 **that our numbers were reduced to less than 10,000 people**
"Humans: From Near Extinction to Phenomenal Success," BBC, 2012, http://www.bbc.co.uk/nature/life/Human.

CONCLUSION

362 **paintings in the caves at Chauvet, in France, and the figurines**
Clottes, "Chauvet Cave (ca. 30,000 B.C.)," Heilbrunn Timeline of Art History; Judith Thurman, "First Impressions," *New Yorker*, June 23, 2008.

363 **99.9 percent of them are identical in every human being**
"Cracking the Code of Life," PBS NOVA, April 17, 2001, http://www.pbs.org/wgbh/nova/body/cracking-the-code-of-life.html; Roger Highfield, "DNA Survey Finds All Humans Are 99.9pc the Same," *Telegraph*, December 20, 2002; University of Utah Genetic Science Learning Center, "Can DNA Demand a Verdict?," http://learn.genetics.utah.edu/content/labs/gel/forensics/.

363 **our 23,000 genes**
"Microbes Maketh Man," *Economist*, August 18, 2012.

363 **millions of proteins**
"Proteomics," American Medical Association, http://www.ama-assn.org/ama/pub/physician-resources/medical-science/genetics-molecular-medicine/current-topics/proteomics.page.

365 **incentives and rules that work to their advantage**
Mancur Olson, *The Rise and Decline of Nations: Economic Growth, Stagflation, and Social Rigidities* (New Haven, CT: Yale University Press, 1984).

369 **"weapons of reason which today arm you against the present"**
Marcus Aurelius, *Meditations* (New York: Penguin, 1964), p. 106. Marcus Aurelius was praised by historians of the Roman empire, including Niccolò Machiavelli and Edward Gibbon.

INDEX

Abdullah, Faisal, 60
abortion, 234, 249–50, 252, 318
acid rain, 333, 344
acquired immune system, 274–77
advertising, 156, 158–60
Afghanistan, 86–87, 170, 177, 180, 194
Afghanistan War, 101, 177, 242
Africa, 19, 131, 163, 169, 180, 187, 194,
    196–98, 304, 350, 353
  population growth in, 166, 167, 194
African Union, 135, 193
"Agent Orange Corn," 267–68
aggregate demand, lowering of, 26
Agha-Soltan, Neda, 59
aging, 254–55
Agrawal, Pramod K., 290
Agricultural Research Service, U.S., 189,
    268
Agricultural Revolution, xvii, 37–38, 49,
    127, 174, 185, 262, 308
agricultural runoff, 302
agricultural waste, 161, 302
agriculture, 10, 106, 188–89, 282, 372
  as affected by global warming, 288
  crop diseases and, 289–90
  loss of jobs in, 22–23, 41
  mechanization of, 21, 39
  organic, 189
  phosphorus and, 145, 190–92
  rise in productivity of, 18, 189
  subsistence, 37
Agriculture Department, U.S., 22, 187,
    200

Ai Weiwei, 60
AIDS, 169, 255, 274
Ailes, Roger, 119
Aken, Scott, 78
Al Jazeera, 61
Al-Qaeda, 132
Alaska, 337
Alberta Basin, 186
alcohol, 258
alfalfa, 262
algae, 189, 296, 302
Algeria, 177
algorithms, financial, 13, 14, 16, 46n, 373
All-Army Conference, 104
Allen, Robert, 93
alphabets, 49–50, 52
alternative energy, 282–84
aluminum, 146
Alzheimer's, 208
Amazon Basin, 304
Amazon rainforest, 193, 335
American Civil Liberties Union, 84, 86
American Express, 82–83
American Medical Association, 260, 261
American Revolution, 52, 99, 106
American Samoa, 312
amoxicillin, 276
amphibians, 339
Amu Darya, 192
Anderson, Michael, 245
Andes, 127, 270, 357
Andorra, 173
Angell, Norman, 133

animal waste, 353
Antarctica, 186, 198, 200, 286, 297, 298, 300, 309–10
Antarctic circumpolar winds, 310
anti-whaling movement, 139
antibiotic-resistant bacteria, 218, 225–28
antibiotics, 200, 225–28, 274–76
antifungals, 200
antiretroviral drugs, 255
antitrust suits, 110, 124
aphids, 290
aquaculture, 200
aquifers, 145, 185, 186, 187–88, 300, 330, 331
Arab League, 135
Arab Spring, 60–61, 100, 138, 176
arable land, 194–95
Aral Sea, 192
architecture, energy consumption reduced by, 282–83
Arctic Climate Impact Assessment, 297
Arctic ice cap, 308
Arctic Ocean, 148, 295–96, 328, 373
Argentina, 263, 271, 288
Aristotle, xx, xxiii
Armstrong, Lance, 247
Arrhenius, Svante, 311
arsenic, 334
artificial cerebellum, 244
artificial intelligence, xiv, 5, 8, 22, 24, 39
artificial life, 218–22
artificial limbs, 242–43
artificial testicle, 247
Asia, 18, 50, 82, 94, 131, 269
assembly lines, 31
asteroids, 339
asthma, 276
AT&T, 82–83
Athens, xx, 177
ATMs, 26
atmosphere, xv, 281
atomic weapons, xxiii, 134
    atmospheric testing of, 139
Aung San Suu Kyi, 58
Australia, 26, 166, 179, 286, 288, 301, 307, 310, 348, 352, 356
Australian Institute for Marine Science, 301
automation, 4–5, 6, 21, 39
automobiles, 10, 39

baby boom generation, 112, 175
Bacillus thuringiensis, 264–65, 267, 270
Bacon, Francis, xx–xxi, 83
bacteria, xxi, 235, 291
    antibiotic-resistant, 218, 225–28
    harmful vs. beneficial, 275
Bahrain, 62
Baidu, 45
Balk, Deborah, 298
Balkans, 304
Ban Ki-moon, 56
bandwith limitation, 63
Bangkok, 299
Bangladesh, 177, 285, 298–99
banks, 26, 125
Bänziger, Marianne, 287
Barnard, Christiaan, 248
Bass, Carl, 32
Becker, Dan, 327
Beddington, John, 299
Bedouins, 182
bee pollination, 268
beef, 228–29
beetles, 290, 337
Beijing, China, 97, 348
Beijing Genomics Institute, 209–10, 214, 235
Belarus, 59
Belgium, 129
Bell, Alexander Graham, 74*n*, 233
Bell, Daniel, xvi
Bellotti, Tony, 290
Bernays, Edward, 157–58, 159–60, 233
Berners-Lee, Tim, 81
"beyond GDP" initiative, 184
big data, 55–56, 206
Bihar, India, 180
Bikini Atoll, 221
Bill of Rights, 99
bin Laden, Osama, 132, 180
biobricks, 214
biochar, 191
biodiversity, 372
bioethics, 213
biofuels, 196–97, 266, 270, 338, 353
biogas, 353
bioinformatics, 216
biological weapons, 137, 221–22
biomass, 353
biopolymers, 266

biotechnology, xiv, 29, 208–11, 235, 316
    and digitalization of life, 211–18
bird flu, 218, 222, 274
Birdwell, Walter Glenn, Jr., 321
birth control, 233
birth rates, 167, 173, 372
Bisphenol A (BPA), 163
Black Death, 18, 169, 285
blackouts, 351, 352
Blaser, Martin, 275, 276
blindness, 244, 253
bliss point, 154
Blood, David, 35
blood transfusions, 247
Body Politic, The (Moreno), 230
Bolívar, Simón, 99
bombers, 135
bonds, bond market, 13, 15, 123, 124, 125
Bonneville Power Administration, 185
books, 50, 133
bookstores, xvii, 44
border adjustments, 347
boreal forests, 337
Borlaug, Norman, 211, 262
Borneo, 335
Bosnia, 130–31, 135
Bostrom, Nick, 239
bottom trawling, 302
Bouazizi, Mohamed, 60
bovine spongiform encephalopathy
    (BSE), 228–29
BP, 148, 295
brain, 38, 46n, 47, 50, 236–39, 275
    artificial, 243–44
brain disease, 207
Brand, Stewart, 71
Branson, Richard, 354
Brazil, 89, 95, 194, 224, 270
    cap and trade in, 348
    GM crops in, 263
Brennan, John O., 77
Bretton Woods Agreement, 97
bridges, 117, 303
Brin, Sergey, 59
Bronze Age, 28–29
Brookings Institution, 135, 178
Brown, James H., 144
Brown, Lester, 150, 194
Brown, Louise, 248
Browner, Carol, 327

Buck, Carrie, 232–33
Buck, Doris, 233
Buck, Emma, 232
Buck, Vivian, 233
Buck v. Bell, 232–33
buckminsterfullerene ("buckeyballs"), 30
Buddhist fundamentalism, 129–30
Burke, Edmund, 326
Bush, George H. W., 345
Bush, George W., 87, 114, 200

C3 plants, 266
C4 plants, 266
Cairo, 57, 61, 133
Calcutta, India, 299
Caldeira, Ken, 330
California, 262, 301, 343, 348
calories, 258–59
    extreme restrictions on, 254
cameras, 83, 87
Campbell, Kurt, 182
Canada, 76, 147, 224, 299, 334, 337, 348
Canary Islands, 182
cancer, 154n, 208, 213, 255, 256–57
Cancer Genome Atlas, 256
canola, 264
cap and trade, 344–48
capital, 27, 33
capital flows, xiv, 10, 13, 35, 125
capital gains tax, 9, 12
capitalism, xxiv, xxviii, 52, 98, 119, 121
    communism vs., 116–17, 158
    as empowering, xxvi, 33
    freedom granted by, 105
    problems of, 33–34, 365, 368–69, 371
carbon, xvii, 189
carbon-based energy companies, 119
carbon capture and sequestration (CCS),
    353–54, 355
carbon credits, 191
carbon dioxide, 11, 147, 225, 232, 280–
    81, 283, 286, 288, 296, 301, 350
    Chinese reduction in, 350–51
    discovery of dangers of, 311–13
    EPA's requirement of reductions in, 327
    insect populations increased by, 290
    plans for lowering, 341–53
    recent reduction in, 333
    tax on, 283, 341–42, 344, 371
    in vegetation, 335, 336

carbon fiber, 29
carbon nanotubes, 29
carcinogens, 257
cardiovascular disease, 153, 154, 256
Carnegie, Andrew, 133
Carter, Jimmy, 116
Cartlidge, John, 13
cassava, 269, 290
Casteel, Clare, 290
Catalonia, 131
cattle, 270, 292, 336, 353
Caucasus, 337
Cayman Islands, 272
CC398, 227
cells, xxi, 207
cement, 146
Central America, 49, 285, 304
Central Asia, 18, 192, 304, 337, 353
Central Europe, 96, 98
cerebellum, artificial, 244
Cerf, Vint, 48n, 74
Cervantes, Miguel de, 52
Chad, 182
Chamber of Commerce, U.S., 76, 112, 318
Chan, Margaret, 226
Chauvet caves, 362
Cheatham, Anne, xvin
Chechnya, 131, 177
checks and balances, 363-64
chemical weapons, 137
chemicals, 161-62
Cheney, Dick, 87, 330
chicken coops, 21-22
chickens, 270, 353
child care, 124
child labor, 139, 358
child mortality, 173, 368
child poverty rate, 118
Chile, 348
China, 18-19, 23, 29, 49, 65, 73, 86, 98, 102-3, 127, 128, 169, 171, 190, 196, 285, 298, 304, 340
   aging population in, 174, 373
   biotechnology in, 209-10, 235-37, 353
   boreal forests of, 337
   building bubble in, 102
   carbon dioxide reduced in, 350-51
   carbon sequestering in, 354
   coal use in, 349-50

   cyberattacks and, 76, 83, 134
   droughts in, 286
   dust and sand storms in, 192-94
   economic growth in, 102, 104, 145-46
   food growth in, 150
   forced abortions in, 234
   fossil fuel subsidies in, 350
   Gini coefficient of, 9
   global warming threat to, 298
   groundwater aquifers in, 145, 165, 185, 186
   income inequality in, 9, 103
   Internet use controlled in, 59-60, 88-89
   irrigation in, 185
   as largest economy, 11
   as leading manufacturing nation, 93
   plastic waste imported by, 161
   population of, 166, 173
   renewable energy commitment of, 343, 348-49
   resources used by, 146
   shale gas in, 332-33
   share of global GDP of, 97
   topsoil degradation in, 183
   tree planting in, 337-38
   U.S. global leadership challenged by, 11, 92, 93
   windmills and solar panels exported by, 10, 327-28, 349, 351
chlorofluorocarbons (CFCs), 309-10
cholera, 163
Chongqing, China, 348
Chow, Carson, 155
Christian fundamentalism, 129-30
chronic respiratory disease, 256
Church, George, 212
cigarettes, 112, 158, 162
Citigroup, 82-83
Citizens United case, 64, 108
civil liberties, 84
civil rights movement, 112, 318
Civil War, U.S., 13, 106, 110, 120, 365
Civil War in France, The (Marx), 108
civil wars, 136
Clarke, Arthur C., 38
Clarke, Richard, 75, 77
clathrates, 296, 297
Clean Water Act, 330
climate change, *see* global warming

climate refugees, 181–82
cloning, 222–25
Clostridium difficile, 276
clouds, 324
coal, 24, 145, 146, 147, 230, 282, 318, 321,
    322–23, 325, 328, 331, 332, 333, 334,
    338, 348, 349–50
  clean technologies for, 343
Coalition for the Urban Poor, 179–80
coastal zones, 298, 303–4
cochlear implants, 208
coconut oil, 145
cod, 199, 302
Cold War, 98, 135, 357
Coleridge, Samuel Taylor, 361
Collapse (Diamond), 150
collateralized debt obligations (CDOs), 17
collective decision making, xxv–xxvi
Collins, Francis, 217–18, 235–36
Colombia, 180, 348
colonialism, 196
colony collapse disorder (CCD), 268
Colorado River, 166, 195
Columbus, Christopher, 18, 50
"Coming Technological Singularity,
    The" (Vinge), 240
commodities, 24, 143, 145–46
Common Sense (Paine), 52, 64
communication, 18
communism, 33, 98–99, 112, 130
  capitalism vs., 116–17, 158
  collapse of, 18, 99
Communist Party, China, 60
Communist Party, USSR, 159
compact discs, 132
compensation, xxviii
complexity, of markets, 15–17
complexity theory, xviii
computational science, 28
computers, xxix, 25, 44, 47, 64, 78, 316
conditional hyper-eating, 171
confederations, 127
Congo, 103n, 180, 196
Congress, U.S., xxv, 63, 76, 104, 106, 112,
    116, 161, 225, 229
Congressional Budget Office, 123
connectome, 207, 236, 237
consciousness, 46n, 47
conservation, 124
conservatives, 121

Constitution, U.S., 99, 111, 113,
    363–64
Consumer Product Safety Commission,
    162–63
consumers, consumption, xiv, xix, 26, 27,
    112, 143, 158, 161, 318
containership revolution, 23–24
continentality, 306
contingency contracts, 107
contraception, 168
contracts, 123, 124
Convention on Biological Diversity,
    200
Convention to Combat Desertification,
    192, 194
Cooke, Jay, 109
Coolidge, Calvin, 158–59
Copernicus, Nicolaus, xx, 51
copper, 24–25, 28, 145, 146
copyright law, 71
coral reefs, 199–200, 237, 281, 300–301,
    302, 339
Corell, Bob, 197
Coriolis effect, 305
corn, 18–19, 145, 188–89, 190, 264, 266,
    268, 269, 270
  decreasing yields of, 288–89
corporations, 95, 104–24
  biotechnology and, 210
  multinational, 46, 113–14, 124, 198,
    210
  as "persons," 107, 113
  political influence of, xxv, 125
  profits of, 106
Cortés, Hernán, 18
cortex, 238
Costa Rica, 57, 348
cotton, 110, 264, 269, 270
Counter-Reformation, 128
coupling, 16
Coursera, 67
Cowdrick, Edward, 158
Cox, Samuel, 73
Crabtree, Robert L., 337
Cray-2, 38
creationism, 217
Creative Destruction of Medicine, The
    (Topol), 206
creativity, xxv
credit, 39

credit crisis, 17, 118
credit derivatives, 15
Crete, 127
Crick, Francis, 212, 223
Critcher, Clayton, 315
Crizotinib, 211
Croatia, 130–31, 135
Crohn's disease, 275
Cronin, Lee, 243
cryptography, 74
Cuban Missile Crisis, 366
culture, globalization of, 132, 143
cyanobacteria, 213
Cyber Command, U.S., 73
cyber-Faustian bargain, 69–70, 79, 88
Cyber Intelligence Sharing and
     Protection Act (CISPA), 88
cybernetics, 216, 245
cybersecurity, 71–78, 82, 83, 134, 367
cyborgs, 239
cyclones, 299, 305

da Gama, Vasco, 18
Dallas, Tex., 291–92
Daly, Herman, 163, 184
dams, 185–86
Darfur, 182
dark nets, 89
Darnovsky, Marcy, 209
Darwin, Charles, xxii, 230, 239
data, 55–56, 81
data mining, 68
data protection law, 82
deafness, 244
death rates, 167, 372
debt, growth of, 34, 122
Decline and Fall of the Roman Empire
     (Gibbon), 52
Deep Packet Inspection (DPI), 81
deepwater drilling, 148
Deepwater Horizon, 148, 281
Defense Advanced Research Projects
     Agency, U.S., 246
Defense Department, U.S., 73, 76, 239
deficits, reducing, 122
deforestation, 153, 194, 336–37
Dekker, Job, 214
Delhi, India, 152
DeLucia, Evan, 290
Demaratus, 73–74

democracy, xxiv, 33, 49, 98, 99, 119,
     139
  change needed in, 365, 368–69,
     370–71
  threats to, 57–67
democratic capitalism, 126
Democratic Party, U.S., 117, 319
Deng Xiaoping, 18, 104, 350
dengue fever, 273, 291
Denmark, 76
derivatives, 15
Descartes, René, xxi, 46, 51
desertification, 149, 287, 306
deserts, 307
designer babies, 208
detention, infinite, 132
developing economies, 5, 11–12
Dhaka, Bangladesh, 152, 179–80, 298,
     299
diabetes, 153, 154, 208, 256
Diabetes Prevention Program, 207
*Diagnostic and Statistical Manual of Mental
     Disorders* (DSM), 47
dialects, 127–28
Diamandis, Peter, 240
Diamond, Jared, 150
Dichter, Ernest, 159
Digital Revolution, 52–54, 204, 212, 215,
     222, 316
digitization:
  of life, 211–18, 243, 258
  of work, 4–5
disability payments, 23
disability rights, 318
Discover Financial, 82–83
diseases, conquering of, xxx
distributed cooperatives, 312
divorce, 172
DNA, xix, 190, 213–14, 249, 250, 272
  cheap analysis of, 215
  discovery of, 207, 212
  sequencing of, xxix
  synthetic, 216–17
  see also genes, genome
Dobson, Andrew, 291
Doha Round, 94
dollar, U.S., 97, 123
Dolly (sheep), 222, 223
dot-com bubble, 14
double clock, xix

Down syndrome, 251
Drake, Edwin, 311
Drake, Thomas, 84
drones, 86–87, 101–2
droughts, 188, 266, 286, 288, 300, 302, 303, 335, 337, 346
dualism, xx–xxi
Duchesne, Ernest, 225
Dust Bowl, 192, 193, 292
dust storms, 192–95
dynamite fishing, 302

e-commerce, 370
e-waste, 161–62
Eagleman, David, 236
Earth, age of, xxii, xxx
East Germany, 126
Eastern Europe, 18, 96, 98, 173
Ebola virus, 274
Ecole Polytechnique Fédérale de Lausanne (EPFL), 238–39
ecological systems, xv, xix, xxi, 143–44
economic efficiency, 40
Economist, 171
ecosystem services, 284
education, 35, 66, 67–68, 123, 124, 125, 129, 173, 174, 179, 363
    of girls, 168, 170, 372
    privatization of, 116
eggs, 21–22, 23
Egypt, 9, 18, 49, 104–5, 127, 182
    global warming threat to, 298
    protests in, 57, 61, 62
Einstein, Albert, 48
Eisenhower, Dwight D., 136
El Niño, 305
Elbe River, 166
elections, U.S.:
    of 1876, 106
    of 2012, 370
electric cars, 352
electricity, 156, 282
    renewable, 282
electricity grids, 351–52
electroencephalography (EEG), 238
electromagnetic spectrum, xxvii, xxx
electronic communications grid, xiv
email, 26, 44, 70, 87
embryonic stem cell research, 252–53
embryos, 251

emergence, xviii, xx, xxix, 5, 217
emergency services, 117
emergent wisdom, xxiv
emerging economies, 5, 11–12, 145–46
emission permits, 345
empires, 127, 131
employment, robosourcing and, 7
"end of history," 98
energy consumption, 11
Energy Department, U.S., 331, 343
energy efficiency, 124
Energy Information Agency, U.S., 343, 350
energy sprawl, 164
engineering, 20, 209
English Civil War, 130, 175
enhanced ploidy, 265
Enigma machine, 74
Enlightenment, xx, 52, 315
entropy, xvi–xviii
environment, 103, 138
environmental protection, 66, 95, 123, 124, 184
Environmental Protection Agency, 163, 283, 327, 330
Enviropigs, 272
epidemiological strategies, 68
epigenetics, 212–13, 216
epilepsy, 208
Epstein, Jonathan, 274
Epstein, Paul, 292
Erdan, Gilad, 181
erythropoietin (EPO), 246–47
Eskimos, 285
estate tax, 9, 111, 122
Estonia, 63
Ethiopia, 183, 195
eugenics, 228–35, 251
Eugenics Record Office, 229
Euphrates River, xvii, 127, 195
Europe, 13, 18, 19, 50, 52, 82, 94, 304
    heat waves in, 286
    Little Ice Age famine in, 285
    topsoil degradation in, 183
    World War II devastation of, 97
European Commission, 210, 222, 266
European Union, 82, 97, 125, 135, 161, 182, 184, 349
Eurozone, 125–26

evolution, xxii, 218, 229, 230
  of flu strains, 218–19
  future course of, 239
  human control of, xiv, xix, 209, 219–20, 373
  rejection of, 217, 231
  short-term thinking and, xxviii
exercise, 154
exports, 11
extensions, 39
externalities, 34
  positive, 34
extinction, 339–41
ExxonMobil, 113–14, 303, 331

Facebook, 45, 47, 56, 57, 59, 172, 242
  privacy issues on, 79–80, 81–82
FaceTime, 54
facial recognition software, 81–82
Faisal, Mohammad al-, 199
false imprisonment, 358
families, 172–74
fascism, 33
fast food, 155, 270
Faust (Goethe), xxiv
FBI, 73, 75, 76, 222
fecal transplants, 276–77
Federalist Papers, 120
feed-in tariffs, 350
Fenton, David, 303
Ferrel cells, 308
Fertile Crescent, 127, 307
fertility, 168, 169, 248–53
fetal DNA, 250
Fidelity Investments, 82–83
Field, Cyrus, 108
Field, David, 108
Field, Stephen, 107, 108, 112
financial services sector, 14, 119
Fineberg, Harvey, 208–9
Finland, 172, 337
fire services, 117
fires, 162
First Amendment, 113
First Opium War, 19
fiscal policy, 124
fish, fisheries, 199–201, 281, 295, 301, 334, 372
FLAVR SAVR, 262–63
flaxseed, 145

Fleming, Alexander, 225
flooding, 286, 287
Florida, 106
flounder, 199
flowback, 329
flu, 68, 218–19, 274
food, 143
  consumption of, 149, 188
  production of, 148–50, 182, 185, 287–88
  shortages of, 149–50, 194
Food and Drug Administration, U.S. (FDA), 154, 210, 225
food prices, 56, 61
  Green Revolution and decline in, 155
  worldwide spike in, 145, 176, 190, 286, 288
food riots, 145
Forbes, 331
Ford, Henry, 31
foreign aid, 168
foreign direct investment, 10
foreign exchange contracts, 15
forest fires, 337
forests, forestry, 281–82, 335–36, 373
fossil fuels, 280–81
  see also specific fuels
Fourteenth Amendment, 107
fracking, 147, 165, 328, 329–31, 333
France, 51, 100, 126, 134, 173, 285
  nuclear power in, 355
  in Security Council, 95
Franco-Prussian War, 108
Frankel, Stuart, 8
Frankenfood, 272
Frankfurter, Felix, 83
Franklin, Rosalind, 212
free competition, 89
free speech, 89, 102
freedom, 138
Freedom of Information Act, 84, 87, 164
Fresh Drinking Water Act, 330
Freud, Sigmund, 156–57, 159
frogs, 339
Frost, Robert, 311
Frozen Planet, The (documentary), 326
fruits, 155
Fukushima, Japan, 355
Fukuyama, Francis, 137–38
Fuller, Buckminster, xvi, 4*n*

functional magnetic resonance imaging (fMRI), 237
fundamentalism, 129–30, 132, 133
    evolution opposed by, 217
fungi, 266, 291
Fust, Johann, 69
"Future Agenda, The," xvi
"Future of Neo-Eugenics, The" (Leroi), 251

G8, 94
G20, 94
Galileo Galilei, xx, 51
Galton, Francis, 230
Galveston, Tex., 299
Ganges Delta, 195
garbage, 161
gas, 165, 189, 321, 325, 329–33
    see also fracking; shale gas
Gates, Bill, 133, 256
Gates, Robert, 73
gay rights, 112, 318
gender roles, 132
General Agreement on Tariffs and Trade (GATT), 18, 94
genes, genome, 68, 215–18, 251, 256–57, 362–63
    splicing of, 262
    transfer of, 213, 270
    see also DNA
genetic analysis, 258
genetic engineering, xxiii, 214, 220, 221, 223, 274
    of animals, 201, 270
    diseases cured by, xix, 247, 254
    of food, 201, 261–69
Genetic Information Nondiscrimination Act, 259
genetics, 216
Geneva Conventions, 139
geoengineering, 355–57
Geological Survey, U.S., 56, 193
Georgia, 107, 187
geothermal energy, 353
Germany, 76, 126, 166
    renewable energy in, 282, 343
    subsidies in, 350
    World War II devastation of, 97
    see also Nazi Germany
Ghazali, Muhammad al-, xx

ghrelin, 276
Gibbon, Edward, 52
Gibson, William, 63
Gillis, Justin, 288
Gini coefficient, 9
girls, education of, 168, 170, 372
glaciers, 284
Gladwell, Malcolm, 20, 61
Gleick, Peter, 187
global change, drivers of, xiii–xv, xxv, 5
global leadership:
    environmental problems and, 280
    in shift to East, xiv
    see also United States, global leadership of
global market crisis (2008), 12, 16, 95, 126, 322
"Global Mind," 46, 53, 55, 88, 89, 101, 136–37, 138, 144, 212, 225
Global Payments, 82–83
global warming, xvii, xxiii–xxiv, xxix, 34, 124, 146, 161, 181, 231–32, 280–359
    and acidification of oceans, 281, 300, 301–2
    computer models of, 324
    crop diseases and pests increased by, 289–90
    deforestation and, 336–37
    deniers of, 283, 314–15, 319–28
    desertification driven by, 287, 306
    diseases increased by, 273, 290–92
    extinctions caused by, 339–41
    false solutions to, 353–59
    flooding and droughts caused by, 286
    food prices affected by, 286, 288
    Hadley cells and, 305–11
    mitigation of vs. adaption to, 303–5
    national security and, 284–85
    new drilling opportunities provided by, 295–96
    recent records in, 286, 293–95
    rising sea levels and, 182, 297–98, 300
    solutions for, 341–53
    as threat to ozone layer, 309
    tribal conflicts exacerbated by, 298–99
    U.S. monitoring of, 324–25
global water cycle, 286–87
globalization, xiv, 5, 18–21, 196
    of culture, 132, 143
glucose monitors, 258

Gnacadja, Luc, 192, 194
Gobi Desert, 193
Goebbels, Joseph, 159, 233
Goethe, Johann Wolfgang von, xxiv
gold, 18
Goldman Sachs, 78
Google, 25, 45, 52, 59, 79, 83, 87
Google Earth, 57
Google Street View, 80
Google Translate, 129
governance, dysfunctionality of, xv
government contracts, 106
GPS devices, 48, 85, 86
GPS stalking, 82
GRAIN, 196
Grand Canyon, 110
Grantham, Jeremy, 145
graphene, 29
grasslands, 149
Great Artesian Basin, 186
Great Barrier Reef, 301
Great Depression, 11, 34, 41, 111, 112,
    145, 167, 260–61, 365
Great Illusion, The (Angell), 133
Great Recession, 11, 12, 34, 118, 322, 327
Greece, 29, 35, 49, 51, 125, 126, 307
Greece, ancient, xx, 73, 260
green plastics, 266
Green Revolution, xxx, 58–59, 62, 149,
    155, 176, 185, 211
greenhouse gases, xxix, 283, 332
    taxes on, 341
    see also carbon dioxide; methane
Greenland, 186, 198, 285, 286, 297, 298,
    300, 305
gross domestic product (GDP):
    as correlated with income of elites,
        142, 143
    problems with, xxvii, 142–43, 171,
        184–85, 372
groundwater, 372
groundwater resources, 145, 165, 183–88,
    281, 300, 330, 331
    global warming and, 288, 332
Guangdong, China, 348
Guangzhou, China, 163–64, 299
Gujarat, India, 150
Gulf of Mexico, 148, 189, 295, 302
Gulf Stream, 305
guns, 32–33, 130, 136

Gutenberg, Johannes, 50, 51, 69
Guterres, António, 180

H1N1, 218
H5N1, 218, 225
hackers, 82, 101
hacktivists, 72, 73
Hadley cells, 305–11
Hai Phong, Vietnam, 299
Haldane, J. B. S., 219–20
Hammond, Kristian, 8
Hansen, James, 293–95, 297–98, 313, 354
Harding, Warren, 111
Hare, Bill, 347
Harl, Neil, 211
Harriman, Tenn., 334
hate groups, 177–78
Hawaii, 127, 200
Hawthorne, Nathaniel, 45, 46
Hawthorne effect, 206–7
Hayes, Richard, 209
Hayes, Rutherford B., 106
health care, 35, 66, 67–68, 118, 123, 124,
    170, 174, 205–8
Health Insurance Portability and
    Accountability Act, 259
healthspans, 254–55
heat stress, 149, 288
heat waves, 286
Helicobacter pylori, 275–76
Henry, Shawn, 75
herbicides, 149, 267, 269
Herzog, Howard, 354
highways, 117, 128
Hindu fundamentalism, 129–30
hippocampus, 244
Hitler, Adolf, 97, 159, 231, 233, 365
HIV, 169, 255, 274
Ho Chi Minh, 100
Ho Chi Minh City, Vietnam, 299
Hobbes, Thomas, 130
Holmes, Oliver Wendell, Jr., 232
Holocaust, xxii
Homeland Security Department, U.S., 73
Homo economicus, 315
Hong Kong, 59, 76, 209, 299
Hoover, Herbert, 159
Hoover Dam, 186, 187
Hopkinson, Neil, 32
horizontal drilling, 147, 333

hospitals, 117
House of Representatives, U.S., xvi, 323, 328
  Science Investigations Subcommittee of, 223
housing bubble, 14, 17
Hubbert, M. King, 146–47
Hubei, China, 348
Huffnagle, Gary, 274
Hull, Cordell, 97
human capital, 124
human cloning, 222–25
Human Connectome Project, 207n
Human Epigenome Project, 213
Human Genome Project, 217–18, 223, 235
  see also genes, genome
Human Microbiome Project, 274
human nature, 362, 364
Human Proteome Project, 212
human rights, 115, 138
hurricanes, 370
Huxley, Aldous, xxiii, 239
Huxley, Julian, 239
hybridization, 211
hydroelectric energy, 353
hydroponic facilities, 199

IBM, 206, 238
Ice Age, xvii, 37, 284, 299
Iceland, 172, 353
identity theft, 82
Illinois, University of, 244
imagined communities, 129
immigration, 231
immune system, 274–77
in vitro fertilization, 252
incentive structures, xxviii
income, xxx, 22–23, 34
income inequality, 8–10, 34, 35
India, 18, 19, 49, 76, 89, 127, 177, 182
  abortion in, 249–50
  agriculture in, 150, 265
  cap and trade in, 348
  coal imported by, 349
  dengue fever in, 273
  electricity grids in, 352
  forced sterilizations in, 235–36
  GDP of, 95
  genetic testing of embryos banned in, 249–50
  global warming threat to, 298
  income inequality in, 9
  kerosene subsidized in, 342
  population growth in, 166
  tree planting in, 338
  water resources in, 145, 165, 185, 288
Indian Ocean, 298
Indonesia, 76, 285, 298, 335
Indus River, xvii, 127, 166
Industrial Revolution, xxi, 19, 29, 37–38, 92, 106, 107, 128, 152, 186, 281, 338
industrialization, 231
industry:
  accidents in, 107
  automation of, 39
  outsourcing of, 5, 12
inequality, 318–19, 372
infant mortality, 118, 168, 169
influenza, 255
information, xvii, 53, 105
information gap, 54
Information Revolution, 98
information technology, 18, 25, 55
infrastructure, 124
Inglis, Chris, 87
inheritance tax, 9, 111, 122
Inner Mongolia, 185, 194
insulin-like growth factor (IGF), 271–72
insurance companies, 68–69, 206, 260–61
intelligence gathering, 134
intelligent design, 217
intercontinental ballistic missiles, 135
interest rates, 15
Intergovernmental Panel on Climate Change (IPCC), 313, 352
International Monetary Fund, 73, 94, 95
Internet, xvii, xxx, 11, 41, 44–47, 53, 68–71, 128, 143, 316, 366–67, 369
  anonymity of, 72
  business models changed by, 44–45
  democracy aided by, 57–62
  education on, 67–68
  as "human right," 54
  journalism on, 66
  languages of, 129
  memory and, 48
  privacy on, 70–72, 78–89
  reading on, 57
  security of, 71–78
  voting and, 63

Internet Use Disorder, 47
Interpol, 73
inventions, 20–21
investment, 7, 125
invisible hand, 52, 105
iPhones, 54, 238
Iran, 170, 180
  cyberattacks and, 77, 83
  Green Revolution in, xxx, 58–59, 62
  Internet use controlled in, 88–89
  organ selling allowed in, 242
  surveillance in, 86
  weapons program of, 136, 335
Iraq, 31, 194
  oilfields of, 350
  refugees from, 180
Iraq War, 101, 132
  drones in, 86, 101
  private contractors in, 102
Ireland, 35
  famine in, 269–70
Irene, Tropical Storm, 346
iron, 71, 145, 146
Iron Age, 28, 29
irreversible phenomena, xvi–xviii
Islam, xx, 130–31, 177, 181, 182
  fundamentalist, 129–30, 132
Israel, 178–79, 181, 182
Istanbul, Turkey, 96
Italy, 5, 35, 126, 159, 173, 307

Jackson, Lisa, 327
Jakarta, Indonesia, 96
James, William, 156
Japan, 19, 76, 96, 112, 126, 128, 137, 182, 337
  cap and trade in, 348
  coal imported by, 349
  fertility rates in, 173
  global warming threat to, 298
  postwar reconstruction of, 97–98
Jay, John, 114
Jefferson, Thomas, xxi, 106, 114, 144
jet streams, 284, 307–8
Jewish fundamentalism, 129–30
jihadists, 177
job training, 123, 124
Johannesburg, South Africa, 96
Johnson, Lyndon, 112
Johnson, Michael, 246

Joplin, Mo., 346
Jordan, 182, 348
Josephson, Matthew, 107
journalism, journalists, xxv, 8, 65–66
Juazeiro, Brazil, 272
Justice Department, U.S., 84–85
Justice Ministry, British, 73
justification theory, 316

Kahn, Robert, 48n, 74
Kaiima, 265
Kansas, 187
Kapital, Das (Marx), 108
Kapor, Mitch, 240
Karachi, Pakistan, 152
Karolinska Institute, 242
Kashmar, Iran, 102
Kass, Leon, 220
Kasyanov, Mikhail, 65
Kazakhstan, 192
Keeling, Charles David, 312–13, 335
Keeling Curve, 312
Kelly, Kevin, 47
Kennedy, John F., 112
Kenya, 180, 269
kerosene, 342
Kessler, David, 154
Keynesian economics, 26
Khan, A. Q., 137
Kim, Jim Yong, 294
Kohl, Helmut, 125
Korea, 128, 196
Korean War, 100
Kosovo, 131
Kübler-Ross, Elisabeth, 315
kudzu, 220–21
Kumtag Desert, 194
Kurdistan, 131
Kurzweil, Ray, 240
Kuznets, Simon, 143, 171, 184–85
Kyoto Protocol, 94, 347–48

La Niña, 305, 326
labor, xiv, 27, 33, 41
  value of technology vs., 8
Labor Department, U.S., 110
labor rights, 6
Lagos, Nigeria, 96, 152
LaHood, Ray, 327
Lamarck, Jean-Baptiste, 230

laptops, 26
Latin America, 9, 100, 169, 190
Latvia, 63
Laughlin, Harry, 229–30, 231, 233
Law of the Sea Treaty, 94
lead, 146, 334
Leeuwenhoek, Antonie van, xxi
legal system, 123, 215–16
leguminous trees, 191
Lenin, Vladimir, 97, 108, 158
Leningrad, siege of, 340
Leroi, Armand, 251
liberals, 120, 121
Liberia, 196
libraries, 50
Library of Congress, 76
Libya, 57, 62
Liebman, Matt, 189
life:
    artificial, 218–22
    digitization of, 211–18, 243, 258
    emergence of, xxx
life insurance, 69
Life on Earth (Wilson), 67
Life Science Revolution, 204, 215, 225,
    248, 257, 261, 316
life sciences, 208
lifespans, xxx, 174–76, 254–55
Limits to Growth, The (Meadows), 144, 146
Lincoln, Abraham, 22, 108, 110, 326, 357
Lippmann, Walter, 156, 157, 159, 160
liquefied natural gas (LNG), 329
liquidity, 15
literacy, xxx, 50, 57, 174
Little Ice Age, 285
liver toxicity, 268
livestock, 193, 256, 274
    antibiotics for, 226–27
    genetic engineering of, 270–73
living space, 152*n*
living standards, 138
loans, low documentation, 321
lobbying, lobbyists, xxv, 14, 105, 107,
    114, 115–16, 119, 162, 225, 228
Lobell, David, 288
Logan, George, 106
London Exchange, 13
long-term thinking:
    markets and, xxviii
    politics and, xxviii–xxix

Lotorobo, Makambo, 196
Lovins, Amory, 352
low-wage countries, outsourcing to,
    6–7, 8
Lubchenco, Jane, 300
Ludd, Ned, 39
Luddite fallacy, 39
Lula da Silva, Luiz Inácio, 335
lung cancer, 112, 211
Luo Yiqi, 185
Luther, Martin, 51
Lyell, Charles, xxii
Lysenko, Trofim, 230, 340

macrolide antibiotics, 276
mad cow disease, 229
Madagascar, 197
Madison, James, 120
Maginot Line, 134
maize, *see* corn
malaria, 255–56
Malawi, 169
Malaysia, 272, 335
Mali, 193
Malthus, Thomas, 144, 150
manioc, 269, 290
Mäntyranta, Eero, 247
manufactured financial products, 13
manufacturing, 10, 11, 116, 161, 186, 282
    dematerialization of, 31
    loss of jobs in, 41
    see also 3D printing
Mao Zedong, xxii, 98, 104
marine biodiversity, 302
market fundamentalism, 117, 120
markets, xiv, 18
    power shifted to, 105, 125
    short-term thinking in, xxviii
    standard model of, 15
marlin, 199, 302
Marshall, Barry, 276
Marshall Plan, 97, 124
Marx, Karl, 108
mass marketing, 156–60, 263
mass migrations, 285
mass production, 31–32
materials science, xiv, 26, 27–28, 29–30
mathematics, mathematicians, xix, 209
Mathews, Debra, 248
Mauna Loa, 312

Mazur, Paul, 157
McConnell, Mike, 75–76
McKibben, Bill, 326
McKinley, William, 109–10
McKinsey, 75
McLean, Malcom, 23–24
McLuhan, Marshall, 39, 46, 48, 50
Mead, Lake, 187
Mead, Margaret, xvi
Meadows, Donella, 144
measles, 256
meat, 153, 191, 336
media, 325–26, 379
    see also newspapers
Medicare, 120, 318
medicine, 170
Mediterranean, 49, 181
Medvedev, Dmitri, 60, 65
Meehl, Gerald, 294–95
mega-deltas, 298
Mekong Delta, 195
memory, 47–48
mental health care, 124
mercury, 334
Mesopotamia, 49
metabolites, 216
meta-journalists, 8
Metcalfe's law, 80
methane, 161, 296, 297, 329–30, 331
methicillin, 227
methyl-mercury, 334
Mexico, 127, 178, 187, 224, 273, 304, 307
    cap and trade in, 348
    electricity grids in, 351–52
Mexico City, Mexico, 96
Miami, Fla., 299
microbial kudzu, 221
microbiome, 216, 274–77
microdrones, 86–87
microscope, xxi
middle class, 133, 138, 153, 179
Middle East, 94, 131, 133, 138, 163, 180,
    194, 335, 350, 351
migration, 176–80
military budgets, 100–101, 102
military industrial complex, 136
milk, 271–72
Milojević, Ivana, xviii
Minerva, 67
mining, 24–26, 282

minorities, 359
misplaced concreteness fallacy, xxvii
Misrata, Libya, 57
missiles, 130
Mississippi River, 189, 300, 302
MIT, 242, 243
Mitchell, George P., 331
mobile devices, 11, 53, 344
molecular manipulation, 27, 29
Molina, Mario, 309
monetary policy, 124
money, 39, 71
Mongol Empire, 18
Monitor Group, 210
monoculture planting, 269, 338
Monsanto, 211, 264, 267
Montreal Protocol, 309
Moore's Law, xxviii, 38, 53, 215, 236, 344
Moreno, Jonathan, 230
Morgan, J. P., 110
Morgan Stanley, 102–3
Morocco, 190, 348
Morrill Land Grant College Act, 22
Morse, Samuel, 45, 74*n*
mortgages, subprime, 16–17, 321–22
Moscow, 57
mosquitoes, 272–73, 291
Mougin, Georges, 199
Moussavi, Mir-Hossein, 58
movies, 56, 133, 156
Mozambique, 196
MRIs, 237
Mubarak, Hosni, 61
Muir Woods, 110
multicellular life, xxx
multinational corporations, 46, 113–14,
    124, 198, 210
multiplayer games, 44
multiple sclerosis, 208, 275
Mumbai, India, 96, 299
Muniruzzaman, A. N. M., 285
Murdoch, Rupert, 119
Murray-Darling River, 166
Murrow, Edward R., 211
music, 44, 133
Muslim fundamentalism, 129–30, 132
Mussolini, Benito, 159
Myanmar, 57, 61, 180
Mycoplasma genitalium, 216
mycorrhizal fungi, 191

Myhrvold, Nathan, 330
myostatin, 247
"mystery algorithm," 13

Naisbitt, John, xvi
nanocomposites, ceramic matrix, 29
nanoparticles, 30
nanotechnology, 216, 243, 244, 258, 373
Nanotechnology Revolution, 29–30
Napoleon I, Emperor of the French, 13
narco-states, 131–32
Narrative Science, 8
NASA, 73
nation-states, 127, 129
    colonial empires of, 131
    decline in power of, 124–33
    overlapping ideas of, 127
nationalism, xiv, 130
National Organ Transplant Act, 242
National Security Agency, 73, 84, 87
natural capital, 184
natural selection, 217
Nazi Germany, 74, 112, 134, 159, 233
Negev, 182
Negroponte, Nicholas, 54
neo-eugenics, 231, 251
neo-evolution, 209
neo-Nazi groups, 177
neocortex, 49
neonicotinoids, 268
Nepal, 182
Netherlands, 106, 241
network effects, 80
network neutrality laws, 89
neurological damage, 334
neurons, 236–37, 363
neuroprosthetics, 245–46
neuroscience, 216, 236
Neuroscience Revolution, 236
neuroscientists, 150
neurotoxins, 268
New Deal, 22, 111, 155, 160, 192
New York, N.Y., 14, 152, 299, 346, 370
New York State, 106
New York Stock Exchange, 14, 16
New York Times, 16, 31, 60, 86, 160, 164,
    242, 245, 288, 337
New Zealand, 348, 353
Newark, N.J., 299
newspapers, xvii, 44, 52, 65, 325

Newton, Isaac, xx
Nicaragua, 57
nickel, 145
Nigeria, 169, 193
Nile River, xvii, 127, 166, 182, 195
Ninety-Five Theses, 51
Niskanen, William, 114
nitrogen, 189–90, 191, 302
nitrous oxide, 333
Nixon, Richard, 85, 112, 119
Njao, Peter, 269
nongovernmental organizations (NGOs),
    135
North Africa, 138, 180, 194–95, 351
North Atlantic Treaty Organization
    (NATO), 97, 135
North Korea, 136, 137
North Polar Ice Cap, 14, 295, 296
Northeast Asia, 128, 137
Northern Sea Route, 295
Norway, 113, 172, 196, 340
    boreal forests of, 337
    carbon sequestering in, 354
Nubian Aquifer, 186
nuclear arms race, xxiii, 85, 317, 357
nuclear power, 354–55
nutrition, 124, 170, 174, 255
Nye, Joseph, 124
Nyerere, Julius K., 168

Obaid, Abdullah al-, 198
Obama, Barack, 136, 178, 259, 327, 328
Obama administration, 228, 266, 327
obesity, 153–56, 171–72, 257–58
Occupy Wall Street, 9, 57, 121
ocean acidification, 281, 300, 301–2
oceans:
    currents in, 284
    temperature of, 200
    warming of, 299–300
Ochalla, Nyikaw, 196
offshoring, technological, 9
oil, 71, 111, 113–14, 146–47, 165, 230,
    282, 318, 321, 325, 330, 332, 338
    carbon dioxide in, 334
    consumption of, 146, 147
    drilling for, 147, 295–96, 328
    embargo on, 344
    peak, 147–48
    price of, 145

oil derivatives, 15
Oklahoma, 187
Olin, John M., 115
Olson, Mancur, 364–65
Olympics, 247, 258
On the Origin of Species (Darwin), 239
one percent, 9–10, 35
OnStar system, 87
open systems, xvii
Operation Shady RAT, 75
optics, xxi
optimism, xxi–xxii, xxiii, xxxi
optogenetics, 216
Oregon, 301
organ donation, 242
organic agriculture, 189
Organic Singularity, 240
Organisation for Economic Co-operation and Development (OECD), 9, 118, 161
Organization of Petroleum Exporting Countries (OPEC), 147, 198
Orthodox Christians, 130
Orwell, George, xxiii, 83, 85
Ostfeld, Richard S., 291
Ottoman Empire, 62, 131
outsourcing, 5–8, 44, 116, 125, 372
    of building blocks of life, 204
    impact on employment of, 9, 12
    resources and, 25
    robosourcing and, 6–7, 17, 45
Overdiagnosed (Welch), 258
overfishing, 302
Overy, Charles, 31
Owen, Adrian, 238–39
Oxford University, 302
oxygen, 312
ozone hole, 309–10, 317

Pacific Ocean, 298
Paine, Thomas, 52, 64
Pakistan, 137, 177, 180, 186
    drones in, 86–87
    flooding in, 286
Palestinian territories, 178–79, 182
palladium, 145
palm oil, 145
Pan Yue, 351
Panama Canal, 110
pandas, 235

Panetta, Leon, 78
Panic of 1873, 109
Papua New Guinea, 335
parahippocampal gyrus, 238
Paralympics, 246
Paris Commune, 108–9, 112
Parkinson's disease, 207–8, 237, 268
Parliament, Myanmar, 58
Pasteur, Louis, 30
Patagonia, 304, 310
patents, 11, 71
Peacock, Warwick, 248
peak oil, 147–48
Pearl Harbor, attack on, 316, 365
Peng, Shaobing, 289
penicillin, 225–26
pensions, 123, 173–74
Penzias, Arno, xvi
Persia, 73
Persian Gulf, 335
Peru, 305, 357
pessimism, xxxi
pest resistance, 270
pesticides, 149, 269
pests, 289–90
pharmaceutical companies, 119, 260
pharmaceuticals, 245–46, 258
Philippines, 298, 336, 353
phishing, 82
Phoenicians, 49
phone hacking, 78
phonetic alphabets, 49
phosphorus, 272, 302
    depletion of, 145, 190–92
photosynthesis, 356
photovoltaic (PV) electricity, 282
pictograms, 49
pigs, 146, 227, 270, 272, 353
Ping, Jean, 193
Pistorius, Oscar, 246
Pizarro, Francisco, 18
plant life, xxx, 153
    diseases of, 269–74
    genetic diversity of, 149
plant protein, 201
Plato, xx, 48
plow, 37
poaching, 339
poison gas, 130, 163
polar jet stream, 308

police, 117
polio vaccine, 211
political campaigns, 117
Pollan, Michael, 148
pollen, 270
pollination, 284
pollution, xiv, 34, 124, 146, 161–66, 283
polysilicon, 10
polyurethane elastomer, 220n
population growth, 61, 151–53, 165–70, 172, 178, 199, 280
   land competition exacerbated by, 193
   pressure on river water from, 195
   social disorder and, 149–50
   surging demand for commodities caused by, 145–46
   as unsustainable, xiv
   in urban areas, 299
Portugal, 35, 352
post-national entities, 131
post-traumatic stress disorder, 101
postal services, 26, 44–45
potatoes, 269–70
Poukens, Jules, 241
poverty, xxx, 34, 118, 167, 318
Powell, Lewis, 112–13
Powell Plan, 114–15, 116, 119
power, necessities for, 123–24
power utilities, 117
pre-fear, 220
precipitation, 149
precision medicine, 205–8
preimplantation genetic diagnosis (PGD), 209, 248, 252
price reduction, 24
Priddy, Albert, 232
Prigogine, Ilya, xvi–xviii
primates, xxx
Principles of Geology (Lyell), xxii
Print Revolution, 52, 62, 64, 92, 127
printing press, 50, 51–52, 57, 62, 66, 89, 128–29
prison population, 117
prisons, 117
privacy, 70–72, 78–89
private schools, 179
production, 158
productivity, 11, 34
progress, xix–xx, xxi–xxii, xxvi, 6
   end of faith in, xxii–xxiii

Progressive Era, 22
Progressive movement, 109–11, 120
Prometheus, 205
property rights, 124, 196
prosthetics, 243–44, 245
PROTECT IP Act, 87–88
protein, 153, 212, 216, 336
proteomics, 68, 216
Protestant Reformation, 50–51
public goods, 35, 41, 66, 123, 367, 372
public interest, xxv
public relations, 157
public safety, 66, 123
public square, 51, 64
public trust, 103
pumps, 165
Pure Food and Drug Act, 110
Putin, Vladimir, 65

"quantified self" movement, 206
quantum computing, 53
quantum properties, 30
quarterly capitalism, xxviii
quarterly democracy, xxviii

radio, 39, 156
Radiofrequency Identification (RFID), 45
railroads, 74n, 106, 128, 260
rain, 288
rain shadows, 306
Rangoon, Burma, 299
rapeseed, 264
ratings agencies, 17
Raymond, Lee, 113–14
Reagan, Ronald, 116, 309, 317
recombinant bovine growth hormone (rBGH), 271
recycling, of water, 188
red blood cells, 246
Red Sea, 181
reductionism, xxi, 217
Reformation, 50–51, 128
Refugee Protocol, 181
refugees, 180–83, 304
regenerative medicine, 216
Rejeski, David, 30
religion, 205
religious freedom, 105

REM sleep, 258
Renaissance, xx
renewable energy, 282–84, 325, 344, 348–49
  subsidies for, 343
Republican Party, U.S., 117
research and development, 123, 124
reservoirs, 187
resources, 25, 27–30, 33
  allocation of, xxiv–xxv, 33
  limits of, 142
  unsustainable consumption of, xiv
Revelle, Roger, 311, 312–13
Revolution of the Spheres (Copernicus), 51
rhesus monkeys, 254
rhizobium bacteria, 191
rice, 185, 235, 269, 340
Rights and Resources Initiative, 196
rinderpest, 256
Rio Grande River, 166
Risen, Jane, 315
Ritaccio, Anthony, 239
Ritalin, 245
RNA, 216
roads, 303, 331
Robber Baron era, 107
Roberts, Dave, 343–44
robosigning, 17
robosourcing, 5, 44, 116, 125, 372
  acceleration of, 6, 138
  of building blocks of life, 204
  of exotic financial instruments, 17
  impact on employment of, 7–8, 9
  in low-wage countries, 7, 12
  outsourcing and, 6–7, 17, 45
  resources and, 25
robot rights, 102
robots, xiv, xxiii, 239
  increasing orders for, 11
  skills needed to operate, 8
Rockefeller, Jay, 322–23
Roland, Sherwood, 309
Roman Empire, 369
Roman Republic, 99
Rome, ancient, xx, 51, 260
Romm, Joseph, 193
Roosevelt, Franklin D., 98, 111, 116, 155, 185, 189, 190, 192, 260–61
Roosevelt, Theodore, 97, 110–11, 233–34

Rose, Charlie, xvi
rubber, 145
Ruedy, Reto, 293
Russia, 18, 60, 65, 114, 334, 353
  agriculture in, 288
  boreal forests of, 337
  cyberattacks and, 83
  Gini coefficient of, 9
  heat wave in, 286
  Internet use controlled in, 88
Russian Revolution, 158
Russo-Japanese War, 110
Rutgers University, 81
Rwanda, 150, 173

Sabin, Albert, 211n
Sack, Kevin, 242
Saffron Revolution, 57
Sagan, Carl, xvi
Sahara Desert, 193, 306
salinization, 195
Salk, Jonas, 211
salmon, 200, 271–72
sandstorms, 193–94
Sandy, Superstorm, 286, 299, 346
Sanger, Margaret, 233, 234
sanitation, 152n, 170, 174, 255
Santa Clara County v. Southern Pacific Railroad Company, 107, 113
São Paulo, Brazil, 96
satellite television, 128, 132
Sato, Makiko, 293
Saudi Arabia, 60, 77, 132, 137, 176, 196, 198–99, 321, 350
savings, 27
Scandinavia, 285, 299
Schalk, Gerwin, 239
Schlenker, Wolfram, 288
Schrems, Max, 82
Schumpeter, Joseph, 206
science, 5, 39, 204–5, 209
scientific method, as reductionist, xxi
Scientific Revolution, xx, xxi, xxix, 19, 51, 62, 69, 123
Scotland, 131, 285, 352
sea lanes, 92, 93
sea levels, 182, 297–98, 300
seawater, desalination of, 198–99
Second Industrial Revolution, xxii, 19
Second Law of Thermodynamics, xvi

Secret Service, U.S., 76
Segre, Julie, 275
selective attention, xxvi–xxvii
selective breeding, 262, 263
self-determinism, 51
self-driving automobile, 25–26
self-governance, 66
self-quantification movement, 259
self-reorganization, xvii
    of Congress, xxv
self-replicating nanobots, 30
self-sourcing, 26
semiautonomous robotic weapons,
    101
Senate, U.S., 73, 323, 374
    Subcommittee on Science of, 223
Seoul, South Korea, 96
separation of powers, 363–64
September 11, 2001, terrorist attacks, 84,
    86, 132
Sepulveda, Francisco, 244
Serbia, 28, 130–31, 135
service jobs, 25, 116
sewage water, 188
sex slavery, 359
sexual values, 132
Shah, Tushaar, 150
Shakespeare, William, 52, 217
shale gas, 147, 165, 328, 331, 332–33
Shanghai, China, 299, 348
sheep, 193–94, 270
shellfish, 334
Shenzhen, China, 152, 209, 348
Shirky, Clay, 63
short-term thinking, xxviii, 12
    in markets, xxviii
    politics and, xxviii–xxix
Sigal, Robert K., 55
Silicon Valley, 20
silk, 220, 235, 270
silver, 18, 145
Singapore, 76, 96
Singh, Manmohan, 114
singularity, 239–41
Sixth Great Extinction, 143
skinheads, 177
Slaoui, Moncef, 244
slavery, 110, 359
sleep monitors, 258
slums, 152

Smalley, Richard, 30
smallpox, 256
smart grid, 351, 352
smartphones, 26, 47, 53, 54, 68, 87,
    257–58
Smith, Adam, 52, 315
smog, 333
social cohesion, 123
social contract, 130, 174, 367–68
Social Darwinism, 230
social discounting, 151
social feedback loops, 121
social media, 26
social networks, 44, 206
Social Security, 23, 26–27, 120, 135,
    261
socialism, 120
Soffer, Arnon, 181
software, 11, 12
Soghoian, Chris, 85
soil, 150
    carbon in, 191
    erosion of, 189
    fertility of, 145, 149, 189
    management of, 197
    moisture in, 287, 296
solar power, 10, 282, 327–28, 342, 343,
    349, 351, 352
Somalia, 131, 180, 285
soot, 329
"Sorcerer's Apprentice, The" (Goethe),
    xxiv
sorghum, 145
South Africa, 89, 348
South America, 99, 269, 305
South Asia, 163, 298
South Korea, 76, 173, 197
    boreal forests of, 337
    cap and trade in, 348
    coal imported by, 349
    nuclear power in, 355
South-North Water Transfer Project,
    186–87
South Sea Company bubble, 106
South Sudan, 131
Southeast Asia, 173, 190, 298
Southern Africa, 307
Southern Europe, 307
Southern Pacific Railroad Company, 107,
    113

Soviet Union, 99, 158
  agriculture in, 192
  biological weapons programs of, 221–22
  in Cold War, 98, 135
  Lysenkoism in, 230
  as nuclear threat, 19, 74n, 98, 357
  as supposed threat to U.S. economic
    dominance, 19, 96
  World War II devastation of, 97
soybeans, 145, 190, 235, 264, 269, 270, 290
space parasol, 355
Spain, 35, 57, 338
species, xiv
  extinction of, 339–41
speech, 49, 52
Spencer, Herbert, 230
sperm, 247, 253, 268
spider silk, 220, 270
spinal pain, 244
spinning mule, 20
Sporns, Olaf, 207n
spruce trees, 336
Sputnik, 23, 96
Square Kilometre Array, 56
"square tomatoes," 263
Stalin, Joseph, xxii, 97, 159
standard deviation, 293–94
standard model of markets, 15
standard of living, xxx
staphylococcus, 227
State of the World's Refugees, 181
states, 127
statism, 120
steam engine, 37, 39, 260
steel, 11, 29, 146
Steinberg, Jeffrey, 209
stem cells, 248
sterilization, 232–35
Sterk, Wolfgang, 348
steroids, 247
Stiglitz, Joseph, 15, 16, 34, 41
stimulus policies, 26, 327
stock market crash of 1929, 111
stock markets, 13, 295, 373
Stone Age, xvii, 28
Stop Online Piracy Act, 87–88
Strategic Air Command, 54
stratosphere, 309, 310–11
strip searches, 85
Strode, Aubrey, 232

stroke, 154
Stuxnet computer worm, 77
subconscious, 157, 160
submarines, 135
subprime carbon assets, 321
subprime mortgages, 16–17, 321–22
subsidies, 342–43, 350
suburban sprawl, 149
Sudan, 180, 193, 195, 197, 335
Suess, Hans, 311
sulfur dioxide, 333, 345, 355, 356
Sun Liping, 103
sunspots, 323
super grids, 351, 352
super salmon, 271–72
supercomputers, 12, 14, 17, 28, 46n, 216
supermarkets, 26
superorganisms, 217
supplementary motor area, 238
supply chains, 4, 75, 148, 298
supranational entities, 135
Supreme Court, U.S., 85
  corporate power entrenched by, 64,
    105, 107
  gene patents allowed by, 210
  lobbying limited by, 107
  New Deal and, 111–12
surveillance drones, 373
surveillance hardware and software,
  86–87
sustainability, xxv, 371, 372
Sustainable Capitalism, 35–36
Svalbard, 340
Swart, Gary, 8
Sweden, 105, 172, 173
  boreal forests of, 337
  forced sterilization in, 234
swine flu, 218
Switzerland, 76, 238–39, 348
symbiosis, 303, 336
synthetic biology, 213–14, 219, 258
synthetic DNA, 216–17
synthetic growth hormone, 270
synthetic nitrogen fertilizers, 338
synthetic viruses, 218
Syr Darya, 192
Syria, 62, 86, 131, 182

Taft, William Howard, 110
Tahrir Square, 57, 61, 62

Taipei, Taiwan, 96
Taiwan, 76
Taklamakan Desert, 194
Taliban, 170, 177
Tambora volcano, 285
Tao, xx
tapioca, 269, 290
taxation, 116, 125, 318
  inequality exacerbated by, 9, 34
  inheritance, 9, 111, 122
  progressive, 34
Tea Party, 118–19
Teapot Dome, 111
techno-optimism, 62
technological capital, 7, 41
technological offshoring, 9
Technological Singularity, 240
technology, 5, 204–5
  power of, 142
  value of labor vs., 8
technology curve, 7
Technology Revolution, 19, 27, 38, 39,
  69, 123
Teilhard de Chardin, Pierre, 27, 46, 207,
  216, 239
Tel Aviv University, 244
telegraph, 39, 45, 74n, 108
telepathy helmets, 239
telephone, 53, 74n, 87
television, 24, 39, 57, 63–65, 67, 119, 128,
  143, 238, 326, 369–70
  commercials on, xxv, 64
  fundamentalism and, 133
  satellite, 128, 132
telomerases, 254–55
temporary work programs, 179
terrorism, 87, 93, 177, 178
test tube babies, 248
testicle, artificial, 247
tetracycline, 227
Texas, 165, 187, 330, 346
text messaging, 44, 87
Thailand, 58, 250, 298, 348
Thames River, 299
Thamus, King of Egypt, 48, 49
Thatcher, Margaret, 309
thermonuclear explosion, 221
Thermopylae, Battle of, 73–74
think tanks, 115, 116
Third Industrial Revolution, 29

Third Reich, xxii
Three Musketeer Principles, 318
three-parent babies, 249
3D printing, 30–33, 219, 241–42, 243
thunderstorms, 305
Tianjin, China, 299, 348
tidal energy, 352
Tigris River, xvii, 127, 195
Tillerson, Rex, 303, 331
tinfoil, 355, 356
titanium, 241
tobacco industry, 112, 162
Toffler, Alvin, xvi
Tokyo, Japan, 14, 152, 299
Tokyo stock market, 295
tomatoes, 262–63
Tongass forest reserve, 110
tools, 37–38
Topol, Eric, 206
topsoil depletion, xiv, 142, 145, 149, 153,
  183–84, 185, 188, 191, 198, 199, 281,
  372
  global warming and, 288, 332
tornadoes, 346
torture, 132
Total Information Awareness (TIA), 87
Tour de France, 246–47
toxic liquid waste, 165
toxic sludge, 334
Toxic Substances Control Act, 163
tracking, 80–81
trade, 125, 347
  interconnectedness and, 7, 18
  restrictions on, 12
  see also globalization
trade routes, 298
transhumanism, 239–40
translation, 129
transportation, 18, 282
Transportation Revolution, 21
Traub, Arielle, 155
tree planting, 337–38
tribal identity, 317
Trinidad Head, Calif., 312
tropical diseases, 255
tropical forests, 335–36
troposphere, 307, 310
Truman, Harry, 261
tuberculosis, 226, 255
tuna, 199, 302

tundras, 336, 337
Tunisia, 9, 60, 61, 170, 176
Turing Test, 240
Turkana, Lake, 196
Turkey, 133, 195, 304, 348
Turki al-Faisal, Prince, 321
Turkle, Sherry, 47
Turkmenistan, 192
Tuscaloosa, Ala., 346
Twitter, 45, 56, 57, 58, 60
Twitter Earthquake Detector, 56
2,4-D (pesticide), 267–68
2000 Millennium Development Goals,
    164
Tyndall, John, 225, 311
type I diabetes, 275
typhoid, 163

Uganda, 197, 269
Ukraine, 288, 348
ulcerative colitis, 275
ulcers, 276
ultraviolet radiation, 310
underemployment, 11, 27, 33
unemployment, xxvi, 27, 177
    eugenics and, 231
    in Great Recession, 12, 34
    in Panic of 1873, 109
    transformation of global economies
        and, 11
unemployment compensation, 260
unions, 116
United Arab Emirates, 170, 197
United Kingdom, 19, 105, 252
    carbon sequestering in, 354
    corporate charters granted by, 106
    cybersecurity in, 76
    Gini coefficient of, 9
    Industrial Revolution in, 20
    privacy laws in, 83
    in Security Council, 95
    temporary work programs in, 179
United Nations, 54, 94, 97, 135, 163, 164,
    176, 181, 187, 192, 194, 302, 336, 337
    Environment Programme of, 192, 335,
        347
    General Assembly of, 317
    Security Council of, 95
United Nations Convention on
    Biodiversity, 340

United States:
    agriculture in, 270
    Al-Qaeda's threat to, 132
    biogas in, 353
    bipartisan divide in, 119–20
    carbon sequestering in, 354
    climate change monitoring by, 324–25
    coal use declining in, 147
    in Cold War, 98, 135, 357
    debt of, 123
    draft abandoned in, 102
    droughts in, 286, 288, 300, 302, 346
    electricity grids in, 351–52
    financial sector in, 13, 14
    genetically modified animals in,
        271–72
    Gini coefficient of, 9
    global warming threat to, 298
    Great Recession in, 11
    groundwater aquifers in, 145, 186
    health care in, 154n, 174, 259–60
    immigration in, 177–78
    income inequality in, 9–10, 34, 35,
        118, 120–21
    Internet in, 63
    irrigation in, 185
    longevity in, 174
    middle class in, 122
    military budget in, 100–101
    organ selling banned in, 242
    population growth in, 167, 178
    Southwest in, 307, 351–52
    tax laws in, 9
    temporary work programs in, 179
    toxic liquid waste in, 165
    tree planting in, 338
    university system of, 100
    urbanization of, 109
United States, degradation of democracy
    in, xv
    avoidance of difficult decisions in,
        119
    and biotechnology, 209
    corporations and, 104–24
    and decline of collective thinking, xxv
    dominance of money and, xxviii
    proponents of, xxv
United States, global leadership of, xiv,
    19, 104, 374
    aid provided by, 100

in Asia, 98
China and, 11, 92, 93
contraception and, 168–69
and global economic output, 96–97
international institutions set up
by, 94
Japan and, 19, 96
as leading manufacturing nation, 93
military strength of, 100–101
overreaching of, 100
Soviet Union and, 19, 96
threats of decline in, 93–94, 95–102,
168–69
universities, 100
urban sprawl, 149, 164, 192
urbanization, 109, 151–53, 155, 174, 176,
231, 372
as unsustainable, xiv
uteri, 253
utility systems, 303
Uttar Pradesh, India, 180
Uzbekistan, 192, 234

vaccines, 170, 218–19
Vatican, 73
Vavilov, Nikolai, 340
vegetables, 155
vegetation, 287
Venezuela, 147, 304, 334
Venice, Italy, 18, 299
Ventner, Craig, 216–17, 218
venture capital, 78
Vermeer, Johannes, xxi
Verne, Jules, xxii, xxiii, xxxi
vertebrates, xxx
Victory Gardens, 191
video games, 66
video rentals, 44
Vietnam, 76, 100, 298, 338, 348
Vietnam War, 100, 102, 318
Vinge, Vernor, 240
viruses, 266, 291
synthetic, 218
visual cortex, 244
vitamin A, 266
voice recognition software, 82
volcanoes, 323–24
Vonnegut, Kurt, 318
voting rights, 173
Voyages of Discovery, 18, 50, 82

wages:
outsourcing and, 12
in Panic of 1873, 109
robosourcing and, 7
Wagner, Robert, 261
walking, 172
Wall Street Journal, 81, 86, 268
Wallace, Helen, 221
Walmart, 10
Walton, Bud, 10
Walton, Sam, 10
Wanyera, Ruth, 269
war, xxx, 133–39, 148
decline in risk of, 133, 134–36
industrialization of, 130
War on Poverty, 112
Washington, George, 99, 110, 144
Washington Consensus, 118
water, xiv, 142, 143, 150, 152*n*, 164, 182,
197, 198, 284, 287–88
competition for, 149, 194–96
natural purification of, 284
pollution of, 163–65
see also groundwater resources
water utilities, 117
water vapor, 286–87, 329
Waterloo, Battle of, 13
Watson, James, 212, 223, 235
Watson system, 206
wave energy, 352
weapons, spread of, 136
weapons of mass destruction, 84, 136–37
Weiss, Ron, 214
Welch, H. Gilbert, 258
wells, 165
Wells, H. G., xxiii, 45
Wells Fargo, 83
Wen Jiabao, 59, 234
West Bengal, India, 180
West Germany, 126
West Nile virus, 291–92
Westphalia, Treaty of, 128, 130
wheat, 269, 270
wheelchairs, 239
Wheeler, David, 246
white flight, 179
Whitehead, Alfred North, xxvii
Wikileaks, 72
Wikipedia, 45
Wilcox, Christie, 292

Wilson, E. O., 67, 217, 317, 339
Wilson, Woodrow, 85, 97, 111, 157, 233
wind power, 282, 342, 343, 352
wind tunnel testing, 31
windmills, 103, 327–28, 349, 351
wisdom of crowds, 99
women:
    empowerment of, 168, 170–71, 372
    on Internet, 47
    smoking by, 158
    in workplace, 170
women's movement, 112
women's rights, 139, 318
Woodhouse, Philip, 196
working conditions, 358
World Bank, 94, 95, 161, 196–97, 294,
    343, 346
World Food Programme, 149
World Health Organization, 153, 155,
    164, 223, 226, 256, 257
World Trade Organization, 94, 347, 371
World War I, xxii, 97, 130, 131, 133, 134,
    145, 163
World War II, xxii, 18, 74, 85, 94, 96,
    97, 99, 126, 130, 136, 145, 151, 163,
    169–70, 191, 261

World Wide Web, xvii, 56, 81
worms, 78
writing, 5, 48
Wyoming, 330

xenophobia, 177
Xerxes, 74

Yadav, Ram Khatri, 288
Yamanaka, Shinya, 253
Yangtze River, 166
Yellow River, xvii, 127, 166
Yemen, 62, 150, 172, 285
Yevtushenko, Yevgeny, 130
Yosemite National Park, 340
YouTube, 58–59
yuan, 123
Yucatan, 339
yucca, 269, 290
Yugoslavia, 130, 135

Zambia, 335
Zhang, David, 285
Zheng He, 18
Zimbabwe, 285
zooxanthellae, 300

ABOUT THE AUTHOR

Former vice president AL GORE is co-founder and chairman of both Generation Investment Management and Current TV. He is also a senior partner at Kleiner Perkins Caufield & Byers, and a member of Apple Inc.'s board of directors. Gore spends the majority of his time as chairman of the Climate Reality Project, a non-profit devoted to solving the climate crisis. Gore was elected to the U.S. House of Representatives four times and to the U.S. Senate twice. He served eight years as vice president. He authored the bestsellers *Earth in the Balance, An Inconvenient Truth, The Assault on Reason,* and *Our Choice.* He is a co-recipient of the 2007 Nobel Peace Prize.

www.algore.com